Human Retrovirus Protocols

METHODS IN MOLECULAR BIOLOGY™

John M. Walker, Series Editor

316. Bioinformatics and Drug Discovery, edited by *Richard S. Larson, 2005*

315. Mast Cells: *Methods and Protocols,* edited by *Guha Krishnaswamy and David S. Chi, 2005*

314. DNA Repair Protocols: *Mammalian Systems, Second Edition,* edited by *Daryl S. Henderson, 2005*

313. Yeast Protocols: *Second Edition,* edited by *Wei Xiao, 2005*

312. Calcium Signaling Protocols: *Second Edition,* edited by *David G. Lambert, 2005*

311. Pharmacogenomics: *Methods and Applications,* edited by *Federico Innocenti, 2005*

310. Chemical Genomics: *Reviews and Protocols,*edited by *Edward D. Zanders, 2005*

309. RNA Silencing: *Methods and Protocols,* edited by *Gordon Carmichael, 2005*

308. Therapeutic Proteins: *Methods and Protocols,* edited by *C. Mark Smales and David C. James, 2005*

307. Phosphodiesterase Methods and Protocols, edited by *Claire Lugnier, 2005*

306. Receptor Binding Techniques: *Second Edition,* edited by *Anthony P. Davenport, 2005*

305. Protein–Ligand Interactions: *Methods and Applications,* edited by *G. Ulrich Nienhaus, 2005*

304. Human Retrovirus Protocols: *Virology and Molecular Biology,* edited by *Tuofu Zhu, 2005*

303. NanoBiotechnology Protocols, edited by *Sandra J. Rosenthal and David W. Wright, 2005*

302. Handbook of ELISPOT: Methods and Protocols, edited by *Alexander E. Kalyuzhny, 2005*

301. Ubiquitin–Proteasome Protocols, edited by *Cam Patterson and Douglas M. Cyr, 2005*

300. Protein Nanotechnology: *Protocols, Instrumentation, and Applications,* edited by *Tuan Vo-Dinh, 2005*

299. Amyloid Proteins: *Methods and Protocols,* edited by *Einar M. Sigurdsson, 2005*

298. Peptide Synthesis and Application, edited by *John Howl, 2005*

297. Forensic DNA Typing Protocols, edited by *Angel Carracedo, 2005*

296. Cell Cycle Control: *Mechanisms and Protocols,* edited by *Tim Humphrey and Gavin Brooks, 2005*

295. Immunochemical Protocols, *Third Edition,* edited by *Robert Burns, 2005*

294. Cell Migration: *Developmental Methods and Protocols,* edited by *Jun-Lin Guan, 2005*

293. Laser Capture Microdissection: *Methods and Protocols,* edited by *Graeme I. Murray and Stephanie Curran, 2005*

292. DNA Viruses: *Methods and Protocols,* edited by *Paul M. Lieberman, 2005*

291. Molecular Toxicology Protocols, edited by *Phouthone Keohavong and Stephen G. Grant, 2005*

290. Basic Cell Culture Protocols, *Third Edition,* edited by *Cheryl D. Helgason and Cindy L. Miller, 2005*

289. Epidermal Cells, *Methods and Applications,* edited by *Kursad Turksen, 2005*

288. Oligonucleotide Synthesis, *Methods and Applications,* edited by *Piet Herdewijn, 2005*

287. Epigenetics Protocols, edited by *Trygve O. Tollefsbol, 2004*

286. Transgenic Plants: *Methods and Protocols,* edited by *Leandro Peña, 2005*

285. Cell Cycle Control and Dysregulation Protocols: *Cyclins, Cyclin-Dependent Kinases, and Other Factors,* edited by *Antonio Giordano and Gaetano Romano, 2004*

284. Signal Transduction Protocols, *Second Edition,* edited by *Robert C. Dickson and Michael D. Mendenhall, 2004*

283. Bioconjugation Protocols, edited by *Christof M. Niemeyer, 2004*

282. Apoptosis Methods and Protocols, edited by *Hugh J. M. Brady, 2004*

281. Checkpoint Controls and Cancer, Volume 2: *Activation and Regulation Protocols,* edited by *Axel H. Schönthal, 2004*

280. Checkpoint Controls and Cancer, Volume 1: *Reviews and Model Systems,* edited by *Axel H. Schönthal, 2004*

279. Nitric Oxide Protocols, *Second Edition,* edited by *Aviv Hassid, 2004*

278. Protein NMR Techniques, *Second Edition,* edited by *A. Kristina Downing, 2004*

277. Trinucleotide Repeat Protocols, edited by *Yoshinori Kohwi, 2004*

276. Capillary Electrophoresis of Proteins and Peptides, edited by *Mark A. Strege and Avinash L. Lagu, 2004*

275. Chemoinformatics, edited by *Jürgen Bajorath, 2004*

274. Photosynthesis Research Protocols, edited by *Robert Carpentier, 2004*

273. Platelets and Megakaryocytes, Volume 2: *Perspectives and Techniques,* edited by *Jonathan M. Gibbins and Martyn P. Mahaut-Smith, 2004*

METHODS IN MOLECULAR BIOLOGY™

Human Retrovirus Protocols

Virology and Molecular Biology

Edited by

Tuofu Zhu, MD

Departments of Laboratory Medicine and Microbiology
University of Washington School of Medicine
Seattle, WA

HUMANA PRESS ✷ TOTOWA, NEW JERSEY

999 Riverview Drive, Suite 208
Totowa, New Jersey 07512

www.humanapress.com

This publication is printed on acid-free paper. ∞
ANSI Z39.48-1984 (American Standards Institute)
Permanence of Paper for Printed Library Materials.

Production Editor: Jennifer Hackworth

Cover design by Patricia F. Cleary

Cover illustration: Figure 4 from Chapter 5, "Isolation and HIV-1 Infection of Primary Human Microglia From Fetal and Adult Tissue," by Kathleen Borgmann, Howard E. Gendelman, and Anuja Ghorpade.

For additional copies, pricing for bulk purchases, and/or information about other Humana titles, contact Humana at the above address or at any of the following numbers: Tel.: 973-256-1699; Fax: 973-256-8341; E-mail: orders@humanapr.com; or visit our Website: www.humanapress.com.

Printed in the United States of America. 10 9 8 7 6 5 4 3 2 1
e-ISBN: 1-59259-907-9
ISSN: 1064-3745

Library of Congress Cataloging-in-Publication Data

Human retrovirus protocols : virology and molecular biology / edited by Tuofu Zhu.
p. cm. -- (Methods in molecular biology ; v. 304)
Includes bibliographical references and index.
ISBN 1-58829-495-1 (alk. paper)
1. Retroviruses--Laboratory manuals. I. Zhu, Tuofu. II. Series.
QR414.5.H785 2005
616.9'188--dc22

200402621:

Preface

Retroviruses are a large and diverse family of enveloped, single-strand RNA viruses characterized by the unique replicative strategy that includes reverse transcription of the virion RNA into linear double-stranded DNA and the subsequent integration of this DNA into the genome of the cells. Human retroviruses are subdivided into three groups: oncoviruses, lentiviruses, and spumaviruses. Oncoviruses have been associated with a variety of cancers. The first human retrovirus, human T-lymphotropic virus type 1 (HTLV-1), was discovered in the late 1970s, and was shown to cause adult T-cell leukemia and chronic neurological conditions. A relative of HTLV-1, HTLV-II, has also been associated with human leukemias. The most significant disease resulting from human retrovirus infection is the acquired immunodeficiency syndrome (AIDS), which is caused by lentiviruses, human immunodeficiency viruses type 1 and type 2 (HIV-1 and HIV-2). Spumaviruses cause no known disease in humans and interest in this category is relatively recent.

Although there are differences in the genetic composition of different types of retroviruses, all carry three basic coding genes in the same order: *gag*, encoding internal structural proteins that form the matrix, the capsid, and the nucleo protein structures; *pol*, encoding the reverse transcriptase and integrase enzymes; and *env*, encoding the surface and transmembrane components of the viral envelope protein. These three genes and their coded proteins are the major focus of studies on the virology, serology, and molecular biology of human retroviruses.

In the last two decades, research on human retroviruses has been progressing at a rapid pace. Not surprisingly, recent studies of human retroviruses have focused largely on HIV-1 and HIV-2. Therefore, most of our present knowledge of human retroviruses is derived from studies on HIV-1. In addition, studies on HIV-1 have significantly advanced our understanding of biology and development of new biotechnologies. The prospect of acquiring a broad view of a field with so many viruses and associated technologies may seem daunting. In hopes of attenuating such concerns, we have organized *Human Retrovirus Protocols: Virology and Molecular Biology* by focusing on methodologies of the virology and molecular biology of human retroviruses. The first of two sections primarily explores methods for the isolation and detection of human retroviruses; the second looks at the interplay between the viruses and the host, focusing on the phenotype of human retroviruses. This division is not intended to be rigid or absolute; the PCR-based assays in the first section are typically molecular techniques.

Part I begins with a chapter on the quantitative isolation of HIV-1 from latently infected resting T-cells from the laboratory that developed this revolutionary technique. Nine chapters follow on methods for the isolation and propagation of HIV, HTLV, and foamy virus from the peripheral blood mononuclear cells (PBMC), blood monocytes, brain tissues, cerebrospinal fluids, semen, vagina, and lymph nodes. The succeeding chapters constitute protocols for detection and quantification of different retrovirus genes, antigens, and antibodies, for which we have placed greater emphasis on "universal" assays for detection of different subtypes of HIV-1 and HIV-2.

Part II begins with biological assays for determining cell tropism of HIV-1 from the laboratory that initially developed the techniques. The next two chapters describe "popular" assays for the determination of co-receptor usage of HIV-1. Three chapters follow on assays recently developed in each author's laboratory for phenotyping HIV-1 infected monocytes and examining HIV-1 fitness with an effective cloning system. The next five chapters are on cloning, construction, and characterization of full-length HIV-1, HIV-2, HTLV-II, and spumaretrovirus. A chapter for assessing drug efficacy follows. *Human Retrovirus Protocols: Virology and Molecular Biology* concludes with two chapters describing new technologies for determining human gene expression with HIV-1 infection by microarrays and assessing genetic polymorphisms in two recently identified HIV-1 co-factors, DC-SIGN and DC-SIGNR.

All chapters offer the detailed steps necessary to carry out the assays, and provide discussion of problems and pitfalls that may be encountered. Most of these protocols can be applied directly, or can be adapted easily.

The editor would like to thank all of the contributors for their high-quality submissions, and the Series Editor, John Walker, for his valuable advice throughout the entire process of producing this volume.

Tuofu Zhu

Contents

Part II. Molecular Biology Methods

Contributors

Awet Abraha • *Division of Infectious Diseases, Department of Medicine, Case Western Reserve University, Cleveland, OH*
Matthew Anderson • *Center for Retrovirus Research, Department of Veterinary Biosciences, Comprehensive Cancer and Cellular and Developmental Biology Graduate Program, The Ohio State University, Columbus, OH*
Stefano Aquaro • *Department of Experimental Medicine, University of Rome "Tor Vergata," Roma, Italy*
Eric J. Arts • *Molecular and Cellular Biology Program, Department of Molecular Biology and Microbiology and Division of Infectious Diseases, Department of Medicine, Case Western Reserve University, Cleveland, OH*
Patricia Bittoun • *CNRS UPR9051, Hôpital Saint-Louis, Paris, France*
Hetty Blaak • *Department of Virology, Erasmus MC, Rotterdam, The Netherlands*
Kathleen Borgmann • *Laboratory of Cellular Neuroimmunology and Center for Neurovirology and Neurodegenerative Disorders and the Departments of Pathology and Microbiology, University of Nebraska Medical Center; Omaha, NE*
Audrey Brussel • *Département des Maladies Infectieuses, Institut Cochin, INSERM U567, CNRS UMR8104, Université René Descartes, Paris, France*
Myhanh Che • *Center for Neurovirology and Neurodegenerative Disorders and the Department of Pathology and Microbiology, University of Nebraska Medical Center, Omaha, NE*
Suzanne M. Crowe • *Macfarlane Burnet Institute for Medical Research and Public Health, Melbourne, Victoria, Australia, and Department of Medicine, Monash University, Clayton, Victoria, Australia*
Tsutomu Daa • *Department of Pathology, Oita University, Oita, Japan*
Florence Damond • *Laboratoire de Virologie, INSERM U552, Hôpital Bichat, Paris, France*
Olivier Delelis • *Département des Maladies Infectieuses, Institut Cochin, INSERM 1567, CNRS UMR 81041, Université René Descartes, Paris, France*
B.D. Desmarais • *Division of Urology, Department of Surgery, Harvard Institutes of Medicine, Boston, MA*

Hedwig Duttmann • Dade Behring Marburg GmbH, Marburg, Germany
Philip K. Ehrenberg • Division of Retrovirology, Henry Jackson Foundation and Department of Molecular Diagnostics and Pathogenesis, Walter Reed Army Institute of Research, Rochville, MD
Philip J. Ellery • Macfarlane Burnet Institute for Medical Research and Public Health, Melbourne, Victoria, Australia, and Department of Medicine, Monash University, Clayton, Victoria, Australia
S.J. Eyre • Division of Urology, Department of Surgery, Harvard Institutes of Medicine, Boston, MA
Jürgen Feldner • Dade Behring Vertriebs-GmbH and Co. OHG, Schwalbach, Germany
Eva Maria Fenyö • Unit of Virology, Division of Medical Microbiology, Department of Laboratory Medicine, Lund University, Lund, Sweden
Susan A. Fiscus • Department of Microbiology and Immunology, University of North Carolina at Chapel Hill, Chapel Hill, NC
Rolf M. Flügel • Research Programm Applied Tumorvirology, German Cancer Research Center Heidelberg, Heidelberg, Germany
Feng Gao • Human Vaccine Institute, Department of Medicine, Duke University Medical Center, Durham, NC
Yong Gao • Division of Infectious Diseases, Department of Medicine, Case Western Reserve University, Cleveland, OH
Howard E. Gendelman • Department of Pathology and Microbiology, Center for Neurovirology and Neurodegenerative Disorders, University of Nebraska Medical Center, Omaha, NE
Anuja Ghorpade • Laboratory of Cellular Neuroimmunology, Center for Neurovirology and Neurodegenerative Disorders, Department of Pathology and Microbiology, Department of Biochemistry and Molecular Biology, University of Nebraska Medical Center, Omaha, NE
Santhi Gorantla • Laboratory of Neuroregeneration, Center for Neurovirology and Neurodegenerative Disorders, Department of Pathology and Microbiology, University of Nebraska Medical Center, Omaha, NE
Paul R. Gorry • Macfarlane Burnet Institute for Medical Research and Public Health, Melbourne, Victoria, Australia, and Department of Medicine, Monash University, Clayton, Victoria, Australia
Patrick L. Green • Center for Retrovirus Research, Departments of Veterinary Biosciences and Molecular Virology, Immunology, and Medical Genetics, Comprehensive Cancer Center and Solove Research Institute, The Ohio State University, Columbus, OH
Marie Gueudin • Unité de Virologie, GRAM EA 2636, CHU Charles Nicolle, Rouen, France

KENJI KASHIMA • *Department of Pathology, Oita University, Oita, Japan*
KATHERINE KEDZIERSKA • *Department of Microbiology and Immunology, The University of Melbourne, Parkville, Victoria, Australia*
ANN A. KIESSLING • *Division of Urology, Department of Surgery, Harvard Institutes of Medicine, Boston, MA*
NEELTJE A. KOOTSTRA • *Department of Clinical Viro-Immunology Sanquin Research at CLB and Landsteiner Laboratory at the Academical Medical Center Amsterdam, The Netherlands*
MICHAEL D. LAIRMORE • *Center for Retrovirus Research and Department of Veterinary Biosciences, Department of Molecular Virology, Immunology ad Medical Genetics, Comprehensive Cancer Center, The Arthur G. James Cancer Hospital and Solove Research Institute, The Ohio State University, Columbus, OH*
RAINER LAUFS • *Institut für Infektionsmedizin, Universitätsklinikum Hamburg-Eppendorf, Hamburg, Germany*
HUANLIANG LIU • *Department of Laboratory Medicine, University of Washington School of Medicine, Seattle, WA (present affiliation: Department of Microbiology and Immunology, University of Miami Miller School of Medicine, Miami, FL)*
MARTIN LÖCHELT • *Abt. Genomveränderung und Carcinogenese, Forschungsschwerpunkt Infektion und Krebs, Deutsches Krebsforschungszentrum, Heidelberg, Germany*
FRANK LÜBBEN • *Dade Behring Marburg GmbH, Marburg, Germany*
ANDRE J. MAROZSAN • *University Department of Microbiology and Immunology, Weill Medical College of Cornell University, New York, NY*
NELSON L. MICHAEL • *Department of Molecular Diagnostics and Pathogenesis, Division of Retrovirology, Walter Reed Army Institute of Research, Rochville, MD*
ANDY MONTGOMERY • *Center for Retrovirus Research and Department of Veterinary Biosciences, The Ohio State University, Columbus, OH*
DAWN M. MOORE • *Molecular and Cellular Biology Program, Department of Molecular Biology and Microbiology, Case Western Reserve University, Cleveland, OH*
ALCINA NICOL • *Immunology Service of Evandro Chagas Research Institute, Oswaldo Cruz Foundation, Rio de Janeiro, Brazil*
ANNA NORDQVIST • *Unit of Virology, Division of Medical Microbiology, Department of Laboratory Medicine, Lund University, Lund, Sweden*
GERARD J. NUOVO • *Department of Pathology, Ohio State University Medical Center, Columbus OH*
CARLO-FEDERICO PERNO • *Department of Experimental Medicine, University of Rome "Tor Vergata," Roma, Italy*

KLAUS-INGMAR PFREPPER • *Department of Research and Development, Mikrogen GmbH, Martinsried, Germany*
SUSANNE POLYWKA • *Institut für Infektionsmedizin, Universitätsklinikum Hamburg-Eppendorf, Hamburg, Germany*
MIGUEL E. QUIÑONES-MATEU • *Department of Virology, Lerner Research Institute, Cleveland Clinic Foundation, Cleveland, OH*
ALI SAÏB • *CNRS UPR9051, Hôpital Saint-Louis, Paris, France*
HANNEKE SCHUITEMAKER • *Department of Clinical Viro-Immunology Sanquin Research at CLB and Landsteiner Laboratory at the Academical Medical Center Amsterdam, The Netherlands*
JANET D. SILICIANO • *Department of Medicine, The Johns Hopkins School of Medicine, Baltimore, MD*
ROBERT F. SILICIANO • *Department of Medicine, The Johns Hopkins School of Medicine and Howard Hughes Medical Institute, Baltimore, MD*
FRANÇOIS SIMON • *Unité de Virologie, GRAM EA 2636, CHU Charles Nicolle, Rouen, France, and Institut Pasteur de Dakar, Dakar, Senegal*
PIERRE SONIGO • *Département des Maladies Infectieuses, Institut Cochin, INSERM U567, CNRS UMR 8104, Université René Descartes, Paris, France*
SECONDO SONZA • *Macfarlane Burnet Institute for Medical Research and Public Health, Melbourne, Victoria, Australia, and Department of Medicine, Monash University, Clayton, Victoria, Australia*
CATHERINE TAMALET • *Laboratory of Virology, Timone Hospital, Marseille, France*
JOËLLE TOBALY-TAPIERO • *CNRS UPR9051, Hôpital Saint-Louis, Centre Hayem, Paris, France*
RYAN M. TROYER • *Division of Infectious Diseases, Department of Medicine, Case Western Reserve University, Cleveland, OH*

I

Methods for Virus Isolation and Detection

1

Enhanced Culture Assay for Detection and Quantitation of Latently Infected, Resting CD4+ T-Cells Carrying Replication-Competent Virus in HIV-1-Infected Individuals

Janet D. Siliciano and Robert F. Siliciano

Summary

Highly active antiretroviral therapy can decrease plasma HIV-1 levels to below the limit of detection. However, HIV-1 persists in latently infected resting-memory $CD4^+$ T-cells carrying an integrated copy of the viral genome. The pool of latently infected cells is extremely stable and represents a major barrier to HIV-1 eradication. Identification and characterization of this reservoir required the development of methods for purifying resting $CD4^+$ T-cells from HIV-1-infected individuals, activating the cells to induce virus production, and detecting and quantitating cells capable of releasing infectious virus. The development of an enhanced viral culture assay to quantitate the number of latently infected cells carrying replication competent virus is described here.

Key Words: HIV-1; latency; resting $CD4^+$ T-cells; reservoir; enhanced viral culture assay; replication-competent virus.

1. Introduction

Analysis of viral load in HIV-1 infection has contributed to our understanding of the pathogenesis and management of the disease. Elegant studies by Wei et al. *(1)* and Ho et al. *(2)* demonstrated that virus production in infected individuals is a highly dynamic process in which there is rapid turnover of both free virus and of productively infected cells. Highly active antiretroviral therapy (HAART) can decrease plasma HIV-1 levels to below the current limit of detection (<50 copies of HIV-1 RNA/mL) *(3)*. This therapy, which involves drugs that specifically inhibit HIV-1 reverse transcriptase and protease, can

From: *Methods in Molecular Biology, Vol. 304: Human Retrovirus Protocols: Virology and Molecular Biology*
Edited by: T. Zhu © Humana Press Inc., Totowa, NJ

prevent new rounds of infection, producing a dramatic drop in plasma virus levels. This decrease reflects the decay of both plasma virus and the cells that produce most of the plasma virus, namely productively infected, activated CD4+ T-cells. In patients treated with HAART, plasma virus levels usually drop below the limit of detection within 2 to 4 mo of initiation of therapy, and thereafter it becomes difficult to culture virus from the blood. Mathematical modeling of the decay of plasma virus and infected peripheral blood mononuclear cells (PBMC) allowed Perelson et al. to make the first rational predictions of the time it would take to eradicate HIV-1 in an infected individual, provided there were no stable viral reservoirs ***(3)***. Although this work raised hopes that eradication of the infection with HAART might be possible, it is now clear that HIV-1 persists in an extremely stable reservoir of latently infected, resting-memory CD4+ T-cells with integrated provirus ***(4–10)***.

Identification, characterization, and quantitation of the major reservoir for latent HIV-1 required the development of a novel assay in which resting CD4+ T-cells were purified to greater than 99.9% purity and then stimulated in a limiting dilution culture assay. Because latently infected cells do not produce virus without cellular activation ***(4,11)***, it was necessary to induce uniform in vitro activation of the resting cells. With this approach, replication-competent HIV-1 could reproducibly be recovered from the resting CD4+ cells of all infected individuals studied ***(5,8–10,12)***. Longitudinal studies of HIV-1-infected adults who maintained long-term suppression of viremia on HAART demonstrated the extraordinary stability of this reservoir. The frequency of latently infected cells in patients on HAART ranged within 0.03–3 infectious units per million (IUPM) resting CD4+ T-cells ***(8–10)***. Statistical analysis, using a random-effects regression model for decay with first-order kinetics, provided an estimate of the decay rate of the pool of latently infected resting CD4+ T-cells. The half-life of decay was 44.2 mo. At this decay rate, eradication of a reservoir of only 10^6 cells would require 73 yr of antiretroviral therapy. Therefore, the culture assay for detecting replication competent HIV-1 in resting CD4+ T-cells has provided the most definitive evidence to date that eradication of HIV-1 infection with antiretroviral therapy alone is unlikely. We describe here the method used to quantitate latently infected cells.

2. Materials

1. Ficoll-Paque™ PLUS (Amersham).
2. 60-mL sterile syringes.
3. Wash media (WM): phosphate-buffered saline (PBS), pH 7.4, 2% heat-inactivated newborn calf serum, 0.1% glucose, 20 U/mL penicillin, 20 μg/mL streptomycin, 12 m*M* HEPES, pH 7.4.
4. Heparin, sodium salt (Sigma): diluted to 1000 U/mL in WM, filter-sterilized, and stored at 4°C.

5. Purified phytohemagglutinin (PHA; Murex): 150 μg/mL stock in WM, filter-sterilized; stock aliquots stored at –80°C (*see* **Note 1**).
6. Complete media (CM): RPMI-1640 with GlutaMAX™ (GibcoBRL), 20 U/mL penicillin, 20 μg streptomycin, 10% heat-inactivated fetal bovine serum (FBS) (*see* **Note 2**).
7. Complete media + IL-2 (CM + IL-2): RPMI-1640 with GlutaMAX, 20 U/mL penicillin, 20 μg/mL streptomycin, 10% heat-inactivated fetal bovine serum, 2% T-cell growth factor (*see* **Note 3**), 100 U/mL recombinant human IL-2.
8. Dynabeads® M-450 sheep anti-mouse immunoglobulin (IgG) magnetic beads (Dynal).
9. Dynabeads M-450 anti-CD8 magnetic beads (Dynal).
10. Dynal MPC-1 magnetic particle concentrator (Dynal).
11. Multiwell™ 24-well sterile tissue culture plate with flat bottom and low evaporation lid (Becton Dickinson).
12. Multiwell 6-well sterile tissue culture plate with flat bottom and low evaporation lid (Becton Dickinson).
13. Mouse anti-human monoclonal antibodies: anti-CD8, clone 3B5 (Caltag); anti-Class II, clone L243 (*see* **Note 4**); anti-CD19, clone SJ25-C1 (Caltag); anti-CD25, clone 3G10 (Caltag); anti-CD14, clone TüK4 (Caltag); anti-CD16, clone 3G8 (Caltag); anti-CD69, clone L78 (Bectin-Dickinson).
14. Fluorochrome conjugated mouse antihuman monoclonal antibodies (Becton-Dickinson): PE-anti-CD4, clone SK3; FITC-HLADR, clone L243.
15. Coulter® HIV-1 p24 Antigen Assay (Beckman Coulter).
16. 50-mL and 15-mL sterile polypropylene centrifuge tubes (Corning).

3. Methods

In this enhanced virus culture assay, three important modifications were made to previously described virus culture assays ***(13–16)***. These modifications allow routine detection and quantification of latently infected resting $CD4^+$ T-cells capable of releasing replication competent virus following in vitro activation. First, the cells that are put into the culture assay are highly purified resting $CD4^+$ T-cells. Resting $CD4^+$ T-cells represent the major long-term reservoir for the virus ***(5,8–10,12)***, and therefore virus isolation is facilitated by starting with purified resting $CD4^+$ T-cells. Resting $CD4^+$ T-cells are obtained from PBMC by initially removing unwanted cell populations ($CD8^+$ T-cells, monocytes, natural killer [NK] cells, B-cells, and activated $CD4^+$ T-cells) using monoclonal antibodies to specific cell surface markers on these populations and magnetic beads conjugated with sheep anti-mouse Ig. Cells that have bound antibodies and beads are subsequently removed with a magnet. Flow cytometry is next used in a second purification step to positively select for resting $CD4^+$ T-cells. This two-step procedure routinely yields a highly purified population of cells that are 99.9% resting $CD4^+$ T-cells.

A second important modification was the use of a procedure that induces uniform activation of resting CD4+ T-cells. This is critical because cellular activation results in a number of changes that allow gene expression from the HIV-1 long terminal repeat (LTR). Previous virus culture assays did not include a specific activation step other than addition of allogeneic lymphoblasts. In our procedure, the addition of both PHA and a 10-fold excess of irradiated, allogeneic PBMC ensures that all latently infected cells in the culture will be uniformly activated *(17)*. The irradiated PBMC are required for high-efficiency activation, possibly because they serve as a source of macrophages and dendritic cells.

A third important modification is that CD4+ lymphoblasts from normal donors are added to amplify virus that is released from latently infected cells. Unlike other culture methods, CD8+ T-cells, which can suppress viral replication, are depleted from the lymphoblasts that are added to the enhanced culture assay.

It is important to note that this assay will detect resting CD4+ T-cells in both pre- and postintegration states of latency *(5,18)*. Viral entry and reverse transcription can occur in resting CD4+ T-cells *(19)*. However, completion of reverse transcription occurs much more slowly in resting CD4+ T-cells than in activated CD4+ T-cells *(19)*. In addition, reverse transcribed HIV-1 cDNA in resting cells does not integrate into the host cell genome owing to a block in nuclear import of the preintegration complex (PIC) *(20)*. The PIC eventually decays and the life cycle of the virus is aborted *(19)*. However, if resting cells in this preintegration stage are activated shortly after infection, the PIC can enter the nucleus and the viral cDNA can integrate into host cell genomic DNA. Production of virus can then take place. Therefore, cells in both pre- and postintegration latency will be measured in this culture assay. In patients who are viremic, the level of latently infected cells can be 100-fold higher than that observed in patients on HAART. When the frequency of latently infected cells is measured before and at various times after initiation of HAART, a biphasic decay in the frequency is observed *(18)*. The rapid intial decay in the first few months of treatment likely represents the loss of cells in the preintegration stage of latency. After the first few months, there is little additional decay owing to the stability of the postintegration form of latency described above. Because the preintegration state of latency is labile, the culture assay can be used to measure postintegration latency in patients who have been on HAART for 6 mo with suppression of viremia to <50 copies/mL.

3.1. Production of CD4+ Lymphoblasts

Because the amount of viral antigen released into the culture supernatant from a single latently infected cell that has been activated in vitro is below the limit of detection of enzyme-linked immunosorbent assays (ELISA), it is necessary to amplify virus through the addition of activated $CD4^+$ blasts to the culture assay. Blasts should be prepared 2–3 d before they are added to the culture assay so that they are fully permissive for viral infection at the time of addition. PBMC are obtained by discontinuous density gradient centrifugation of heparinized blood from an HIV-1 negative donor, and cells are subsequently activated with PHA. One hundred twenty mL of blood (approx $120–300 \times 10^6$ PBMC) is drawn into two 60-cc sterile syringes, each containing filter-sterilized heparin (10 U/mL of blood). The syringes are inverted one to three times to prevent clotting of the blood. Thirty milliliters of blood is carefully layered onto 15 mL Ficoll in each of four 50-mL sterile polypropylene centrifuge tubes. PBMC are separated from red blood cells and granulocytes by centrifuging the tubes at 470*g* in a Sorvall H-1000B swinging bucket rotor for 25 min at room temperature. Fifteen to 20 mL of the top layer of plasma is carefully removed so as not to disturb the PBMC at the interface of the two layers. The band of PBMC from each tube is carefully removed with a sterile 10-mL pipet and placed in separate fresh 50-mL centrifuge tubes. WM is added to each of the four tubes to a final volume of 45 mL, and the tubes are centrifuged at 470*g* for 15 min. WM is completely removed by aspiration, and the cell pellets are resuspended in 45 mL WM. The tubes are then centrifuged at 470*g* for 8 min. After removal of the WM by aspiration, the cell pellet is resuspended in 30 mL CM+IL-2 to which PHA has been added to a final concentration of 0.5 μg/mL. Typical yields of PBMC are $150–300 \times 10^6$ cells; therefore, the concentration ranges from 5 to 10×10^6 cells/mL. Once resuspended, the cells are added to a T75 tissue culture flask and the flask is incubated in an upright position for 2 d at 37°C in a humidified, 5% CO_2 incubator.

Because $CD8^+$ T-cells can suppress virus replication, the resulting lymphoblasts are depleted of $CD8^+$ T-cells using Dynal anti-CD8 magnetic beads. Briefly, the PHA-stimulated cells are counted, centrifuged to pellet the cells, and resuspended in 10 mL WM. One hundred microliters Dynal anti-CD8 beads/10^7 cells are aliquoted into a 15-mL sterile polypropylene centrifuge tube. It is critical that the beads that are to be added to the cells are washed once in WM to remove alcohol present in the bead solution. This is done by adding 10 mL of WM to the appropriate volume of beads and then placing the 15-mL tube into the Dynal MPC magnet. The liquid is removed by aspiration. The cell suspension is quickly added to the beads so that the beads do not dry. Cells and beads are mixed by gently pipetting up and down twice. The cells are incubated with the beads for 20–25 min at 4°C, with gentle rocking. Incubation with beads for longer than 25 min will greatly decrease cell viability. The Dynal magnet is then used to pull out the bead-coated $CD8^+$ T cells. The cells that do not bind to the magnet are collected. The beads are gently washed once to collect any cells that may be nonspecifically trapped within the beads. Again, the cells that do not bind to the magnet are collected, combined with the cells from the first pass, pelleted by centrifugation, and resuspended in CM + interleukin (IL)-2. Suspensions of $CD4^+$ T lymphoblasts prepared in this way are added to the virus cultures on d 2 and d 9 (*see* **Note 5**). Blasts added to the culture on d 2 are resuspended at a concentration of 0.67×10^6 cells/mL, whereas blasts added on d 9 are resuspended at 1×10^6 cells/mL.

3.2. Preparation of Irradiated PBMC

In order to culture virus from latently infected, resting $CD4^+$ T-cells, it is very important to uniformly activate all resting cells in culture. This is accomplished through the addition of both PHA and irradiated PBMC isolated from an HIV-1-negative donor. Irradiated PBMC are prepared the day that the culture assay is set up (d 1; *see* **Note 5**). For efficient activation, a 10-fold higher number of irradiated PBMC than purified resting $CD4^+$ T-cells are needed. Therefore, it is necessary to collect 120–180 cc blood from an uninfected donor, depending on how many patient cells will be put into culture. After blood is drawn into heparinized syringes, it is transferred into 50-mL sterile polypropylene tubes and cells are inactivated by γ-irradiation with 5000R in a cesium source irradiator. Irradiated PBMC are then isolated by discontinuous density gradient centrifugation and washed as described in **Subheading 3.1.** The irradiated PBMC are resuspended in CM + IL-2 at a concentration of 2.5×10^6 cells/mL. PHA is added to a final concentration of 1 μg/mL and the tube is immediately placed on ice until use to prevent macrophages from adhering to the walls of the centrifuge tube.

3.3. Isolation and Purification of Resting $CD4^+$ T-Cells From Infected Individuals

All experiments with blood and cells from HIV-positive individuals are carried out in a biological safety cabinet following standard BSL-3 protocols. One hundred eighty cc of blood is drawn from an HIV-1-positive individual into three 60-cc syringes containing heparin, as described above. PBMC are collected by discontinuous density gradient centrifugation and washed as described in **Subheading 3.1.** In the enhanced viral culture assay, d 0 refers to the day that blood is drawn from the infected patient (*see* **Note 5**). PBMC are resuspended in 30 mL CM, added to a T75 tissue culture flask, and placed, lying flat, at 37°C in a humidified, 5% CO_2 incubator. Incubation at 37°C allows monocytes to adhere to the tissue culture flask, thereby removing them from the population of cells that will be further purified. The length of incubation time can vary from 2 h to 18 h. Monocyte-depleted PBMC are collected from the flask, an aliquot is counted, and the remainder of the cells collected by centrifugation. The supernatant is removed, and the cells are placed on ice. The cells are incubated with saturating concentrations of monoclonal antibodies to

CD8 (30 μL/10^7 cells), CD19 (7 μL/10^7 cells), Class II (3 μL/10^7 cells), CD69 (6 μL/10^7), CD25 (3 μL/10^7 cells), CD14 (3 μL/10^7 cells), and CD16 (2 μL/10^7 cells) on ice for 25 min, with occasional mixing. Unbound antibodies are removed by washing the cells twice in ice-cold WM. It is very important that incubation of cells with antibody and subsequent washes are carried out at 4°C to prevent endocytosis of cell surface antigens bound to antibody. Cells with bound antibody are removed from the population by incubation with sheep anti-mouse Dynal magnetic beads. Prior to adding cells, beads are washed as described in **Subheading 3.1.** Cells and beads are incubated with rocking at 4°C for 25 min. The Dynal magnet is used to pull out bead-coated cells as described in **Subheading 3.1.** An aliquot of the cells not removed by the magnet is counted; the remaining unbound cells are collected by centrifugation and placed on ice.

Further purification of resting CD4+ T-cells is achieved by flow-cytometric sorting for small lymphocytes that express CD4 but not HLA-DR. To prepare cells for sorting, partially purified cells are labeled with saturating amounts of PE-conjugated anti-CD4 (67 μL/10^7 cells) and FITC-conjugated anti-HLADR (67 μL/10^7 cells) for 25 min on ice, with occasional mixing. To prevent photobleaching of the fluorochrome-labeled antibodies, the ice bucket is covered with aluminum foil and the fluorescent light in the laminar flow hood turned off. The labeled cells are washed twice in ice-cold WM to remove unbound antibodies, and the pellet resuspended in ice-cold WM to a final concentration of 2×10^7 mL. Cells are kept on ice in the dark until they are sorted. The sorting is done in a BSL-3 facility with aerosol protection. During the sorting, activated CD4+ T cells are removed based on cell size and expression of HLA-DR. Sorted cells are collected by centrifugation, counted, and put into the culture assay (*see* **Subheading 3.4.**). This two-step method of purification routinely yields a population of highly purified resting CD4+ T-cells that is >99% free of contaminating activated CD4+ T-cells.

3.4. Enhanced Virus Culture Assay

A limiting dilution culture assay is set up to detect and quantitate the frequency of latently infected resting CD4+ T-cells. Day 1 of the enhanced viral culture assay is the day that the purified resting CD4+ T-cells are put into culture and activated with PHA and irradiated PBMC (*see* **Note 5**). Purified resting cells (*see* **Subheading 3.3.**) are carefully counted and resuspended to 1×10^6 cells/mL in CM + IL-2 (Tube A). The number of cells in Tube A will vary depending on how many resting CD4+ T-cells are recovered from the sort. Five 15-mL centrifuge tubes are labeled B–F; 2.2 mL of CM + IL-2 is added to each of them. Serial fivefold dilutions are made by first taking 550 μL of cells from Tube A and adding the cells to Tube B. The cell suspension is gently mixed with a 2-mL pipet and 550 µL is removed and added to Tube C. Three additional fivefold serial dilutions are made in the same manner. Therefore, tubes A–F contain 1×10^6 cells/mL (Tube A), 200,000 cells/mL (Tube B), 40,000 cells/mL (Tube C), 8000 cells/mL (Tube D), 1600 cells/mL (Tube E), and 320 cells/mL (Tube F), respectively. One mL from each of these tubes is added to the appropriate tissue culture well (discussed later).

The limiting dilution assay with fivefold serial dilutions of patient cells is set up as follows: All tissue culture plates are set up in duplicate. Both 6-well and 24-well tissue culture plates are used, with the higher cell concentrations plated in the 6-well plate. The maximum number of resting $CD4^+$ T-cells that are put into one well of a 6-well plate is 1×10^6. Two individual wells of a 6-well tissue culture plate are labeled "1A" and "2A." This plate is labeled the 1×10^6 plate. To prevent contamination of virus from one well into another, wells are separated by an empty well. In addition to the resting $CD4^+$ T-cells (1 mL of cells from Tube A), each well also has 1×10^7 irradiated PBMC + PHA (*see* **Subheading 3.2.**) and 3 mL CM + IL-2, for a final volume of 8 mL/well. If more than 2.5×10^6 resting $CD4^+$ cells are obtained from sorting, additional wells of a 6-well plate are set up, each containing 1×10^6 resting $CD4^+$ T cells. Five such wells, considered together, provide for the next highest cell dose in the dilution series.

Cell concentrations below 1×10^6 are cultured in 24-well tissue culture plates. Wells labeled B–G are set up on duplicate plates, taking care not to place cells in wells that are directly adjacent to each other. Well G serves as a negative control; no patient cells are added to this well. Each well of the 24-well tissue culture plate is set up with 2.5×10^6 irradiated PBMC + PHA (1 mL; *see* **Subheading 3.2.**), and 1 mL of cells from the appropriate dilution tube. For example, 1 mL of cells from Tube B cells will be put into well B of each plate. Control well G has irradiated PBMC and 1.0 mL of CM + IL-2. The final volume of cells and media in the wells of a 24-well plate is 2 mL.

Plates are stacked, four plates per stack, wrapped in aluminum foil, and placed at 37°C in a humidified 5% CO_2 incubator.

On d 2 (*see* **Note 5**), PHA-containing supernatants are removed from all wells of the tissue culture plates, being very careful not to remove any cells. It is important to remove PHA because prolonged exposure will cause cell death. A different pipet must be used for each well to prevent cells and/or virus from one well contaminating another well. Typically, 6 mL is removed from each well of the 6-well plates and 1.5 mL is removed from each well of the 24-well plates. In order to ensure that all the PHA has been removed, the same volume of CM + IL-2 is carefully added to all wells. Cells are allowed to settle for two hours at 37°C, and then the supernatants are once again carefully removed. Blasts are now added to the culture. Each well of a 6-well plate receives 4×10^6

blasts in 6 mL of CM + IL-2. Each well of a 24-well plate receives 1 × 10^6 blasts in 1.5 mL CM + IL-2.

It is important to monitor the pH of the culture media by visual inspection of the plates every day. Typically, the media will turn yellow by d 5 of the assay owing to the proliferation of both blasts and patient cells. Therefore, 3 mL of media are removed from each of the large wells and 0.75 mL is removed from all small wells; all wells are replenished with the same volume of fresh CM + IL-2. On d 7, all wells are split in half. Cells and media are gently mixed by pipetting and half of each well is discarded into a bleach bucket. A different pipet is used for each well. A second addition of blasts, prepared two days earlier (*see* **Subheading 3.1.**), is added to all wells. Plates are further incubated at 37°C in a humidified incubator with 5% CO_2. On d 10, wells are again fed with fresh media, as on d 5. On d 14, the cultures are tested for the presence of HIV-1 antigen using the Coulter HIV-1 p24 Antigen Assay kit per the manufacturer's instructions. Two hundred microliters of culture supernatant is removed from each well for the analysis. A third addition of blasts may be added to the culture and the plates further incubated for an additional week in the same manner if p24 results are negative.

3.5. Statistical Analysis

Infected cell frequencies are determined by the maximum likelihood method *(21)* and were expressed as IUPM resting $CD4^+$ T-cells. The precision of the measurement depends on the number of replicates tested at each cell dilution. In general, the yield of purified resting $CD4^+$ T-cells is only few million cells; thus the assay is generally set up as a duplicate fivefold dilution series. For assays set up in this format, frequencies can be determined from a tabulated set of calculated frequencies representing the maximum likelihood estimate for each possible outcome in such a dilution series. This table may be obtained from the ACTG. Frequencies for common outcomes are given in **Table 1**. The confidence intervals for individual determinations are ± 0.7 log IUPM.

4. Notes

1. Once a stock tube of 150 µg/mL PHA has been thawed, it can be stored at 4°C for up to 7 d.
2. Multiple lots of FBS are tested to find one that will support optimal growth of mitogen-stimulated primary human T-cells. A proliferation assay with $CD4^+$ blasts and media containing a specific lot of FBS is set up in the following manner: $CD4^+$ blasts are prepared as described in **Subheading 3.1.** The blasts are resuspended in CM at a concentration of 5 × 10^5 cells/mL. Using a multichannel pipettor, 0.1 mL of cells is added to the appropriate number of wells of a 96-well, round-bottom tissue culture plate. Then 0.1-mL aliquots of RPMI-1640 + 20% of the FBS lots to be tested are added to each well. The plate is incubated at 37°C in

Table 1
Limiting Dilution Analysis of the Frequencies of Latently Infected Cells

Input number of cells/well[a]	Number of wells positive for virus growth at the indicated number of input cells[b]															
25,000,000	1	2	2	2	2	2	2	2	2	2	2	2	2	2	2	2
5,000,000	0	0	1	2	2	2	2	2	2	2	2	2	2	2	2	2
1,000,000	0	0	0	0	1	2	2	2	2	2	2	2	2	2	2	2
200,000	0	0	0	0	0	0	1	2	2	2	2	2	2	2	2	2
40,000	0	0	0	0	0	0	0	0	1	2	2	2	2	2	2	2
8000	0	0	0	0	0	0	0	0	0	0	1	2	2	2	2	2
1600	0	0	0	0	0	0	0	0	0	0	0	0	1	2	2	2
320	0	0	0	0	0	0	0	0	0	0	0	0	0	0	1	2
Frequency (IUPM)c	**0.023**	**0.053**	**0.12**	**0.32**	**0.51**	**1.6**	**3.2**	**8.1**	**16**	**41**	**82**	**206**	**421**	**1121**	**2503**	**5608**

[a]Purified resting $CD4^+$ T-cells are cultured in duplicate fivefold dilution series as described in **Subheading 3.4.** For the 25 million and 5 million cell dilutions, multiple wells with 1 million cells/well are set up and treated as a group.

[b]For duplicate dilution series. This table indicates the most common patterns. For analysis of other patterns, tabulated frequencies are available from the ACTG Virology Core.

[c]Based on maximum likelihood analysis **(21)**. IUPM, infectious units per million.

a humidified, 5% CO_2 incubator for 48 h. One microcurie of ^{3}H-thymidine in 50 µL CM is added to each well. The plates are further incubated at 37°C for 24 h. Labeled cells are collected onto a filter using a cell harvestor. Once dry, the filter is counted to determine the amount of radioactivity incorporated into the cells. Lots giving maximal proliferation are selected.

3. T-cell growth factor is a general name given to human IL-2 containing supernatants of mitogen-activated PBMC. The supernatants are prepared from PBMC in the following manner: two leukopaks are collected from two different donors by trained individuals in a hemaphoresis laboratory. PBMC are isolated by density gradient centrifugation and washed as described in **Subheading 3.1.** Typically, approx 1.5×10^9 cells from each donor are obtained. The cells from each donor are cultured together in 500 mL RPMI-1640 with Glutamax, 2.5% heat-inactivated AB human serum (Gemini), 20 µg/mL streptomycin, and 20 U/mL penicillin in 150 cm^2 tissue culture flasks, 70–75 mL/flask. The cells are incubated overnight at 37°C in a humidified incubator with 5% CO_2. The next day, PHA and PMA (Phorbol 12-myristate 13 acetate; Sigma) are added to each flask to a final concentration of 2 µg/mL and 5 ng/mL, respectively. The cells are further incubated for 4 h at 37°C. The supernatants from each flask are gently poured out and the flasks rinsed with 15 mL prewarmed WM, three times/flask. Then, 75 mL RPMI-1640 with Glutamax, 2.5% heat-inactivated AB human serum, 20 µg/mL streptomycin, and 20 U/mL penicillin is added to each flask. Cells are cultured for 40 h at 37°C. Supernatants from the activated cells are collected, centrifuged at 470*g* to remove cells, filtered through 0.45-µm Nalgene filters, and stored in 45-mL aliquots at –20°C. Supernatants are assayed in a proliferation assay (*see* **Note 2**). The concentration that gives 50–75% maximal proliferation is added to CM.
4. Hybridoma clone L243 IgG was produced by the Monoclonal Antibody Core of the Johns Hopkins University Center for AIDS Research. The hybridoma line L243 was weaned for growth in serum-free medium (CD Hybridoma Medium; Invitrogen) and then grown in an Integra culture flask (IBS, Switzerland) to densities greater than 10^8/mL. Supernatants containing mg/mL quantities of IgG were collected every 3 to 4 d and pooled. The pooled supernatants were dialyzed against PBS. The antibody was titered using freshly prepared PBMC.
5. Day 0 of the culture assay is the day that blood is drawn from an HIV-positive individual. The day that purified resting $CD4^+$ T-cells are put into culture with irradiated PBMC and PHA is d 1. PHA is removed and lymphoblasts are added to the culture on d 2.

Acknowledgments

This work was supported by grants from the NIH (AI43222), the Doris Duke Charitable Foundation, and the Howard Hughes Medical Institute.

References

1. Wei, X., Ghosh, S. K., Taylor, M. E., Johnson, V. A., Emini, E. A., Deutsch, P., et al. (1995) Viral dynamics in human immunodeficiency virus type 1 infection. *Nature* **373,** 117–122.
2. Ho, D. D., Neumann, A. U., Perelson, A. S., Chen, W., Leonard, J. M., and Markowitz, M. (1995) Rapid turnover of plasma virions and CD4 lymphocytes in HIV-1 infection. *Nature* **373,** 123–126.
3. Perelson, A. S., Essunger, P., Cao, Y., Vesanen, M., Hurley, A., Saksela, K., Markowitz, M., and Ho, D. D. (1997) Decay characteristics of HIV-1-infected compartments during combination therapy. *Nature* **387,** 188–191.
4. Chun, T.-W., Finzi, D., Margolick, J., Chadwick, K., Schwartz, D., and Siliciano, R. F. (1995) Fate of HIV-1-infected T cells in vivo: Rates of transition to stable latency. *Nat. Med.* **1,** 1284–1290.
5. Chun, T.-W., Carruth, L., Finzi, D., Shen, X., Digiuseppe, J. A., Taylor, H., et al. (1997) Quantitation of latent tissue reservoirs and total body load in HIV-1 infection. *Nature* **387,** 183–188.
6. Chun, T. W., Stuyver, L., Mizell, S. B., Ehler, L. A., Mican, J. M., Baseler, M., et al. (1997) Presence of an inducible HIV-1 latent reservoir during highly active antiretroviral therapy. *Proc. Natl. Acad. Sci. USA* **94,** 13,193–13,197.
7. Wong, J. K., Gunthard, H. F., Havlir, D. V., Zhang, Z. Q., Haase, A. T., Ignacio, C. C., et al. (1997) Reduction of HIV-1 in blood and lymph nodes following potent antiretroviral therapy and the virologic correlates of treatment failure. *Proc. Natl. Acad. Sci. USA* **94,** 12,574–12,579.
8. Finzi, D., Hermankova, M., Pierson, T., Carruth, L. M., Buck, C., Chaisson, R. E., et al. (1997) Identification of a reservoir for HIV-1 in patients on highly active antiretroviral therapy. *Science* **278,** 1295–1300.
9. Finzi, D., Blankson, J., Siliciano, J. D., Margolick, J. B., Chadwick, K., Pierson, T., et al. (1999) Latent infection of CD4+ T cells provides a mechanism for lifelong persistence of HIV-1, even in patients on effective combination therapy. *Nat. Med.* **5,** 512–517.
10. Siliciano, J. D., Kajdas, J., Finzi, D., Quinn, T. C., Chadwick, K., Margolick, J. B., et al. (2003) Long term follow-up studies confirm the extraordinary stability of the latent reservoir for HIV-1 in resting CD4+ T cells. *Nat. Med.* **9,** 727–728.
11. Chun, T. W., Justement, J. S., Lempicki, R. A., Yang, J., Dennis, G., Jr., Hallahan, C. W., et al. (2003) Gene expression and viral prodution in latently infected, resting CD4+ T cells in viremic versus aviremic HIV-infected individuals. *Proc. Natl. Acad. Sci. USA* **100,** 1908–1913.
12. Persaud, D., Pierson, T., Ruff, C., Finzi, D., Chadwick, K. R., Margolick, J. B., et al. (2000) A stable latent reservoir for HIV-1 in resting CD4(+) T lymphocytes in infected children. *J. Clin. Invest.* **105,** 995–1003.
13. Ho, D. D., Moudgil, T., and Alam, M. (1989) Quantitation of human immunodeficiency virus type 1 in the blood of infected persons. *N. Engl. J. Med.* **321,** 1621–1625.

14. Connor, R. I., Mohri, H., Cao, Y., and Ho, D. D. (1993) Increased viral burden and cytopathicity correlate temporally with CD4+ T-lymphocyte decline and clinical progression in human immunodeficiency virus type 1-infected individuals. *J. Virol.* **67,** 1772–1777.
15. Spina, C. A., Prince, H. E., and Richman, D. D. (1997) Preferential replication of HIV-1 in the CD45RO memory cell subset of primary CD4 lymphocytes in vitro. *J. Clin. Invest.* **99,** 1774–1785.
16. Aids Clinical Trials Group (1997) Quantitative PBMC microculture assay. In *Virology Manual*, National Institute of Allergy and Infectious Disease, Bethesda, MD, pp. 61–66.
17. Hermankova, M., Siliciano, J. D., Zhou, Y., Monie, D., Chadwich, K., Margolick, J. B., et al. (2003) Analysis of HIV-1 gene expression in latently infected resting CD4+ T lymphocytes in vivo. *J. Virol.* **77,** 7383–7392.
18. Blankson, J. N., Finzi, D., Pierson, T. C., Sabundayo, B. P., Chadwick, K., Margolick, J. B., et al. (2000) Biphasic decay of latently infected CD4+ T cells in acute HIV-1 infection. *J. Infect. Dis.* **182,** 1636–1642.
19. Pierson, T. C., Zhou, Y., Kieffer, T., Ruff, C. T., Buck, C., and Siliciano, R. F. (2002) Molecular characterization of preintegration latency in HIV-1 infection. *J. Virol.* **76,** 8518–8531.
20. Bukrinsky, M. I., Sharova, N., Dempsey, M. P., Stanwick, T. L., Bukrinskaya, A. G., and Stevenson, M. (1992) Active nuclear import of human immunodeficiency virus type 1 preintegration complexes. *Proc. Natl. Acad. Sci. USA* **89,** 6580–6584.
21. Myers, L. A., McQuay, L. J., and Hollinger, F. B. (1994) Dilution assay statistics. *J. Clin. Micro.* **32,** 732–739.

2

Isolation, Propagation, and Titration of Human Immunodeficiency Virus Type 1 From Peripheral Blood of Infected Individuals

Hanneke Schuitemaker and Neeltje A. Kootstra

Summary

HIV-1 can be isolated from peripheral blood mononuclear cells and is easily propagated on primary cells in vitro. Here we describe the method for bulk isolation of the HIV-1 quasispecies and a limiting dilution virus isolation protocol by which single coexisting clones can be obtained. In addition, methods for propagation and titration of HIV-1 are provided.

Key Words: HIV isolation; bulk isolate; HIV variant; $TCID_{50}$; stock preparation.

1. Introduction

To study biological properties of the human immunodeficiency virus type 1 (HIV-1) in relation to the clinical course of infection, the in vitro preservation of phenotypical characteristics of the virus that may be relevant in vivo is essential. Even in the early days of HIV-1 research, it was clearly recognized that passage through immortalized T-cell lines was successful with only some of the viruses *(1)*. In addition, adaptation to T-cell lines changed the viral phenotype *(2)*. Thus, primary HIV-1 isolates should be isolated and propagated on primary peripheral blood mononuclear cells (PBMC), as this provides the most optimal preservation of the biological phenotype of the virus. This chapter describes the isolation, propagation, and titration of HIV-1 bulk isolates and biologically cloned HIV variants.

2. Materials

1. Iscove's modified Dulbecco's medium (IMDM; BioWhittaker).
2. Heparin.

From: *Methods in Molecular Biology, Vol. 304: Human Retrovirus Protocols: Virology and Molecular Biology*
Edited by: T. Zhu © Humana Press Inc., Totowa, NJ

3. Trisodium citrate dihydrate (TNC, Merck).
4. Phosphate-buffered saline (PBS).
5. Fetal calf serum (FCS; Hyclone).
6. Recombinant interleukin-2 (rIL-2; Chiron Benelux).
7. Penicillin (Pen), 100 U/mL (GibcoBRL).
8. Streptomycin (Strep), 100 U/mL (GibcoBRL).
9. Phytohaemagglutinin (PHA; Wellcome).
10. Heparinized peripheral blood or buffy coat.
11. Ficoll-Isopaque (Pharmacia).
12. Sterile 10-mL pipets.
13. Sterile 25-mL pipets.
14. Sterile 50-mL plastic tubes.
15. Pasteur pipets.
16. PFH medium: PBS supplemented with 5% FCS and 10 U/mL heparin.
17. PFT medium: PBS supplemented with 5% FCS, 10% TNC.
18. IF medium: IMDM supplemented with 5% FCS.
19. PHA medium: IMDM supplemented with 10% FCS, 1 μg/mL PHA, Pen/Strep.
20. Interleukin (IL)-2 medium: IMDM supplemented with 10% FCS, Pen/Strep, 10 U/mL rIL-2.
21. 0.2% Triton X-100 solution: PBS supplemented with 0.2% Triton X-100.
22. 96-well flat-bottom microtiter plates.
23. Multichannel pipet.
24. Sterile pipet tips.

3. Methods

The methods described below outline (1) the isolation of peripheral blood mononuclear cells from patient blood and buffy coat, (2) the virus isolation from patient PBMC in bulk, (3) HIV isolation under limiting dilution conditions to obtain biological virus clones, (4) the propagation of primary HIV isolates and variants, and (5) the titration of virus stocks (*see* **Note 1**).

3.1. Isolation of PBMC From Patient Blood and Buffy Coats

Transfer heparinized venous patient blood to plastic 50-mL tubes (maximum of 25 mL patient blood per tube) and dilute with an equal volume of PFH medium. When citrate has been used as an anticoagulant, add an equal volume of PFT medium. When PBMC are isolated from a buffy coat, transfer the buffy coat to a 250-mL flask and add PFT medium to a final volume of 150 mL. Gently load 25 mL of diluted blood on top of 12.5 mL of a Ficoll Isopaque solution with a density of 1.077 g/mL in a 50-mL tube. Centrifuge at 760*g* for 20 min at room temperature. Collect the cell band on top of the Ficoll layer. Wash the harvested cells twice by adding IMDM to a total volume of 50 mL and centrifuge at 425*g* for 10 min. PBMC can now either be cryopreserved in

IMDM containing 10% FCS and 10% dimethyl sulfoxide (DMSO) or directly used (*see* **Note 2**).

3.2. Isolation of Human Immunodeficiency Virus on Donor PBMC

3.2.1. Selection of Donor PBMC

PBMC may vary in their susceptibility to infection with HIV. To optimize the efficiency of the isolation procedure, it may therefore be useful to select optimal susceptible donor PBMC.

3.2.2. Exclusion of PBMC From Donors With a 32-Basepair Deletion in CCR5 Gene

HIV-1 susceptibility is most obviously determined by the CCR5 genotype and associated β-chemokine production levels. The CCR5 gene encodes for the coreceptor for R5 HIV variants. To achieve optimal susceptible target cells during virus isolation, it is recommended to exclude buffy coats from donors who are homozygous or heterozygous for the 32-basepair (bp) deletion in CCR5 that results in a premature stop codon and the absence of functional CCR5 on the cell membrane. The presence of this 32-bp deletion is rapidly determined with a polymerase chain reaction (PCR) *(3)*. Genomic DNA can be most conveniently isolated from donor PBMC using the Qiagen blood kit (Qiagen, Hilden, Germany). Subsequently, 100 ng of DNA can be used for PCR analysis with primers (sense, position 612 to 635 in CCR5, 5'-GATAGGTACCTGGCTGTCGTCCAT-3'; antisense, position 829 to 850 in CCR5, 5'-AGATAGTCATCTTGGGGCTGGT-3') flanking the described 32-nucleotide deletion in the CCR5 gene. Samples are amplified with 1 U of Taq polymerase (Promega, Madison, WI) in the buffer provided by the manufacturer, with a final $MgCl_2$ concentration of 3 mmol/L. Conditions for PCR reaction are: 5 min of denaturation at 95°C; 30 cycles of 1 min at 95°C, 1 min at 56°C, and 2 min at 72°C; and 5 min of elongation at 72°C in a Perkin-Elmer Cetus DNA thermal cycler 480 (Perkin-Elmer, Foster City, CA). The PCR product lengths are 238 bp for the wild-type allele and 202 bp for the mutant allele. DNA samples that show both bands are obviously from donors with a CCR5 D32 heterozygous genotype. PCR products can be analyzed by 2% agarose gel electrophoresis and ethidium bromide staining.

3.2.3. Measurement of HIV Susceptibility of PBMC

To determine relative HIV susceptibility of donor PBMC, a cell-free virus stock of a R5 HIV isolate is titrated in quadruplicate on PBMC from different donors in microtiter plates. PBMC from donors in which the highest $TCID_{50}$ value is achieved are selected for further use. For a detailed description of the $TCID_{50}$ assay, *see* **Subheading 3.5.**

3.3. Isolation of Primary HIV-1 on PBMC

Primary HIV-1 can be isolated either in bulk or under limiting dilution conditions to obtain biological clonal virus variants. For both methods, PBMC from a healthy blood donor volunteer are used as target cells. These donor PBMC are stimulated with PHA medium for 2 to 3 d, starting at a cell density of 5 × 10^6/mL. PHA-stimulated cells are then centrifuged at 425*g* and resuspended in IL-2 medium at a cell density of 1 × 10^6/mL.

3.3.1. Isolation of Bulk HIV (the "Quasi-Species")

Mix 5 mL of the PHA-stimulated donor PBMC suspension with 1 to 3 mL of isolated patient PBMC at a cell density of 1 × 10^6 in IL-2 medium in a 15-mL tube (*see* **Note 3**). Spin the cells down and incubate the cell pellet in a 37°C water bath for 1 h prior to culture in a 25-cm^2 flask. Alternatively, put the cells directly in the 25-cm^2 flask and culture the cells overnight in the 37°C incubator standing under a 45° angle, allowing the cells to settle and be in close contact in a corner of the flask. Afterwards, place the flask in an upright position and further incubate the cells. Four days after the start of the co-cultivation, cells are collected, centrifuged, and resuspended in the original volume of fresh IL-2 medium. Collected supernatant is tested for the presence of p24 gag antigen reflecting virus production (*see* **Note 4**). On d 7, cells are centrifuged and the supernatant is again tested for the presence of p24 antigen. To half of the cell pellet, 5 × 10^6 fresh 2–3 d PHA-stimulated donor PBMC are added to a total volume of 8 mL IL-2 medium and cultures are maintained. The other half of the pellet can be cryopreserved in IMDM with 10% DMSO. Medium changes are repeated on d 11, 18, and 25. The addition of fresh PHA-stimulated cells is repeated on d 14, 21, and 28. In general, cultures are continued for 4 to 5 wk. Supernatant is stored in small aliquots (1–1.5 mL/ampule) at –80°C.

3.3.2. Isolation of Biological HIV Clones

Molecular studies have demonstrated that in HIV-1-infected individuals at each moment a large number of related but different virus variants may coexist. Virus isolation as described in **Subheading 3.2.1.** will result in the rapid outgrowth of one or few of the infectious HIV-1 variants present in the patient's PBMC with the most fit phenotype under these isolation conditions. To obtain a more complete picture of the diversity of the HIV quasispecies present in an individual in vivo, a virus isolation protocol was developed that allows for the isolation of multiple HIV-1 variants from a single PBMC sample, avoiding overgrowth and loss of slowly replicating variants (**Fig. 1**) *(4)*.

Biological virus clones can be obtained by cocultivation of patient PBMC with PHA-stimulated healthy blood donor PBMC (donor PHA-PBMC) under

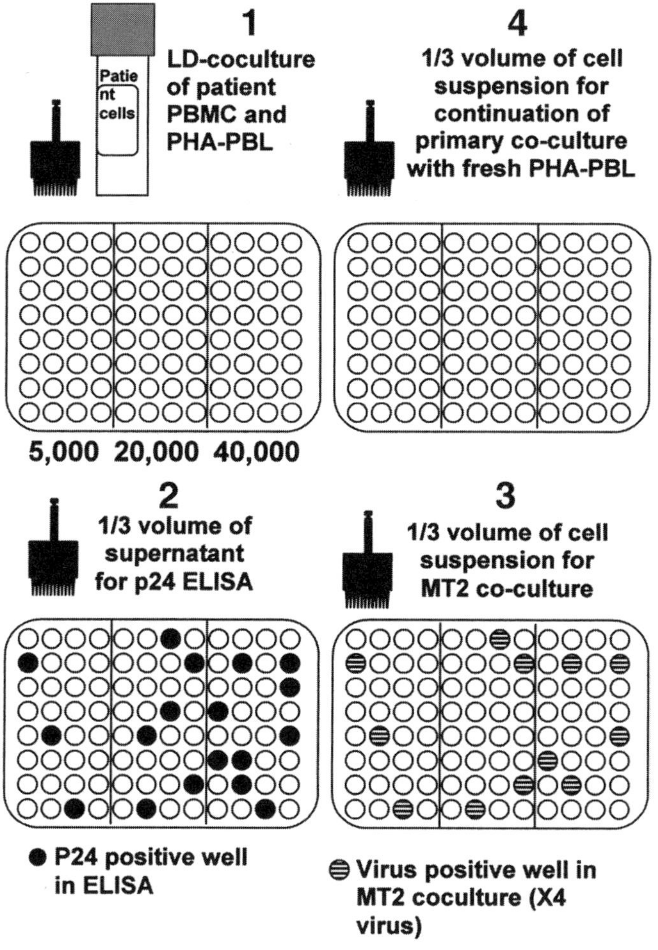

Fig. 1. Isolation of biological HIV-1 clones is achieved by co-culture of limiting dilutions of HIV-1-infected patient peripheral blood mononuclear cells (suggested cell numbers in figure are 5000, 20,000, and 40,000 patient cells per well; 32 wells per patient cell number) with phytohemagglutinin (PHA)-stimulated healthy donor peripheral blood lymphocytes (PBL). Step 1 is the initiation of the co-culture, step 2 is the harvesting of 50 μL supernatant for analysis of virus production in a p24 enzyme-linked immunosorbent assay to be performed weekly, step 3 is the transfer of 50 μL of the cell suspension to a MT2 co-culture to monitor the presence of X4 viruses that will induce large syncytia in MT2 cells, step 4 is the transfer of 50 μL of the cell suspension to fresh PHA-stimulated healthy-donor PBL for continuation of the primary culture. From this plate, steps 2, 3, and 4 are repeated at d 14. At d 21, steps 2, 3, and 4 are repeated from the primary co-culture plate that was established at d 14, and so forth. *See* text for further details.

limiting dilution conditions (*see* **Note 4**). Patient PBMC are isolated by Ficoll density gradient centrifugation as described above and diluted in medium to cell concentrations of 5×10^4, 10×10^4, 20×10^4, and 40×10^4 cells/mL IL-2 medium. These cell suspensions are then seeded in 96-well microtiter plates, 100 μL/well and 32 to 48 wells per cell concentration (*see* **Note 5**). Subsequently, 10^5 PHA-stimulated healthy donor PBMC in 50 mL IL-2 medium are added to each well and cultures are incubated in an incubator at 37°C, with a humidified atmosphere and 5% CO_2. Every week, 50 μL culture supernatant is transferred from all wells, row by row, to wells of a new 96-well plate, using a multichannel pipet. With the same pipet tips, the remaining cells in that row of the microtiter plate are resuspended and 50 μL is transferred from each well of that row to the corresponding row in a new 96-well microtiter plate. Subsequently, 10^5 freshly prepared PHA-stimulated healthy donor PBMC in 100 μL IL-2 medium are added to propagate the culture. Microcultures are continued in an incubator as described above. To each well of the original 96-well plate, add 50 μL of the initial PBMC co-culture and 100 μL of 10^5 MT2 cells in log phase. These cultures will be microscopically studied with a ×40 magnification at d 5 after start of the MT2 co-culture for the presence of syncytia, which indicates that the HIV variant present in the co-culture has the capacity to use HIV coreceptor CXCR4.

To reveal virus production in the original PBMC co-cultures, the presence of p24 antigen in the supernatants is tested using a p24 specific antigen capture enzyme-linked immunosorbent assay (ELISA) (*see* **Note 3**). Prior to performing the ELISA, infectious HIV that may be present in the supernatant should be inactivated by adding 50 μL of 0.2% Triton X-100 solution.

The whole procedure is repeated at d 14, 21, and 28. The frequency of infected PBMC is estimated using the formula for Poisson distribution ($u = -\ln F_0$, in which F_0 is the fraction of negative cultures). Isolates can be considered clonal if 37% or less of the cultures become positive, which corresponds to ≤0.5 infected cells/well.

3.4. Propagation of Primary HIV

Once virus replication is detected by p24 ELISA (*see* **Note 3**), the virus should be stored as quickly as possible to minimize the number of passages, as this may lead to adaptation of the virus to in vitro conditions. Virus can be stored as cell-free supernatant at –80°C, or as infected viable frozen PBMC at –80°C or in liquid nitrogen.

3.4.1. Expansion of Primary HIV From Cell-Free Supernatant

To expand HIV from cell-free supernatant, PHA-stimulated PBMC are prepared as described above (*see* **Subheadings 3.1.** and **3.2.**). For each inoculation, a cell suspension with an absolute cell number of 8×10^6 cells is transferred to a 15-mL tube. Cells are centrifuged (425*g*, 10 min at room temperature) and the supernatant is decanted. The cell pellet is then resuspended in 1–1.5 mL HIV-positive supernatant from the thawed ampule and incubated for 2 h in a 37°C shaking water bath. Subsequently, 7 mL IL-2 medium is added and the cell suspension is transferred to a 25-cm^2 tissue culture flask and cultured at 37°C. Every 3–4 d, 100 µL is sampled from the supernatant to monitor for virus production in a p24 antigen capture ELISA. At d 4, 11, 18, and 25, half of the medium is replaced with fresh IL-2 medium. At d 7, 14, and 21, half of the cell suspension is removed and 4×10^6 fresh PHA-stimulated PBMC are added. When the optical density (OD) of the sample in the p24 ELISA is four times the OD of the negative control, the supernatant is aliquoted in 1–1.5-mL samples and stored at –80°C until further use.

3.4.2. Expansion of Primary HIV From Viable Frozen HIV-Infected PBMC

The propagation of HIV from HIV-infected viable frozen PBMC is identical to the procedure described for virus isolation from HIV-infected patient PBMC (*see* **Subheading 3.3.1.**).

3.5. Determination of Infectious Titer of HIV in Supernatant

To establish the titer of a virus stock, serial 1:5 dilutions of the stock are made in IL-2 medium. For each virus, six 15-mL tubes with 2 mL IL-2 medium each are prepared. Then 0.5 mL from the original virus stock is transferred to the first 15-mL tube with 2 mL IL-2 medium. After thorough mixing, 0.5 mL is transferred from this tube to the next tube with 2 mL IL-2 medium. This is repeated until six dilutions have been made.

From 2-d PHA-stimulated PBMC, a cell suspension of 10^6/mL IL-2 medium is prepared. This suspension is then seeded in a 96-well plate, 100 µL per well, 4 columns (32 wells) per virus titration. Subsequently, undiluted virus stock is added to 4 wells in row A of the plate, 50 µL/well. The 1:5 dilution of the virus stock is added to 4 wells in row B, the 1:25 dilution of the virus stock is added to 4 wells in row C, and so on. There is room for three virus titrations on one 96-well plate. Cultures are incubated at 37°C with 5% CO_2 in a humidified atmosphere. On d 7 and 10, 75 µL supernatant is harvested using a multichannel pipet and transferred to a new plate to be inactivated with 75 µL/well of a 0.2% Triton X-100 solution. On d 7, subsequently 75 µL fresh IL-2 medium is added to each well, to propagate the cultures. Cultures are terminated after supernatant sampling on d 10.

The 50% tissue culture infectious dose ($TCID_{50}$) in the supernatant is calculated as follows:

–log $TCID_{50}$ endpoint = –log a– [(b – 0.5) × log c], in which a = the lowest virus dilution in the assay (e.g., 10^{-1}); b = sum of the percent p24 positive cultures at each dilution; and c = the dilution factor.

4. Notes

1. Working with replication-competent HIV-1 requires a laboratory at biological safety level 2 (BSL-2).
2. Working with primary cells from blood donor and HIV-infected patients may result in the frequent introduction of mycoplasma. Not only should precautions be taken to prevent infection of permanent cell lines that are being used in the laboratory, but it is also recommended to add cyproxin to the cultures, to inhibit replication of mycoplasma.
3. An inhibitory effect of CD8 cells on HIV replication has been reported. When an attempt is made to isolate virus from an HIV-infected individual with low virus load, it is recommended to deplete CD8 cells from patient PBMC and from PHA-stimulated PBMC. The latter should be performed after PHA stimulation, as the absence of CD8 cells seems to interfere with PHA stimulation.
4. As an alternative for the p24 antigen capture ELISA, a reverse transcriptase (RT) assay can be performed to detect virus production.
5. It is possible to isolate HIV-1 variants from specific peripheral-blood T-cell subsets. For this purpose, the cells of choice should be sorted on a Fluorescence Activated Cell Sorter (FACS) first and then used for virus isolation according to one of the protocols described earlier.

References

1. Popovic, M., Sarngadharan, M. G., Read, E., and Gallo, R. C. (1984) Detection, isolation and continuous production of cytopathic retroviruses (HTLV-III) from patients with AIDS and pre-AIDS. *Science* **224,** 497.
2. Sawyer, L. S. W., Wrin, M. T., Crawford-Mikza, L., Potts, B., Wu, Y., Weber, P. A., et al. (1994) Neutralization sensitivity of human immunodeficiency virus type 1 is determined in part by the cell in which the virus is propagated. *J. Virol.* **68,** 1342–1349.
3. De Roda Husman, A. M., Koot, M., Cornelissen, M., Brouwer, M., Broersen, S. M., Bakker, M., et al. (1997) Association between CCR5 genotype and the clinical course of HIV-1 infection. *Ann. Intern. Med.* **127,** 882–890.
4. Schuitemaker, H., Koot, M., Kootstra, N. A., Dercksen, M. W., De Goede, R. E. Y., Van Steenwijk, R. P., et al. (1992) Biological phenotype of human immunodeficiency virus type 1 clones at different stages of infection: progression of disease is associated with a shift from monocytotropic to T-cell-tropic virus populations. *J. Virol.* **66,** 1354–1360.

3

Isolation of Human Immunodeficiency Virus Type 1 From Peripheral Blood Monocytes

Paul R. Gorry, Secondo Sonza, Katherine Kedzierska, and Suzanne M. Crowe

Summary

Human immunodeficiency virus type 1 (HIV-1) can infect circulating peripheral blood monocytes and resting $CD4^+$ T lymphocytes despite sustained suppression of plasma viremia to undetectable levels. These persistently infected cell populations pose a barrier for virus eradication by highly active antiretroviral therapy (HAART), and are a significant reservoir of HIV-1 contributing to viral rebound following cessation or failure of HAART. This chapter provides a protocol for isolating replication-competent HIV-1 from peripheral blood monocytes of HIV-1-infected individuals, including those with sustained plasma HIV-1 RNA levels below 50 copies/mL, by co-culture with CD8-depleted, phytohemagglutinin-activated donor peripheral blood mononuclear cells. In our laboratory, this protocol has the sensitivity to achieve a success rate of positive HIV-1 isolation in approx 70% of cases. The study of HIV-1 strains harbored by peripheral blood monocytes of patients undergoing HAART will contribute to the understanding of viral persistence in cellular reservoirs that impede effective HAART.

Key Words: Monocyte; macrophage; human immunodeficiency virus; HIV-1; monocyte isolation; HAART; CD8-depleted PBMC; co-culture.

1. Introduction

The ability to isolate HIV-1 strains from different cell types assists in elucidating the phenotypic and/or genetic properties of particular viral strains that enable them to persist in various cellular compartments. However, success rates in isolating virus vary greatly depending on the source of tissue and method employed *(1)*. Because fewer monocytes are thought to harbor HIV-1 in vivo compared with T-lymphocytes and tissue macrophages, and because monocytes are less permissive to productive virus replication compared with the

From: *Methods in Molecular Biology, Vol. 304: Human Retrovirus Protocols: Virology and Molecular Biology*
Edited by: T. Zhu © Humana Press Inc., Totowa, NJ

latter cell types, the isolation of HIV-1 from peripheral blood monocytes requires techniques with a greater level of sensitivity. However, our laboratory and other investigators have shown that isolation of HIV-1 from peripheral blood monocytes is readily achievable even from patients undergoing highly active antiretroviral therapy (HAART) for periods of up to 80 wk, with undetectable plasma viral loads ***(2–4)***.

Although patterns of virus isolation vary from patient to patient, HIV-1 can generally be detected in culture supernatants by 14 to 21 d of co-culture of monocytes with CD8-depleted, phytohemagglutinin (PHA)-activated peripheral blood mononuclear cells (PBMC). The levels of HIV-1 recovered from monocytes are typically very low compared with levels normally isolated from other cell types such as PBMC, and tissues such as lymph node, spleen, and brain ***(1)***. Subsequently, the detection of virus in culture supernatants is often achievable only by measurement of HIV-1 RNA by reverse transcriptase (RT)-polymerase chain reaction (PCR), especially when isolating virus from monocytes of patients undergoing HAART with suppressed plasma viremia.

There is no apparent correlation between the period of viral suppression by HAART and the ability to recover infectious virus from patients' monocytes. Although it is well established that resting-memory $CD4^+$ ($CD45RO^+$) T lymphocytes constitute a significant latent reservoir of HIV-1 in individuals receiving HAART ***(5–10)***, it is unlikely that monocytes represent a latently infected cellular reservoir, because monocytes have a relatively short period of circulation within blood before differentiation into specific tissue macrophages. In fact, our studies and those of other investigators have demonstrated the presence of circular viral DNA and spliced RNA in HIV-1-infected monocytes, albeit at very low levels ***(2,11)***, suggesting that their infection is recent and transcriptionally active rather than latent. Futhermore, findings of genetic evolution of HIV-1 in peripheral blood monocytes indicates HIV-1 replicates in monocytes in vivo in patients with ***(11)*** and without HAART ***(12)***.

The protocol for isolating HIV-1 from peripheral blood monocytes described in this chapter involves three major procedures: (1) isolation of peripheral blood monocytes from PBMC of HIV-1-infected individuals; (2) preparation of CD8-depleted, PHA-activated donor PBMC; and (3) co-culture of patient monocytes with donor CD8-depleted, PHA-activated PBMC.

2. Materials

2.1. Materials Required for Isolation of Peripheral Blood Monocytes From PBMC of HIV-1-Infected Individuals

1. 30 mL of anticoagulated patient blood (*see* **Notes 1** and **2**).
2. Supplemented Iscove's medium (Iscove's medium containing penicillin [100 U/

mL], streptomycin [100 μg/mL], L-glutamine [25 μg/mL], and 10–20% [v/v] heat-inactivated AB+ human serum) (*see* **Note 3**).
3. Phosphate-buffered saline (PBS).
4. PBS, calcium- and magnesium-free (PBS-CMF).
5. Heat-inactivated fetal calf serum (FCS).
6. Ficoll-Hypaque (Amersham-Pharmacia, cat. no. 17-1440-03).
7. Anti-CD14 mouse monoclonal antibody (Mab) conjugated to phycoerythrin (CD14-PE; Becton Dickinson, cat. no. 347497).
8. Anti-CD3 mouse Mab conjugated to fluorescein isothiocyanate (CD3-FITC; Becton Dickinson, cat. no. 349201).
9. 150-cm^2 plastic tissue culture plates (e.g., Nunc, cat. no. 168381).
10. 50-mL conical centrifuge tubes (e.g., Corning, cat. no. 430828).
11. 5-mL polypropylene round-bottom tubes (Becton Dickinson, cat. no. 35-2063).
12. Cell scrapers (e.g., Nunc, cat. no. 179707).
13. 3% (v/v) ultrapure formaldehyde.

2.2. Materials Required for Preparation of CD8-Depleted, PHA-Activated Donor PBMC

1. Buffy coat packs from healthy HIV-1-seronegative individuals (obtained from the blood bank).
2. Supplemented Iscove's medium (*see* **Subheading 2.1.**).
3. Supplemented RPMI medium (RPMI-1640 medium containing penicillin [100 U/mL], streptomycin [100 μg/mL], L-glutamine [25 μg/mL], and 10% [v/v] heat-inactivated FCS).
4. Ficoll-Hypaque (Amersham-Pharmacia, cat. no. 17-1440-03).
5. PBS.
6. Interleukin (IL)-2 (e.g., Roche, cat. no. 799 068).
7. PHA (e.g., Murex Diagnostics, cat. no. HA15).
8. 50-mL conical centrifuge tubes (e.g., Corning, cat. no. 430828).
9. 15-mL conical tubes (e.g., Corning, cat. no. 430790).
10. 75-cm^2 tissue-culture flasks (e.g., Nunc, cat. no. 156472).
11. Anti-CD8 magnetic beads (Dynal, cat. no. 111.08).
12. Magnetic particle concentrator (Dynal, cat. no. 120.21).
13. Sample mixer (e.g., Dynal, cat. no. 159-03).

2.3. Materials Required for Co-Culture of Patient Monocytes With Donor CD8-Depleted, PHA-Activated PBMC

1. Supplemented Iscove's medium (*see* **Subheading 2.1.**).
2. Freshly purified peripheral blood monocytes from HIV-1-infected individuals (from **Subheading 3.1.**).
3. CD8-depleted, PHA-activated donor PBMC (from **Subheading 3.2.**).
4. IL-2 (e.g., Roche, cat. no. 799 068).
5. 24-well tissue culture plates (e.g., Costar, cat. no. 3524).

6. 1.0-mL cryovials (e.g., Nunc, cat. no. 377224).
7. HIV-1 p24 antigen detection kit (e.g., Perkin Elmer, cat. no. NEK050B001KT).
8. Amplicor HIV-1 Monitor v1.5 kit (Roche Diagnostics, cat. no. 1118390) (*see* **Note 4**).

3. Methods

3.1. Isolation of Peripheral Blood Monocytes From PBMC of HIV-1-Infected Individuals

The isolation of peripheral blood monocytes requires (1) the preparation of the PBMC fraction from peripheral blood, (2) the isolation of monocytes from the total PBMC, and (3) testing the purity of the isolated monocytes by flow cytometry.

3.1.1. Preparation of PBMC Fraction From Peripheral Blood

Peripheral blood mononuclear cells are isolated from the patient blood by first diluting the blood 1:2 with PBS-CMF and gently overlaying 30 mL of the diluted blood onto 15 mL of Ficoll-Hypaque in two 50-mL conical centrifuge tubes. The tubes are then centrifuged at 700*g* for 20 min at room temperature with the centrifuge brake turned off. Keeping the brake turned off during this step is necessary to ensure that efficient separation of the red and white cell fractions occurs. The buffy coat fraction is then collected into a fresh 50-mL tube and washed once by resuspension in 50 mL of PBS-CMF followed by centrifugation at 500*g* for 7 min at room temperature. The cells are then washed a further two times by resuspension in 50 mL of PBS-CMF and centrifugation at low speed (300*g*) for 7 min at room temperature to remove platelets. The resulting PBMC are counted with a hemocytometer, and resuspended at 2×10^7 cells/mL in prewarmed (37°C) supplemented Iscove's medium.

3.1.2. Isolation of Monocytes From Total PBMC

The resuspended cells are added to 150-cm^2 tissue culture plates at a density of no more than 3×10^8 cells per plate (usually 10 to 15 mL per plate), and incubated for 1 to 2 h in a humidified CO_2 incubator at 37°C. The adherence properties of monocytes allows them to remain adherent to the plastic during this step, whereas nonmonocytes will remain nonadherent. After this incubation, plates are examined with an inverted phase contrast microscope to check that they contain adherent cells. The nonadherent cells are removed by repeated gentle washing and aspiration using warm (37°C) PBS. PBS is aspirated using a plastic pipet attached to a suction system. This process is repeated six to eight times, during which the culture plates are repeatedly examined for floating cells using an inverted phase contrast microscope. After a quick wash with cold (4°C) PBS-CMF to remove any remaining PBS containing magnesium and calcium, the adherent monocytes are detached by incubating the cells with 20 mL of cold (4°C) PBS-CMF for 15 min on ice. The cells are gently lifted with a cell scraper and isolated monocytes are collected into 50-mL centrifuge tubes containing 2.5 mL of FCS. The plates are washed twice with cold (4°C) PBS-CMF, and the remaining cells gently lifted and collected into the same 50-mL tubes. A final check under the inverted phase contrast microscope will ensure that plates do not still contain adherent monocytes. The cells within the tube are centrifuged at 500g for 7 min at 4°C. Monocytes are resuspended in 10 mL of supplemented Iscove's medium and a cell count performed using a hemocytometer.

3.1.3. Testing the Purity of the Isolated Monocytes by Flow Cytometry

To test the purity of the freshly isolated monocytes, 2×10^5 cells are placed into 5-mL polypropylene round-bottom tubes (one tube for specific Mab, and one tube for control Mab), washed once with cold PBS-CMF, centrifuged at 500*g* to pellet the cells, and PBS-CMF aspirated, leaving approx 100 µL. Following this, the cells are incubated with 1 µg of anti-CD14-PE and anti-CD3-FITC for 30 min on ice. An isotype-matched control Mab conjugated to FITC or PE must be included in each assay. The cells are then washed with cold (4°C) PBS-CMF, fixed with 3% (v/v) ultrapure formaldehyde, and analyzed by flow cytometry.

3.2. Preparation of CD8-Depleted, PHA-Activated Donor PBMC

The next procedure in the protocol to isolate HIV-1 from peripheral blood monocytes is the preparation of CD8-depleted, PHA-activated donor PBMC that are to be used to co-culture with the patient monocytes. This procedure requires (1) the preparation and PHA-activation of PBMC from buffy coat packs obtained from healthy, HIV-1-seronegative donors, and (2) the depletion of $CD8^+$ cells from the PBMC.

3.2.1. Preparation of PBMC From Buffy Coat Packs

The buffy coat is diluted 1:2 with PBS, and 30 mL gently overlaid onto 15 mL of Ficoll-Hypaque in each of several (usually four) 50-mL conical centrifuge tubes. Following this step, the method for isolating PBMC from the buffy coat is the same as for isolating PBMC from patient's peripheral blood, which is outlined in **Subheading 3.1.1.** The only difference is that after the low-speed centrifugation steps to remove platelets, the cells are counted and resuspended in supplemented RPMI medium at a concentration of 2×10^6 cells/mL, and then added to 75-cm^2 tissue culture flasks. The PBMC are then stimulated with

PHA (5 µg/mL) and incubated for 3 d in a humidified CO_2 incubator at 37°C (*see* **Note 5**).

3.2.2. Depletion of $CD8^+$ T Cells From PBMC

After activation of the PBMC with PHA, a cell count should be performed using a hemocytometer and PBMCs pooled from different donors at a ratio of 1:1. The cells are centrifuged at 500*g* for 7 min and resuspended in 5 mL supplemented RPMI medium. The $CD8^+$ T-cells (approx 10% of total PBMC) are depleted using anti-CD8 beads according to the manufacturer's instructions. First, the anti-CD8 beads are washed by transferring the required number of beads (100 µL of beads per 1×10^7 PBMC) into a 15-mL tube and resuspending them in 1 mL of supplemented RPMI medium. The tube is placed on a magnetic particle concentrator for 3 min and then the medium is gently removed using a sterile transfer pipet. One mL of supplemented RPMI medium is added and the process of washing the beads is repeated. The anti-CD8 beads are mixed with the PBMC suspension in a 15-mL tube and incubated on a rotary sample mixer for 30 min at 4°C. The tube is placed on a magnetic particle concentrator for 3 min and the CD8-depleted PBMCs collected in a new tube. This step should remove >95% of the $CD8^+$ T cells. If desired, a second round of CD8 depletion to remove any remaining $CD8^+$ T cells can be performed but is generally not required. This second time, the PBMCs are incubated with only 10 µL of prewashed beads per 1×10^7 cells, and then the method proceeds as for the first round of depletion. The CD8-depleted PBMCs are transferred into a 15-mL tube, centrifuged at 500*g* for 7 min, the medium aspirated, and cells resuspended at a concentration of 1×10^6/mL in supplemented Iscove's medium containing 10 U/mL IL-2.

3.3. Co-Culture of Patient Monocytes With Donor CD8-Depleted, PHA-Activated PBMC

The final procedure in this protocol utilizes the purified cell populations outlined in **Subheadings 3.1.** and **3.2.** in co-culture experiments to isolate replication-competent HIV-1 from the patient monocytes.

The freshly isolated monocytes (described in **Subheading 3.1.**) are resuspended in supplemented Iscove's medium at a density of 5×10^5 cells/mL. One mL of this suspension (i.e., 5×10^5 monocytes) is placed into wells of 24-well plates and the cells are allowed to adhere for 1 h at 37°C in a humidified CO_2 incubator. This is followed by a thorough wash to remove any residual contaminating nonadherent cells (*see* **Note 6**). One mL of the CD8-depleted, PHA-activated PBMCs from HIV-1-negative donors at a concentration of 1×10^6 cells/mL (prepared in **Subheading 3.2.**) is added to each well containing 5×10^5 monocytes (i.e., at a PBMC to monocyte ratio of 2:1) and the cultures incubated at 37°C in a humidified CO_2 incubator for 3 wk (*see* **Notes 7** and **8**). The culture medium is replaced twice weekly. This is done by transferring 1 mL of the culture supernatant into 15-mL conical tubes, centrifuging at 700*g* for 10 min (to clarify viral supernatant) and resuspending the cell pellet in 1 mL of fresh supplemented Iscove's medium containing 10 U/mL IL-2 before adding back to the co-cultures. The viral supernatant is filtered through 0.22-μm filters, and stored in 1-mL cryovials at –70°C. The virus stocks are tested by quantifying HIV-1 p24 antigen (using a commercial p24 kit) or, if levels of HIV-1 are low, by HIV-1 RNA (using the Amplicor HIV-1 Monitor v1.5 kit) according to the manufacturer's instructions (*see* **Note 4**).

4. Notes

1. Ethylenediamine tetraacetic acid (EDTA) or acid citrate dextrose (ACD) are suitable anticoagulants that can be used in patient blood samples. The use of heparin is avoided because of the inhibitory effect that heparin has on RT-PCR reactions. It is perhaps unlikely that the detection of HIV-1, produced by cells that have been cultured for several weeks, by RT-PCR will be impeded by the presence of heparin in the initial blood sample. However, its use as an anticoagulant in these studies should nonetheless be avoided if possible.
2. Procedures described in this chapter involve the use of human blood, cells obtained from HIV-1-seropositive individuals and HIV-1 virus stocks. Appropriate biosafety procedures must be observed while working with human blood, whereas BSL-3 biosafety practices must be performed at all times while handling virus stocks and HIV-1-infected cells.
3. As an alternative to Iscove's medium, supplemented RPMI medium may also be used (*see* **Subheading 2.2.**). All incubations are performed in a humidified 5% CO_2 incubator at 37°C unless otherwise noted. Media should be heated to 37°C, except where specified. All solutions must be endotoxin-free and sterile, and proper sterile techniques are required. Culture supernatants, reagents, and media need to be tested to exclude the presence of endotoxin, as lipopolysaccharide (LPS) contamination has been shown to affect HIV-1 replication in monocyte/macrophages *(13)*. Each new batch of AB+ human serum should be tested for toxicity and sterility and the ability to support monocyte culture.
4. In some isolation attempts, very low concentrations of HIV-1 may be recovered; in these cases, the concentration of p24 antigen in culture supernatants may be below the limit of detection of commercial kits. The occurrence of undetectable p24 antigen in culture supernatants is more frequent in isolations from monocytes of patients undergoing HAART with plasma HIV-1 RNA levels below detection, compared to untreated viremic patients or patients failing HAART. In these cases it is advisable to quantify HIV-1 RNA in culture supernatants using the Amplicor HIV-1 Monitor v1.5 kit because of the ultrasensitivity of this assay. However, because of the higher cost associated with measuring HIV-1 RNA, it may be prudent to first test for the presence of HIV-1 by measurement of p24

antigen, and then quantify HIV-1 RNA if necessary.

5. PBMCs from two different HIV-1-negative donors are required for the most successful amplification of primary HIV-1 isolates. PBMC from different donors are cultured separately for the first 3 d after isolation and then pooled prior to CD8 depletion and incubation with monocytes isolated from HIV-1-infected individuals. It is recommended that at least 1×10^7 PBMCs from each donor are stimulated on the day of isolation.
6. The purity of the isolated monocytes is very important for these studies, because HIV-1 harbored by any contaminating T-cells may be amplified in the co-culture. Thorough washing to remove nonadherent cells is usually sufficient to minimize T-cell contamination. The use of AB+ human serum instead of FCS enhances the adherence of monocytes to plastic, and adherence may be enhanced further if necessary by increasing the concentration of AB+ human serum from 10% (v/v) to 20% (v/v). The additional adherence step described in **Subheading 3.3.** increases the purity of the monocyte population to approx 99% *(2)*. The level of T-cell contamination in this purified monocyte population can be determined down to a sensitivity of 1 contaminating T cell in 10^4 monocytes by analysis of TcR mRNA using RT-PCR *(2)*.
7. In this protocol care is taken to avoid the presence of LPS in all solutions, media, and media supplements because of the ability of LPS to modulate HIV-1 replication in monocytes/macrophages. However, it should be noted that methods to isolate HIV-1 from patient monocytes employed by other investigators utilize LPS stimulation of the monocytes to activate virus replication *(4)*. We have also found that LPS stimulation of patient monocytes, or stimulation with other agents including M-CSF or tumor necrosis factor (TNF)-α prior to co-culture with CD8-depleted PHA-activated PBMC, may increase the amount of virus isolated, or decrease the number of monocytes required for successful virus isolation. However, successful virus isolation can be achieved in approx 70% of cases without any prior stimulation step *(2)*.
8. A common concern during the monocyte-PBMC co-culture is the relatively high density at which the cells are cultured. The co-culture starts at an initial density of 1.5×10^6 cells/mL (ie., 5×10^5 monocytes with 1×10^6 PBMC), but the presence of IL-2 in the culture medium promotes the proliferation of the PBMC during the course of the 3 wk that the cells are cultured. This results in the cell density increasing significantly, and commonly causes the culture medium to turn a yellow shade. In practice, however, this does not appear to have any adverse effect on cell viability nor the ability to successfully isolate HIV-1 from monocytes.

References

1. Gorry, P. R., Bristol, G., Zack, J. A., Ritola, K., Swanstrom, R., Birch, C. J., et al. (2001) Macrophage tropism of human immunodeficiency virus type 1 isolates

from brain and lymphoid tissues predicts neurotropism independent of coreceptor specificity. *J. Virol.* **75,** 10,073–10,089.

2. Sonza, S., Mutimer, H. P., Oelrichs, R., Jardine, D., Harvey, K., Dunne, A., et al. (2001) Monocytes harbour replication-competent, non-latent HIV-1 in patients on highly active antiretroviral therapy. *AIDS* **15,** 17–22.
3. Crowe, S. M. and Sonza, S. (2000) HIV-1 can be recovered from a variety of cells including peripheral blood monocytes of patients receiving highly active antiretroviral therapy: a further obstacle to eradication. *J. Leukoc. Biol.* **68,** 345–350.
4. Lambotte, O., Taoufik, Y., de Goer, M. G., Wallon, C., Goujard, C., and Delfraissy, J. F. (2000) Detection of infectious HIV in circulating monocytes from patients on prolonged highly active antiretroviral therapy. *J. Acquir. Immune Defic. Syndr.* **23,** 114–119.
5. Finzi, D., Blankson, J., Siliciano, J. D., Margolick, J. B., Chadwick, K., Pierson, T., et al. (1999) Latent infection of CD4+ T cells provides a mechanism for lifelong persistence of HIV-1, even in patients on effective combination therapy. *Nat. Med.* **5,** 512–517.
6. Finzi, D., Hermankova, M., Pierson, T., Carruth, L. M., Buck, C., Chaisson, R. E., et al. (1997) Identification of a reservoir for HIV-1 in patients on highly active antiretroviral therapy. *Science* **278,** 1295–1300.
7. Chun, T. W., Stuyver, L., Mizell, S. B., Ehler, L. A., Mican, J. A., Baseler, M., et al. (1997) Presence of an inducible HIV-1 latent reservoir during highly active antiretroviral therapy. *Proc. Natl. Acad. Sci. USA* **94,** 13,193–13,197.
8. Wong, J. K., Hezareh, M., Gunthard, H. F., Havlir, D. V., Ignacio, C. C., Spina, C. A., and Richman, D. D. (1997) Recovery of replication-competent HIV despite prolonged suppression of plasma viremia. *Science* **278,** 1291–1295.
9. Persaud, D., Pierson, T., Ruff, C., Finzi, D., Chadwick, K. R., Margolick, J. B., et al. (2000) A stable latent reservoir for HIV-1 in resting CD4(+) T lymphocytes in infected children. *J. Clin. Invest.* **105,** 995–1003.
10. Siliciano, J. D., Kajdas, J., Finzi, D., Quinn, T. C., Chadwick, K., Margolick, J. B., et al. (2003) Long-term follow-up studies confirm the stability of the latent reservoir for HIV-1 in resting CD4+ T cells. *Nat. Med.* **9,** 727–728.
11. Zhu, T., Muthui, D., Holte, S., Nickle, D., Feng, F., Brodie, S., et al. (2002) Evidence for human immunodeficiency virus type 1 replication in vivo in CD14(+) monocytes and its potential role as a source of virus in patients on highly active antiretroviral therapy. *J. Virol.* **76,** 707–716.
12. Fulcher, J. A., Hwangbo, Y., Zioni, R., Nickle, D., Lin, X., Heath, L., et al. (2004) Compartmentalization of human immunodeficiency virus type 1 between blood monocytes and CD4+ T cells during infection. *J. Virol.* **78,** 7883–7893.
13. Kornbluth, R. S., Oh, P. S., Munis, J. R., Cleveland, P. H., and Richman, D. D. (1989) Interferons and bacterial lipopolysaccharide protect macrophages from productive infection by human immunodeficiency virus in vitro. *J. Exp. Med.* **169,** 1137–1151.

4

Isolation, Propagation, and HIV-1 Infection of Monocyte-Derived Macrophages and Recovery of Virus From Brain and Cerebrospinal Fluid

Santhi Gorantla, Myhanh Che, and Howard E. Gendelman

Summary

Mononuclear phagocytes (MP: monocytes, dendritic cells, and tissue macrophages) are host cells for the human immunodeficiency viruses types 1 and 2. MPs are both the first lines of defense and vehicles for viral dissemination in the infected human host. Viral infection of MP can affect the disease directly during interstitial pneumonitis and HIV encephalitis. Both revolve around MP secretions of immune regulatory and neurotoxic factors. Clearly, laboratory models that mimic disease need to include primary human MP infected with viral isolates obtained from diseased tissues. Over the past two decades our laboratory has developed state-of-the-art methods for isolation and propagation of monocytes from peripheral blood. This technology directly supports work at the University of Nebraska Medical Center as well as research performed throughout the United States, including the laboratories of Drs. Mario Stevenson, William Tyor, David Volsky, Loyda Melendez, and Mary-Jane Potash, among others. The importance of these cells as targets for virus and reservoirs of persistent infection are discussed.

Key Words: Monocytes; macrophages; elutriation; HIV-1 infection; mycoplasma; endotoxin contamination.

1. Introduction

Cells of mononuclear phagocyte lineage (MP: monocytes, dendritic cells, and tissue macrophages) are primary cellular targets for HIV infection and viral dissemination *(1–4)*. They also serve as reservoirs of virus throughout body compartments including, most notably, peripheral blood, lung, and the central nervous system (CNS) *(5–7)*. Macrophage-tropic HIV-1 can be propagated directly onto monocyte-derived macrophages (MDM) from infected body

From: *Methods in Molecular Biology, Vol. 304: Human Retrovirus Protocols: Virology and Molecular Biology*
Edited by: T. Zhu © Humana Press Inc., Totowa, NJ

fluids and tissues including plasma, blood, lung, brain, and cerebrospinal fluids ***(8,9)***. Monocytes can be maintained in culture for more than 3 mo and thus provide susceptible targets for the propagation of virus from patient samples. Permissiveness of macrophages to productive HIV infection depends on the differentiation state of the cells ***(10)***. Viability and differentiation of monocytes isolated from peripheral blood are maximized when cultivated with recombinant human macrophage colony stimulating factor type 1 (referred to as MCSF). Such control conditions also permit increased susceptibility to HIV-1 infection ***(9)***.

Viral recovery from infected brain samples or cerebrospinal fluid directly onto MDM is appropriate as MP disseminate virus to the CNS and are productively infected during the disease. Moreover, viral production is often continuous in MP. Macrophage indicator systems represent adherent cell targets where virus-induced cytopathicity is readily observed. Isolation of monocytes by countercurrent centrifugal elutriation allows the isolation of a large number of relatively pure cell populations ***(11)***. Alternatively, isolation of cells by adherence or gradient systems diminishes cell purity and may change the cell phenotype ***(12)***. In this chapter, we describe the isolation of monocytes as well as quality control measures for achieving optimal cell culture conditions free of microbial contamination for the purpose of studies of HIV-host cell interactions as they may occur during disease.

2. Materials

2.1. Isolation and Propagation of Human Monocytes

1. Leukocyte-enriched whole blood (leukopak), from HIV-1,-2 and hepatitis B seronegative donors, is obtained from a blood bank and processed within 6 h of collection.
2. Lymphocyte separation medium (LSM).
3. Phosphate-buffered saline (PBS), filter-sterilize through 0.2-µm filter unit or autoclave for sterilization.
4. Elutriation, J6-MI centrifuge with a JE-5.0 elutriation system (Beckman, Fullerton, CA).
5. Cell counter and size analyzer (Beckman Coulter, Fullerton, CA).
6. Isoton diluent (Beckman Coulter).
7. Percoll gradient (1.087 g/mL): 12.5 mL of Percoll, 5.5 mL of sterile distilled water, 2 mL of 10X PBS.
8. Dulbecco's minimal essential medium (DMEM) without L-glutamine
9. Human AB serum (HS), heat inactivated at 56°C for 30 min.
10. L-glutamine.
11. Gentamicin.
12. Ciprofloxacin.
13. Recombinant human MCSF.

14. Monocyte medium without MCSF: DMEM, 10% HS, 2 m*M* L-glutamine, 50 µg/mL gentamicin, 10 µg ciprofloxacin. Filter-sterilize using 0.2 µm filter unit. Store at 4°C. Discard after 2 wk.
15. Monocyte medium with MCSF: DMEM, 10% HS, 2 m*M* L-glutamine, 50 µg/mL gentamicin, 10 µg ciprofloxacin, 1000 U/mL rhMCSF-1. Filter-sterilize using 0.2-µm filter unit. 16. Store at 4°C. Discard after 2 wk.
16. Trypan blue solution, 0.4 % in saline or PBS.

Aliquot reagents 8, 9, 11, and 12 for use in 500-mL bottles of DMEM. Reagents are stored at –20°C and freeze–thaw cycles are minimized.

2.2. Propagation of HIV-1 From Brain Tissue

1. Kevlar gloves or cut-resistant gloves.
2. Scalpels and forceps, sterile.
3. Monocyte medium with and without MCSF (as described in **Subheading 2.1.**).

2.3. Reverse Transcriptase (RT) Assay

1. Solution A (disruption buffer): 100 m*M* Tris-HCl, pH 7.9, 300 m*M* KCl, 10 m*M* dithiothreitol (DTT), 0.1% NP40.
2. Solution B (RT reaction mix): 50 m*M* Tris-HCl, pH 7.9, 150 m*M* KCl, 5 m*M* DTT, 15 m*M* $MgCl_2$, 0.05% NP40, 10 µg/mL Poly A, 0.25 U/mL oligo dt pd$(T)_{12-18}$, 4 µCi/mL ^{3}H-thymidine (add ^{3}H-thymidine just before use; ^{3}H-thymidine is radioactive and must be handled carefully according to radiation safety procedures). Solution A and solution B can be stored as aliquots at –20°C for up to 6 mo.
3. 10% trichloroacetic acid (TCA).
4. Standard RT enzyme.
5. Cell harvester.
6. Liquid scintillation counter.

2.4. Screening for Mycoplasma and Endotoxin Contamination

1. Pyrotell *Limulus* Amebocyte Lysate (LAL) (Associates of Cape Cod, Inc., East Falmouth, MA).
2. Control standard endotoxin (CSE) (Associates of Cape Cod).
3. Pyrotubes (Associates of Cape Cod).
4. LAL reagent water (Associates of Cape Cod).
5. Triton X-100.
6. Tributylphosphate.
7. Mycoalert (Cambrex Bio Science Rockland, Rockland, ME).
8. Luminometer.

3. Methods

3.1. Isolation of Monocytes From Peripheral Blood

Countercurrent centrifugal isolation is performed on leukocyte-enriched blood samples of approx 100 mL. The yield of monocytes by this method is 1.5

to 2 billion cells. An alternative method to isolate monocytes from smaller volumes (<50 mL) of blood is by Percoll gradient separation or by adherence. In this section we describe all three methods for monocyte isolation.

3.1.1. Donor Screening

Donors are screened for HIV-1 and 2 and hepatitis B and C. Leukopaks are obtained only from seronegative individuals for monocyte isolation. The donors are screened for ABO Rh and antibodies against hepatitis B surface (HBS) or core antigen (HCV), hepatitis C virus (HCV), HIV, human T lymphotropic virus (HTLV), cytomegalovirus (CMV), and syphilis per the institutional regulations.

Donors are chosen without heart, lung, and liver disease. Those with anemia regardless of cause and pregnancy are excluded. Donors undergo a physical examination by a physician before being qualified for leukophoresis. On the day the procedure is performed, the donor's hematocrit and platelet count are checked and must be $\geq$38% and 150×10^{10}/mL, respectively.

3.1.2. Countercurrent Centrifugal Elutriation

The procedure is carried out in a biological safety cabinet (BSC) under sterile conditions (*see* **Note 1**); biosafety level 2 (BSL-2) practices are required for handling human specimens, given the risks associated with their use. BSL-2 and BSL-3 practices are followed per National Institutes of Health guidelines; these include wearing laboratory coats and gloves and decontaminating liquid waste with 50% bleach. All materials are autoclaved before being discarded.

3.1.2.1. Preparation of Elutriation Setup

1. Prepare the elutriation setup before starting to process the blood sample.
2. Assemble the chambers and the rotor per the instrument instruction manual.
3. Connect the input tubing and the output tubing to the rotor chamber.
4. The input tubing is connected through a peristalitic pump that controls the flow rate. To the other end of the input tube connect a three-way stopcock, which is used to control the input of either PBS or cells pumped from two separate reservoirs (the manual provides detailed instructions for the setup). The output tubing goes directly into a waste container while collecting washes, or a sterile bottle while collecting the cells (*see* **Note 2**).
5. Sterilize the chamber and the tubing by pumping 95% ethanol (approx 200 mL) through the system (*see* **Note 3**).
6. Fill PBS in both reservoirs and flush the system with PBS to remove air bubbles. Use at least 400 mL of PBS.
7. Flush PBS through the system by increasing the pump setting to maximum. This is to ensure that the setup can handle the pressure. Check the system for any air bubbles and prevent them by flushing the system again with more PBS.
8. Now proceed to processing the leukopak.

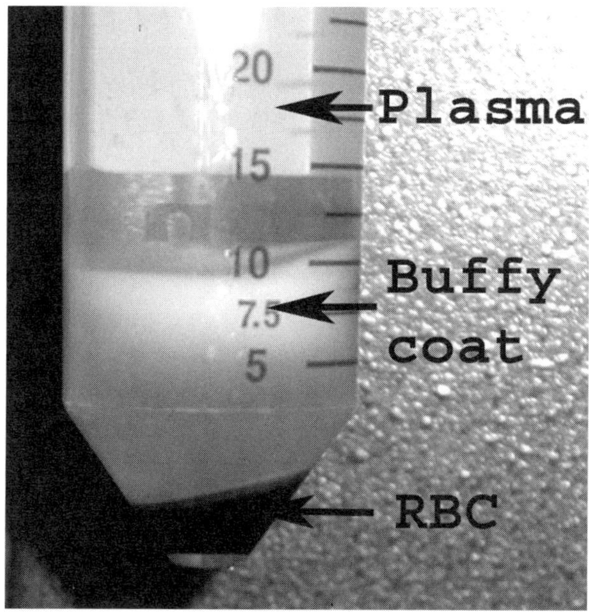

Fig. 1. Lymphocyte separation medium (LSM) gradient. Separation of whole blood using LSM and showing a buffy coat at plasma and media interface.

3.1.2.2. Isolation of Monocytes

1. Transfer the blood sample from the leukopak into a 250-mL centrifuge tube. Use sterile scissors to cut the leukopak. Dilute the blood sample by adding enough PBS to make up to a total of 250 mL. Pipet up and down to mix gently.
2. Aliquot 10 mL of LSM into each of the 10 50-mL centrifuge tubes.
3. Layer 25 mL of diluted blood sample carefully onto the LSM without disturbing the interface. To maintain the interface, the tube is held at a 45° angle.
4. Centrifuge at 600*g* at room temperature for 20 min without brake.
5. After centrifugation, transfer the tubes from the centrifuge to the BSC without disturbing the gradient. Red blood cells settle at the bottom and the peripheral blood mononuclear cells (PBMC) form a buffy coat layer at the LSM and plasma interface. This is shown in **Fig. 1**. Aspirate the top layer containing diluted plasma and platelets directly above the buffy coat.
6. Collect the buffy coat containing PBMC from each of the gradient tubes and transfer the recovered cells to a fresh 250-mL centrifuge tube.
7. Make the volume up to 250-mL by adding PBS. Mix well by gently pipetting up and down.
8. Centrifuge for 10 min at 400*g* at room temperature.
9. Decant supernatant and resuspend the cell pellet in 250 mL of PBS and repeat **step 8**. Finally, suspend the cells in 25 mL of PBS.

10. Replace one of the reservoirs in the elutriation setup with the tube containing the cell suspension.
11. With PBS flow open (controlled with the stopcock), turn on the pump at a flow rate of 40 mL/min. Let PBS flow through the system.
12. Turn on the centrifuge and when the rotor reaches 1960 rpm (*see* **Note 4**), stop PBS flow and load the cells by turning the stopcock. After all the cells are loaded, turn the stopcock back to allow PBS flow. Wait for 10–15 min to allow the cells to separate by size within the rotor chamber.
13. Turn the pump to a flow rate of 45 mL/min and wait for 2 min to remove the remaining red blood cells.
14. Turn the pump to 50 mL/min and collect peripheral blood lymphocytes (PBL).
15. Collect aliquots of cell suspension (10 μL) at each speed setting of the pump. Dilute the cell aliquot with 10 mL of isoton diluent. Count and analyze the cells using an electronic counting device (Beckman Coulter cell counter and analyzer), which also gives a profile indicating the cell size. Calibrate the cell analyzer with latex beads of known size before analyzing the cells. Lymphocyte profile will be that of approx 6–8 μm-diameter cells. Use the cell counter according to the manufacturer's instructions (*see* **Note 5**).
16. Turn the pump up to 55 mL/min to collect more PBL. Count the cells; if it shows >500 cells/mL increase the pump flow by 5 mL/min at a time until all the PBL are eluted.
17. When a profile of monocytes starts showing (refer to the cell analyzer manual; the profile of cells should be same as that obtained for cells of 8–10 mm diameter), stop the centrifuge, set the pump to 200 mL/min, and collect all the monocytes.
18. Centrifuge tubes with PBL or monocytes at 400*g* for 10 min at room temperature. Decant the supernatant and resuspend the cells in 25 mL of PBS.
19. Use a small aliquot (10 mL) to count the cells and calculate the total number of cells.
20. To check the purity of PBL and monocyte populations, make a cell smear on a glass slide using a small aliquot from each cell suspension and stain with hematoxylin and eosin (H&E). Observe under the light microscope. **Figure 2** shows H&E-stained monocytes and PBL separated by this procedure. Monocytes have distinct kidney-shaped nuclei and can be easily identified.

3.1.3. Percoll Gradients

1. Dilute blood sample 1:1 with PBS.
2. In 15-mL conical tubes take 4 mL of Percoll, adjusted to 1.087 g/mL density, and layer 8 mL of diluted blood carefully over the Percoll gradient without disturbing the interface.
3. Centrifuge the tubes at 600*g* at room temperature for 30 min without brake.
4. Red blood cells will settle in the bottom and PBMC will form a buffy coat layer at the interface of the Percoll (bottom layer) and the plasma (top layer). Aspirate the plasma carefully up to the buffy coat layer without disturbing it. Transfer the buffy coat into a fresh 15-mL conical tube and dilute with 10 mL PBS.

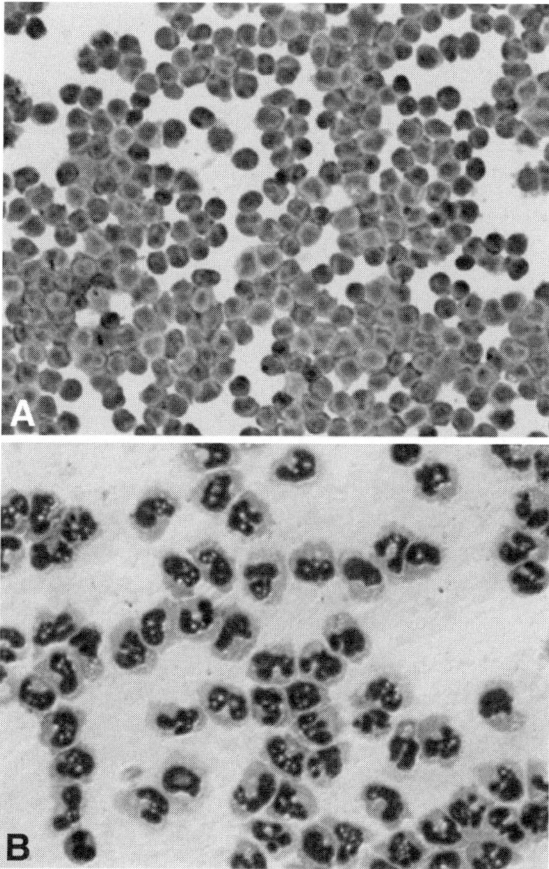

Fig. 2. Lymphocytes and monocytes isolated by elutriation. Cell fractions of peripheral blood lymphocytes **(A)** and monocytes **(B)** isolated by countercurrent centrifugal elutriation are stained with H&E, original magnification ×40.

5. Spin the tubes at 400*g* for 5 min and decant the supernatant.
6. Suspend the cells in 10 mL of PBS and repeat **step 5**.
7. Finally, suspend the cells in 5 mL of monocyte PBS and count the cells by using trypan blue and a hemocytometer.

3.1.4. Adherence

1. Dilute the blood sample 1:1 with PBS.
2. In 15-mL conical tubes take 4 mL of LSM and layer 8 mL of diluted blood without disturbing the interface.
3. Centrifuge the tubes at 600*g* for 20 min at room temperature without brake.
4. Collect the buffy coat as described in **Subheading 3.1.2.2., step 5**.
5. Dilute with 10–15 mL PBS and centrifuge the tubes at 400*g* for 10 min.

Table 1
Cell Number and Volume of Medium To Be Used According to Size of Culture Dish

Culture dish	Cell no. ($\times 10^6$)	Media added d 3 (mL)
75-cm^2 flask	30/flask in 20 mL	15
6-well plate	3/well in 3 mL	2
24-well plate	0.75/well in 0.75 mL	0.75
48-well plate	0.25/well in 0.25 mL	0.25
96-well plate	0.1/well in 0.1 mL	0.1

6. Resuspend the pellet in 10 mL PBS by pipetting up and down gently and repeat **step 5**.
7. Suspend the pellet in 10 mL monocyte medium without MCSF. Take a 10-µL aliquot to count the cells using trypan blue and a hemocytometer.
8. Adjust the cell suspension to 2–3 × 10^6 cells/mL and add an appropriate volume of cell suspension to the culture dish. Take 20 mL in a 75 cm^2 flask, 5 mL/well in a 6-well plate, 2 mL/well in a 24-well plate, 1 mL/well in a 12-well plate, or 200 µL/well in a 96-well plate. Choose the size of the culture dish according to the cell number and also how it will be used in further experiments (*see* **Note 6**).
9. Incubate the culture at 37°C with 5% CO_2 and 95% humidity for 2–3 h.
10. After 2–3 h cells will adhere to the plastic. Aspirate the supernatant containing nonadherent cells without disturbing the adherent cells. Wash the cells at least 4 times by adding fresh monocyte medium without MCSF without disturbing the adhered cells, swirl the flask, and aspirate the medium to remove as many lymphocytes as possible.
11. Replace with the same amount of fresh monocyte medium with MCSF. Add medium to the sides of the culture dish to prevent cells from coming off of the surface. Incubate the culture under the same conditions (**step 9**).
12. On d 3 and d 5, feed the cells with fresh monocyte medium containing MCSF by doing a half-exchange as described in **Subheading 3.2.**, **step 5**.
13. On d 7, cells can be used for HIV-1 infection.

3.2. Monocyte Propagation

1. Suspend monocytes in monocyte medium with MCSF at 1 × 10^6 cells/mL. Gently mix the cells by pipetting to obtain uniform suspension. Plate the cells immediately after elutriation.
2. Plate an appropriate volume of cells for type of plate or culture dish (*see* **Table 1**).
3. Incubate at 37°C with 5% CO_2 and 95% humidity.
4. On d 3, add additional monocyte medium with MCSF (as shown in **Table 1**) and incubate under the same conditions. By this time cells adhere to the bottom of the culture dish, forming a monolayer.

Table 2
Amount of Virus Used for Infection According to Size of Culture Dish

Culture dish	Volume (mL) HIV (MOI of 0.01)	Media added after 4 h (mL)
75-cm^2 flask	10	20
6-well plate	3	2
24-well plate	1	0.5
48-well plate	0.25	0.25
96-well plate	0.1	0.1

MOI, multiplicity of infection.

5. Half-exchange with fresh medium every 2 d as follows: remove half of the total volume of medium from the culture, without disturbing the cells (*see* **Note 7**), and replace with an equal volume of monocyte medium with MCSF.
6. From d 7, use monocyte medium without MCSF (*see* **Note 8**).

3.3. HIV Infection of MDM

All procedures handling infected specimens should be carried out in at least BSL-2 containment facilities. BSL-3 is recommended for handling high-titer virus preparations.

3.3.1. Laboratory HIV-1 Isolates

Viral production stays in an exponential state depending on the size of the culture. In a 6-well plate it will be from approx d 7 to d 21. In a 75-cm^2 flask there is a higher cell number; therefore, the viral production will be in an exponential state for up to 1 mo. The size of the culture to be used for infection depends on the amount of viral stock needed.

1. Culture monocytes in an appropriate dish as described in **Subheading 3.2.**
2. On d 7, remove all medium from the culture and infect with HIV stock diluted in monocyte medium without MCSF at 0.01 multiplicity of infection (MOI). Refer to **Table 2** for the volume of medium containing the virus to be used for each type of culture dish (*see* **Notes 9** and **10**).
3. Incubate at 37°C with 5% CO2 and 95% humidity for 4 h.
4. Add monocyte medium without MCSF (*see* **Table 2** for the volume), without removing the virus, and incubate overnight under the same conditions.
5. The following morning, remove all the infection medium from the culture and add the same amount of fresh monocyte medium without MCSF.
6. On d 10, collect culture supernatant during half-exchange of the media (as described in **step 5** of **Subheading 3.2.**) instead of discarding. Reserve 30–50 μL for RT assay and store the rest at 4°C (*see* **Note 11**).

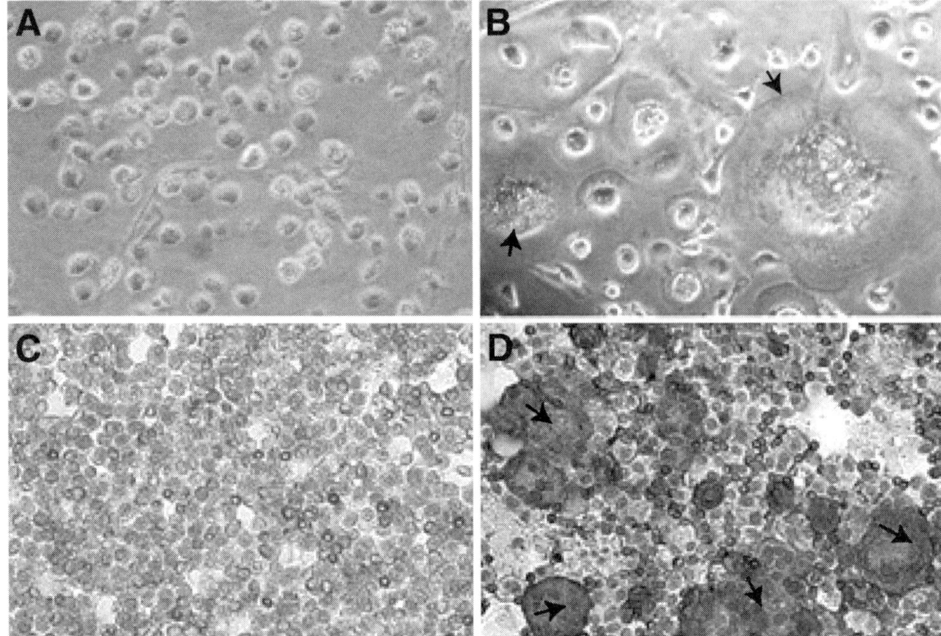

Fig. 3. Cytopathicity of HIV-1-infected monocyte-derived machrophages (MDM). MDM (**A**) and HIV-1_{ADA} infected MDM (**B**) are cultivated in 6-well plates. Cytopreparations of MDM (**C**) and HIV-1_{ADA} infected MDM (**D**) were immunostained with antibodies against HIV-1 p24. Virus-positive cells are black, original magnification ×20. Arrows show multinucleated virus infected cells.

7. Repeat **step 6** on d 12, 14, 16, 18, 20, 22, and 25 (*see* **Note 12**).
8. Obtain the RT results (perform RT assay as described in **Subheading 3.4.**) and pool the supernatants with peak RT values.
9. Centrifuge at 300*g* for 15 min at 4°C, remove the supernatant, and aliquot into cryovials for storage either at –80°C or in a liquid nitrogen tank (*see* **Note 13**).

HIV-infected MDM form multinucleated giant cells (fusion of infected and uninfected cells, **Fig. 3B**). HIV-infected MDM express HIV-p24 (**Fig. 3D**).

3.3.2. Isolation of HIV From Infected Cerebrospinal Fluid (CSF)

1. Monocytes cultured in a 6-well plate are obtained as described in **Subheading 3.2.**
2. On d 7, add 1 mL of diluted patient CSF/well for infection or directly inoculate 100 μL of fluid into 1 mL total culture fluid volume.
3. Incubate the cells at 37°C with 5% CO_2 and 95% humidity for 24 h.
4. After 24 h incubation add an additional 4 mL of fresh monocyte medium without MCSF.

5. Every other day, collect the samples by replacing one half of culture media (per **Subheading 3.3.1.**). Samples are tested for RT activity and stored as described.

3.3.3. Isolation of HIV From Infected Brain Tissue

1. Prepare monocyte culture for infection as described in **Subheading 3.2.** On d 7, cocultivation of infected brain tissue and MDM is initiated.
2. HIV-infected brain tissue obtained by autopsy is processed in a biological safety cabinet following at least BSL-2 practices. Basal ganglia or frontal cortex are optimal brain regions for viral rescue. Virus recovery is optimal if tissue is processed immediately after autopsy. Brain tissue can be frozen but freeze–thaw cycles diminish viral yields.
3. Wear cut-resistant Kevlar gloves during tissue slicing.
4. Transfer tissue into a sterile Petri dish, then wash and remove excess blood with PBS. Using scalpels, chop the tissue into 2–3-cm^2 pieces. Remove any excess blood vessels and connective tissue with forceps.
5. Remove the PBS and cover the tissue pieces with monocyte medium without MCSF. Slice the tissue into 1–2-mm^2 pieces using scalpels.
6. Aspirate medium from the monocyte culture and transfer equivalent amount of media containing the brain pieces. No more than 30–40% of the monolayer should be covered with the tissue pieces.
7. Incubate the culture over night at 37°C with 5% CO_2 and 95% humidity.
8. The following day (d 8), remove tissue pieces along with the medium using a pipet without disturbing the cells. Replace with the equal amounts of monocyte media without MCSF.
9. From d 10, do half-exchange and collect samples every 2 d as described in **Subheading 3.3.1.** Pool the samples that contain peak RT activity and store (**Subheading 3.3.1.**).

3.4. RT Assay

1. Pipet 10 µL of solution A into an appropriate number of wells in a 96-well round-bottom plate. Assay in quadruplicate. A blank (media), standards, and unknown samples are included in the assay. Use different dilutions of the standard, ranging from 0.1 to 10 U/10 µL. Dilute the standards immediately before use.
2. Add 10 µL of each sample to appropriate wells and mix by swirling the plate.
3. Incubate the plate for 15 min at 37°C with 95% humidity.
4. Add 25 µL of solution B to each well. Solution B contains radioactive thymidine, appropriate cautions need to be taken while handling the radioactive material.
5. Incubate the reaction plate for 18–24 h at 37°C with 95% humidity.
6. After the incubation add 50 µL of ice-cold 10% TCA to each well and keep the plate at 4°C for at least 15 min.
7. Collect the precipitate by harvesting the plate using a cell harvester.
8. Place the filter paper disks separately into vials containing the scintillation fluid and count the radioactivity.
9. Calculate the amount of RT activity in the unknown samples from the standard curve made by using different dilutions of commercial standard RT.

3.5. Screening for Endotoxin and Mycoplasma

Endotoxin and mycoplasma contaminations are serious problems associated with cell culture (*see* **Note 14**). If they are left undetected, it is hard to conclude a negative result affected by these contaminations. HIV propagated on human cells, cell culture supernatants, and any homemade cell culture reagent are routinely tested for endotoxin and mycoplasma. All other cell culture reagents procured directly from the companies undergo equivalent quality control measures before being supplied.

3.5.1. Endotoxin Assays

1. Use pyrotubes for dilution of standards and the assay. The procedure must be performed in a biological safety cabinet (*see* **Note 15**).
2. Prepare the various dilutions of CSE in reagent water: 100 pg/mL, 50 pg/mL, 25 pg/mL, and 12.5 pg/mL.
3. Transfer 100 µL of each dilution into fresh pyrotubes for the assay, to prepare the standard curve. Use 100 µL of reagent water for the blank.
4. Make at least two dilutions of the sample to be tested and take 100 µL of undiluted and diluted samples for the assay.
5. Reconstitute LAL with 2 mL of reagent water (*see* **Note 16**). Add 100 µL of LAL reagent to the assay tubes with the samples. Vortex the tube immediately after adding LAL. Once the tube is set back in the test-tube rack, do not move the tube.
6. After adding LAL to all the tubes, incubate at 37°C for 60 min. Any disturbance to the tubes during the incubation will cause false negatives.
7. After 60 min observe each tube carefully for the gel clot. Take the tubes out of the rack one by one with minimal disturbance for observation.
8. A positive (+) reaction appears as a gel clot. A positive/negative (±) reaction appears as a gel that breaks, or turbid precipitate. A negative (–) reaction has no gel formation.

3.5.2. Mycoplasma

1. Take 2 mL of the sample to be tested for mycoplasma detection assay.
2. HIV must be inactivated before the assay. For inactivation, add 3 µL of tribatylphosphate and 20 µL of Triton X-100 (to get a final concentration of 1% Triton X-100) to 2 mL of virus and incubate at 37°C overnight.
3. The following day, perform mycoplasma detection according to the kit manufacturer's instructions (*see* **Note 17**).

4. Notes

1. Cell isolation and cell culture procedures must be carried out strictly under sterile conditions in a BSC. All pipets, centrifuge tubes, and tips must be sterile, and the packages should be opened only inside the BSC. The working area of the BSC should be wiped with 75% ethanol for sterilization.

2. Do not insert the tubing directly into either PBS or cell suspension. Connect a sterile aspirating pipet to the tubing and insert the pipet into the reservoir. Do not let the reservoirs become empty while the pump is on, as air bubbles can enter the system. Connect another sterile aspirating pipet to the output tubing and put it in the collection bottle.
3. Before and after using the elutriation system, it is very important to take proper precautions to clean and maintain the system to avoid clogging and to maintain sterile conditions.
4. The flow rate, rotor speed to elute a particular cell type, is determined by the nomogram given in the instruction manual provided with the elutriation system. In the protocol, the rotor speed is mentioned in rpm as obtained from the nomogram to retain particles of 6 μm or larger at a flow rate of 40 mL/min.
5. Cell analysis should be done fast without time delay as the pump cannot be stopped during elutriation. Otherwise, there can be loss of cells.
6. Removing adhered monocytes either by cell scraping or trypsinization is not recommended for reculture, as the physiology of monocytes can get affected.
7. Monocytes are adherent cells, so the supernatants can be collected carefully without disturbing the cells. If there are any loose cells present, spin the supernatant at 400*g* for 5 min, collect the supernatant, and transfer the cells back into the corresponding well.
8. Human monocytes require MCSF for differentiation. After 6–7 d into culture, the cells begin making their own MCSF, and supplementation is no longer required (hence the change to monocyte medium without MCSF).
9. Incubating the cells with HIV in a minimal volume for the first 4–5 h facilitates maximum HIV-cell interaction to enhance infection levels.
10. Monocytes cultured between 6–9 d can be used for HIV infection. HIV-macrophage infection is highly dependent on cellular differentiation state. Between d 6 and d 7 of culture, the cells are differentiated enough to obtain successful viral infection.
11. Two consecutive cultures showing RT activity indicates successful viral infection.
12. The infected cultures should be closely monitored. The supernatants can be collected as long as there is an intact monolayer without complete loss of cells. An active cell culture is also indicated by the pH change to an acidic side (yellow) after 24 h of medium change.
13. Storing viral aliquots in liquid nitrogen is recommended for long-term usage to retain the infectivity.
14. Excellent tissue culture practices are extremely important, both for the safety of the personnel and for preventing endotoxin contamination. Human sweat, dust, and the like, are potential sources of endotoxin.
15. Throughout the assay, it is important to note that improperly performed procedures can lead to false negatives, and that false positives can occur as a result of the introduction of endotoxin during the assay. Sterility and careful handling of reagents is crucial.

16. Do not make up LAL in advance. Reconstituted LAL left to sit can cause both false-negatives and false-positives.
17. Samples infected with mycoplasma will give a final ratio greater than 1, as detected by the luminometer.

References

1. Gendelman, H. E., Narayan, O., Molineaux, S., Clements, J. E., and Ghotbi, Z. (1985) Slow, persistent replication of lentiviruses: role of tissue macrophages and macrophage precursors in bone marrow. *Proc. Natl. Acad. Sci. USA* **82,** 7086–7090.
2. Gendelman, H. E., Narayan, O., Kennedy-Stoskopf, S., Kennedy, P. G., Ghotbi, Z., Clements, J. E., et al. (1986) Tropism of sheep lentiviruses for monocytes: susceptibility to infection and virus gene expression increase during maturation of monocytes to macrophages. *J. Virol.* **58,** 67–74.
3. Gendelman, H. E. and Meltzer, M. S. (1989) Mononuclear phagocytes and the human immunodeficiency virus. *Curr. Opin. Immunol.* **2,** 414–419.
4. Zhu, T., Mo, H., and Wang, N. (1993) Genotypic and phenotypic characterization of HIV-1 in patients with primary infection. *Science* **261,** 1179–1181.
5. Stevenson, M. and Gendelman, H. E, (1994) Cellular and viral determinants that regulate HIV-1 infection in macrophages. *J. Leukoc. Biol.* **56,** 278–288.
6. Zhu, T. (2002) HIV-1 in peripheral blood monocytes: an underrated viral source. *J. Antimicrob. Chemother.* **50,** 309–311.
7 Persidsky, Y. and Gendelman, H. E. (2003) Mononuclear phagocyte immunity and the neuropathogenesis of HIV-1 infection. *J. Leukoc. Biol.* **74,** 691–701.
8. Heinzinger, N., Baca-Regen, L., Stevenson, M., and Gendelman, H. E. (1995) Efficient synthesis of viral nucleic acids following monocyte infection by HIV-1. *Virology* **206,** 731–735.
9. Kalter, D. C., Nakamura, J. A., Turpin, J. A., Baca, L. M., Hoover, D. L., Dieffenbach, C., et al. (1991) Enhanced HIV replication in macrophage colony-stimulating factor-treated monocytes. *J. Immunol.* **146,** 298–306.
10. Gendelman, H. E., Orenstein, J. M., Martin, M. A., Ferrua, C., Mitra, R., Phipps, T., et al. (1988) Efficient isolation and propagation of human immunodeficiency virus on recombinant colony-stimulating factor 1-treated monocytes. *J. Exp. Med.* **167,** 1428–1441.
11. Figdor, C. G., Van, E. W., Leemans, J. M., and Bont W. S. (1984) A centrifugal elutriation system of separating small numbers of cells. *J. Immunol. Methods* **68,** 73–87.
12. Bennett, S. and Breit, B. S. (1994) Variables in the isolation and culture of human monocytes that are of particular relevance to studies of HIV. *J. Leukoc. Biol.* **56,** 236–240.

5

Isolation and HIV-1 Infection of Primary Human Microglia From Fetal and Adult Tissue

Kathleen Borgmann, Howard E. Gendelman, and Anuja Ghorpade

Summary

Glial inflammation, principally involving astrocytes and microglia, underlies the pathogenesis of a broad range of neurodegenerative disorders, including, most notably, human immunodeficiency virus (HIV-1)-associated dementia. Indeed, for the latter, disease mechanisms are attributed to viral infection and activation of microglia and perivascular macrophages and their resultant neurotoxic activities. Although monocyte-derived macrophages have served as models for microglia, they are limited both qualitatively and quantitatively in their immune responses and susceptibility to viral infection. Thus, the acquisition of primary human microglial cells is critical for laboratory studies of human neurological disease. In this chapter, we provide detailed methods of isolation, cultivation, characterization, HIV-1 infection, and experimental applications of primary human fetal and adult microglial cells, with particular emphasis on studies of HIV-1 neuropathogenesis.

Key Words: Neuroimmunology; microglia; human immunodeficiency virus; human microglia; microglial activation; neuroinflammation; HIV-1-associated dementia.

1. Introduction

In 1932, del Rio-Hortega gave the first suggestions for the origins of microglia by stating that, unlike the neural and other glial elements of the central nervous system (CNS), they were of mesenchymal origin *(1)*. The observations that microglial cells apparently invaded the brain from the meninges and that they had features similar to these of blood leukocytes suggested to del Rio-Hortega that they were probably blood-borne cells arising "from polyblasts or embryonic cells of the meninges" *(1)*. Until recently and within the past 20 yr, microglia were a relatively underrecognized, widely distributed cell popula-

From: *Methods in Molecular Biology, Vol. 304: Human Retrovirus Protocols: Virology and Molecular Biology*
Edited by: T. Zhu © Humana Press Inc., Totowa, NJ

tion within brain parenchyma constituting about 1% to 2% of all cells. Yet, in the mid-1980s, the microglial cell "came out of the closet" to become the cell of the brain decade ***(2)***. The biology and function of microglia is central to a variety of diseases of the nervous system. Microglia and brain macrophages have been recognized to play crucial roles in important diseases such as viral infections, autoimmunity, and neurodegenerative disorders ***(3–8)***. Human immunodeficiency virus (HIV)-associated dementia (HAD), multiple sclerosis, and Alzheimer's disease are examples where understanding the role of microglia promises to hold essential information concerning disease pathogenesis ***(8–10)***. To researchers who study the disease process involved in the CNS neuroinflammatory disorders and HAD in particular, it is imperative to use primary human microglial cells as in vitro tools to address questions regarding pathogenesis and mechanisms and to uncover therapeutic targets. In this chapter, we describe the protocols utilized for obtaining primary human microglial cells. Human fetal tissue has been the more commonly used source of human microglia because of advantages in yield and quality of cells and the ease of preparation and manipulation. Human adult microglia, although more difficult to obtain, are valuable systems for confirmatory studies, as several of the diseases that involve microglial activation are of the adult brain.

The procedures involved in the preparation of microglia at both the fetal and adult developmental stages are based on the principal of tissue dissociation, followed by enzymatic digestion with or without a gradient to remove debris, and then purification of microglia by preferential adhesion. The term "preferential adhesion" is coined to indicate the high preference and speed with which microglial cells adhere to tissue culture-treated plastic surfaces. Although both types of cells use the principle of preferential adhesion, there is an important technical distinction in this process. In fetal tissue preparations, microglia are obtained as the nonadherent population released from a confluent monolayer of mixed glial cells over a period of time and then purified by preferential adhesion. In adult microglial cells, it is indeed the microglial population that will preferentially adhere in the primary culture followed by removal of nonadherent, nonmicroglial cells. The procedures for HIV-1 infection of microglial cells are similar as those applied to human monocyte-derived macrophages. Yet the susceptibility of microglia to viral infection is different from those of monocyte-derived macrophages (MDM), perhaps as a result of their terminally differentiated quiescent status in the brain parenchyma. In this chapter, we describe, in detail, the preliminary dissociation and purification of human microglia from fetal and adult brain tissue and their infection with macrophage-tropic HIV-1.

2. Materials

2.1. Fetal Tissue

1. Hank's balanced salt solution (HBSS), 1X with/without (w/o): calcium chloride, magnesium chloride, magnesium sulfate, and phenol red.
2. Petri dishes 150 × 20 mm size, disposable, sterile, polystyrene.
3. Forceps, stainless steel, sterile, 5.5 in long.
4. Serological pipets: 1, 2, 5, 10, and 25 mL sizes, sterile, individually wrapped.
5. Trypsin 2.5% thawed 1X previously and aliquoted into 5-mL amounts and stored at –20°C.
6. Heat-inactivated fetal bovine serum (FBS).
7. Pasteur pipet, 7-in, sterile.
8. Microglia medium: Dulbeco's modified Eagle medium (DMEM), 10% FBS, 20 m*M*/mL L-glutamine, 1% penicillin/streptomycin/neomycin, 1000 U/mL macrophage colony stimulating factor (MCSF) (Genetics Institute).
9. 70-μm filter, sterile, individually wrapped.
10. Cell-culture-treated plates or flasks: sterile, individually wrapped.
11. Conical tubes, 50-mL, polypropylene, sterile.
12. Trypan blue stain 0.4%.
13. Hemocytometer and cover slips.

2.2. Adult Tissue

1. HBSS, 1X w/o: calcium chloride, magnesium chloride, magnesium sulfate, and phenol red.
2. Conical tubes, 50 mL, polypropylene, sterile.
3. Forceps, stainless steel, sterile, 5.5 in long.
4. Scalpel, sterile and individually wrapped (size 10).
5. Strainer spoon, sterile.
6. Petri dishes 150 × 20 mm size, disposable, sterile, polystyrene.
7. Kevlar gloves.
8. Surgical gloves, sterile.
9. Razor blades, one-sided blade, sterile.
10. 250-mL flask with lid, sterile.
11. DNAse: 20 mg lyophilized, diluted in the above HBSS at 2 mg/mL, aliquoted into 5-mL aliquots and stored at –20°C.
12. Trypsin 2.5% thawed 1X previously, aliquoted into 5-mL amounts and stored at –20°C.
13. Rotary shaker water bath.
14. Parafilm.
15. Sodium dodecylsufate (SDS) 10%, premade solution for decontaminating anything that may corrode in NaOH.
16. NaOH, 2 *N*, premade stock solution for decontaminating glassware.

17. 95% alcohol in squirt bottle.
18. 100% bleach in squirt bottle.
19. Heat-inactivated fetal bovine serum (FBS).
20. Serological pipet: 1, 5, 10, and 25 mL sizes, sterile, individually wrapped.
21. Spinal needles: 18G, 3.5 inches, sterile, individually wrapped.
22. Syringe: 60 mL, with leur lock, sterile, individually wrapped.
23. Screen cups, for cell dissociation with 200 μm, 100 μm, 60 μm mesh screens, assembled, sterile.
24. Beakers, 250 mL, sterile.
25. Centrifuge tubes, for ultracentrifuge, 175 mL, sterile.
26. Centrifuge tube adaptors.
27. Ultracentrifuge.
28. Adult microglia medium: DMEM 1X, w/high glucose, w/L-glutamine, w/sodium pyruvate, w/pyridoxine hydrochloride, 10% heat-inactivated FBS, 2% HEPES buffer solution, 0.1% gentamicin.
29. Trypan blue stain 0.4%.
30. Hemocytometer and cover slips.
31. Cell culture-treated plates or flasks: sterile, individually wrapped, 48-well.

2.3. Reverse Transcriptase Assay

1. Solution A (disruption buffer): 100 m*M* Tris-HCl, pH 7.9, 300 m*M* KCl, 10 m*M* DTT, 0.1% NP40.
2. Solution B (RT reaction mix): 50 m*M* Tris-HCl, pH 7.9, 150 m*M* KCl, 5 m*M* DTT, 15 m*M* $MgCl_2$, 0.05% NP40, 10 μg/mL Poly A, 0.25 U/mL oligo dt pd$(T)_{12-18}$, 4 μCi/mL ^{3}H-thymidine (add ^{3}H-thymidine just before use; ^{3}H-thymidine is radioactive and must be handled carefully according to radiation safety procedures). Solution A and solution B can be stored as aliquots at –20°C up to 6 mo.
3. 10% Trichloroacetic acid (TCA).
4. Standard RT enzyme.
5. Cell harvester.
6. Liquid scintillation counter.

3. Methods

3.1. Fetal Tissue

The methods outlined below are divided into the following sections: (1) obtaining fetal tissue specimens in accordance to institutional, state, and National Institutes of Health (NIH) guidelines; (2) cleaning and enzymatic dissociation; (3) preparation of a single-cell suspension, culture, and maintenance; and (4) characterization and experimental applications. We and others have previously described this method in multiple publications ***(11–16)***.

3.1.1. Obtaining Fetal Tissue Specimens

In the recent past, obtaining fetal brain tissue specimens has emerged as a controversial endeavor depending on the location in the United States. Regardless, tissue should be obtained with strict guidelines that follow appropriately documented informed consent. Individual institutions and state guidelines should be followed in addition to NIH ethical guidelines. Tissue should be stored in HBSS at 4°C and transported back to the laboratory in wet ice. All tissues should be recorded in a logbook without any additional identifiers. All procedures outlined below should be performed in a minimum of biosafety level 2 (BSL-2) type of tissue culture facility given the risks associated with use of human tissue.

3.1.2. Cleaning and Enzymatic Dissociation

Overall, the tissue specimens obtained are very clean and contain little meninges and blood vessels. However, it is important to ensure that the enzymatic digestion is performed on brain tissue free of other contamination. Also, tissue should be carefully washed to remove any blood vessels. As it is easier to clean the tissue, wash the brain tissue in HBSS w/o phenol red. Transfer the tissue from the tube to a Petri dish and use a forceps to get rid of any adherent meninges. Gestational age ideal for microglial isolation is early to mid second trimester (*see* **Note 1**).

1. Pipet the tissue to a fresh Petri dish with a 5.0-mL pipet. Simultaneously (tissue will usually dissociate with only one pass through the pipet tip), triturate it gently by pipetting up and down through a 5.0-mL pipet. This will dissociate the bigger pieces of tissue.
2. Incubate it with trypsin in the above-mentioned covered Petri dish (diluted in HBSS w/o Ca^{2+} and Mg^{2+} at 37°C in the incubator) for 30–45 min (final volume of tissue: total trypsination volume is 1:5 and the final trypsin concentration is 0.25%).
3. At the end of the incubation period, transfer the tissue with a pipet to a conical tube containing cold heat-inactivated FBS (final concentration of FBS ≅ 10%). Triturate gently with a 5-mL pipet a few times. Allow it to stand for 2–3 min on ice (*see* **Note 2**).

3.1.3. Preparation of Single-Cell Suspension, Culture, and Maintenance

1. Add cold HBSS (ratio of 40 mL HBSS for every 10 mL of tissue). Triturate gently to mix. Centrifuge at 500*g* for 10 min. Decant off the supernatant and repeat wash with the same volume of cold HBSS two times.
2. Resuspend the tissue in 30 mL microglia medium (*see* composition described earlier). Triturate with a 5.0-mL pipet followed by a 2.0- and a 1.0-mL pipet. At the end of these steps, triturate with a 7-in Pasteur pipet. Trituration should not be so harsh that it would cause more cell death. Allow it to stand for about 5 min or until tissue settles down.
3. Collect the supernatant in a fresh tube without collecting any tissue fragments. Add more fresh microglia medium and repeat **step 2**. Repeat **step 2** several times

until there is very little tissue left at the bottom of the tube. The volume of the supernatant collected should not exceed 150–200 mL. If need be, re-use the cell suspension instead of fresh medium for trituration.

4. Filter the single cell suspension (*see* **Note 3**) through a 70-μm filter and then count the cell yield. As per our experience, 5–7 cc of tissue should give about 300–400 million cells. Seed in T-75 flasks at a density of 80–100 million cells/ total 25 mL medium ($3.2–4 \times 10^6$ cells/mL).
5. Incubate undisturbed for a week (*see* **Note 4**). At d 7, very carefully replace the entire 25 mL medium and incubate for additional 7 d. Floating cells will be seen on an adherent layer of astrocytes at the end of this incubation period. Purify by preferential adhesion as described later.

3.1.4. Purification of Microglia by Preferential Adhesion

1. Gently withdraw the supernatant from the primary flasks (T-75 flasks) on d 14, 21, and 28. Collect in a fresh tube (*see* **Note 5**).
2. Centrifuge at 500*g* at room temperature for 10 min. Decant off as much of the supernatant as possible by pipetting with a 5-mL pipet and resuspend cells in a minimal volume (*see* **Note 6**).
3. Count the microglia. Only the bigger, ameboid or round sometimes-granulated cells should be counted. The other smaller or tiny cells should be disregarded, as they are not microglia and would be washed off during processing. Usually, the cells are 100% viable; and the density of the cells at this juncture should be about $1–2.0 \times 10^6$/mL.
4. Culture the microglia from the concentrated cell suspension at appropriate density (1×10^6/mL stock concentration). You will not need to precoat any culture dishes. The density is as follows: 50,000 cells/50 μL/well for 96-well plates and 8-mm chamber-tech slides; 100,000 cells/75–100 μL/well for 24-well plates (Co-Star) (placed as a drop in the center prior to wash below); 450–500,000 microglia/500 μL/well for 6-well plates (for PCR/conditioned media experiments) (*see* **Note 7**).
5. Allow the cells to adhere for 2–4 h (*see* **Note 8**). Add fresh microglia medium from the edge of the well and not directly in the center. Remove the medium, still leaving some medium, so that the cells do not dry. Final volume for 96-well plates is 200 μL/well; for 8-mm chamber-tech slides, 400 μL/well; and for 24 well plates, 500 μL/well. Wash a few wells at a time taking care that they do not dry.
6. Incubate further for 7 d. Supplement thereafter with culture medium without any MCSF addition. MCSF only jump-starts the differentiation procedure. Regardless, media should be replaced twice a week. Cells are ready for experiments. **Figure 1** demonstrates the morphology of cells in mixed culture (**Fig. 1A**) and after preferential adhesion (**Fig. 1B**). Cells harbor bipolar processes and also numerous fine fimbriae-like projections (**Fig. 1C,D**). Immunocytochemically, cells are highly positive for CD68, vimentin, and seldom for HLA-DR (**Fig. 1**). HIV-1 infection can be performed as early as 4 d as described below.

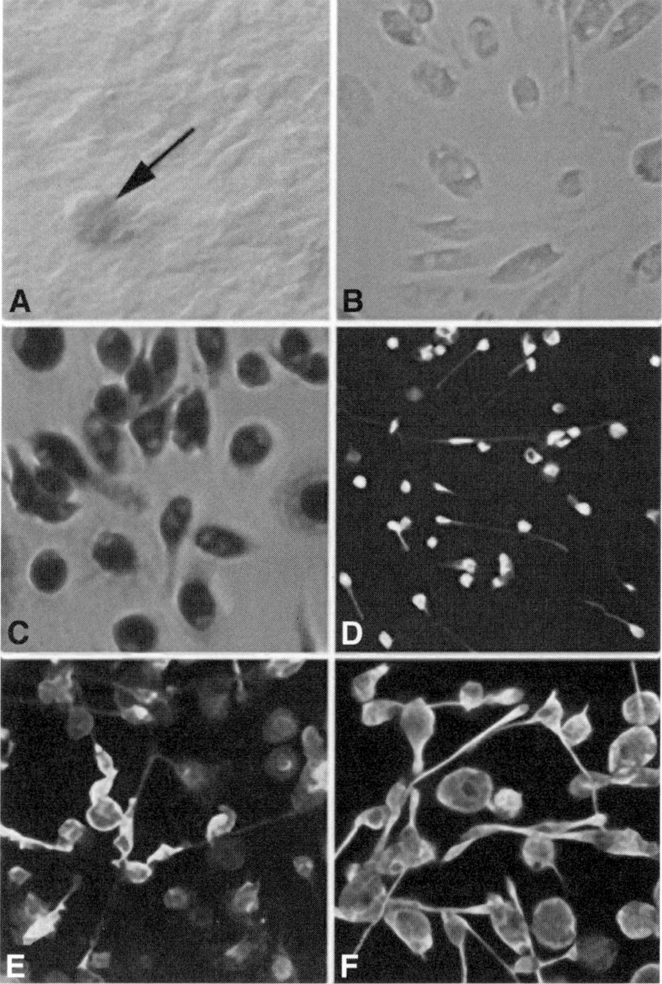

Fig. 1. Characterization of human fetal microglial cultures. Human fetal microglia are isolated from a mixed dissociated brain tissue culture (**A**). Microglial cells are released from a dense monolayer of mixed glia (arrow in **A**). Upon purification of the floating population as described by preferential adhesion, microglial cells differentiate to develop polar processes and in some instances multiple fine fimbriae-like projections (**B**). These morphological features are enhanced with Wright's staining (data not shown) on reduction of 3-(4,5-dimethyl-2-thiazolyl)-2,5-diphenyl-2H-tetrazolium bromide (MTT), a chromogen, to form purple zymosan crystals by metabolically active microglia (**C**). These cells are all immunoreactive to the mononuclear phagocyte marker CD68 (**D**) and a subpopulation expressed elevated levels of human major histocompatibility (MHC) class II (HLA-DR) as shown in **E**. Vimentin, a mononuclear phagocyte-specific intermediate filament protein, readily stains human microglial cells with strong immunoreactivity (**F**). Original magnifications ×200 for **A–C**, **E**, and **F**, ×100 for **D**.

3.1.5. Characterization and Experimental Application

Human fetal microglial cells prepared under these conditions are >99% pure by CD68 staining. In all cultures, maintain a few wells in a chamber-tech slide or a 48-well plate for immunocytochemistry analysis of microglial cells *(13–15)*. Antibodies to CD68 (a mononuclear phagocyte marker), glial fibrillary acidic protein (GFAP, an astrocyte marker), and mouse immunoglobulin (Ig)G (for nonspecific control) should be used in all cases. In addition, other markers specific for the research purposes can be used successfully for these cells using immunofluorescence or immunocytochemistry analysis with peroxidase color development. Vimentin is highly expressed, typical of mononuclear phagocytes, and a subpopulation of these cells can be activated and express HLA-DR (**Fig. 1**).

3.2. Adult Tissue

The methods below outline the process of isolating adult microglia from rapid autopsy tissue by (1) obtaining tissue, (2) cleaning and enzymatic disassociation, (3) mechanical dissociation, (4) Percoll gradient, and (5) plating, care, and use. The following protocol was developed with modifications to other previously published methods *(17–21)*.

3.2.1. Obtaining Adult Autopsy Tissue Specimens

Tissue is obtained in a number of ways, depending on each institution's rules and regulations. However, issues of safety, sterility, and time must be considered.

3.2.1.1. Donor Eligibility

1. The donor should be free of infectious disease (HIV, tuberculosis, viral hepatitis, Creutzfeldt-Jakob disease, undocumented dementia, and so on), as they pose risk to those processing the tissue.
2. The donor should be free from head trauma, prolonged hypoxia, or prolonged respirator brain death, as the viability of the microglia could be significantly reduced.
3. The donor should be free of sepsis and/or meningitis to avoid contamination of cells post isolation.
4. The postmortem interval (PMI) should be less than 24 h and preferably close to 4 h *(22)*.

3.2.1.2. Harvesting of Tissue

1. Frontal cortex and corpus collossum brain tissue should be harvested in a sterile manner using presterilized saw blades, forceps, scalpels, and gloves. A total of 20–60 g of tissue is ideal for significant cell retrieval.
2. Tissue should be collected into a preweighed sterile 50-mL conical tube containing 15–20 mL of ice-cold, sterile HBSS.

3. The tissue should be transported back to the processing area on wet ice and processing should begin as soon as possible.

3.2.2. Cleaning and Enzymatic Dissociation of Tissue

During the washing and chopping procedure, blood, blood vessels, and connective tissues are removed to improve the effectiveness of the enzymatic digestion and to prevent contamination of the final cell culture.

The methods from here to the end should be performed under BSL-2 precautions, including but not limited to: lab coat, exam gloves, sterile laminar flow hood, bleach bucket, and autoclaving, and proper disposal of biohazard trash. Also, for the safety of the processors and the purity of the tissue, the work area and instruments used are decontaminated.

3.2.2.1. Washing and Debris Removal

1. Weigh the 50-mL conical tubes containing the tissue and subtract the preweight from the new weight to get the actual tissue weight. Record this weight for processing purposes *(see* **Note 9**).
2. Rinse the exterior of the 50-mL conical tubes with alcohol and place the tubes in the culture hood.
3. Wash each tissue by filling the tube with HBSS and gently shaking and inverting it to mix the tissue with the HBSS. The tissue should move freely inside the tube for proper removal of blood residue. The tissue should be allowed to settle and the excess liquid should be decanted off into the bleach bucket. This process should be repeated 10–15 times or until the decanted liquid is clear.
4. Place the tissue in a sterile Petri dish, cover with HBSS, and use sterile forceps to remove any meninges and blood vessels *(see* **Note 10***).*
5. Using a sterile scalpel, cut tissue into approx 2.5-cm^3 pieces. During this washing and chopping procedure, continue to remove blood vessels and connective tissue.
6. Place tissue in a fresh 50-mL conical tube and wash two to three times in HBSS. Avoid picking up any loose connective tissue or blood vessels.
7. Wash again in a 50-mL conical tube two to three times by gentle shaking and inverting of the tubes with HBSS.
8. Carefully place the tissue chunks into another sterile Petri dish, cover with HBSS, and cut into approx 1.5 cm^3 pieces.
9. Place tissue in a fresh 50-mL conical tube and wash two to three times in HBSS. Avoid picking up any loose connective tissue or blood vessels.
10. Place tissue into another sterile Petri dish, cover in HBSS. Change into Kevlar gloves covered with sterile surgical gloves and chop tissue into approx 0.2 cm^3 pieces *(see* **Note 11***).*
11. Gently wash this tissue a few times in the Petri dish with HBSS by adding and removing liquid with a 25-mL pipet. Be careful not to lose any of the tissue during the washes.

Table 1
Volume of Reagents Required for Trypsination Dependent on Tissue Weight

Tissue (g)	HBSS (mL)	Trypsin 2.5% (mL)	DNAse 2mg/mL (mL)
10	20	2	1.5
20	50	3	2
30	80	5	3.5

12. Transfer the tissue pieces into the 250-mL flask with the strainer spoon. Use a 25-mL pipet to rinse the Petri dish and transfer any tissue pieces stuck to the dish.
13. Tilt the 250-mL flask at an angle and allow the tissue to settle; remove as much of the HBSS as possible without losing any tissue.

3.2.2.2. Enzymatic Dissociation With Trypsin

1. Refer to **Table 1** and add the appropriate volumes of HBSS, trypsin and DNase to the flask. These are dependent on the original weight that was recorded in **Subheading 3.2.2.1., step 1**.
2. Put on the cap, seal with parafilm, and swirl the contents gently to mix. Put in a rotary shaker water bath at 37°C and 150 rpm for 1 h.

3.2.2.3. Decontamination of Instruments, Disposables, and Working Surface

1. Nondisposable, metal instruments (forceps and strainer spoons) should be covered in 10% SDS for a minimum of 1 h and then rinsed thoroughly before being resterilized.
2. Disposable plasticware (Petri dishes and 50-mL conical tubes) should be rinsed with bleach and thrown into biohazard trash. Any debris, connective tissue, or blood vessels should be incubated in 10% bleach water before disposal.
3. Sharps should be rinsed in bleach and then disposed of in a biohazard sharps container.
4. The entire working surface should be rinsed with 95% alcohol.
5. The materials for the next section should be placed in the hood before turning on the ultraviolet (UV) light for the remaining incubation time. This step further sterilizes the reagents, materials, and working surface for the next step.

3.2.3. Mechanical Dissociation of Tissue

The trypsin digestion breaks many of the bonds holding the tissue together. However, there are still intact tissue pieces in the creamy trypsin-DNase mixture. With repetitive mechanical agitation these pieces will disassociate further into a more homogenous mixture. After straining to remove large pieces, connective tissue, and blood vessels, the single-cell suspension will be ready for the Percoll gradient.

3.2.3.1. Repetitive Agitation With Pipets and Spinal Needle

1. After digestion, turn off UV light and remove flasks from the water bath. Clean the exterior with alcohol to keep the working surface sterile.
2. Add 2.0 mL of FBS to each flask and swirl gently.
3. Divide contents of each flask evenly into two or three 50-mL conical tubes
4. Centrifuge tubes for 10 min at room temperature and 500*g*.
5. Record the volume of the pellet. Remove the supernatant, being careful not to disturb the tissue pellet, as it is not solid and will move down the length of the tube.
6. Resuspend the tissue pellet in an equal amount of HBSS. For example: with 7 mL tissue, add 7 mL HBSS for a final total volume of 14 mL.
7. Triturate through a 25-mL pipet aggressively to break up tissue chunks until the tissue passes freely through the 25-mL pipet three times. Be careful not to aerosolize (create bubbles in) the solution.
8. Triturate through a 10-mL pipet. Do this until the tissue passes freely through the 10-mL pipet three times. Be careful not to aerosolize the solution.
9. Triturate though a spinal needle using a 60-mL syringe. Do this until the tissue passes freely through the spinal needle three times. When expelling the solution from the syringe run it down the tube's side to avoid aerosolizing the solution (*see* **Note 12**).

3.2.3.2. Straining

1. Filter cell solution through a 60-μm mesh into a sterile 250-mL beaker. The solution will not flow through the mesh easily. It must be constantly agitated by running it over the mesh with a 10-mL pipet. Furthermore, it will need to be rinsed with small amounts of HBSS until all tissue but connective tissue and blood vessels has gone through the mesh (*see* **Note 13**).
2. Take the 60-μm filtered cell solution and filter it again though a 100-μm mesh into another 250-mL beaker.
3. Take this 100-μm filtered cell solution and filter it again though a 200-μm mesh into another 250-mL beaker.

3.2.4. Purification Through Percoll Gradient

The Percoll separates the different cell types and debris in the cell solution into distinct layers that can then be harvested separately, leading to a pure microglia culture.

3.2.4.1. Establishing the Percoll Gradient

1. Refer to **Table 2** for the appropriate final volume for the cell suspension (volume = grams tissue × 10.5 mL) and add enough HBSS to the filtered cell solution to adjust it to that volume. Again, the volumes are dependent on the tissue weight recorded in **Subheading 3.2.1., step 1**.
2. Divide the cell suspension between the appropriate number of tubes (10 g = 1 tube, 11–20 g = 2 tubes, 21–29 g = 3 tubes, 30–35 g = 4 tubes).

Table 2
Volumes of Reagents for Percoll Gradient Dependent on Original Tissue Weight

Grams tissue example	Total Percoll (mL) = tissue (g) × 4.5 (mL)	Total cell suspension (mL) = tissue (g) × 10.5 (mL)	mL Percoll/tube = total Percoll/ # tubes	mL cell susp./ tube = total cell susp/ # tubes	# of tubes
15	67.5	157.5	33.75	78.75	2
25	112.5	262.5	37.5	87.5	3

3. Add the appropriate volume of Percoll (volume = grams tissue × 4.5 mL) to each tube, cap, and agitate gently.
4. Weigh the centrifuge tubes, along with the adaptors, and balance to within 0.01 g (*see* **Note 14**).
5. Centrifuge for 45 min at 10,000*g* and 4°C. The brake should not be used to slow the centrifugation, as it will disturb the gradient.

3.2.4.2. Decontamination of Instruments, Disposables, and Working Surface

1. Nondisposable, metal instruments (strainer cups and meshes) should be covered in 10% SDS for a minimum of 1 h and then rinsed thoroughly before being resterilized.
2. Nondisposable glassware (250-mL flasks and 250-mL beakers) should be covered with 2 *N* NaOH for a minimum of 2 h and then rinsed thoroughly before being resterilized.
3. Disposable plasticware (50-mL conical tubes and 60-mL syringe) should be rinsed with bleach and thrown into biohazard trash. Any debris, connective tissue, or blood vessels should be incubated in 10% bleach water before disposal.
4. Sharps (spinal needles) should be rinsed in bleach and then disposed of in a biohazard sharps container.
5. The entire working surface should be rinsed with 95% alcohol.
6. The materials for the next section should be placed in the hood before turning on the UV light for the remaining centrifugation time. This step further sterilizes the reagents, materials, and working surface for the isolation step.

3.2.4.3. Harvesting the Percoll Gradient

During centrifugation the Percoll gradient separates the cell suspension into five layers: the debris layer, a hazy interface between the debris and cellular layer, the cellular layer, the red blood cell layer, and HBSS at the very bottom. The microglia settle in the middle of the cellular layer.

1. After spin completes, gently remove the tubes. You should see distinct layers in the tube.
2. Turn off the UV light, alcohol the exterior of the centrifuge tubes, and place them in the hood.

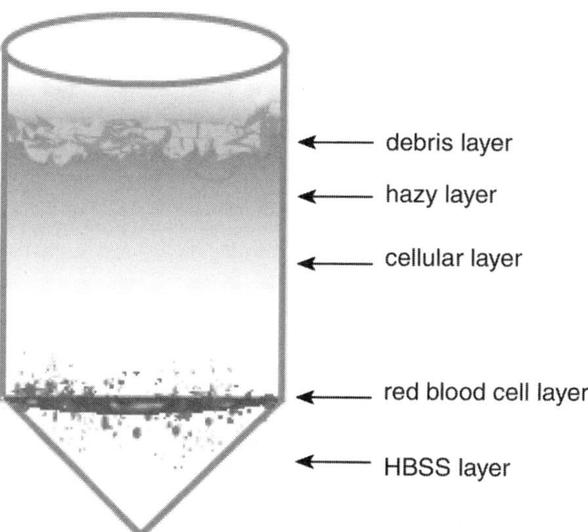

Fig. 2. Percoll gradient. This is a schematic representation of the gradient as observed after human adult dissociated tissue is centrifuged. Respective layers are labeled in the drawing. The viable cell layer is harvested as the source of viable glial cells.

3. Refer to **Fig. 2** to visualize the layers of the Percoll gradient. Remove the debris layer using a 10-mL pipet. It is thick and will move into the pipet in a continuous section as long as the pipet is kept in contact with the surface of the debris layer. Be careful not to drip debris from the pipet back into the tube, as this will disturb the gradient. Discard the debris into the bleach bucket. Use a clean pipet each time you return to remove debris.
4. Once the thick debris layer is removed, pinpoint as much of the white debris chunks that may be floating loosely in the upper hazy cellular layer with a 10-mL pipet and discard them into the bleach.
5. Collect the cellular layer containing the microglia. It lies between the hazy remains of the debris layer and the red blood cell layer. Divide it equally into six 50-mL conical tubes with no more than 15 mL per tube. Do not drip during this removal and be very careful not to include any of the red blood cell layer (*see* **Note 15**).
6. To each 50-mL conical tube containing cell suspension, resuspend in a total volume of 45 mL with HBSS.
7. Spin for 10 min at room temperature at 500*g*.
8. Remove and discard supernatant. At this point the pellets are small and firm; thus, the supernatant can easily be removed (*see* **Note 16**).
9. Resuspend each pellet in about 5 mL of HBSS and triturate well, as the pellet is firm and good washes remove dead cells and debris. Combine three tubes into one tube and add HBSS up to 45 mL in each tube.
10. Repeat **steps 7–9** until all tubes are combined into one.

11. Spin one final time at room temperate at 500*g* for 10 min.
12. Resuspend the single pellet in 5 mL adult microglia media and triturate gently but thoroughly, as a single cell suspension is necessary for an accurate cell number count. Avoid aerosolizing the solution as cells will become entrapped in the bubbles and the cellular count will be affected.
13. Count cells with standard method using a four-cell hemocytometer. Record the concentration.
14. Discard the red blood cell layer and remaining contents in the ultra centrifuge tube and decontaminate empty tube and cap by covering them with 10% SDS for at least 1 h.

3.2.5. Plating, Care, and Use of Adult Microglia Culture

The adult microglia can be plated into a variety of culture plates and flasks. However, the final experimental use should be considered in the plating of these cells, as they are very difficult to passage. The cells become firmly attached to the culture surface and many, if not all, cells are lost in trypsination.

3.2.5.1. Initial Plating of Adult Microglia Cultures

1. Microglia should be plated in tissue culture-coated plates at a density of between 0.5 and 0.6×10^6 cells/cm^2, initially in a small amount of adult microglia media (100 μL media per 100,000 cells).
2. Cultures should be left for a few days without being disturbed to allow the cells to attach and begin differentiation. The media should be doubled after 2 d for proper nourishment, but none should be removed.

3.2.5.2. Purification of Adult Microglia Cultures Through Preferential Adhesion

1. After 1 wk, the cells should begin to grow processes and should be fairly well attached. There will be debris and floating cells in the well that should be removed by a couple of complete media exchanges. The wash media removed from the well should be collected in a 50-mL conical tube.
2. The wash media can be centrifuged at room temperature, 500*g* for 10 min, and then resuspended in new media. A hemocytometer count may reveal sufficient cell numbers to replate these cells *(see* **Note 17***)*.
3. Following purification, adherent cultures should receive a half-medium exchange twice a week with adult microglia media.

3.2.5.3. Differentiation and Characterization

The cells exhibit ameboid morphology earlier in culture and differentiate normally into small bipolar cells with finite processes extending to 0.0–2-fold of the cell length on either side. This bipolar phenotype lasts from 3 d to about 2 wk in culture; however, individual preparations can vary in the time frame. After 3 wk in culture, microglial cells appeared well differentiated, and multiple fibrous extensions in addition to the polar processes can be visualized under phase-contrast. Some preparations appear granulated and retain these granules throughout the period of cultivation despite the differentiation of cells from ameboid into bipolar cells. On average, pure microglia preparations are differentiated and ready for experimental use 10 to 15 d postcultivation, or 3 to 10 d postpurification. **Figure 3** demonstrates live cell views of microglial cells prior to purification and after the removal of debris and contaminating populations.

All microglial preparations should be checked with immunocytochemical labeling with standard markers (*see* **Note 18**). At a mean time period of 2 to 3 wk, cells are fixed with a 1:1 mixture of ice-cold acetone:methanol for 20 min and allowed to dry. Cells are incubated with antibodies to CD68 (MP marker), GFAP (astrocyte marker), factor VIII (endothelial cells marker), fibronectin (fibroblast marker), galactocerebroside (gal-C, oligodendrocyte marker), mouse IgG (nonspecific control) and 4',6-diamidino-2-phenylindole, dilactate (DAPI [nucleic marker]) according to manufacturer's specifications. Reactivity to cell-specific markers is detected with appropriate secondary antibodies conjugated to fluorescent markers. A typical culture shows a majority of cells positive for CD68. Occasionally, a cluster of GFAP-positive astrocytes is observed; however, this event is very rare. Adherent microglial preparations are consistently <99% devoid of factor VIII and fibronectin, indicating a lack of contaminating endothelial cells and fibroblasts. Preliminary immunocytochemical characterization of pure adult microglial cells is shown in **Fig. 4**.

3.3. Infection of HIV on Microglia

All procedures handling infectious HIV or HIV-infected specimens should be carried out in a BSL-3 facility or a BSL-2 facility with BSL-3 practices.

3.3.1. Infection With HIV-1

1. Culture microglia in an appropriate dish for experimental purposes as described in **Subheadings 3.1.4.** (fetal) and **3.2.5.1.** (adult).
2. When microglia are sufficiently differentiated (*see* **Subheadings 3.1.4.6.** [fetal] and **3.2.5.3.** [adult]), remove all media from the culture and infect with HIV stock diluted in appropriate media at 300,000 cpm (as measured by RT assay [*see* **Subheading 3.3.2**]) per 0.5–1 × 10^6 cells in a sufficient volume to cover and nourish the cells, 50 µL/100,000 cells (*see* **Note 19**). If multiple HIV-1 isolates are used in an experiment, it is important to normalize the infection levels by RT measurements in the same assay to avoid variable levels of HIV-1 particles for different strains.
3. Incubate at 37°C with 5% CO_2 and 95% humidity overnight.
4. The following morning, remove all infection medium from the culture and add the required amount of the appropriate medium for chosen plating method.
5. Monitor the infection every 2–3 d by looking for cytopathic effects (**Fig. 5**) and/or by performing a RT assay on reserved media (30–50 µL) from postinfection medium exchanges (*see* **Note 20**).

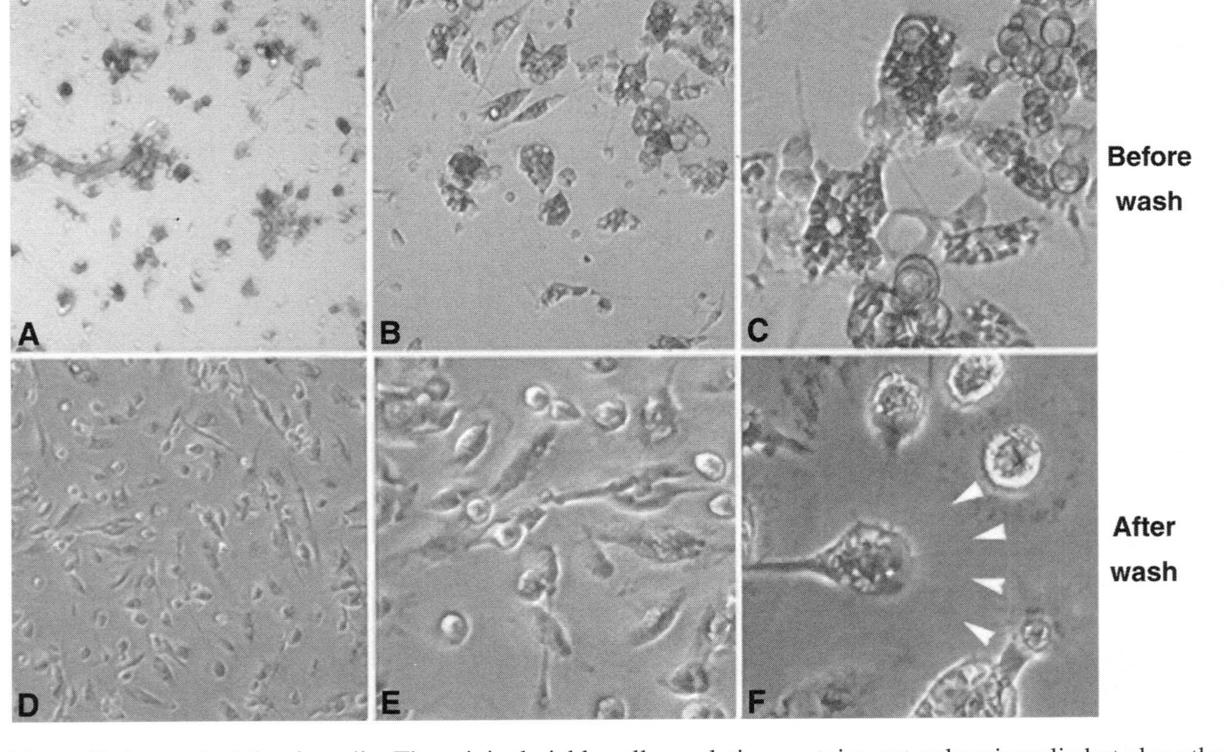

Fig. 3. Live cell views of adult microglia. The original viable cell population contains not only microglia but also other debris and contaminating populations. **A–C** demonstrate this mixed culture with increasing magnification. **A** demonstrates the granular debris and other smaller ameboid cells that can be withdrawn in the wash. **B** and **C** demonstrate, with higher magnification, that under the debris and contaminating cells, microglial cells begin to differentiate within 4 to 5 d after cultivation. Cells are washed on d 4 or 5 depending on the morphological differentiation of microglial cells in the adherent population. **C–D** demonstrate with increasing magnification the purified, differentiated microglial cells. Under high power and phase-contrast numerous fine processes can be visualized (shown by arrows in **F**). Original magnifications, X100 for **A** and **C**; ×200 for **B** and **D** and ×400 for **C** and **E**.

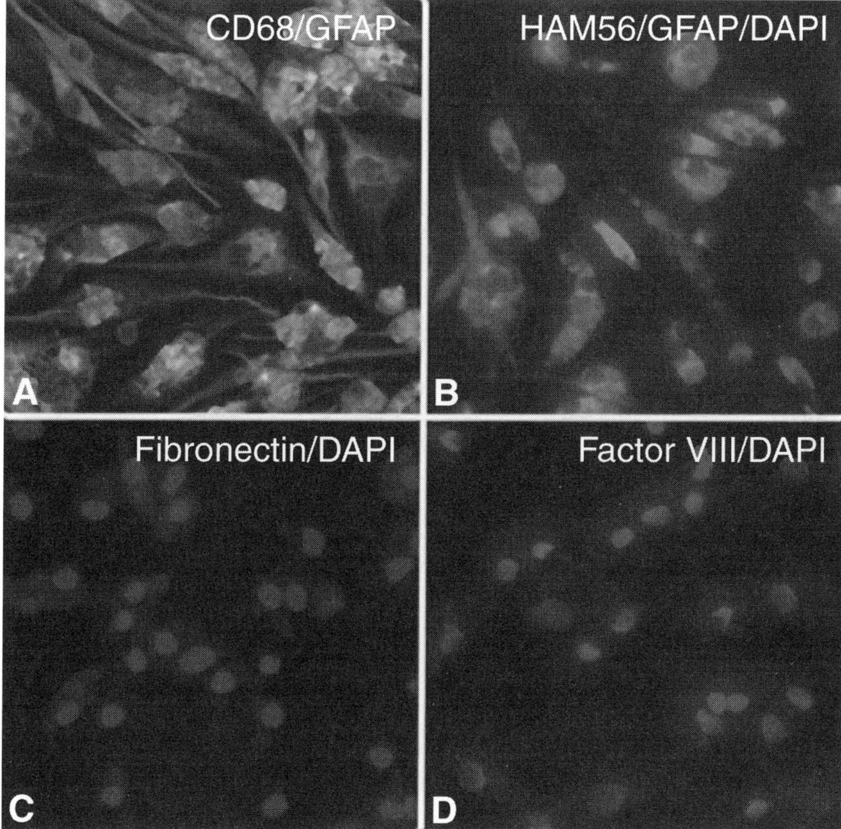

Fig. 4. Immunocytochemical characterization of primary human adult microglia. Adherent monolayers of microglia were fixed with acetone/methanol as described previously and immunostained for mononuclear phagocyte-specific marker (CD68), astrocyte-specific marker (GFAP), endothelial cells (fibronectin and factor VIII). Immunoreactivity was visualized with appropriate secondary antibodies conjugated to flourescein isothiocyanate conjugate (FITC) or rhodamine. DAPI was used to visualize cellular nuclei. **(A)** CD68/GFAP; **(B)** CD68/GFAP/DAPI (nuclear); **(C)** Fibronectin/DAPI (nuclear); **(D)** Factor VIII /DAPI (nuclear). Original magnifications ×200.

6. Microglia can be used in experiments once cytopathic effects are visualized or when a rise in RT is noticed, usually about 7 d post infection (**Fig. 5**). Typical cytopathic effects, such as formation of multinucleated giant cells, are demonstrated in **Fig. 5**. If the experimental plan involves further incubation of the microglia, the experiment should be started with the initial signs of infection, so the microglia survive through the entire experiment.

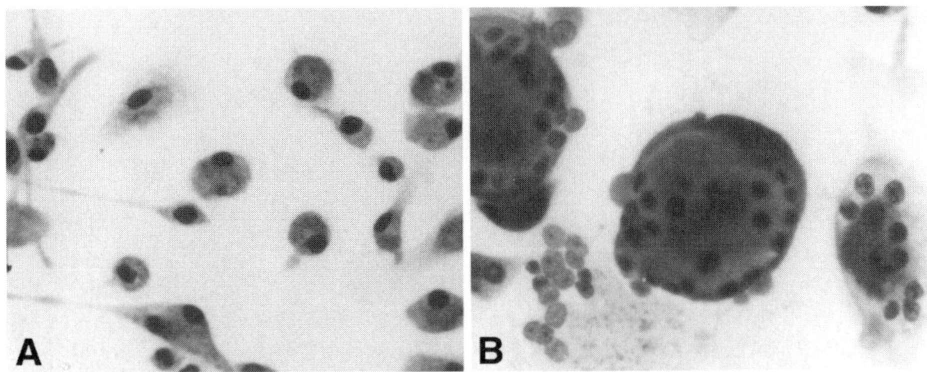

Fig. 5. HIV-1-infection and cytopathic effects in human microglia. Human fetal and adult microglial cells are infected with HIV-1_{ADA} with very little variation from the human monocyte-derived macrophage infection procedure. **(A)** Uninfected, bipolar, well-differentiated control cells. Upon viral infection, cells demonstrate ballooning and other cytopathic effects, formation of multinucleated giant cells, and halo around the giant cells as demonstrated in **B**. Original magnification ×200.

3.3.2. Reverse Transcriptase Assay

1. Pipet 10 µL of solution A into an appropriate number of wells in a 96-well round-bottomed plate. Assay is done in quadruplicate. A blank (medium), standards, and unknown samples are included in the assay. Use different dilutions of the standard, ranging from 0.1 to 10 U/10 µL. Dilute the standards immediately before use.
2. Add 10 µL of each sample to appropriate wells and mix by swirling the plate.
3. Incubate the plate for 15 min at 37°C with 95% humidity.
4. Add 25 µL of solution B to each well. Solution B contains radioactive thymidine, so take appropriate cautions while handling the radioactive material.
5. Incubate the reaction plate for 18–24 h at 37°C with 95% humidity.
6. After the incubation, add 50 µL of ice-cold 10% TCA to each well and keep the plate at 4°C for at least 15 min.
7. Collect the precipitate by harvesting the plate using a Skatron Harvester or any cell harvester.
8. Place the filter paper disks separately into scintillation vials containing the scintillation fluid and count the radioactivity using a liquid scintillation counter.

Calculate the amount of RT activity in counts per minute (cpm) of the unknown samples from the standard curve made by using different dilutions of commercial standard RT enzyme.

4. Notes

1. Human fetal brain tissue can be used for the preparation of microglia as well as astrocytes and neurons ***(23,24)***. Because neurogenesis precedes gliogenesis during development, early to mid-first trimester specimens yield better purity of neuronal cells with relatively less glial cell contamination. Microglia and astrocytes can be successfully prepared from late first-trimester or second-trimester gestational age specimens.
2. At the end of the digestion, tissue should form a tight pellet. If this does not happen, and it continues to float halfway through the conical tube, it is recommended that trituration be repeated until it can form a well-defined pellet.
3. The single-cell suspension, at this point, will still harbor minute tissue remnants. It is important to filter these out; otherwise, the mixed culture may have a high number of astrocytes mixed with microglial cells after 14 d of culture. In this case, it is very difficult to obtain highly pure preparations of microglia.
4. Do not change the media, and do not take the cells out to look under the inverted scope for at least 4 d.
5. Obviously, the days of collection are not definitively defined; for example, if medium starts to turn yellow sooner you may need to collect earlier, i.e., d 12 post explantation. The caveat is that you need to check the cultures to make sure the microglia are not consuming the growth medium and causing a pH change.
6. We prefer the adhesion properties of Co-Star and Nunc tissue culture plates.
7. Allow the tube to stand for a few minutes. Some medium will still be left in the tube that would accumulate at the bottom. Resuspend the cells in the same medium that has pooled at the bottom.
8. The time is not definitively defined. If the preparation is highly enriched with microglia (>95% pure), 3–4 h is acceptable. If the preparation is not highly pure, and contains >20% other cells, washing at 1–2 h is recommended or purify cells using αCD45 microbeads (Miltenyi Biotec, Auburn, CA, cat. no. 130-045-801, per manufacturer's instructions).
9. At this point, the solutions needed later in the methods protocol should be thawed in a 37°C water bath to avoid unnecessary delays. The solutions are: DNase, trypsin, FBS, HBSS, Percoll, and culture media.
10. The lid of the Petri dish can be used to keep the instruments sterile as one switches from washing to cleaning and cutting and back. The lid also serves as a discard area for blood vessels, connective tissue, and meninges removed during cleaning. This keeps the debris associated with the waste away from the tissue and helps one to avoid putting it in the new 50-mL conical tube for the following wash.
11. During the razor blade chopping step, both the Kevlar and the sterile surgical gloves need to be tight against the fingertips. If there are bubbles at the fingertips they can be submerged into the tissue and are a potential source of contamination. Other types of blades and multiple types of handles have been tried and this method remains the fastest and most effective way of finely chopping the tissue before trypsination.

12. The spinal needle often becomes clogged with unremoved portions of meninges. The easiest way to remove the clog is to remove it with sterile forceps. The needle may also be changed but the tissue suspension inside the needle will be lost.
13. If a significant portion of the tissue will not pass through the mesh after multiple washes, a sterile glass pestle may be used to manually break up the small pieces with circular moves and gentle pressure. Attention must be paid to avoid pushing the mesh through the cup holder, especially with the fine 200-μm mesh. Often, the mesh catches minute blood vessels and connective tissue. The forcing of these types of tissue through the mesh should be avoided. Preform Percoll gradient calculations prior to straining to avoid rinsing too much HBSS and over-diluting the cell suspension.
14. Once the Percoll and cell suspension have been combined, a delicate ratio has been established that allows for the *in situ* formation of the gradient during centrifugation. During balancing, solution should be moved from one centrifuge tube to another rather than adding small amounts of HBSS that would offset this necessary ratio.
15. When collecting the cellular layer, no more than 15 mL of the Percoll gradient should be put into each 50-mL conical tube. The gradient must be diluted sufficiently (1:3) with HBSS to allow the microglia to pellet. Furthermore, it is not beneficial to try to collect too close to the debris and red blood cell layers. Although one may see a small increase in the number of microglia collected, a large amount of debris and red blood cell contamination will be observed.
16. The color and size of the pellet observed at this point can be very misleading. Many pellets may have a sight red color and show little red blood cell contamination, and large pellets may have debris. The only sure way of knowing the cellular make of the pellet is to count them.
17. The cells recovered from washes often have different cellular properties from pure microglia cultures. Astrocyte and endothelial cells have been observed in these cultures when not present in the initial plating.
18. In planning the experimental layout in plates and/or flasks, an additional plate with four to eight wells can be cultured solely for the purpose of cell characterization. This way the composition of each donor is recorded, as differences do exist owing to large donor variation.
19. A typical viral stock has a cpm measured by RT assay of 30,000 to 60,000 cpm. Thus, some variability exists in the infection range as a ratio to cell number. The specific cpm may need to be adjusted for viral strains and stock infectability.
20. Two consecutive media exchanges showing RT activity indicates successful viral infection.

References

1. del Rio-Hortega, P. (1932) in *Cytology and Cellular Pathology of the Nervous System* (Penfield, W., ed.), Hoeber, New York: pp. 481–584.
2. Gehrmann, J., Matsumoto, Y., and Kreutzberg, G. W. (1995) Microglia: intrinsic immuneffector cell of the brain. *Brain Res. Brain Res. Rev.* **20,** 269–287.

3. Streit, W., Graeber, M., and Kreutzberg, G. (1988) Functional plasticity of microglia: a review. *Glia* **1,** 301–307.
4. Griffin, W. S., Sheng, J. G., Royston, M. C., Gentleman, S. M., McKenzie, J. E., Graham, D. I., et al. (1998) Glial-neuronal interactions in Alzheimer's disease: the potential role of a 'cytokine cycle' in disease progression. *Brain Pathol.* **8,** 65–72.
5. Watkins, L. R. and Maier, S. F. (1999) Implications of immune-to-brain communication for sickness and pain *Proc. Natl. Acad. Sci. USA* **96,** 7710–7713.
6. Raivich, G., Bohatschek, M., Kloss, C. U. A., Werner, A., Jones, L., and Dreutzberg, G. W. (1999) Neuroglial activation repertoire in the injured brain: graded response molecular mechanisms and cues to physiological function. *Brain Res. Rev.* **30,** 77–105.
7. Williams, K. C. and Hickey, W. F. (2002) Central nervous system damage, monocytes and macrophages, and neurological disorders in AIDS. *Annu. Rev. Neurosci.* **25,** 537–562.
8. Minagar, A., Shapshak, P., Fujimura, R., Ownby, R., Heyes, M., and Eisdorfer, C. (2002) The role of macrophage/microglia and astrocytes in the pathogenesis of three neurologic disorders: HIV-associated dementia, Alzheimer disease, and multiple sclerosis. *J. Neurol. Sci.* **202,** 13–23.
9. Gonzalez-Scarano, F. and Baltuch, G. (1999) Microglia as mediators of inflammatory and degenerative diseases. *Annu. Rev. Neurosci.* **22,** 219–240.
10. Benveniste, E. N. (1997) Role of macrophages/microglia in multiple sclerosis and experimental allergic encephalomyelitis *J. Mol. Med.* **75,** 165–173.
11. Dickson, D. W., Lee, S. C., Mattiace, L. A., Yen, S. H., and Brosnan, C. (1993) Microglia and cytokines in neurological disease, with special reference to AIDS and Alzheimer's disease. *Glia* **7,** 75–83.
12. Ghorpade, A. (1997) in *The neurology of AIDS* (Gendelman, H. E., Lipton, S. A., Epstein, L., and Swindells, S., eds.), Chapman and Hall, New York, pp. 86–96.
13. Ghorpade, A., Nukuna, A., Che, M., Haggerty, S., Persidsky, Y., Carter, E., et al. (1998) Human immunodeficiency virus neurotropism: an analysis of viral replication and cytopathicity for divergent strains in monocytes and microglia. *J. Virol.* **72,** 3340–3350.
14. Ghorpade, A., Xia, M. Q., Hyman, B. T., Persidsky, Y., Nukuna, A., Bock, P., et al. (1998) Role of the beta-chemokine receptors CCR3 and CCR5 in human immunodeficiency virus type 1 infection of monocytes and microglia. *J. Virol.* **72,** 3351–3361.
15. Ghorpade, A., Persidskaia, R., Suryadevara, R., Che, M., Liu, X. J., Persidsky, Y., et al. (2001) Mononuclear phagocyte differentiation, activation, and viral infection regulate matrix metalloproteinase expression: implications for human immunodeficiency virus type 1-associated dementia. *J. Virol.* **75,** 6572–6583.
16. Peterson, P. K., Hu, S., Salak-Johnson, J., Molitor, T. W., and Chao, C. C. (1997) Differential production of and migratory response to beta chemokines by human microglia and astrocytes. *J. Infect. Dis.* **175,** 478–481.

17. Lue, L. F., Brachova, L., Walker, D. G., and Rogers, J. (1996) Characterization of glial cultures from rapid autopsies of Alzheimer's and control patients. *Neurobiol. Aging* **17,** 421–429.
18. De Groot, C. J. A., Montagne, L., Janssen, I., Ravid, R., Van Der Valk, P., and Veerhuis, R. (2000) Isolation and characterization of adult microglial cells and oligodensrocytes derived from postmortem human brain tissue. *Brain Res. Protocols* **5,** 85–94.
19. Yong, V. W. and Antel, J. P. (1997) in *Protocols for Neural Cell Culture* (Fedoroff, S. and Richardson, A., eds.), Humana Press, Totowa, NJ. pp. 157–172.
20. Ghorpade, A., Bruch, L., Persidsky, Y., et al. (2005) Development of a rapid autopsy program for studies of brain immunity. *J. Neuroimmunol.*, in press.
21. Ghorpade, A., Persidsky, Y., Swindells, S., et al. (2005) Investigations of neuroinflammatory responses from microglia recovered from HIV-1-infected and seronegative subjects. *J. Neuroimmunol.*, in press.
22. Schuenke, K. and Gelman, B. B. (2003) Human microglial cell isolation from adult autopsy brain: brain pH, regional variation, and infection with human immunodeficiency virus type 1. *J. Neurovirol.* **9,** 346–357.
23. Suryadevara, R., Holter, S., Borgmann, K., Persidsky, R., Labenz-Zink, C., Persidsky, Y., et al. (2003) Regulation of tissue inhibitor of metalloproteinase-1 by astrocytes: Links to HIV-1 dementia. *Glia* **44,** 47–56.
24. Zheng, J., Thylin, M., Ghorpade, A., Cotter, R., Persidsky, Y., and Gendelman, H. E. (1998) CXCR4 mediates neuronal dysfunction by HIV-1 infected macrophage secretory product: Importance for HIV-1 associated dementia. *Soc. Neurosci. Abs.* **24,** 776.776 (Abstr).

6

Isolation of Human Immunodeficiency Virus Type 1 From Semen and Vaginal Fluids

Ann A. Kiessling

Summary

Semen and vaginal fluids transmit HIV infection. The virus is present as cell-free particles and as infected cells. Isolation of infectious virus from both genital tract fluids poses unique problems. Vaginal fluids are heavily contaminated with normal bacterial flora, and seminal plasma is cytotoxic to peripheral blood mononuclear cells. Adaptations of routine laboratory procedures have been developed to largely overcome these problems, allowing the culture and characterization of genital-tract HIV.

Key Words: Semen; vaginal fluids; PBMCs; RT activity; HIV.

1. Introduction

Semen and vaginal fluids present unique challenges to the recovery of infectious HIV. Vaginal fluids are heavily contaminated with bacterial flora and seminal fluid is remarkably toxic to peripheral blood mononuclear cells *(1)*. Methods to at least partially overcome these limitations have been developed.

Each genital fluid can be divided into a cellular fraction and a fluid fraction. Semen is formed at ejaculation and is a mixture of components from several organs and glands of the male genitourinary tract. Spermatozoa develop within the seminiferous tubules of the testis and are highly immunogenic because they are not produced until after puberty. They are protected from the male immune system by a combination of anatomic and immunologic barriers provided principally by the Sertoli cells, which line the basement membrane surrounding the tubules *(2,3)*. Leukocytes are absent from the seminiferous tubules, of an immune-competent male, but are present in the interstitial areas surrounding the seminiferous tubules, which also contain lymphatic ducts and blood capillar-

From: *Methods in Molecular Biology, Vol. 304: Human Retrovirus Protocols: Virology and Molecular Biology*
Edited by: T. Zhu © Humana Press Inc., Totowa, NJ

ies. Sperm exit the seminiferous tubules by way of a series of short ducts termed the rete testis, which join the initial segment of the epididymis. The epididymis is a specialized sperm storage organ consisiting of long, coiled tubules that carry sperm to the ductus (vas) deferens. The epididymis is replete with leukocytes in the interstitial areas surrounding the tubules, which are lined internally by a single layer of unique, ciliated, columnar epithelial cells that actively secrete and absorb proteins and other molecules *(4)*. Leukocytes are present in the epithelium of the epididymal lumen, and appear to contribute to tissue organization during development *(5)*. The ductus (vas) deferens is a more muscular tubule that connects the epididymis in the scrotum with the ejaculatory ducts in the region of the prostate near the base of the penis. An outpouching of the ductus deferens near the ejaculatory ducts forms a small sperm storage area, the ampulla. Leukocytes are found within the epithelial cell layer that lines the lumen of the ductus deferens and the ampulla. Should interstitial leukocytes in the testis or epididymis be HIV-infected, nascent virus could enter the tubules with the sperm, but only in the epididymis and ductus deferens could infected leukocytes enter the tubular lumen.

At ejaculation, sperm are squeezed from the epididymis, ductus deferens, and ampulla into the ejaculatory ducts, immediately followed by contractions of the prostate and seminal vesicles, forcing their glandular secretions (seminal plasma) into the ejaculatory ducts with the sperm. Little is known about leukocyte populations within the secretory lumen of the prostate and seminal vesicles, although it is clear that prostatitis, a common affliction in men, may be accompanied by an increase in leukocytes in prostatic secretions *(6)*. An average semen specimen contains on the order of 10^8 spermatozoa, 10^6 to 10^7 nonsperm cells, including leukocytes and immature spermatozoa, and 2 to 5 mL of seminal plasma. Density gradient centrifugation can separate semen into seminal plasma, nonsperm cells, and sperm themselves.

Vaginal fluids are constitutively expressed by cervical and vaginal mucosal cells. The exact composition of the fluids is influenced by the menstrual cycle, but they contain a mixture of mucosal secretions, leukocytes, and epithelial cells, plus normal vaginal bacterial flora. Regionalized sampling studies have indicated that HIV may be more highly expressed in the upper vagina, including the cervix, than in the lower vagina *(7)*. It is difficult to obtain infectious virus from small, regional samples of vaginal secretions. Installation of 10 mL of isotonic solution in a circular motion around the vaginal vault, followed by gentle aspiration, yields a cervicovaginal lavage (CVL) that can be separated into fluids and cells by centrifugation.

2. Materials and Reagents

2.1. Virus Culture

1. Fresh, heparinized human blood, 40 to 100 cc.
2. Ficoll-Hypaque, density 1.077 gm/mL.
3. Nycodenz.
4. Hemocytometer.
5. Standardized suspension of glass beads.
6. Microscope with 20× objective.
7. Centrifuge equipped with aerosol-safe swinging bucket rotor.
8. 50-mL sterile, conical tubes.
9. 15-mL sterile, conical tubes.
10. RPMI-1640 culture medium.
11. Glutamine solution.
12. Fetal bovine serum (FBS).
13. Sterile tissue culture pipets.
14. Dimethylsulfoxide (DMSO) in 1-mL sealed glass ampules.
15. Cryotubes for liquid nitrogen storage (e.g., Nunc).
16. Phytohemagglutinin solution.
17. Interleukin (IL)-2 solution (recombinant human, R&D Systems).
18. Trypan blue solution, 0.25% (Sigma).
19. Sterile phosphate-buffered saline (PBS, Mg^{2+} and Ca^{2+}-free).
20. T-25 tissue-culture flasks.
21. 24-well tissue-culture plates.
22. Humidified CO_2 incubator.
23. Sterile plugged tissue culture pipets.
24. Sterile bulbed disposable pipets.
25. Pipettor, manual or electric.
26. Small tubes or welled plates.
27. Laminar flow hood (minimum class 2, type A biosafety hood).
28. Gloves (latex, vinyl, or nitrile).
29. Lab coat with cuffed sleeves.
30. Bleach (1% household bleach) in a spray bottle and a beaker.
31. 70% isopropanol or ethanol in a spray bottle.
32. Sterile culture-tested deionized water.
33. 0.45-μm, sterile syringe filters.
34. 10-cc syringes.
35. Penicillin.
36. Streptomycin.
37. Tube racks.
38. Sharps biohazard container.

39. Biohazard bag.
40. Freezing canister for –70 to –80°C freezer that will provide cooling at approx 1°/min in the cryovials. May be fashioned from insulating materials such as styrofoam, or purchased commercially.
41. Liquid nitrogen storage tank.

2.2. Reverse Transcriptase Assay

1. Prepare the following:
   a. 10X assay buffer.
   b. Embryo-tested water (sterile, Sigma).
   c. Ethylene glycol, enzyme-grade (Fisher).
   d. 1 *M* KCl.
   e. 25% NP-40 solution.
   f. 50 mg/mL bovine serum albumin (BSA) solution.
   g. 1 *M* Tris-HCl, pH 8.9, at 25°C.
   h. 0.22-µm filter.
   i. Combine components to form a final solution that is: 200 m*M* Tris-HCl, 200 m*M* KCl, 1 mg/mL BSA, 0.5% NP-40, and 35% ethylene glycol. Sterile filter and store at –20°C (will not freeze). Solution is stable for up to 6 mo.
2. PolyA·oligod$T_{12–18}$ 250 µg/mL in sterile water.
3. 0.5 *M* $MgCl_2$, sterile-filtered.
4. Lysis buffer: 0.25% NP40, 10 m*M* dithiothreitol, 50% glycerol in sterile water (stored at –20°C, will not freeze).
5. Whatman GF/C glass fiber filters.
6. Sterile 96-well plates with gummed plate sealers.
7. TCAPE: 10% trichloroacetic acid with 0.1 *M* KH_2PO_4, and 0.01 *M* ethylenediamine tetraacetic acid (EDTA).
8. PPE: 0.1 *M* KH_2PO_4, 1% w/v potassium pyrophosphate, and 0.001 *M* EDTA 0.1 *M* HCl without and with 0.01 *M* KH_2PO_4.
9. Spectophotometry-grade acetone.
10. Polyethylene glycol (MW 8000), 30% w/v in sterile water.
11. 2.5 mL gas-tight Hamilton syringe with pipet tip adapter and repeating dispenser.
12. Liquid scintillation counter and fluor.

3. Methods

3.1. Peripheral Blood Mononuclear Cell Culture

Peripheral blood mononuclear cells (PMBC) are recovered in relatively high yield from freshly drawn anticoagulated venous blood by centrifugation onto a cushion of density medium that does not support denser leukocytes, such as granulocytes. Once recovered, the PBMCs can be cultured immediately, or cryopreserved in liquid nitrogen for future use. On the order of 1 to 2 million PBMCs are recovered per milliliter of blood. All blood-handling procedures are carried out with strictly sterile technique within the laminar flow hood, the biosafety glass in place and the blower on. Lab coats should have sleeves cuffed at the wrist. Two gloves should be worn, one under the cuffed sleeve and one over the cuffed sleeve. All disposable, nonsharp, plastic- and glassware are discarded in a biohazard bag placed within the hood. All disposable "sharps," such as plastic pipets and pipet tips, are discarded in a sharps container within the hood. No scissors, needles, or glass Pasteur pipets are used in the hood.

3.1.1. Preparation of PBMC Culture Medium

FBS is heat-inactivated at 56°C for 20 min, either by the manufacturer or following arrival in the laboratory. There is considerable lot-to-lot variation in the ability of FBS to support PBMC culture. It may be necessary to test more than one lot to obtain consistent culture results. FBS may be stored frozen in aliquots at –20 to –80°C for up to 1 yr, or according to the expiration date provided by the manufacturer. Some lots may be stored successfully at 4°C for up to 1 mo.

Glutamine, prepared from powder or purchased as a stock solution, is aliquoted and stored frozen at –20 to –80°C according to the manufacturer's instructions and expiration date. A concentrated stock solution (1000X) of penicillin and streptomycin is prepared by dissolving 200,000 U of penicillin powder and 200,000 mg streptomycin powder in 2 mL of sterile, deionized, tissue-culture-tested water; aliquots of 0.5 mL are stored at –20°C for up to 1 yr, or according to manufacturer's instructions. HEPES buffer, 1 *M* solution, pH 7.4, is prepared from HEPES powder, or purchased as a sterile tissue culture stock solution. It is stable as a sterile solution at 4°C for up to 1 yr, or according to manufacturer's expiration date.

The basic medium for PBMC culture is RPMI-1640 supplemented with 2 m*M* glutamine, one-fifth volume FBS, (14% v/v final concentration), 25 m*M* HEPES, 100 U/mL penicillin, and 100 mg/mL streptomycin. The basic medium is routinely prepared in 100- to 500-mL batches, sterile-filtered, and stored at 4°C for up to 1 mo.

Growth medium (GM) for PBMC culture is basic medium supplemented with 1.0 to 1.5 ng/mL recombinant IL-2. GM is also stable at 4°C for up to 2 wk. IL-2 maintains the viability of the PBMCs as well as supporting division of some T-cells.

The addition of phytohemagglutinin (3 to 5 μg/mL) for 48 to 72 h to stimulate lymphocytes increases the susceptibility of the PBMCs to infection by HIV-1.

3.1.2. Recovery of PBMC

Carefully remove the rubber stopper from the heparinized blood collection tube (*see* **Note 1**). With either a sterile bulbed disposal pipet or a stoppered

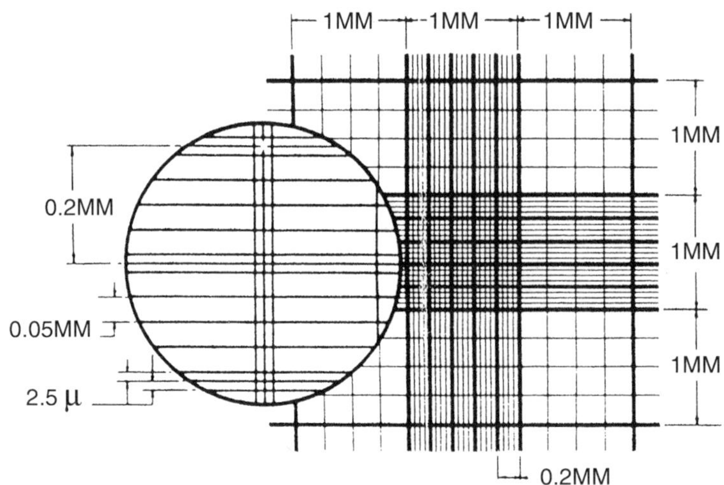

Fig. 1. Hemocytometer. The volume of each corner square containing 16 smaller squares is 1×10^{-4} mL.

tissue culture pipet attached to a pipettor, transfer two tubes of blood, 10–12 cc/tube, to a 50-mL conical tube. Dilute with one-half volume of room temperature PBS; mix well. Final volume will be approx 30 to 35 mL of diluted blood mixture.

Underlay the diluted blood with Ficoll-Hypaque. Draw up 10 to 12 mL of Ficoll-Hypaque into a plugged tissue culture pipet. Holding the 50-mL conical tube containing the diluted blood at a slight angle (e.g., 10 to 20 degrees), place the tip of the pipet near the bottom of the tube and slowly release the Ficoll solution. The goal is to form a sharp interface between the Ficoll and the blood mixture by avoiding swirling of the Ficoll solution. Cap the 50-mL tube and place upright into the centrifuge bucket; cap the bucket. When all buckets are prepared, move gently into the centrifuge. Centrifuge at 500 to 1000*g* for 30 min at room temperature with the centrifuge brake off.

Standardize the hemocytometer. Following cleaning with 70% alcohol, dry the hemocytometer and its cover slip. With the cover slip in place, load both sides of the hemocytometer with the standardized suspension of glass beads. Using the ×20 microscope objective, count the beads in the four large corner squares of the hemocytometer (**Fig. 1**). Include beads touching two of the four sides of each square. Determine the average bead count per square. Multiply the average by 10,000. Repeat with the other side of the hemocytometer. Average the count for the two sides and compare with the bead suspension manufacturer's specifications. Plus or minus 10% variation is acceptable; if variation is greater, repeat the glass bead counting procedure. If variation is still greater, replace either the cover slip or the entire hemocytometer.

Remove the 50-mL conical tube from the centrifuge bucket within the laminar flow hood. The red cells should be in the bottom with a clear interface below the Ficoll-Hypaque. The denser leukocytes, e.g., granulocytes and neutrophils, will be within the clear Ficoll-Hypaque layer. The PBMC layer will be faintly visible at the top of the Ficoll cushion. If the upper and lower Ficoll interfaces are not sharp, the tube may need to be centrifuged longer.

Double-gloved and using sterile technique, remove the 50-mL conical tube cap, and with a sterile, plugged pipet, carefully "vacuum" up the PBMCs with a steady back-and-forth motion over the top of the Ficoll layer. Avoid collecting the Ficoll. Transfer the PBMC from each centrifuge tube to a fresh 50-mL conical tube. With practice, the PBMC layer can be collected in a volume of 5 mL or less. Dilute the PBMCs with 1–2 volumes of PBS; centrifuge at 500*g* for 10 min; repeat two more times. Discard the supernatant blood plasma, Ficoll solution, and PBS washes in the bleach beaker.

Resuspend the PBMC pellet in 2 to 5 mL of room-temperature basic medium. Add 100 µL of the cell suspension to 100 µL of the trypan blue solution in a small tube or multiwell plate. Count the trypan blue-cell suspension with the hemacytometer. Trypan blue is a vital dye; it is not taken up by live cells, only by dead ones. Therefore, live cells appear colorless, or slightly pink, and dead cells appear blue. No more than 2% of the PBMC should be dead.

3.1.3. PBMC Cryopreservation

Prepare the cryprotectant solution by adding 1 mL of DMSO dropwise to 9 mL of FBS or 50% FBS in RPMI 1640. The solution may become warm. When it cools to room temperature, sterile-filter. The FBS-DMSO cryoprotectant may be stored frozen in l-mL aliquots for up to 1 yr, or according to the expiration date of the FBS. Label the cryotubes.

PBMC to be cryopreserved should be pelleted, excess PBS removed, and the cell pellet resuspended by dropwise addition of cryoprotectant that has been precooled to 4°C. Add sufficient cryoprotectant for a final PBMC concentration of 15–20 × 10^6 cells/mL. Use gentle swirling to resuspend the PBMC, and immediately transfer 0.5 to 0.8 mL of the cell suspension to the prelabeled cryotubes. Place in a container that will support a cooling rate of –1°C/min at –70° to –80°C. Exposure to the cryoprotectant for more than a few minutes before freezing will reduce cell viability. After at least 90 min at –70 to –80°C, quickly transfer the cryovials to liquid nitrogen for storage. It is important that the frozen cells not warm up above –60°C during this transfer procedure (*see* **Note 2**).

3.1.4. PBMC Culture

Fresh PBMC should be resuspended to a concentration of 1×10^6 viable cells/mL in 1 mL of GM in wells of a 24-well culture plate. Reserve the corner wells for 2 cc sterile water for humidity control. Add phytohemagglutinin (PHA), 3–5 µg/mL, to each well, and culture at 37°C, 8% CO_2 in air for 48–72 h.

Cryopreserved PBMC should be thawed in a 37°C water bath and added immediately to five volumes of GM, centrifuged at 500*g* for 10 min to remove the DMSO, and resuspended at 2×10^6 cells/mL of fresh GM for cell viability count in the hemacytometer. Adjust cell concentration to 1×10^6 viable cells/mL for plating in the 24-well culture plate. Stimulate with phytohemagglutinin and culture as for fresh PBMC.

3.2. Infectious HIV From Semen

Semen specimens can be collected by a variety of methods, including (1) masturbation into a sterile container, such as a urine container; (2) at intercourse using double condoms, the inner one designed to maintain cell viability, such as Male Factor Pak, or the Milex condom; or (3) by masturbation into the viability condom. Several studies have shown that sperm counts are higher when semen specimens are collected at intercourse, suggesting that such specimens may be more "complete" and more representative of sexually transmitted HIV burden. There are several advantages to transporting specimens in a fertility condom, such as ease of temperature control, keeping the specimen intact, and the ability of the study participant to collect the specimen in private.

Infertility clinics require cleansing of the penis, foreskin (if present), and urethral meatus with soap and water prior to collecting semen specimens to avoid contamination of the sperm; such cleansing procedures may, however, actually remove sources of HIV transmissible during sexual intercourse. Infertility clinics also advise men to abstain from ejaculation for 3 d, but no more than 10 d, before producing a semen specimen for fertility treatment. The relevance of this to sexual transmission of HIV is entirely unknown, and frequent ejaculation, known to stress the prostate and sometimes release "pockets" of infection, may lead to higher HIV burden in semen. For studies of infectious HIV, semen specimens should be produced under the most "usual" circumstances possible, with information about length of abstinence provided with the specimen. Circumcision, evidence of genitourinary tract inflammation, and history of genitourinary tract disease are obtained by questionnaire and urologic exam performed near the time of specimen collection.

Several lines of investigation have revealed that HIV in seminal plasma is not produced by the infected cells in semen ***(8–10)***, observations that explain the discordance observed between detection of cell-free and cell-associated semen virus *(7)*.

3.2.1. Semen Fractionation

1. High-molecular-weight seminal vesicle proteins confer viscosity on the ejaculate; they are cleaved by prostatic proteases, including prostate-specific antigen, which leads to thinning ("liquefaction") of the specimen. Semen specimens from some men fail to liquefy, thought to be due to abnormalities in prostate enzymes. Following 30–60 min at room temperature in a laminar flow hood, the specimen is transferred to a 15-mL sterile polypropylene conical tube with a wide-mouth plastic transfer pipet and diluted with an equal volume of PBS containing penicillin (100 U/mL) and streptomycin (100 μg/mL). An aliquot of the diluted specimen (approx 10%) is fixed with semen fixative for the determination of leukocytes ***(11)***.
2. Sperm and nonsperm cells in the diluted specimen are counted with a hemacytometer. Focusing on the center square that is gridded and subdivided into 25 smaller squares (**Fig. 1**), both motile and nonmotile sperm are counted in five of the smaller squares, the sum of which is multiplied by 5×10^4 to obtain the sperm concentration/mL in the diluted specimen. Nonsperm cells (NSC) are counted in the four squares used to count PBMC (**Fig. 1**); the average cell count is multiplied by 10^4 to calculate the concentration of NSC in the diluted specimen. Multiplying the concentration by the total volume of the diluted specimen provides an estimate of the number of NSC available for co-culture with PBMC to recover infectious, cell-associated HIV. Because the morphology of lymphocytes is similar to some immature sperm forms (*see* **Note 3**), such as round spermatids, it is not possible to accurately distinguish leukocytes and monocytes from immature sperm without staining for a leukocyte antigen, usually CD45 ***(11)***.
3. The diluted semen specimen is underlaid with a 19% cushion of Nycodenz in PBS with penicillin and streptomycin. Two mL of Nycodenz solution is pulled into a sterile stoppered pipet. Holding the conical tube at a slight angle, the pipet tip is advanced to the bottom of the tube and the Nycodenz solution slowly dispensed. The goal is to create a sharp interface between the diluted semen and the Nycodenz cushion. The tube is capped, placed into a centrifuge bucket with a cap, and centrifuged at 1000*g* for 20 min. NSC layer atop the cushion, the sperm pellet to the bottom of the tube.
4. With sterile technique and double-gloved, supernatant seminal plasma is removed to within approx 1 cm of the NSC layer, filter-sterilized through a 0.8-μm filter, and divided into 0.5-mL aliquots for culture or for storage at –80°C.
5. The NSC layer is removed with a wide-bore, bulbed, transfer pipet, transferred to a fresh 15-mL conical tube, and diluted with 4 to 5 vol of PBS with penicillin, streptomycin and nystatin (100 to 200 U/mL). NSC are recovered by centrifugation at 500*g* for 10 min, the supernatant removed and discarded in the bleach beaker, and the process repeated twice more.
6. Washed NSC are resuspended in GM at approx 2×10^6 cells/mL, diluted with trypan blue, and counted as described for PBMC (*see* **Subheading 3.1.2.**). Those not used immediately for culture of infectious virus may be cryopreserved as described for PBMC (*see* **Subheading 3.1.3.**).

3.2.2. Seminal Plasma Virus

The profound cell-killing effect of seminal plasma on cultured PBMC* has severely hampered recovery of infectious cell-free semen HIV (*see* **Note 4**). The following protocol partially, but not completely, alleviates this problem.

Day 1. Sterile-filtered, diluted seminal plasma is added to PHA-stimulated PBMCs at concentrations of 10% and 20% in triplicate wells of a 24-well plate. Thus, 0.9 mL of the seminal plasma preparation (*see* **Subheading 3.2.1.**) is spread among 6 wells, each containing 1.0 mL with 2×10^6 PBMC. Following 6 h of culture, cells from each well are transferred into sterile, 15-mL conical tubes, diluted with 1 mL fresh, warm GM, centrifuged at 500*g* for 10 min and the culture medium discarded in the bleach beaker. The cell pellets are resuspended in fresh, warm GM and re-plated in a 24-well plate (*see* **Note 5**).

Day 2. Following an additional 24 h of culture, 0.5 mL of the culture medium is removed from each well and replaced with 0.5 mL of fresh, PHA-stimulated PBMC containing 1×10^6 cells. The culture is fed and/or replenished with fresh PHA-stimulated PBMC every 3 to 4 d, for up to 6 wk (*see* **Note 6**).

Day 5 or 6; Day 12–14; Day 19–22. Remove 0.5 mL culture medium without disturbing the cells, freeze at –80°C for assay of HIV RT. Replace with 0.5 mL fresh GM.

Day 8–10; Day 15–18. Remove 0.5 mL culture medium without disturbing the cells, freeze for HIV RT assay, replace with 0.5 mL fresh GM containing 0.5×10^6 PHA-stimulated PBMC. Assay the collected cell culture supernatants for RT following 3 wk of culture; if positive, prepare stocks of supernatant virus and infected cells. If not positive, continue culture for an additional 3 wk before declaring the culture negative for virus.

Day 22–26; Day 30–34. Remove 0.5 mL culture medium without disturbing the cells, freeze for HIV RT, replace with 0.5 mL fresh GM containing 0.5×10^6 PHA-stimulated PBMC on d 17 or 18. Carefully remove 0.5 mL culture medium, freeze at –80°C for RT assay; replace with 0.5 mL fresh GM containing 0.5×10^6 PHA-stimulated PBMC.

Day 26–30; Day 34–38. Remove 0.5 mL culture medium without disturbing the cells, freeze at –80°C for assay of HIV RT. Replace with 0.5 mL fresh GM.

Day 41 or 42. Terminate the culture and assay the collected culture supernatants for reverse transcriptase activity (*see* **Note 7**).

3.2.3. Semen Cell Virus

Day 1. Washed NSC, 1×10^6 in 0.5 mL, are added to 0.5-mL cultures of 1×10^6 PHA-stimulated PBMC, resulting in a final cell culture of 2×10^6 cells in 1 mL of GM fortified with 100–200 U/mL nystatin. The co-cultures are fed and/or replenished with fresh PHA-stimulated PBMC every 3–4 d for up to 6 wk.

Day 4 or 5; Day 11–13; Day 18–21. Remove 0.5 mL of culture medium without disturbing the cell layer; freeze at –80°C for HIV RT assay; replace with fresh GM without the nystatin.

Day 7–9; Day 14–17. Remove 0.5 mL of culture medium, freeze at –80°C for HIV RT assay; replace with fresh GM containing 1×10^6 PHA-stimulated PBMC. Assay the collected cell culture supernatants for RT following 3 wk of culture; if positive, prepare stocks of supernatant virus and infected cells. If not positive, continue culture for an additional 3 wk before declaring the culture negative for virus.

Day 21–25; Day 29–33. Remove 0.5 mL culture medium without disturbing the cells, freeze for HIV RT, replace with 0.5 mL fresh GM containing 0.5×10^6 PHA-stimulated PBMC on d 17 or 18. Carefully remove 0.5 mL culture medium, freeze at –80°C for RT assay; replace with 0.5 mL fresh GM containing 0.5×10^6 PHA-stimulated PBMC.

Day 25–29; Day 33–37. Remove 0.5 mL culture medium without disturbing the cells, freeze at –80°C for assay of HIV RT. Replace with 0.5 mL fresh GM.

Day 41 or 42. Remove 0.5 mL culture medium without disturbing the cells, freeze at –80°C for RT assay; replace with 0.5 mL fresh GM.

Day 43 or 44. Terminate the culture and assay the collected culture supernatants for RT activity (*see* **Note 8**).

3.3. Infectious HIV in Cervicovaginal Lavage Fluids

Women should not be menstruating and need to refrain from douching and unprotected sex (assumed, but reminded) for 48 h prior to collection of cervicovaginal lavage (CVL) fluids.

Women recline on an exam table with feet in stirrups and undergo a routine gynecologic exam. A speculum is inserted into the vagina and the vagina and cervix are inspected for signs of menstruation, inflammation, and other irregularities, all of which are noted. If the speculum occupies more than one third of the depth of the vaginal vault, it is withdrawn slightly.

A 14–18-gage angiocath (without the stylet) is fitted to a 10-mL sterile plastic syringe filled with sterile PBS. The PBS is directed in a gentle stream all around the vaginal vault, including the cervix and the cervical os. The buffer will pool in the posterior fornix; it is reaspirated into the syringe and the process repeated twice more. The volume of the final CVL is measured. Amounts above 10 mL are assumed to be vaginal fluids; if less than 10 mL is recovered, it is assumed that there are negligible starting fluids in the vagina.

3.3.1. Fractionation of CVL Fluids

Fluids are fractionated within 1 h of collection.

1. First, a 0.5-mL aliquot of the CVL is added to semen fixative to immunostain for leukocytes. Penicillin (100 U/mL), streptomycin (100 µg/mL), and nystatin (100 U/mL) are added to the CVL.
2. The CVL is separated into supernatant fluid and cell pellet by centrifugation at 500*g* for 10 min. The supernatant is withdrawn to within 1 cm of the pellet, sterile-filtered through 0.8-µm filters, and

3. The cell pellet is washed by suspension in 5 vol of fresh PBS with antibiotics followed by centrifugation twice more. The final pellet is resuspended in 2 mL of GM, 20 µL of which is diluted with trypan blue and counted as for PBMC (*see* **Subheading 3.1.2.**); the percentage of dead cells is noted. Those cells not used for cell culture may be cryopreserved, as described for PBMC (*see* **Subheading 3.1.2.**).

3.3.2. CVL Fluid Virus

Day 1. Sterile-filtered CVL fluid is added to PHA-stimulated PBMC in GM at concentrations of 10% and 20% in triplicate wells of a 24-well plate. Thus, 0.9 mL of CVL fluids (**Subheading 3.3.1.**) is spread among 6 wells each containing 1.0 mL GM with 2×10^6 PBMC supplemented with 50 µg/mL nystatin.

Day 3 or 4; Day 10–12; Day 17–20. Remove 0.5 mL culture medium without disturbing the cells; freeze at –80°C for assay of HIV RT. Replace with 0.5 mL fresh GM.

Day 7–8; Day 14–17. Remove 0.5 mL culture medium without disturbing the cells, freeze for HIV RT assay; replace with 0.5 mL fresh GM containing 0.5×10^6 PHA-stimulated PBMC. Assay the collected cell culture supernatants for RT following 3 wk of culture; if positive, prepare stocks of supernatant virus and infected cells. If not positive, continue culture for an additional 3 wk before declaring the culture negative for virus.

Day 22–26; Day 30–34. Remove 0.5 mL culture medium without disturbing the cells, freeze for HIV RT; replace with 0.5 mL fresh GM containing 0.5×10^6 PHA-stimulated PBMC on d 17 or 18. Carefully remove 0.5 mL culture medium, freeze at –80°C for RT assay; replace with 0.5 mL fresh GM containing 0.5×10^6 PHA-stimulated PBMC.

Day 26–30; Day 34–38. Remove 0.5 mL culture medium without disturbing the cells; freeze at –80°C for assay of HIV RT. Replace with 0.5 mL fresh GM on d 41 or 42. Remove 0.5 mL culture medium without disturbing the cells, freeze at –80°C for RT assay; replace with 0.5 mL fresh GM.

Day 41 or 42. Terminate the culture and assay the collected culture supernatants for RT activity.

3.4. HIV Reverse Transcriptase Assay

The synthetic template-primer, PolyA·oligodT$_{12–18}$, directs the incorporation of ^{3}H-thymidine by HIV RT. Using high specific activity tritiated thymidine triphosphate, and overnight incubation, assay conditions *(12–15)* can be adjusted to detect on the order of 10^4 virus particles.

1. Culture supernatant, 1.1 mL, is centrifuged at 800*g* for 10 min to remove cellular debris. Taking care not to disturb the pellet, 1.0 mL of the supernatant is transferred to fresh microfuge tubes, and 0.5 mL of PEG solution is added. Tubes are capped and stored at 4°C overnight (*see* **Note 9**).

2. PEG-precipitated virus is pelleted by centrifugation at 10,000*g* in a microcentrifuge for 10 min. Taking care to not disturb the pellet, the supernatant is carefully decanted off and the rim of the tube blotted on a Kimwipe. Add 50 µL lysis buffer. Keep cold for 20 to 30 min.
3. Prepare assay mix with a final concentration of: 1.1X assay buffer, 30 µg/mL PolyA·oligodT$_{12-18}$, 0.2 to 0.5 nmol ^{3}HTTP, and 6 m*M* $MgCl_2$. Add the $MgCl_2$ just before use
4. Transfer 10 µL of the virus lysate (agitate the pellet by pulling up and down in the pipet three to five times before transferring the 10 µL) into one well of a 96-well, round-bottom, sterile plate. Add 50 µL of assay mix to each well with the repeating Hamilton syringe, taking care to not touch the virus lysate in the bottom of each well.
5. Seal the plate and incubate at 37°C for at least 2 h, or overnight.
6. Prepare filters to collect reaction products: saturate the Whatman glass fiber filters with PPE and dry thoroughly. Cut trapezoids approx 2 cm long, 1.5 cm at the wide end, and 0.5–0.7 cm at the narrow end. Treated filters may be stored at room temperature for at least 1 yr. Number filters with India ink.
7. Carefully remove the plate sealer. Place each numbered assay filter upright, narrow end down, in the corresponding well of the 96-well plate, taking care to advance the filter to the bottom of the well, and allow the assay solution to wick into the filter ***(12,13)***. When the wells appear dry, transfer the filters to a beaker with cold TCAPE in an ice bucket. Use approximately 5 mL of TCAPE/filter. After 15 min, carefully decant the TCAPE into the radioactive waste. Replace with fresh, cold TCAPE; repeat once more.
8. Cover the bottom of a Buchner funnel with two layers of filter paper (e.g., Whatman no. 1); dampen with cold 0.1 *M* HCl. Attach to gentle vacuum, or water aspirater, according to radioactive disposal practices. Arrange the filter papers around the Buchner funnel so none are overlapping. Overlay with one more filter paper. Rinse the filters twice with cold 0.1 *M* HCl plus phosphate by filling the Buchner funnel and allowing it to drain to dryness. Repeat with two washes of cold 0.1 *M* HCl, followed by acetone if desired. There is almost no residual radioactivity at the Buchner funnel wash step; the washes are designed to reduce background.
9. Oven dry the filters to dryness (approx 15 min in a 65°C oven). Place them in numbered order into scintillation vials, add fluor appropriate for tritium, and count in a scintillation counter.
10. Background radioactivity for assays with only lysis buffer (varies from 200 to 600 counts per minute) is subtracted from samples (*see* **Note 10**).

3.5. Preparation of Virus Stocks

Cultures that yield two successive rising titers of virus are deemed productively infected and suitable for deriving a virus stock. Because each round of infection produces virus with new mutations, virus stocks derived as early in the culture process as possible will be the most representative of the virus quasispecies in the patient.

Nascent virus may stay associated with the plasma membrane of the host cell for several hours, or even days. For this reason, cells need to be shaken to release the virus into the culture supernatant.

At d 3 or 4 of culture after replacement of fresh medium (longer culture times are associated with an increasing percentage of noninfectious virus in the culture medium), the entire culture, including cells stuck to the bottom of the plate, is removed to a sterile 15-mL conical tube, and shaken by rotating end to end and in a circle. Mild frothing may occur.

Following approx 10 min of agitation, the cells are pelleted at 1000*g* for 15 min and the supernatant removed. An aliquot is saved for RT measurement; the remainder is adjusted to 40 to 50% FCS, sterile-filtered through a 0.45-µm filter, and stored in 0.5- to 1.0-mL aliquots. Infectivity is more stable if the stocks are stored in liquid nitrogen, but infectious virus can usually be recovered from stocks stored at –80°C for at least 1 yr.

4. Notes

1. Blood may be anticoagulated by other common anticoagulants, such as EDTA and acid citrate dextrose (ACD). Blood-bank blood may also be used, but PBMC isolated from blood more than 3 d old do not stimulate as well as fresh PBMC.
2. Several methods and devices for cryopreservation of cells are available. Controlled-rate freezers, using either liquid nitrogen or alcohol, may result in more reliable and reproducible numbers of live cells following thawing. Other, far less expensive, devices, e.g., "Mr. Frosty," are commercially available. If vials are to be stored on canes in liquid nitrogen refrigerators, a device that provides for a –1°C cooling rate that accommodates vials already loaded onto canes can be fashioned from insulating materials such as Styrofoam. The probe of an electronic thermometer inserted into a cryovial with freezing medium only can be used to check, and adjust, the freezing rate.
3. Distinguishing mononuclear cells from immature sperm cells in semen is more difficult than distinguishing granulocytes and other peroxidase-positive leukocytes. A common test to distinguish leukocytes from immature sperm cells is the Endtz test, which employs a solution of benzidine that undergoes a color-reaction in the presence of peroxidase. The advantage of a colorimetric test for leukocytes is that it provides a quick identification of some of the nonsperm cells in semen specimens. The Endtz test has a number of drawbacks, including the toxicity of benzidine, and the subtle color change from pink to pink-brown in the presence of peroxide. The Endtz test can be replaced by the AEC (aminoethyl carbazole, a diazo dye) staining kit available from Zymed. It forms a red-brown precipitate when present in an assay containing peroxide and peroxidase. Diluting an equal volume of semen into a 2X concentration of AEC (according to kit instructions) will result in a readily detectable red-brown precipitate in peroxi-

dase-positive leukocytes within 2 to 5 min. Although the AEC stain is a more sensitive and safer test than the Endtz test, it does not distinguish monocytes from immature sperm, which requires detection of common leukocyte antigen, CD45. A mixture of mouse monoclonal antibodies against human CD45 that is commonly used to detect lymphocytic infiltrates in aldehyde-fixed pathology sections is LCA (Dako), which is more specific for lymphocytes than granulocytes in fixed semen specimens ***(11)***. A different mouse monoclonal antibody, Hle-1 (Becton Dickinson), is used for flow-cytometry detection of leukocytes, and stains all leukocytes in both fixed blood and fixed semen specimens ***(11)***. Using both immunostaining methods will distinguish lymphocytes/monocytes from peroxidase-positive leukocytes and immature germ cells in semen.

4. Eliminating fetal calf serum from the culture medium has been reported to eliminate the toxicity of seminal plasma to PBMC, but in agreement with other studies, we have found that it only delays the toxicity ***(16)***, and in the case of some specimens, there is no benefit to eliminating the serum.
5. Seminal plasma at concentrations up to 10% v/v of culture medium is not toxic to H9 cells ***(17)***. This provides an opportunity to more easily recover HIV species from seminal plasma that is infectious to T lymphocytes.
6. Some investigators have reported that depleting the CD8 cells from the PBMC populations recovered from donor blood markedly decreases the cytotoxic effects of seminal plasma ***(18)***. In our experience this step is somewhat useful in some cases, but not in all.
7. Detection of HIV p24 protein was commonly used to assay culture fluids for HIV release. Several kits were developed by Abbott Laboratories, but all have now been discontinued. Kits are still available from PerkinElmer.
8. RT-polymerase chain reaction (PCR) assays for HIV RNA are more sensitive than reverse transcriptase assays, but far more expensive. If the RT step is allowed to proceed overnight, bracket-nested PCR of the resulting cDNAs can detect 10–100 RNA copies with ethidium bromide gel staining.
9. An alternate method of concentrating virus from cell supernatants is centrifugation at 100,000*g* for 90 min. This provides more reproducible results and a cleaner virus pellet, but is less practical for screening large numbers of samples.
10. The 96-well plate, batch filter washing method described here minimizes liquid radioactive waste and the time involved in washing assay filters; but it can be replaced by more conventional arrays of individual assay tubes and individual filters on a vacuum manifold.

References

1. Okamoto, M., Byrn, R., Eyre, R. C., Mullen, T., Church, P., and Kiessling, A. A. (2002) Seminal plasma induces programmed cell death in cultured peripheral blood mononuclear cells. *AIDS Res. Hum. Retroviruses* **18,** 797–803.
2. Pentikainen, V., Erkkila, K., and Dunkel, L. (1999) Fas regulates germ cell apoptosis in the human testis in vitro. *Am. J. Physiol.* **276,** E310–E316.
3. McClure, R. F., Heppelmann, C. J., and Paya, C. V. (1999) Constitutive Fas ligand gene transcription in Sertoli cells is regulated by Sp1. *J. Biol. Chem.* **274,** 7756–7762.

4. Kiessling, A. A., Crowell, R., and Fox, C. (1989) Epididymis is a principal site of retrovirus expression in the mouse [published erratum appears in *Proc. Natl. Acad. Sci. USA* 1989 Dec;86(24):10133]. *Proc. Natl. Acad. Sci. USA* **86,** 5109–5113.
5. Mullen, T. E., Jr., Kiessling, R. L., and Kiessling, A. A. (2003) Tissue-specific populations of leukocytes in semen-producing organs of the normal, hemicastrated, and vasectomized mouse. *AIDS Res. Hum. Retroviruses* **19,** 235–243.
6. Tanner, M. A., Shoskes, D., Shahed, A., and Pace, N. R. (1999) Prevalence of corynebacterial 16S rRNA sequences in patients with bacterial and "nonbacterial" prostatitis. *J. Clin. Microbiol.* **37,** 1863–1870.
7. Coombs, R., Reichelderfer, P., and Landay, A. (2003) Recent observations on HIV type-1 infection in the genital tract of men and women. *AIDS* **17,** 455–480.
8. Eyre, R., Zhung, G., and Kiessling, A. (2000) Multiple drug resistance mutations in human immunodeficiency virus in semen but not blood of a man on antiretroviral therapy. *Urology* **55,** 591.
9. Byrn, R. A., Zhang, D., Eyre, R., McGowan, K., and Kiessling, A. A. (1997) HIV-1 in semen: an isolated virus reservoir [letter]. *Lancet* **350,** 1141.
10. Paranjpe, S., Craigo, J., Patterson, B., Ding, M., Barroso, P., Harrison, L., et al. (2002) Subcompartmentalization of HIV-1 quasispecies between seminal cells and seminal plasma indicates their origin in distinct genital tissues. *AIDS Res. Hum. Retroviruses* **18,** 1271–1280.
11. Kiessling, A. A., Yin, H. Z., Purohit, A., Kowal, M., and Wolf, B. (1993) Formaldehyde-fixed semen is suitable and safer for leukocyte detection and DNA amplification. *Fertil. Steril.* **60,** 576–581.
12. Kiessling, A. A. and Goulian, M. (1976) A comparison of the enzymatic responses of the DNA polymerases from four RNA tumor viruses. *Biochem. Biophys. Res. Commun.* **71,** 1069–1077.
13. Kiessling, A. A. and Goulian, M. (1979) Detection of reverse transcriptase activity in human cells. *Cancer Res.* **39,** 2062–2069.
14. Rey, M. A., Spire, B., Dormont, D., Barre-Sinoussi, F., Montagnier, L., and Chermann, J. C. (1984) Characterization of the RNA dependent DNA polymerase of a new human T-lymphotropic retrovirus (lymphadenopathy associated virus). *Biochem. Biophys. Res. Commun.* **121,** 126–133.
15. Hoffman, A. D., Banapour, B., and Levy, J. A. (1985) Characterization of the AIDS-associated retrovirus reverse transcriptase and optimal conditions for its detection in virions. *Virology* **147,** 326–335.
16. Fiore, J., La Grasta, L., DiStefano, M., Buccoliero, G., Pastore, G., and Angarano, G. (1997) The use of serum-free medium delays, but does not prevent, the cytotoxic effects of seminal plasma in lymphocyte cultures: implications for studies on HIV infection. *Microbiologica* **20,** 339–344.
17. Rasheed, S., Li, Z., and Xu, D. (1995) Human immunodeficiency virus load. Quantitative assessment in semen from seropositive individuals and in spiked seminal plasma. *J. Reprod. Med.* **40,** 747–757.
18. Zhang, H., Dornadula, G., Beumont, M., Livornese, L., Jr., Van Uitert, B., Henning, K., et al. (1998) Human immunodeficiency virus type 1 in the semen of men receiving highly active antiretroviral therapy [see comments]. *N. Engl. J. Med.* **339,** 1803–1809.

7

Isolation and Quantification of HIV From Lymph Nodes

Catherine Tamalet

Summary

This chapter describes a standardized microculture technique adapted from a peripheral blood mononuclear cells (PBMC) microculture assay to isolate and quantify HIV from lymph nodes. This quantitative lymph node microculture estimates the number of infectious cells per million lymph node mononuclear cells. The assay, as described below, is performed in two 24-well tissue culture plates using six fivefold dilutions. Each sample of patient cells is cocultivated with phytohemagglutinin-stimulated normal donor PBMC for 21 d. The supernatant from each well is assayed for HIV p24 antigen production by the standard HIV p24 enzyme-linked immunosorbent assay (ELISA). HIV-1 infectious titers are expressed as tissue culture infective dose ($TCID_{50}/10^6$ cells) according to Reed and Muench.

Key Words: HIV isolation; viral burden; infectious HIV titer; lymph node mononuclear cells; lymph node cell co-culture; quantitative microculture.

1. Introduction

The major role of lymphoid tissue as tissue reservoir during the entire course of HIV-1 infection has been recognized since the onset of the HIV epidemic *(1–3)*. Higher virus titers are usually found in the lymph nodes than in the peripheral blood. Therefore the detection of infectious viruses from the lymph nodes can provide useful information for evaluating the efficiency of new antiretroviral drugs aimed at suppressing viral replication. The feasibility of a lymph node cell culture method has been established ***(4–6)***: it is derived from the semiquantitative blood lymphocyte co-culture procedure *(7)* that is a more successful method for HIV-1 isolation than the primary culture. The lymph node mononuclear cells (LNMC) are isolated from lymph nodes using a Ficoll-Hypaque gradient (*see* **Note 1**). The semiquantitative assay includes co-culture and identification of the 50% endpoint dilution on the basis of HIV-1 p24 anti-

From: *Methods in Molecular Biology, Vol. 304: Human Retrovirus Protocols: Virology and Molecular Biology*
Edited by: T. Zhu © Humana Press Inc., Totowa, NJ

gen production. Of course, the absence of palpable lymph nodes is a limiting factor. Moreover, the indication of the lymph node cell co-culture should be extremely restricted because lymph node sampling requires an invasive procedure (*see* **Note 2**).

2. Materials

1. RPMI-1640 medium with 5% L-glutamine (GibcoBRL, Scotland) at 4°C.
2. Fetal bovine serum (FBS) (GibcoBRL, Grand Island, NY) at –20°C; should be heat-inactivated before use.
3. Interleukin (IL)-2 human (h IL-2) 200 U/mL (Bohringer Mannheim Gmbh, Mannheim Germany) at –20°C.
4. Penicillin-streptomycin (10,000 IU/mL–10.000 µg/mL) (GibcoBRL, Grand Island, NY) at –20°C.
5. Phytohemagglutinin (PHA) (M form) lyophilized (GibcoBRL, Grand Island, NY) at 4°C.
6. Complete (growth) RPMI 1640 medium: RPMI-1640 with L-glutamine supplemented with 20% FBS, 20 IU/mL IL-2, 100 U/mL penicillin, and 100 µg/mL streptomycin.
7. Stimulation RPMI-1640 medium: complete RPMI-1640 containing 2 µg/mL PHA.
8. Dulbecco's phosphate-buffered saline (PBS) without Ca^{2+} and Mg^{2+}.
9. Ficoll-Hypaque: Ficoll-Paque solution or other commercially available lymphocytic separation medium (LSM).
10. HIV-1 p24 antigen enzyme-linked immunosorbent assay (ELISA) kits (Innogenetics, N.V., Ghent, Belgium).

3. Methods

The methods described below outline the different steps of the lymph node cells co-culture. These include (1) preparation of donor peripheral blood mononuclear cells (PBMC); (2) preparation of patient lymph node cells; (3) semiquantitative HIV-1 co-culture assay; and (4) determination of the infectious titer.

3.1. Donor Cells

Donor-cell processing is described in **Subheadings 3.1.1.** and **3.1.2.**, including sampling and preparation of lymphocytes from a seronegative donor, then stimulation of these cells before carrying out the co-culture.

3.1.1. Preparation of Donor Cells

1. Collect 80 mL of blood from a seronegative healthy donor and place in heparin, ACD, or CPD anticoagulated vacutainers.
2. Dilute each blood tube (1:1) with RPMI within 6 h of blood collection.
3. Carefully layer two volumes of diluted blood over one volume of LSM or Ficoll-Hypaque into a 50-mL centrifuge tube. Do not mix the sample with the LSM.

4. Centrifuge at 400*g* for 30–40 min at room temperature.
5. Draw off the upper layer of plasma using a Pasteur pipet.
6. Collect the band of lymphocytes at the plasma/LSM interface with a fresh Pasteur pipet.
7. Transfer the cells from each gradient to centrifuge tubes and wash in 15 mL of RPMI at 250*g* for 10 min.
8. Pool the cells harvested from each tube and transfer them to a fresh 50-mL centrifuge tube. Wash twice in 45 mL of RPMI at 200*g* for 10 min. Prior to each wash, resuspend the cells in a small volume (5 mL) before filling up with RPMI.
9. After washing, resuspend the pellet in a small volume of stimulation medium containing PHA and count the cells with a hemocytometer.
10. Bring the cell suspension to a concentration of 1.0×10^6 cells/mL with stimulation medium and divide the volume obtained among several tissue culture flasks.

3.1.2. Stimulation of Donor Cells

1. Incubate the flasks at 37°C in a humidifed 5% CO_2 incubator for 2 or 3 d.
2. Prior to addition to co-culture, sediment PHA-stimulated donor cells at 400*g* for 10 min at room temperature and enumerate them after resuspending in a small volume of growth medium.
3. Add growth medium to obtain a donor-cell suspension at a concentration of 2×10^6 cells/mL. Approximately 6×10^7 donor lymphocytes are required for this semiquantitative co-culture assay.

3.2. Patient Lymph Node Cells

The lymph node is obtained from surgical excision (*see* **Note 3**) and is minced inside the safety laboratory. The steps described below outline the procedure used for preparing the lymph node cell suspension (*see* **Note 4**).

1. Centrifuge the cell suspension at 400*g* for 30–40 min at room temperature.
2. Resuspend the pellet in 6 mL of RPMI.
3. Layer the lymphoid cell suspension over 1:2 vol of lymphocyte separation medium into a centrifuge tube.
4. Centrifuge at 400*g* for 30–40 min at room temperature.
5. Draw off the upper layer using a Pasteur pipet.
6. Collect the whitish band of mononuclear cells using a fresh Pasteur pipet.
7. Transfer the cells to a centrifuge tube and wash three times in 15 mL of RPMI at 400*g* for 10 min.
8. Either co-cultivate the cells immediately or store the pellet in liquid nitrogen.
9. If co-cultivated, re-suspend the pellet in 3 mL of growth medium and count the cells with a hemocytometer.
10. Bring the cell suspension to a concentration of 1.0×10^6 cells/mL with growth medium. A minimal amount of 4.0×10^6 cells is required for semiquantitative co-culture.

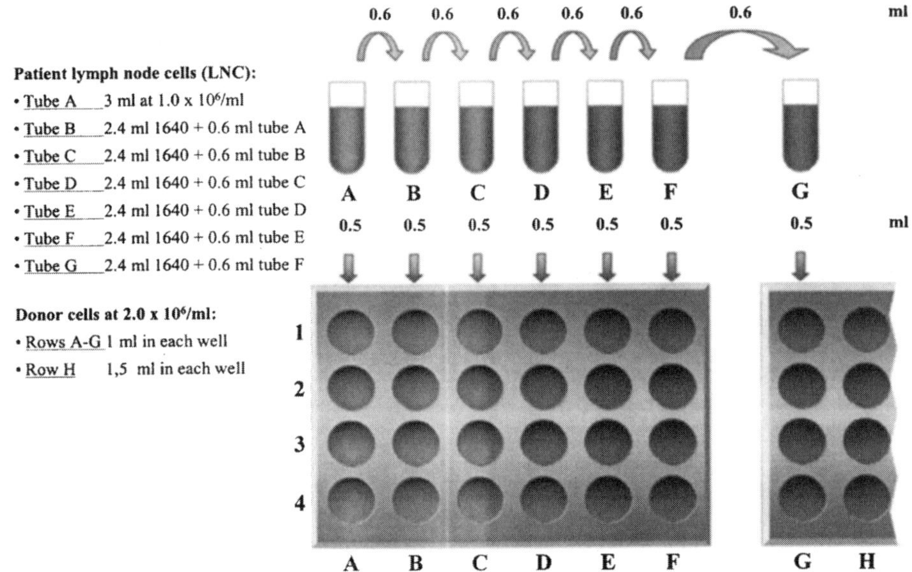

Fig. 1. Quantitative microculture plate.

3.3. Semiquantitative Co-Culture

The next steps in this process involve the preparation of co-culture plates, the maintaining of co-culture and the reading of the plates.

3.3.1. Preparation of Co-Culture Plates

1. Using seven sterile tubes (**Fig. 1**), prepare fivefold dilutions from the cell suspension at a starting concentration of 1.0×10^6 lymph node cells/mL. Add 3.0 mL of cell suspension at a concentration of 1.0×10^6 cells/mL to tube A.
2. Add 2.4 mL of growth medium to each of the tubes B–G.
3. Transfer 0.6 mL from tube A to tube B. Mix thoroughly. Change pipet and transfer 0.6 mL from tube B to tube C. Continue making fivefold dilutions to obtain concentrations from 1.0×10^6 cells to 6.4×10^1 cells/mL from tube A to tube G.
4. Using two 24-well tissue culture plates (**Fig. 1**), add 1.0 mL of the PHA-stimulated donor cell suspension to the concentration of 2.0×10^6 cells in the four wells of rows A–G and 1.5 mL in each well of row H (donor lymphocyte control).
5. Add 0.5 mL of lymph node cell suspension to the consecutive dilutions of 1.0×10^6, 2.0×10^5, 40×10^4, 8.0×10^3, 3.2×10^2, and 6.4×10^1 cells/mL, each dilution in quadruplicate (*see* **Note 5**), from wells of row A to wells of row G, respectively. Change pipets at each dilution. The final volume in each well is 1.5 mL (*see* **Note 6**).

3.3.2. Maintaining Cultures

1. Incubate the two plates at 37°C in a humidified 5% CO_2 incubator. Keep the plates covered to avoid evaporation and contamination.
2. Carefully remove 0.75 mL of the supernatant from each well and replenish with 0.75 mL of fresh growth medium twice a week. Change pipet tips for each well (*see* **Note 7**).
3. Add 3.0×10^5 PHA-stimulated donor cells to each well on d 7 and 14.
4. Maintain the cultures as long as the number of antigen-positive wells is increasing within 21 d.

3.3.3. Reading Plates

1. Three times per week, detect the cytopathic effects (ballooning cytoplasm, giant cells, syncytia), which are generally preceded by p24 antigen production (*see* **Note 8**).
2. Once a week, collect culture supernatant from each well and store at –70°C for HIV-1 p24 antigen production assay. A well is considered positive if the result exceeds the cutoff on two consecutive determinations.
3. Carry out the decontamination procedure adequate for working with HIV-1-infected cells.

3.4. Measurement of Infectious Titer

The Reed and Muench method can be used to quantify the infectivity of the lymph node cells suspension *(8)* (*see* **Note 9**). At the peak of HIV-1 p24 antigen, the dilution at which 50% of the wells are found antigen-positive can be assessed as shown in the following example:

LNC suspension

P24 Ag	5^0	5^{-1}	5^{-2}	5^{-3}	5^{-4}	5^{-5}	5^{-6}
1	+	+	+	+	+	–	–
2	+	+	+	+	–	–	–
3	+	+	+	+	–	–	–
4	+	+	+	–	–	–	–

None of these dilutions gives a 50% dose; this dose lies between dilutions 5^{-3} and 5^{-4}. An interpolated value is calculated from the cumulative numbers of wells that are antigen-positive or -negative. This interpolated value is given in the following formula:

$$\frac{\text{\% wells antigen positive at dilution just above 50\%} - 50\%}{\text{\% wells antigen positive just above 50\%} - \text{\% wells antigen-positive at dilution just below 50\%}}$$

Table 1
Cumulative Values

Dilution rate	Wells positive for p24 antigen	Wells negative for p24 antigen	Total positive	Total negative	Infection ratio	%
5^{-2}	4	0	8	0	8/8	100
5^{-3}	3	1	4	1	4/5	80
5^{-4}	1	3	1	4	1/5	20
5^{-5}	0	4	0	8	0/8	0

In this example, the interpolated value is =

$$\frac{80-50}{80-20} = \frac{30}{60} = 0.5$$

The cumulative values are calculated as shown in **Table 1**.

The $TCID_{50}$ titer lying between 5^{-3} and 5^{-4} is $5^{3.5}/5.10^5$ cells, i.e., 560 $TCID_{50}/1.10^6$ cells. If converting the result in log 10, the titer is expressed as $10^{2.75}$ $TCID_{50}/1.10^6$ cells (*see* **Note 10**).

4. Notes

1. An alternative method has been reported in which it is not necessary to subject the cells to a Ficoll-Hypaque gradient ***(9)***.
2. All patients must be informed of the aims of the study and the side effects of the surgical procedure, and sign a written consent form approved by the institutional ethics committee before biopsy. Few postoperative side effects have been observed. The complication rate (local infections) is approx 3%.
3. An alternative noninvasive technique is lymph node aspiration ***(10)*** using a 21- or 23-gage needle with a 10-mL syringe. However, this attractive procedure did not make it possible to achieve a successful semiquantitative co-culture in all the patients in whom it was attempted.
4. It is highly desirable that the patient presents a palpable lymph node with a mean surface of approx 0.6–0.8 cm in diameter. The lymph node is surgically excised under local anesthesia and in most cases is in the axillary or cervical region. The lymph node is placed into a flask and covered by RPMI-1640 medium. It must be carried out within 3 h. Bring the flask to a laminar flow hood of the laboratory in a biological safety cabinet. The lymph node is potentially infectious and therefore must be handled with care. Transfer it to a 100-mm Petri dish. Using scalpel and small scissors, finely mince the tissue. Add drops of RPMI during mincing to keep it moist. Separate and discard all clotted blood and connective tissue. Bring the volume of RPMI to 8 mL. Pipet several times to disperse cell clumps and make a cell suspension that can easily pass through the opening of a 10-mL sterile pipet.

5. Using from five to eight replicates at each dilution is more accurate, but truly consuming in lymph node cells.
6. Take care not to touch the edge of the well or the medium in order to avoid contaminating the tip.
7. Each well must be considered as a culture dish in itself. It is mandatory to change the pipet tips for each well whenever one removes the medium even for the same wells of the same dilution step.
8. In the positive co-culture, p24 antigen production is detected first and precedes the cytopathic effect.
9. The virus titer and estimation of 50% endpoints can also be calculated according to the Spearman-Kärber formula.
10. The $TCID_{50}$/mL titer may vary between different intra assays up to 0.7 log 10 on an identical lymph node cell stock. In order to reach a significant difference between two samples, ≥1 log is required.

Acknowledgments

The author thanks Professor J. M. Huraux and Professor C. Rouzioux for reading the manuscript and for their helpful comments.

References

1. Pantaleo, G., Graziosi, C., Butini, L., Pizzo, P. A., Schnittman, S. M., Kotler, D. P., and Fauci, A. S. (1991) Lymphoid organs function as major reservoirs for human immunodeficiency virus. *Proc. Natl. Acad. Sci. USA* **88,** 9838–9842.
2. Fox, C. H., Tenner-Racz, K., Racz, P., Firpo, A., Pizzo, P. A., and Fauci, A. S. (1991) Lymphoid germinal centers are reservoirs of human immunodeficiency virus type 1 RNA. *J. Infect. Dis.* **164,** 1051–1057.
3. Pantaleo, G., Graziosi, C., Demarest, J., Butini, L., Montroni, M., Fox, C., et al. (1993) HIV infection is active and progressive in lymphoid tissue during the clinically latent stage of disease. *Nature* **362,** 355–358.
4. Lafeuillade, A., Tamalet, C., Pellegrino, P., Tourres, C., Yahi, N.,Vignoli, C., et al. (1993) High viral burden in lymph nodes during early stages of HIV-1 infection. *AIDS* **11,** 1527–1528.
5. Tamalet, C., Lafeuillade, A., Yahi, N., Vignoli, C., Tourres, C., Pellegrino, P., et al. (1994) Comparison of viral burden and phenotype of HIV-1 isolates from lymph nodes and blood. *AIDS* **8,** 1083–1088.
6. Tamalet, C., Lafeuillade, A., Fantini, J., Poggi, C., and Yahi, N. (1997) Quantification of HIV-1 viral load in lymphoid and blood cells: assessment during four-drug combination therapy. *AIDS* **11,** 895–901.
7. Rouzioux, C., Puel, J., Agut, H., Brun-Vezinet, F., Ferchal, F., Tamalet, C., et al. (1992) Comparative assessment of quantitative HIV viraemia assays. *AIDS* **6,** 373–377.
8. Dulbecco, R. and Ginsberg, H. S. (1988) The nature of viruses. In: *Virology*. Lippincott Williams & Wilkins, Philadelphia: pp. 22–25.

9. Gartner, S. and Popovic, M. (1990) Virus isolation and production. In: *Techniques in HIV research* (Aldovini, A. and Walker, B. D., eds.), New York: Stockton Press, pp. 53–71.
10. Meylan, P. R. A., Burgisser, P., Weyrich-Suter, C., and Spertini, F. (1996) Viral load and immunophenotype of cells obtained from lymph nodes by five needle aspiration as compared with peripheral blood cells in HIV-infected patients. *J. Acquir. Immune Defic. Syndr.* **13,** 33–47.

8

Isolation of Human Immunodeficiency Virus Type 2 Biological Clones From Peripheral Blood Lymphocytes

Hetty Blaak

Summary

Infectious variants of human immunodeficiency virus (HIV) can be isolated from peripheral blood mononuclear cells (PBMC) of infected individuals by propagating the virus in co-cultures with healthy donor PBMC. Standardized culture protocols have been designed specifically for the isolation of HIV type 1 (HIV-1) and proven effective for the isolation of virus from virtually all HIV-1-infected individuals. For the isolation of HIV-2, however, standard HIV-1 culture protocols have been only partially effective. While suitable for the isolation of HIV-2 from PBMC of individuals in advanced stages of disease, they appeared largely inadequate for the isolation of HIV-2 from asymptomatically HIV-2-infected individuals.

This chapter describes a variant HIV isolation procedure with culture conditions adapted to the isolation of HIV-2 from PBMC of "healthy" HIV-2-infected individuals, i.e., individuals with high $CD4^+$ T-cell numbers and no detectable viral RNA in plasma. By using a limiting dilution format, several biological clones representing individual HIV variants can be obtained from the PBMC of an infected individual. In addition, the frequency of PBMC infected with HIV-2 variants capable of propagating in vitro can be estimated.

Key Words: HIV-2; PBMC; virus isolation; co-culture; biological clone; virus stock; infectious unit.

1. Introduction

Isolation of HIV from peripheral blood mononuclear cells (PBMC) is achieved by co-culturing (pro)virus-containing patient cells with healthy donor PBMC, previously stimulated with phytohemagglutinin (PHA) *(1,2)*. The co-culture is carried out in the presence of interleukin (IL)-2 to stimulate both

From: *Methods in Molecular Biology, Vol. 304: Human Retrovirus Protocols: Virology and Molecular Biology*
Edited by: T. Zhu © Humana Press Inc., Totowa, NJ

donor and patient cells, and HIV is released from the patient cells and propagated in the healthy donor cells. Standardized HIV-1 isolation protocols take into account the influence on virus propagation of several culture conditions, e.g., the patient PBMC to negative PBMC ratio, the total cell concentration, the cell density at the bottom surface of culture vessels and the age and freshness of the PHA-stimulated cells ***(1,3–7)***. $CD8^+$ T-cells of HIV-infected individuals, especially of so-called long-term nonprogressors, produce virus-inhibitory factors that may interfere with virus-production in vitro (e.g., CAF, β-chemokines, α-defensins) ***(8–10)***. Removal of the $CD8^+$ T-cells prior to co-culture enhances the success rate of virus isolation in these cases ***(5,11,12)***.

To determine the frequency of patient PBMC carrying infectious virus (or virus able to propagate in vitro), replicates of a number of dilutions of patient cells are cultured in quantitative microculture assays ***(6,13–15)***. From the number of patient cells per well and the proportion of negative wells, the frequency of cells carrying infectious virus can be estimated ***(16,17)***. Generally, this frequency is indicated as infectious units per million PBMC, or IUPM. Besides the possibility of quantifying the cellular infectious load, the microculture assay also enables the generation of so-called biological clones, virus variants propagated from one single infected cell. Thus this technique offers the opportunity to study properties of different virus variants isolated from the PBMC of one infected individual. Moreover, it enables isolation of HIV variants with relatively low fitness, which might not be detected in cultures where they have to compete with other, more fit variants.

Standard HIV-1-isolating techniques appeared not adequate for isolating HIV-2 from PBMC of asymptomatically infected individuals with high $CD4^+$ T-cell numbers ***(18–20)***. This may be related to the fact that these individuals have very low, often undetectable, plasma viremia ***(20–22)***. The low plasma viremia may reflect a low frequency of cells with replication-competent virus, and/or an overall low replication capacity of HIV-2, and/or a more adequate immune response, e.g., the presence of more efficient cytotoxic T lymphocytes or a higher production of viral inhibitory factors. Based on these possibilities, an enhanced variant of standard HIV-1 microculture procedures has been designed to isolate HIV-2 from PBMC of individuals with undetectable plasma RNA load ***(23)***. The enhanced culture conditions include (1) use of high numbers of patient cells, (2) use of freshly isolated patient cells, (3) use of freshly isolated and freshly stimulated healthy donor cells, (4) depletion of $CD8^+$ T-cells from patient cells, and (5) depletion of $CD8^+$ T-cells from healthy donor cells. As it may be preferred to use the same protocol also for isolation of HIV-2 from individuals in more advanced stages of disease, the chapter additionally describes how to adapt the culture format to make the protocol suitable for isolation of HIV-2 from individuals in all stages of disease.

2. Materials

All used materials should be sterile.

1. Buffy coats from healthy blood donors. Use within 24 h.
2. Patient blood. Use without delay.
3. Ficoll-Paque PLUS (Amersham Biosciences).
4. Phosphate-buffered saline (PBS).
5. 0.13 *M* Trisodium citrate (TSC), adjust to pH 7.0 with 4 *N* HCl. Once opened, store at 4°C.
6. Iscove's modified Dulbecco's medium (IMDM) (Biowhittaker).
7. Fetal calf serum (FCS) (Fetalclone; Hyclone). Heat-inactivate at 56°C for 30 min.
8. PHA (Murex). Reconstitute in IMDM (0.2 mg/mL) and store aliquots of 0.5 mL at –20°C. Prevent multiple rounds of freezing and thawing.
9. Polybrene (Sigma-Aldrich). Reconstitute in IMDM (800 µg/mL) and store aliquots of 0.5 mL at –20°C.
10. Recombinant interleukin-2 (rIL-2) (Proleukin; Chiron). Reconstitute in IMDM (4000 U/mL; *see* **Note 1**) and store aliquots at –20°C. Prevent multiple rounds of freezing and thawing.
11. Constituted media
    a. Complete IMDM: IMDM, 10% FCS, 100 U/mL penicillin, 100 µg/mL streptomycin. Can be stored up to 1 mo at 4°C.
    b. PHA medium: complete IMDM, 1 µg/mL PHA. Can be stored up to 1 wk at 4°C.
    c. Culture medium: complete IMDM, 20 U rIL-2, 4 µg/mL polybrene (*see* **Note 2**). Can be stored up to 1 wk at 4°C.
    d. Freeze medium: IMDM, 20% dimethyl sulfoxide (DMSO), 30% FCS. Can be stored up to 1 wk at 4°C.
12. "Mr. Frosty" (1°C freezing container; Nalgene).
13. Isopropanol.
14. CD8 magnetic beads + Dynal magnetic particle concentrator (Dynal MPC) (Dynal Biotech).
15. HIV-2 detection assay (e.g., SIV core antigen enzyme-linked immunosorbent assay [ELISA] [Coulter Beckman]; Lenti-RT activity kit [Cavidi Tech AB]).

3. Methods

The isolation of HIV-2 biological clones from patient PBMC is a multistep procedure altogether taking 7–9 wk. First, the size and the format of the assay are determined (**Subheading 3.1.**). Two to three days prior to the start of co-culture, target cells are isolated from HIV-negative healthy blood donors and stimulated with PHA (**Subheadings 3.2.–3.4.**). When the target cells are properly activated, patient PBMC are prepared (**Subheading 3.5.**), both patient PBMC and PHA-blasts depleted of $CD8^+$ T-cells (**Subheading 3.6**), and the co-culture is initiated (**Subheadings 3.7** and **3.8.**). The co-culture is maintained for 5 wk, during which cultures are refreshed every seventh day and virus pro-

Table 1
Summary of Day-to-Day Actions in Co-Culture Procedure

Day	Action	Subheading(s)
–2 or –3	Prepare PHA-blasts	**3.2.** (**3.3.3.–4.**), **3.4.**
0	Initiatiate virus isolation co-culture	**3.5.–3.8.**
3 or 4	Prepare PHA-blasts	**3.2.** (**3.3.3.–4.**), **3.4.**
7	Refresh virus isolation co-culture	**3.6.**, **3.9.**
11 or 12	Prepare PHA-blasts	**3.2.** (**3.3.3.–4.**), **3.4.**
14	Refresh virus isolation co-culture	**3.6.**, **3.9.**
18 or 19	Prepare PHA-blasts	**3.2.** (**3.3.3.–4.**), **3.4.**
21	Refresh virus isolation co-culture	**3.6.**, **3.9.**
	Start growth of virus stocks from wells positive at d 7/14	**3.6.**, **3.11.**
25 or 26	Prepare PHA-blasts	**3.2.** (**3.3.3.–4.**), **3.4.**
28	Refresh virus isolation co-culture	**3.6.**, **3.9.**
	Refresh virus stock cultures from wells positive at d 7/14	**3.6.**, **3.11.**
	Start growth of virus stocks from wells positive at d 21	**3.6.**, **3.11.**
33	Prepare PHA-blasts	**3.2.** (**3.3.3.–4.**), **3.4.**
35	Terminate virus isolation co-culture	**3.9.**
	Freeze virus stocks from wells positive at d 7/14	**3.11.**
	Refresh virus stock cultures from wells positive at d 21	**3.6.**, **3.11.**
36	Start growth of virus stocks from wells positive at d 28/35	**3.6.**, **3.11.**
40 or 41	Prepare PHA-blasts	**3.2.** (**3.3.3.–4.**), **3.4.**
43	Freeze virus stocks from wells positive at d 21	**3.11.**
	Refresh virus stock cultures from wells positive at d 28/35	**3.6.**, **3.11.**
50	Freeze virus stocks from wells positive at d 28/35	**3.11.**

In addition, monitor virus production in wells every week starting on d 7. PHA, phytohemagglutinin.

duction is monitored (**Subheading 3.9.**). The frequency of patient cells carrying infectious virus can be estimated based on the proportion of negative wells (**Subheading 3.10.**). A selection of positive wells is further propagated to grow virus stocks (**Subheading 3.11.**). A schematic summary of the day-to-day actions involved in the complete isolation and propagation procedure as shown in **Table 1**.

Table 2
Guidelines to Determine Culture Format Based on Plasma RNA Load

RNA load (copies/mL)	"Expected" IUPM range[a]	Total patient cells needed for culture	Microculture plates
<500	0.01–1	$50–100 \times 10^6$	24-well
500–1000	0.1–5	$10–50 \times 10^6$	24-well/96-well
1000–10,000	1–50	$2–10 \times 10^6$	96-well
10,000–100,000	10–500	$0.5–2 \times 10^6$	96-well
>100,000	100–>1000	$0.1–0.5 \times 10^6$	96-well

[a]The values are estimates based on the author's experience; deviations may occur. IUPM, infectious units per million.

3.1. Choosing Format of Assay

To obtain biological HIV-2 clones and to estimate the frequency of PBMC containing infectious virus, several dilutions of patient cells are cultured in multiple replicates. The total number of patient cells cultured should at least be high enough to contain one infectious unit (resulting in one biological clone), but preferably more, e.g., 5–10. At the same time, in at least some of the tested dilutions, the proportion of negative wells should be high enough to maximize the probability that viruses in positive wells are indeed clones. To increase the chance of success at the first isolation attempt, prior knowledge of infectious virus load is useful. As a guideline for the infectious PBMC load, viral RNA load in plasma can be used (*see* **Table 2**) (*see* **Note 3**). Using plasma RNA as a guideline, testing four to six PBMC twofold dilutions is generally sufficient. The less that is known about the patient's disease status, however, the higher the number of dilutions needed. A minimum of 24 replicates per cell dilution is sufficient to reliably estimate the frequency of infected cells ***(16)***, but because only one or two dilutions will result in "highly probable" biological virus clones, 32 or 48 replicates are preferable.

3.2. Target Cells: Isolation of PBMC of Healthy Donors From Buffy Coats

Two to three days prior to start of co-culture, PBMC from healthy donors are isolated from buffy coats and freshly stimulated with PHA (**Subheading 3.4.**). If necessary, however, PBMC can be frozen and stored for later use (**Subheading 3.3.**) (*see* **Note 4**). PBMC from two different healthy donors are needed for each co-culture, which are kept separate throughout the procedure (*see* **Note 5**).

3.2.1. Preparations

1. Prepare a solution containing 10% of 0.13 *M* TSC in PBS (PBS/10% TSC).
2. Prepare complete IMDM.
3. Prewarm solutions (PBS/10% TSC, complete IMDM, Ficoll) to room temperature.

4. For each buffy coat fill six 50-mL tubes with 12.5 mL Ficoll.
5. Pour 50 mL PBS/10% TSC into a sterile 75-cm^2 culture flask.

3.2.2. Isolation of PBMC

1. Transfer each buffy coat (50 mL) from the bag to a flask containing PBS/10% TSC.
2. Mix blood and PBS/10% TSC.
3. Adjust the volume to 150 mL with PBS/10% TSC.
4. Carefully layer 25 mL of diluted blood on top of every 12.5-mL Ficoll-Paque layer.
5. Spin 20 min at 700*g* with the *brake off*.
6. Collect layers of white blood cells in fresh 50-mL tubes. Pool two layers per tube but keep blood donors separate.
7. Adjust volume to 50 mL with PBS/10% TSC.
8. Spin 10 min at 400*g*.
9. *Carefully* decant the supernatant (cell pellet is loose).
10. Resuspend the cells in 5 mL of PBS/10% TSC.
11. Adjust the volume to 50 mL with PBS/10% TSC.
12. Spin 10 min at 250*g* (*see* **Note 6**).
13. Carefully decant the supernatant.
14. Resuspend the cells in 3 mL complete IMDM.
15. Pool the contents of three tubes. Keep blood donors separate.
16. Count the cells.
17. If cells are directly used, proceed with **Subheading 3.4.**

3.3. Target Cells: Freezing and Thawing of Cells

3.3.1. Freezing—Preparations

1. Fill Mr. Frosty to the fill line with isopropanol.
2. Prechill Mr. Frosty at 4°C.
3. Prepare freeze medium. Keep on ice.

3.3.2. Freezing and Storage of PBMC

1. Adjust cell concentration to 20–100 × 10^6 cells/mL with complete IMDM.
2. Place the cells on ice for 30 min.
3. Add an equal volume of ice-cold freeze medium to the cells drop by drop while gently mixing.
4. Aliquot the cells in freeze vials (10–50 × 10^6/mL).
5. Place the vials containing cells immediately in a Mr. Frosty and place at –70°C (*see* **Note 7**).
6. After 1 to 3 d, transfer vials for long-term storage in liquid nitrogen or –135°C freezer.

3.3.3. Thawing—Preparations

It is important to be well prepared so that every action can be performed as quickly as possible, thereby minimizing the time cells are warmed up in 10% DMSO.

1. Prepare ice-cold complete IMDM.
2. Prepare one 50-mL tube for each (or maximally two) vial(s). Keep donors separate.

3.3.4. Thawing of Cryopreserved PBMC

1. Thaw vials containing PBMC in a 37°C water bath, until only a small ice clump remains.
2. Transfer the contents of the vial to a 50-mL tube.
3. Add ice-cold complete IMDM, at first drop by drop with a gradually increasing rate, and shake tube while adding the medium. Dilute cells in DMSO at least 10 times.
4. Spin cells 10 min at 400*g*.
5. Discard supernatant.
6. Resuspend cells in 1–5 mL complete IMDM.
7. If applicable, pool contents of tubes containing cells from the same donor.
8. Count the cells.
9. Proceed with **Subheading 3.4.**

3.4. Target Cells: PHA Stimulation of Healthy Donor PBMC

1. Prepare PHA medium and prewarm to room temperature.
2. Determine how many cells are to be stimulated (*see* **Note 8**).
3. If only a proportion of the donor PBMC is used, transfer the appropriate number of cells to fresh tubes. Keep donors separate.
4. Spin cells 10 min at 400*g*.
5. Discard the supernatant.
6. Resuspend the cells in 2–5 mL PHA medium.
7. Adjust the cell concentration to 5×10^6/mL with PHA medium.
8. Transfer the cells of each donor to an appropriately sized culture flask at a cell surface density of 2 to 2.5×10^6 cells/cm^2.
9. Place flasks horizontally in incubator at 37°C/5% CO_2 for 2 or 3 d.
10. *Optional* (only when fresh PBMC were used): Freeze and store surplus cells for backup purposes.

3.5. Preparing Patient PBMC

On the day the co-culture is initiated, patient PBMC are prepared. Virus can be isolated from freshly isolated or from cryopreserved PBMC (*see* **Note 9**). The isolation of patient PBMC from blood (*see* **Subheading 3.5.1.**) follows basically the same procedure as for buffy coats from healthy donors with the exception that the starting material is whole blood, generally collected in sodium heparin vacutainers. If cryopreserved patient PBMC are used, proceed with **Subheading 3.5.2.**

3.5.1. Isolation of PBMC From Patient Blood

1. For each 10–12.5 mL of blood that has been collected, fill one 50-mL tube with 12.5 mL Ficoll.
2. Fill an appropriately sized tube or culture flask with PBS/10% TSC in a volume half that of the collected blood.
3. Transfer the patient blood from vacutainer(s) to tube or culture flask containing PBS/10% TSC. If applicable, pool vacutainers from one patient.
4. Rinse vacutainers with PBS/10% TSC and add to the blood suspension.
5. Adjust the volume so that the blood to PBS/10%TSC ratio is at least 1:1 and the end volume is a multiple of 25.
6. Carefully layer 25 mL of blood suspension on top of 12.5 mL Ficoll in 50-mL tubes.
7. Spin 20 min at 700*g*, brake off.
8. Collect the PBMC layers in a fresh 50-mL tube. If applicable, pool two layers (of the same patient) per tube.
9. Adjust the volume to 50 mL with PBS/10% TSC.
10. Spin 10 min at 400*g*.
11. Carefully discard the supernatant.
12. Resuspend the cells in 3 mL PBS/10% TSC.
13. Adjust volume to 50 mL with PBS/10% TSC.
14. Spin 15 min at 250*g*.
15. Carefully discard the supernatant.
16. Resuspend cells in 3 mL PBS.
17. Count the cells.
18. Proceed with **Subheading 3.6.**

3.5.2. Thawing of Cryopreserved Patient PBMC

1. Follow the procedures as described in **Subheadings 3.3.3.** and **3.3.4., steps 1–5**.
2. Resuspend the cells in 3 mL PBS.
3. Count the cells.
4. Proceed with **Subheading 3.6.**

3.6. CD8 Depletion of Patient PBMC and PHA-Blasts

Prior to co-culture the $CD8^+$ T-cells are removed from patient cells and PHA-blasts to increase the chance of successful virus isolation (*see* **Note 10**).

3.6.1. Preparations

1. Prepare culture medium. Prewarm to 37°C.
2. Prepare a sterile, ice-cold PBS/2% FCS solution.

3. Transfer to a 15-mL tube 75 µL sterile "Dynabeads CD8" (approx 10×10^6 beads) for every 20×10^6 PBMC to be depleted (see **Note 11**).
4. Wash beads three times with PBS/2% FCS using a Dynal magnetic device (Dynal MPC).
5. Resuspend the beads to the original volume in ice-cold PBS/2% FCS.

3.6.2. Harvest of PHA-Stimulated Healthy Donor Cells

1. Analyze the cell culture both macro- and microscopically (*see* **Note 12**).
2. Transfer cells to 50-mL tubes. Keep cells from different donors separated.
3. Spin 10 min at 250*g*.
4. Discard the supernatant.
5. Resuspend the cells in PBS.
6. Pool cells from the same donor in one tube.
7. Count the cells.
8. Determine how many PHA-blasts are to be depleted of $CD8^+$ T-cells (*see* **Note 8**).
9. Take half the number of total required cells from one donor and transfer to a fresh tube.
10. Take half the number of total required cells from the other donor and add to the tube.

3.6.3. Depletion of $CD8^+$ Cells

1. Spin PHA-blasts and patient PBMC (from **Subheading 3.5.**) 10 min at 400*g*.
2. Discard supernatants.
3. Resuspend cells in ice-cold PBS/2% FCS and adjust cell concentration to 20×10^6/mL (but minimal volume of 1 mL).
4. Add the washed beads to the cells.
5. Incubate at 4°C for 30 min while rotating (*see* **Note 13**).
6. Adjust volume to 5 to 10 mL with PBS/2% FCS.
7. Place the tube in the Dynal MPC for 1–2 min.
8. Transfer the supernatant to a new tube, being careful not to transfer any of the beads.
9. Spin the cells 10 min at 400*g*.
10. Resuspend the cells in culture medium.
11. Count the cells.
12. Proceed with **Subheading 3.7.** (for patients with undetectable RNA load) and/or **Subheading 3.8.** (for patients with >1000 copies/mL).

3.7. Co-Culture: Isolation of HIV-2 From Individuals With Undetectable Plasma RNA Load

When patients have undetectable plasma load, $50–100 \times 10^6$ CD8-depleted cells are expected to yield several biological clones (*see* **Table 2**). The next protocol is based on a situation where 50×10^6 are available. Tested are four twofold dilutions, each in 24-plo, in 24-well plates. The protocol needs to be adjusted based on the amount of cells truly available (*see* **Note 14**).

3.7.1. PHA-Blasts

1. Adjust the CD8-depleted PHA-blast cell suspension to 2×10^6/mL with culture medium. For this specific format the total amount of cells needed is 100×10^6.
2. Fill each well with 0.5 mL (i.e., 1×10^6 cells) PHA-blast suspension.
3. Wrap the plates in foil and place in 37°C/5% CO_2 incubator until patient cells are prepared.
4. Immediately proceed with **Subheading 3.7.2.**

3.7.2. Patient Cells

For each patient cell concentration, a cell suspension of 25 mL is prepared.

1. Fill 3 fresh 50-mL tubes with 25 mL of culture medium.
2. Adjust the CD8-depleted patient cell suspension to 50 mL with culture medium. This tube contains the maximum cell concentration, in this case 1.0×10^6 cells/mL.
3. Transfer 25 mL of the cell suspension to the next tube containing 25 mL culture medium.
4. Resuspend and transfer 25 mL to the next and so on to the fourth tube. The four resulting concentrations are 1 million, 500,000, 250,000, and 125,000 cells/mL.
5. Take the culture plates containing PHA-blasts from the incubator.
6. Add 1 mL of patient cell dilution per well containing PHA-blasts. Fill up one 24-well plate with each cell dilution.
7. Wrap the plates in foil and place them in a 37°C/5% CO_2 incubator.
8. *Optional:* Use remaining cells (in this case approx 3×10^6) for fluorescence-activated cell sorter (FACS) analysis to assess the results of the CD8-depletion procedure.

3.8. Co-Culture: Isolation of HIV-2 From Individuals With More Than 1000 Copies RNA/mL

The total amount of patient cells required depends on the viral load of the patient (*see* **Table 2**). The next protocol is an example for individuals with 1000–10,000 RNA copies per mL of plasma, where 2 to 10×10^6 CD8-depleted cells are sufficient. In this example, 3×10^6 CD8-depleted PBMC are cultured in six twofold dilutions, 32 replicates per dilution (*see* **Note 15**). A 96-well format is preferred when fewer than 10×10^6 patient cells are cultured.

3.8.1. PHA-Blasts

1. Adjust the CD8-depleted PHA-blast cell suspension to 1.5×10^6/mL culture medium. For this specific format the total amount of cells needed is 30×10^6.
2. Fill each well of flat-bottom 96-well plates with 100 µL PHA-blast suspension.
3. Wrap the plates in foil and place in 37°C/5% CO_2 incubator until patient cells are prepared.
4. Immediately proceed with **Subheading 3.8.**

3.8.2. Patient Cells

1. Fill five tubes with 3.5 mL of culture medium.
2. Adjust the CD8-depleted patient cells to 7 mL with culture medium. The maximum cell concentration is now 428,000 cells/mL.
3. Transfer 3.5 mL of the cell suspension to the next tube, containing 3.5 mL culture medium.
4. Resuspend, and transfer 3.5 mL to the third tube, and so on until 6 cell concentrations are prepared.
5. Take the two plates containing PHA-blasts from the incubator.
6. Add 100 µL of the patient cell suspensions to the wells, two cell concentrations per plate.
7. The resulting cell concentrations are 42,800–21,400–10,700–5350–2675–1340 cells per well.
8. Wrap the plates in foil and place in 37°C/5% CO_2 incubator.

3.9. Prolongation of Co-Cultures

Co-cultures are maintained for 5 wk and refreshed and tested for virus production every seventh day, at which time points fresh CD8-depleted PHA-blasts must be ready. The procedures for 24-well and 96-well cultures are described separately for d 7 in **Subheadings 3.9.2.** and **3.9.3.**, respectively.

3.9.1. Preparations: Days 4/5, 11/12, 18/19, 25/26, and 33

1. Isolate PBMC from buffy coats from two donors, as described in **Subheading 3.2.**
2. Stimulate healthy donor PBMC with PHA, as described in **Subheading 3.4.**

3.9.2. 24-Well Cultures: Day 7

1. Prepare the PHA-blasts (deplete CD8$^+$ T-cells) as described in **Subheading 3.6.**
2. Make a PHA-blast suspension of 1.5×10^6/mL of culture medium.
3. Add 1 mL PHA-blast suspension to each well of fresh 24-well plates.
4. Take the 7-d-old co-cultures from the incubator.
5. Remove 500 µL of the culture supernatants (try not to remove cells).
6. Transfer 100–150 µL to fresh 96-well plates and discard the remainder.
7. Store the supernatant at –20°C for virus detection assays (*see* **Note 16**).
8. Resuspend the cells and transfer 500 µL to the plates containing fresh PHA-blasts.
9. Wrap the plates in foil and place them in the 37°C/5% CO_2 incubator.
10. Discard the old culture plates.
11. Test the cultures supernatants to identify the HIV-2-producing wells.

3.9.3. 96-Wells Cultures: Day 7

1. Prepare the PHA-blasts (deplete CD8$^+$ T-cells) as described in **Subheading 3.6.**
2. Make a PHA-blast suspension of 1×10^6/mL culture medium.
3. Add 135 µL of PHA-blast suspension per well in fresh flat-bottom 96-well plates.

4. Take the 7-d-old co-cultures from the incubator.
5. Remove 65 µL of culture supernatants and transfer to fresh 96-well plates.
6. Store supernatants at –20°C for virus detection assays (*see* **Note 16**).
7. Resuspend the cells and transfer 65 µL of cell suspension to the plates containing PHA-blasts.
8. Wrap the plates in foil and place them in the 37°C/5% CO_2 incubator.
9. Discard the "old" plates.
10. Test the culture supernatants to identify the HIV-2-producing wells.

3.9.4. Day 14

1. Repeat the procedure of **Subheading 3.9.2., steps 1–10** (24-well cultures) or **Subheading 3.9.3., steps 1–9** (96-wells cultures).
2. Test supernatants for virus production to confirm earlier results and to identify new positive wells.

3.9.5. Days 21 and 28

1. Repeat the procedure of **Subheading 3.9.2., steps 1–9** (24-well cultures) or **Subheading 3.9.3., steps 1–8** (96-well cultures). *Do not discard* "old plates."
2. Positive wells in the "old" plates are used to generate virus stocks (see **Subheading 3.11.**).
3. Test supernatants for virus production to confirm earlier results and to identify new positive wells.

3.9.6. Day 35

1. Take the 35-d-old cultures from the incubator.
2. Transfer 100–150 µL (24-well cultures) or 65 µL culture (96-well cultures) supernatants to 96-well plates.
3. Determine whether new wells are HIV-2-positive.
4. If so, transfer the well(s) to a larger vessel as described in **Subheading 3.11.** If not, discard plates.

3.10. Calculation of Infectious Units Per Million (IUPM)

The number of virus-positive wells depends on the number of patient PBMC added to the wells and the frequency of infectious units (i.e., cells carrying virus capable of replicating in vitro). The probability for a well to acquire an "infectious unit" is assumed to follow a Poisson distribution. Using a maximum likelihood method the frequency of infectious units in a patient PBMC sample can be estimated from the proportion of wells that remain negative (*see* **Note 17**).

3.11. Co-Culture: Propagation of Virus Stocks From Positive Wells

Starting from d 21, surplus cultures in the "old" plates are used to propagate virus stocks (*see* **Note 18**). Selecting wells from cultures with the lowest patient cell concentrations maximizes the probability that wells contain "true" biological clones (*see* **Note 19**).

3.11.1. Transfer Positive Wells to Larger Culture Vessels

1. Add 3×10^6 fresh 2- or 3-d-old CD8-depleted PHA-blasts in 3 mL of culture medium to T25 flasks, one flask for each HIV-2-positive well to be propagated.
2. Transfer the cell suspensions remaining in the "old" culture plates (500 µL for 24-well cultures and 65 µL for 96-well cultures) to the T25 flasks (*see* **Note 20**).
3. Place the flasks slanted in a 37°C/5% CO_2 incubator.

3.11.2. Refreshment of Virus Stock Cultures (d 7)

1. Add 5×10^6 2- or 3-d-old CD8-depleted PHA blasts in 5 mL of culture medium to the cultures in flasks.
2. Return flasks (this time upright) to a 37°C/5% CO_2 incubator.

3.11.3. Storage of Virus Stocks (d 14)

Cultures now contain 7–8 mL of virus stock and can be stored.

1. Determine virus content in culture supernatants (*see* **Note 21**).
2. Transfer contents of culture flasks to a 15-mL tube.
3. Spin 10 min at 400*g*.
4. Aliquot the culture supernatant and store at –70°C.
5. *Optional:* Store viable co-cultured cells for backup purposes (as in **Subheading 3.4.3.**).
6. *Optional:* Store dry cell pellet for DNA isolation procedures.
7. Determine the infectious titer of the virus stocks.

4. Notes

1. According to the manufacturer's instructions, one vial contains 18×10^6 IU (international units) rIL-2 per mL after reconstitution in 1.2 mL; in other words, one vial contains 21.6×10^6 IU. In our hands, this amount agrees with 1×10^6 "true" units. It is advisable to test the amount of rIL-2 inducing half-maximal proliferation of PBMC in one's own culture system.
2. The effect of polybrene has been described to be either positive or neutral *(2,5)*. Because there is no deleterious effect and polybrene might especially make a difference for variants with relatively low replication capacity, polybrene is added.
3. If RNA load is not known, CD4 numbers may serve as a crude indicator. Roughly, when CD4 counts are above 500/µL, the infectious load is likely to be lower than 10 IU/10^6 CD4 cells, and when CD4 counts are below 200/µL, the infectious load is often higher than 100 IU/10^6 CD4 cells.
4. Both fresh and frozen healthy donor PBMC can be used for HIV isolation, but use of fresh cells has been described to increase HIV-1 isolation success rate and virus yields, especially for PBMC from asymptomatically infected individuals *(3,4)*. The use of fresh cells is therefore recommended for isolation of HIV-2 from infected individuals with undetectable viremia, even though it is time-consuming. Frozen material may sometimes be preferred, however, because it creates the possibility of characterizing PBMC (e.g., CCR5 genotype, susceptibility

to HIV infection) prior to use and the possibility of using the same (characterized) cells throughout the virus isolation procedure and/or in virus isolations to be carried out at a later time.

5. To minimize the risk of failure of virus isolation owing to (relative) insusceptibility of the target cells (for instance, because of a $CCR5^{\Delta32/\Delta32}$ genotype), always use PBMC from two individual healthy donors. Blood and isolated PBMC from the different donors are kept separate until after the PHA stimulation procedure. This way, a safety net is created in case (one of the) PBMC are found to be unusable (e.g., because of failed stimulation or contamination with bacteria/fungi) and alloactivation is prevented. Alloactivation should be kept at a minimum, as it induces an upregulation of β-chemokines and may have a net negative effect on production of HIV variants sensitive to β-chemokines *(24,25)*. For the same reason, target cell mixtures never contain PBMC from more than two donors.
6. Thrombocytes do not appear to interfere with virus production and, to minimize the loss of lymphocytes, further washing is not recommended. If, however, PBMC are to be stored, one or two additional washes (repeat **steps 10–13**) may be considered to reduce clumping of PBMC after the process of thawing.
7. Alternatively, a controlled-rate freezing apparatus can be used or, less accurate but adequate, vials can be wrapped in tissues, placed in a Styrofoam box, which is then sealed with tape and placed at –70°C.
8. About 30% of the cells will be lost as a result of thawing and/or toxic effects of PHA. A further 10–30% (the proportion of $CD8^+$ T-cells in healthy donor PBMC) is expected to be lost after depletion of $CD8^+$ T-cells. To be on the safe side, stimulate at least four times the total amount of cells needed for co-culture and start the CD8-depletion procedure with two times the total amount of cells needed. (If PBMC from two donors are used, take into account that both will constitute only half of the total cells needed.)
9. In limiting diluting cultures, the minimum number of patient cells resulting in a positive HIV-1 culture have been described to be 5–500 times lower for fresh cells compared to cryopreserved cells *(13)*, suggesting a reduction in infectious PBMC load upon cryopreservation. When optional, use fresh PBMC for the isolation of HIV-2 from individuals with undetectable viremia.
10. Depletion of $CD8^+$ T-cells from patient PBMC has been shown to enhance isolation success rate and virus yield for asymptomatically infected HIV-1 *(2,5)*, and may therefore be essential in the isolation of HIV-2 from aviremic individuals. For isolation of HIV-2 from individuals with advanced-stage disease, depletion of $CD8^+$ T-cells might be omitted; however, for comparison of aviremic and viremic individuals use of the same protocol is preferable.

    $CD8^+$ T-cells are removed from the PHA-blasts to reduce possible and variable production of inhibitory factors and to increase the number of target cells without increasing the cell concentration and cell surface density.
11. According to the manufacturer's instructions, 75 μL of beads can efficiently deplete $CD8^+$ T-cells from 10×10^6 PBMC assuming that maximally 25% of the PBMC are $CD8^+$. For the present purpose, complete elimination of $CD8^+$ T-cells

is not required and 75 μL beads per 20×10^6 PBMC is sufficient. This way, often less than 1% $CD8^+$ cells will remain, even though less complete depletion may be achieved when blood donors have high percentages of $CD8^+$ T-cells (and for depletion of $CD8^+$ T-cells from AIDS patients a higher bead:PBMC ratio may be needed).

12. The medium should be yellowish orange and large clumps of cells visible by eye. Microscopic inspection will show that the majority of cells are enlarged, representing stimulated T-cells (blasts). Make sure that there are no contaminating fungi, bacteria, or yeast present in the culture.
13. When no rotating apparatus is available, gently mix tubes by hand every few minutes.
14. When fewer cells than in the example are available, the same protocol is used, resulting in lower cell concentrations being tested. It may be considered to increase the number of target cells so that the total cell concentration remains 0.8–1.3×10^6 cells/mL. Do not reduce the number of replicates per concentration, as 24 is already minimal. If more cells than in the example are available, increase the number of replicates per cell concentration and/or the number of concentrations, rather than the cell concentrations. Addition of too many patient cells per well (in our experience, between 1×10^6 and 1.3×10^6) will not result in a proportional increase in positive wells. Explanations for a relatively lower yield compared to lower cell inputs may be crowding and/or or a less optimal target cell:patient cell ratio. For patients with RNA load between 500 and 1000 copies the same protocol can be used with fewer cells and lower tested cell concentrations. Alternatively, a 96-well format (**Subheading 3.8.**) may be favored depending on the amount of cells available (e.g., 10×10^6 cells).
15. When fewer cells are available the same protocol can be used, resulting in lower cell concentrations. Alternatively, fewer cell concentrations (e.g., 4 or 5) and/or fewer replicates (minimally 24) per concentration can be tested. When more cells are available, increase the number of replicates per concentration (e.g., 48) and/or test higher cell concentrations (though not more than 100,000 cells/well). When isolating virus from individuals with a higher load, the same protocol can be used with fewer patient cells and lower cell concentrations per well. If RNA load is not known, test more concentrations.
16. Suitable commercially available HIV-2-detecting kits are a p26 SIV core antigen ELISA kit from Coulter Beckman and an RT-activity detection kit from Cavidi Tech. Both assays are very sensitive but expensive and "in-house" alternatives may be preferred. However, high sensitivity may be essential for the detection of variants with impaired replication kinetics. To reduce costs, samples are first tested in pools after which wells from positive pools are tested individually. At d 14–35, take care not to include wells already known to be positive into the pools. Positivity is to be confirmed in the samples of the subsequent week.
17. At a given frequency of infectious units in PBMC (ϕ), a fixed number of patient cells added to a well (x) will contain on average λ infectious units ($\lambda = \phi \cdot x$). The probability for a well to acquire k infectious units depends on λ and is described

by the following formula: $P(X = k) = e^{-\lambda}. \lambda k/k!$ (Poisson distribution). The probability for a well to remain negative is therefore described by $P(X = 0) = e^{-\lambda}$. For example, a patient has an infectious load of 50 IUPM ($\phi = 50/10^6$). When 20,000 cells are added per well ($x = 20{,}000$), the average number of infectious units per well, or λ, is 1. The probability of receiving no infectious unit is, in this situation, e^{-1} or 0.37 for each well. In other words, the average proportion of negative wells in cultures with $\lambda = 1$ (or $\phi = 1/x$) is 37%. Based on the proportions of negative wells in the different tested patient PBMC dilutions (each with its own λ), the frequency of infectious units in PBMC (ϕ) can be estimated using the (jackknifed version of the) maximum likelihood method ***(16)***. In this method, the frequency of infectious units with the highest likelihood to result in the observed proportion of negative wells is calculated and represents an estimate of the true frequency ***(17)***.

18. The majority of positive wells will be positive between d 7 and 14 of co-culture. Starting growth of virus stocks as soon as possible will minimize the time in culture and thereby the chance to acquire mutations.
19. Positive wells can be considered clonal when the probability of receiving two or more infectious units is negligible. The chance that a positive well will not contain more than 1 IU increases with decreasing λ. However, the number of positive wells will decrease with decreasing λ, which is not desirable from the virus isolation point of view. As a compromise viruses are selected from cultures of patient cell dilutions where *at least* 70% of the wells are negative. An average of 70% of wells is expected to be negative when $\lambda = 0.36$ ($\lambda = -\ln[0.7]$). In that same situation, the probability of 1 infectious unit per well = $P(X = 1) = e^{-0.36} \times 0.36^1/1! = 0.25$ (the probability of 2 IU per well = $P(X = 2) = e^{-0.36} \times 0.36^2/2! = 0.045$) or, when 30/100 wells are positive, an average of 25/30 may be expected to be clonal.
20. For slow growing viruses, the cell suspensions of 96-well cultures (65 µL) may also be first transferred to 1 mL culture medium containing 1×10^6 PHA-blasts in a 24-well well. Using this protocol, stocks are maintained for a minimum of 21 d: at d 7 transfer cells to a 25-cm^2 flask and add 2×10^6 PHA-blasts in 2 mL culture medium, at d 14 add 5×10^6 cells in 5 mL, and at d 21 freeze and store supernatants.
21. When virus production in one or more stocks is relatively low, refresh one extra time instead of freezing: remove 5 mL of cell suspension and add 5×10^6 fresh CD8-depleted PHA-blasts in 5 mL culture medium. If larger stocks than 7–8 mL are desired, propagate the culture by transferring to a 75-cm^2 flask and add 12×10^6/12 mL PHA blasts, and so on.

References

1. Castro, B. A., Weiss, C. D., Wiviott, L. D., and Levy, J. A. (1988) Optimal conditions for recovery of the human immunodeficiency virus from peripheral blood mononuclear cells. *J. Clin. Microbiol.* **26,** 2371–2376.
2. Gallo, D., Kimpton, J. S., and Dailey, P. J. (1987) Comparative studies on use of fresh and frozen peripheral blood lymphocyte specimens for isolation of human immunodeficiency virus and effects of cell lysis on isolation efficiency. *J. Clin. Microbiol.* **25,** 1291–1294.

3. Farzadegan, H., Imagawa, D., Gupta, P., Lee, M. H., Jacobson, L., Saah, A., et al. (1990) The effect of fresh lymphocytes on increased sensitivity of HIV-1 isolation: a multicenter study. *J. Acquir. Immune Defic. Syndr.* **3,** 981–986.
4. Balachandran, R., Thampatty, P., Rinaldo, C., and Gupta, P. (1988) Use of cryopreserved normal peripheral blood lymphocytes for isolation of human immunodeficiency virus from seropositive men. *J. Clin. Microbiol.* **26,** 595–597.
5. Ulrich, P. P., Busch, M. P., el Beik, T., Shiota, J., Vennari, J., Shriver, K., Vyas, G. N. (1988) Assessment of human immunodeficiency virus expression in cocultures of peripheral blood mononuclear cells from healthy seropositive subjects. *J. Med. Virol.* **25,** 1–10.
6. Dimitrov, D. H., Melnick, J. L., and Hollinger, F. B. (1990) Microculture assay for isolation of human immunodeficiency virus type 1 and for titration of infected peripheral blood mononuclear cells. *J. Clin. Microbiol.* **28,** 734–737.
7. Hollinger, F. B., Bremer, J. W., Myers, L. E., Gold, J. W., and McQuay, L. (1992) Standardization of sensitive human immunodeficiency virus coculture procedures and establishment of a multicenter quality assurance program for the AIDS Clinical Trials Group. The NIH/NIAID/DAIDS/ACTG Virology Laboratories. *J. Clin. Microbiol.* **30,** 1787–1794.
8. Walker, C. M. and Levy, J. A. (1989) A diffusible lymphokine produced by CD8+ T lymphocytes suppresses HIV replication. *Immunology* **66,** 628–630.
9. Cocchi, F., DeVico, A. L., Garzino-Demo, A., Arya, S. K., Gallo, R. C., and Lusso, P. (1995) Identification of RANTES, MIP-1 alpha, and MIP-1 beta as the major HIV-suppressive factors produced by CD8+ T cells. *Science* **270,** 1811–1815.
10. Zhang, L., Yu, W., He, T., Yu, J., Caffrey, R. E., Dalmasso, E. A., et al. (2002) Contribution of human alpha-defensin 1, 2, and 3 to the anti-HIV-1 activity of CD8 antiviral factor. *Science* **298,** 995–1000.
11. Tersmette, M., de Goede, R. E., Al, B. J., Winkel, I. N., Gruters, R. A., Cuypers, H. T., et al. (1988) Differential syncytium-inducing capacity of human immunodeficiency virus isolates: frequent detection of syncytium-inducing isolates in patients with acquired immunodeficiency syndrome (AIDS) and AIDS- related complex. *J. Virol.* **62,** 2026–2032.
12. Walker, C. M., Moody, D. J., Stites, D. P., and Levy, J. A. (1986) CD8+ lymphocytes can control HIV infection in vitro by suppressing virus replication. *Science* **234,** 1563–1566.
13. Gupta, P., Enrico, A., Armstrong, J., Doerr, M., Ho, M., and Rinaldo, C. (1990) A semiquantitative microassay for measurement of relative number of blood mononuclear cells infected with human immunodeficiency virus. *AIDS Res. Hum. Retroviruses* **6,** 1193–1196.
14. Ho, D. D., Moudgil, T., and Alam, M. (1989) Quantitation of human immunodeficiency virus type 1 in the blood of infected persons. *N. Engl. J. Med.* **321,** 1621–1625.
15. Schuitemaker, H., Koot, M., Kootstra, N. A., Dercksen, M. W., de Goede, R. E., van Steenwijk, R. P., et al. (1992) Biological phenotype of human immunodeficiency virus type 1 clones at different stages of infection: progression of disease is associated with a shift from monocytotropic to T-cell-tropic virus population. *J. Virol.* **66,** 1354–1360.

16. Strijbosch, L. W., Buurman, W. A., Does, R. J., Zinken, P. H., and Groenewegen, G. (1987) Limiting dilution assays. Experimental design and statistical analysis. *J. Immunol. Methods* **97,** 133–140.
17. Myers, L. E., McQuay, L. J., and Hollinger, F. B. (1994) Dilution assay statistics. *J. Clin. Microbiol.* **32**, 732–739.
18. Simon, F., Matheron, S., Tamalet, C., Loussert-Ajaka, I., Bartczak, S., Pepin, J. M. , et al. (1993) Cellular and plasma viral load in patients infected with HIV-2. *AIDS* **7,** 1411–1417.
19. Albert, J., Naucler, A., Bottiger, B., Broliden, P. A., Albino, P., Ouattara, S. A., et al. (1990) Replicative capacity of HIV-2, like HIV-1, correlates with severity of immunodeficiency. *AIDS* **4,** 291–295.
20. De Cock, K. M., Adjorlolo, G., Ekpini, E., Sibailly, T., Kouadio, J., Maran, M., et al. (1993) Epidemiology and transmission of HIV-2. Why there is no HIV-2 pandemic. *JAMA* **270,** 2083–2086.
21. Berry, N., Ariyoshi, K., Jaffar, S., Sabally, S., Corrah, T., Tedder, R., and Whittle, H. (1998) Low peripheral blood viral HIV-2 RNA in individuals with high CD4 percentage differentiates HIV-2 from HIV-1 infection. *J. Hum. Virol.* **1,** 457–468.
22. Damond, F., Gueudin, M., Pueyo, S., Farfara, I., Robertson, D. L., Descamps, D., et al. (2002) Plasma RNA viral load in human immunodeficiency virus type 2 subtype A and subtype B infections. *J. Clin. Microbiol.* **40,** 3654–3659.
23. Blaak, H., Boers, P. H. M., Schutten, M., van der Ende, M. E., and Osterhaus, A. D. M. E. (2004) HIV-2-infected individuals with undetectable plasma viremia carry replication-competent virus in peripheral blood lymphocytes. *J. AIDS* **36,** 777–782.
24. Wang, Y., Tao, L., Mitchell, E., Bravery, C., Berlingieri, P., Armstrong, P., et al. (1999) Allo-immunization elicits CD8+ T cell-derived chemokines, HIV suppressor factors and resistance to HIV infection in women. *Nat. Med.* **5,** 1004–1009.
25. Moriuchi, H., Moriuchi, M., and Fauci, A. S. (1999) Induction of HIV-1 replication by allogeneic stimulation. *J. Immunol.* **162,** 7543–7548.

9

Isolation and Confirmation of Human T-Cell Leukemia Virus Type 2 From Peripheral Blood Mononuclear Cells

Michael D. Lairmore and Andy Montgomery

Summary

Human T-cell leukemia virus type 2 (HTLV-2) was first isolated from leukemia patients, but has been found to be endemic among asymptomatic groups worldwide, including certain American Indian tribes *(1)*. The virus infection is associated with a low incidence of disease among infected subjects, but has been found in patients with neurologic disorders and contributes to bacterial sepsis in AIDS patients *(2,3)*. Polymerase chain reaction (PCR) and virus isolation techniques revealed that a high percentage of HTLV seroreactivity among intravenous drug users and blood donors in the United States is caused by HTLV-2 *(4)*. Among serologic methods, enzyme-linked immunosorbent assays (ELISA) using whole virus preparations or in combination with recombinant and synthetic peptides are used as a primary screen for the infection. Antigen-capture systems have increased the sensitivity and accuracy in verification of HTLV-2 culture systems. The verification of HTLV-2 infection and detection of new strains of related viruses has been enhanced by employing virus-isolation methods using primary lymphocytes. Lymphocyte culture methods have also been used to test transformation properties of the virus and create stably expressing cell lines. This chapter briefly summarizes the biology of HTLV-2 infection and disease and details methods to isolate and verify the virus in lymphocyte cultures.

Key Words: Human T-cell leukemia virus type 2; HTLV-2; human T-cell leukemia virus type 1; HTLV-1; lymphoma; retrovirus; cell culture; antigen; replication; isolation; detection.

1. Introduction

The first human retrovirus described, human T-cell leukemia virus type 1 (HTLV-1), is considered the cause of adult T-cell leukemia/lymphoma (ATLL) and is associated with chronic inflammatory diseases such as HTLV-1 myelopathy/tropical spastic paraparesis (HAM/TSP) *(5)*. Human T-cell leukemia

From: *Methods in Molecular Biology, Vol. 304: Human Retrovirus Protocols: Virology and Molecular Biology*
Edited by: T. Zhu © Humana Press Inc., Totowa, NJ

virus type 2 (HTLV-2) was later isolated from a patient with hairy cell leukemia ***(6)*** and has subsequently been determined to be endemic among persons sharing many of the same risk factors as those with HTLV-1 infection ***(7–12)***. The virus infection is also endemic among certain American Indian groups in North and South America and among at-risk groups worldwide ***(3,7)***. Detailed studies of the pathogenesis of HTLV-2 infection have not been reported, in part, due to the low incidence of disease among infected subjects. However, the infection is associated with neurologic disorders and contributes to bacterial sepsis in AIDS patients. With the development of polymerase chain reaction (PCR) techniques it was discovered that a high percentage of HTLV seroreactivity among intravenous (IV) drug users and blood donors in the United States is caused by HTLV-2 ***(14–16)***. However, gaps remain in the literature regarding disease associations, transmission routes, and epidemiologic features of HTLV-2 infections.

Serologic surveys to detect antibodies against HTLV-1 and HTLV-2 remain a primary tool for virologists and epidemiologists interested in population-based studies of HTLV infection. The most widely used system of antibody detection is the enzyme-linked immunosorbent assay (ELISA), using whole-virus preparations for antigen preparations or in combination with recombinant and synthetic peptide-based immunoassay systems. The development of ELISA-based antigen-capture systems to detect small quantities of HTLV-1 and HTLV-2 core antigens has increased the sensitivity and accuracy of HTLV culture systems and largely replaced measurement of reverse transcriptase activity to verify cultures that express these highly cell-associated viruses. The use of PCR-based methods has dramatically enhanced the ability to detect HTLV-2 proviral DNA and has allowed significant improvements in virus detection compared to the labor-intensive procedures of virus isolation. The verification of HTLV-2 infection and detection of new strains of related viruses, however, has also been through virus isolation ***(11,17,18)***. Long-term culture of lymphocytes has been employed to test transformation properties and antigen expression of HTLV-2. This is owing, in part, to the fact that leukocytes derived directly from infected subjects are often negative when tested by conventional methods (e.g., immunofluorescence). This chapter briefly summarizes the biology and clinical features of HTLV-2 infection and disease. Virus isolation methods to detect HTLV-2 infection, to isolate new stains of the virus, and for basic comparative studies have expanded the knowledge base of this important human retrovirus.

The HTLV family shares structural features in common with all retroviruses, including the viral genes for group-specific antigens (*gag*), reverse transcriptase (*pol*), and envelope *(env)* proteins. However, HTLV-1 and HTLV-2 contain unique pX region genes, which code for important regulatory (i.e., Tax and Rex) and accessory proteins (e.g., p28 of pX ORF II of HTLV-2) at the 3' end of the provirus (**Table 1**). HTLV-2 shares 60% amino acid identity with HTLV-1

Table 1
Human T-Lymphotropic Virus Type 2 (HTLV-2) Genomic Organization

Gene	Function	Protein product (kD)
Long terminal repeat (*LTR*)	Regulation	Noncoding
Group-specific antigen (*gag*)	Core proteins	p53 precursor
		p19 matrix
		p24 capsid
		p15 nucleocapsid
Protease (*pro*)	*gag* cleavage	p14 protease[a]
Polymerase (*pol*)	Replication	p62 transcriptase[a]
		p32 integrase[a]
Envelope *(env)*	Envelope	p61-68 precursor
		gp46 surface
		gp21 transmembrane
pX ORF IV	Regulation	p37 Tax (transactivation)
pX ORF III	RNA regulation	p24, p26[b] Rex (regulator of splicing)
pX ORF II	[a]	p28[c,d]
pX ORF I	[a]	p10[c]
pX ORF V	[a]	p 11[c]

[a]Defined for HTLV-1, not defined for HTLV-2.
[b]Rex phosphorylation determines size.
[c]Predicted protein product.
[d]May regulate viral RNA through posttranscriptional mechanism (P. Green, personal communications).

and shares genome structures and in vitro biological properties. As a result, HTLV-2 remains an important comparative model for the study of the deltraretroviruses. Examination of the genome sequence of the deltaretrovirus family of retroviruses that includes HTLV-1, HTLV-2, simian T-lymphotropic virus (STLV), and bovine leukemia virus (BLV) reveals conserved organization of genes within the 3' end of these viruses. Like HTLV-1, proteins encoded in the pX region of HTLV-2 are likely to be essential for viral replication during the natural infection ***(19)***.

Retroviruses initiate their replication by binding to specific host cellular receptors through complementary viral envelope components. Following penetration of the host cell, single-stranded viral RNA with associated nucleoproteins and reverse transcriptase is released to allow synthesis of a DNA copy of the viral RNA. The resulting DNA copy (provirus) typically integrates into the host genome and initiates synthesis of viral RNA synthesis including copies of the genome. Repeated nucleotide sequences (terminal repeat segments) at either end of the viral genome, in concert with viral and host genes, control viral replication. During productive phases of retrovirus replication, progeny viruses bud from host cell membranes (often obtaining host proteins in the process) and are released to infect other cells. Transcriptional blocks of virus replications, defective viral genomes, and highly regulated viral encoded proteins regulate the degree of virus expression within the host cell.

This ability of this group of retroviruses to activate cellular genes allows HTLV-1 and HTLV-2 to overcome cellular controls and cause inappropriate expression of particular genes (e.g., host transcriptional factors), which contribute to lymphocyte proliferation and eventual transformation of the cell ***(20)***. These viruses are difficult to detect by conventional methods (e.g., *in situ* hybridization, immunofluorescence) within infected tissues. The predominant cell target of HTLV-1 appears to be the $CD4^+$ lymphocyte, while $CD8^+$ lymphocytes appear to be preferentially transformed by HTLV-2. The basis for the differential cell tropism is not understood, but may be related to post entry mechanisms ***(21)***.

HTLV-2, like HTLV-1 is highly cell associated and is transmitted from infected mothers to their offspring, primarily via breast milk. Virus culture and PCR studies of cord blood samples suggest that HTLV-1 and HTLV-2 are not transmitted *in utero*. Injection of infected blood is a common means of transmission of both viruses. In blood, the virus is associated with the cellular fractions that account for virtually all transmission of the infection. Parenteral exposure to the virus from needle sharing is a predominant mode of transmission of the virus among drug abusers ***(1,22–25)***. Procedures used for virus isolation and subsequent confirmation of viral antigens of HTLV-2 need to take into account the cell-associated nature of this group of viruses. Particular attention is also required in cell target preparation and in methods to test for viral antigen or nucleic acid in lymphocyte cultures.

2. Materials

1. Complete RPMI-1640 supplemented with 15% fetal bovine serum (FBS), 1% streptomycin/penicillin, and 1% glutamine (complete RPMI, cRPMI).
2. Human interleukin (hIL)-2 added at 10 U/mL as either purified or recombinant IL-2.
3. Normal uninfected human peripheral blood mononuclear cells (PBMC) obtained and maintained in complete RPMI-1640 media.
4. PBMC used for co-culture; prestimulated for 4 d with hIL-2 (10 U/mL) and phytohemagglutinin (PHA, 2 μg/mL) to activate for optimal infectivity prior to co-culture with test subject cells.
5. Viral core antigen production to test cell culture supernatants by enzyme-linked immunosorbent assay (ELISA) (Zeptomatrix, Buffalo, NY) .
6. Western blot lysis buffer: 1% deoxycholic acid, 0.1% sodium dodecyl sulfate (SDS), 1% Triton X-100, 150 m*M* NaCl, 50 m*M* Tris-HCl, 20 μg/mL leupeptin, 20 μg/mL aprotinin, 1 m*M* sodium orthovanadate, 1 m*M* PMSF.

7. HTLV-2 polyclonal antiserum.
8. Genomic DNA extracted from cell cultures obtained by commercial affinity column.
9. HTLV-2 specific oligonucleotide primers.

3. Methods

3.1. Isolation of HTLV-2

The detection of HTLV-2 by virus isolation is dependent on activation of T lymphocytes and cell culture conditions that allow for the long-term growth of lymphocytes. The virus is transmitted by cell-to-cell transfer, and successful infection of cell cultures with cell-free virus preparations have generally been unsuccessful. Because of these biologic factors, virus isolation procedures require that test subjects' PBMC or tumor material be stimulated (e.g., by mitogens) and monitored for the presence of viral reverse transcriptase, antigens, or nucleic acids. In addition, coculturing with appropriate uninfected target cells (e.g., mitogen-stimulated PBMC) will enhance the ability of detection of HTLV-2 from infected persons.

The development of sensitive and specific antigen capture assays has greatly improved the ability to detect HTLV-2 from infected cell cultures. These assays detect soluble antigens (primarily major core antigens) in cell culture supernatants using specific antibodies to capture antigen. The bound antigen–antibody complex is detected colorimetrically. By comparison of the optical density readings of test samples with standard preparations, the quantitity of viral antigen within the culture can be estimated. Alternate procedures to detect the presence of HTLV-2 in cell cultures include immunofluorescence assay (IFA) and avidin–biotin complex methods that allow visualization of viral antigens from the surface of infected cells (**Table 2**).

Infection of uninfected, but mitogen activated human PBMC is an effective method to amplify the number of HTLV-2 infected cells.

1. This is performed by co-culture with test subjects PBMC or lymphocytes with target unifected PBMC prestimulated for 4 d with hIL-2 (10 U/mL) and phytohemagglutinin (PHA, 2 μg/mL) ("activated") in cRPMI at a 1:1 ratio (*see* **Notes 1** and **2**).
2. Co-culture is performed in 24-well plates or 25-cm^2 flasks in complete RPMI. Care must be taken not to cross-contaminate 24-well plates; separation of individual subject cultures in separate plates is recommended.
3. Supernatants are then tested for viral core antigen by antigen capture assay at 3–4 d intervals through 14 d of culture. Samples may be frozen in 1-mL aliquots and tested in batches.
4. Typically, positive cultures are detected by antigen capture assay within 2–3 d, but may not become positive until 7–14 d postculture. Cell clumping is common in early cell cultures due to mitogen stimulation and cell outgrowth of transformed cells (*see* **Note 3**).

Table 2
Methods to Detect Human T-Lymphotropic Virus Type 2 (HTLV-2) Antigens in Cell Culture

Methods	Comments
Immunofluorescence assay	Simple to perform for detection of HTLV from infected cells. Cost-effective, labile signal.
Avidin-biotin complex assay	Easily performed for detection of HTLV from in fected cells. Stable signal; allows slide record.
Antigen capture assay	Excellent sensitivity for HTLV-2 core antigen in cell culture supernatants. Will cross-react with HTLV-1 core antigen. Microtiter plate to test multiple samples.

5. Cultures negative for viral antigen or by PCR after 2 wk are generally considered negative and rarely produce positive results. Cells should be monitored for viable cell counts by standard dye procedures and split when cell growth exceeds 1.5×10^6 cells/mL. Cell splitting should result in cell concentrations of no less than 5×10^5 cells/mL to maintain cell-to-cell contact in the culture.

3.2. Antigen Capture Assay to Confirm HTLV-2 Core Antigen in Cultures

Viral core antigen production is used to test cell culture supernatants by ELISA (Zeptomatrix) according to the manufacturer's protocol (*see* **Notes 4** and **5**).

1. Collect cell supernatant after centrifugation at 1200 rpm for 10 min to remove cell debris. Samples (1 mL) may be tested directly or frozen at –20°C at tested when multiple samples are to be tested to reduce costs.
2. Virus antigen production may be detected as early as 2–3 d in culture. Cultures with low numbers of infected cells may take 7–14 d to rule out as negative.
3. Serial passage of the cells with supernatant samples are taken at each split (every 3–4 d) using fresh complete RPMI media.

3.3. Western Blot Assay for Confirmation of HTLV-2 Proteins From Cell Cultures

Western blot analysis is an alternative to evaluate the expression of cell-associated viral proteins and to confirm antigen capture or PCR results.

1. 5×10^6 HTLV-2-infected cells are lysed in standard Western blot buffer and 20 µg total protein separated by 10% SDS-polyacrylamide gel electrophoresis (SDS-PAGE).
2. After transfer to nitrocellulose, viral proteins are detected by an anti-HTLV-1 or -2 human antiserum using commercially available secondary antibodies and detection systems.

3. HTLV-2 proteins are visualized with appropriate secondary antibodies conjugated to horseradish peroxidase (HRP) using chemiluminescence. Control positive (e.g., Mo-T-cell line) and negative cell lysates (e.g., Jurkat T-cells) should be included to rule out nonspecific bands.

3.4. Polymerase Chain Reaction for Detection for HTLV-2 From Cell Cultures

HTLV-2 antigen expression is limited from freshly isolated subject cells as a result of an active immune response and perhaps transcriptional silencing of the provirus in vivo. As a result the direct detection of viral proteins from cells or tissues is difficult. Therefore, methods used to detect viral nucleic acids have been employed to detect the presence of HTLV from infected cells (**Table 3**).

Techniques to detect relatively large copy numbers of viral nucleic acid include both Southern blot assays and *in situ* hybridization. Each assay is dependent on the sensitivity and specificity of the probe used to detect proviral DNA, as well as the methods and tissues used in the preparations. *In situ* hybridization uses radiolabeled probes that consist of portions of the viral genome (e.g., single-stranded DNA), which are applied to specifically treated slides of cells or tissues. This technique has the major advantage of morphological identification of infected cells; however, it is technically difficult and less sensitive than gene amplification methods (e.g., PCR). Southern blot procedures are similar to *in situ* hybridization in preparation of labeled probes; however, the detection is based on identification of viral nucleic acids from blotted membranes. Restriction enzymes with unique cutting sites are used to allow predictions of the integration of HTLV from lymphoma cells or cell cultures. The ability of Southern blot to detect monoclonal integration of HTLV-I in tumor cells has been employed diagnostically to predict the status of ATLL patients. Direct Southern blot of HTLV, like *in situ* hybridization, is technically difficult and is not sensitive to detect HTLV in low numbers of infected cells (e.g., seropositive but asymptomatic patients). Thus, PCR has become the method of choice to confirm the presence of HTLV-2 nucleic acid from cell cultures.

The application of PCR for the detection of HTLV-2 has allowed more sensitive and specific detection of the virus. The use of specific primers and probes have allowed investigators to discriminate the nucleic acids of HTLV-1 from closely related retroviruses like HTLV-2. The procedure has the major advantage of amplification of the viral signal. However, because of the sensitivity of the assay, strict adherence to control procedures is necessary to avoid cross-contamination. The development of PCR procedures for HTLV-1 and HTLV-2 has allowed investigations to determine the prevalence and potential disease associations of HTLV-2 ***(1,8,26,27)***. In addition, PCR can be used to directly diagnose individuals with HTLV-2 infection without the need for virus isolation.

Table 3
Diagnostic Methods For Detection of Human T-Lymphotropic Virus (HTLV) Nucleic Acids in Cell Cultures

Method	Procedure	Comment
Southern blot assay	DNA from HTLV-2 cells used with labeled complementary probes typically of long terminal repeats	Sensitivity determined by probes and conditions of assay. Method to estimate copy number in cell lines
In situ hybridization	HTLV-2 DNA or RNA in tissues, complementary probes	Excellent to determine cell type expressing virus. Less sensitive directly from patient for RNA *in situ* owing to low level of transcriptionally active cells
Polymerase chain reaction	DNA or RNA from HTLV-2 cells	Excellent sensitivity, useful to differentiate from HTLV-1, contamination a potential problem

1. DNA for PCR can be prepared from cells simply by lysing in the presence of detergents or by using more rigorous methods of phenol extraction.
2. Primer sequences for HTLV-2 are readily available ***(1,8,26,28,29)*** and can be purchased as oligonucleotides from commercial sources. Amplified products are detected after separation in agarose or acrylamide gels followed by transfer and probing by hybridization with labeled probes. Specific detection of HTLV-2 can be based on primer or probe design depending on preference of detection (*see* **Notes 6–8**).
3. Appropriate controls should include both negative and positive cell controls, molecular weight standards, sensitivity controls (serially diluted positive controls), and specificity controls (related but different viral cell lines) (*see* **Note 6**).
4. Controls and test samples should be prepared in a consistent manner to prevent variability in the methods of cell preparation and to avoid contamination of test samples.

4. Notes

1. Cultures derived from human PBMC must be handled with universal biosafety precautions, i.e., assumed to carry blood-borne pathogens. In addition, long-term cell cultures used to produce HTLV-2 antigen should also be tested for the presence of other human viruses including the herpes viruses, Epstein-Barr virus, and the like, which may contaminate PBMC-derived cultures. Pretesting of blood

samples from donors is recommended to eliminate common blood-borne pathogens, but this does not preclude the use of universal precautions.

2. Cultures that become rapidly (within 24 h) acidic or basic should be considered suspicious for the presence of bacterial, mycoplasma, or fungal contamination and appropriately tested. Cultures should be observed daily to check for cell growth and to monitor potential contamination.
3. Standard viable cell counts are useful to monitor the growth of HTLV-2 infected cell cultures. Cell clumping may occur and is expected until mitogenic stimulation and following transformation by HTLV-2. Cultures should be maintained between 1×10^6 and 1.5×10^6 cells/mL for optimal cell-to-cell contact and growth.
4. The viral antigen capture assay used to measure cell culture supernatants by ELISA can be used to test for HTLV-1 antigen production, but will not differentiate these viruses. Monitoring cultures for the presence of reverse transcriptase is of limited value because of the low levels produced by this class of retrovirus. In addition, the measurement of this retrovirus enzyme is not specific for HTLV-2 and will also detect the presence of other retroviruses (e.g., HIV) and potentially detect endogenous host polymerases.
5. Unlike the situation for HIV-infected persons, these HTLV-2 antigen capture assays have not been useful to detect viral antigen directly from patients' bodily fluids (e.g., serum). This limitation of antigen detection directly from patients is caused by several factors, including complexing of antigens with patients' antibodies, but is principally due to the low levels of HTLV-2 antigens because of the cell-associated nature of the viral infection.
6. Laboratory space used for all procedures involving DNA extraction should be separated physically from rooms used for amplification of the target DNA to prevent contamination of primer pairs and probes.
7. Several important caveats must be kept in mind in the application of PCR to detect HTLV-2 infection. The procedure detects the DNA of HTLV-2 and does not provide information concerning the replication capacity or infectious nature of the virus infection. Because of potential cross-contamination problems and detection of endogenous retrovirus sequences, the criteria of a positive PCR test should ideally include detection of HTLV-2 using primers from at least two separate portions of the genome.
8. A variety of alternative PCR methods (e.g., real-time PCR) have been developed to test for HTLV-1 and HTLV-2 from patient samples and cell cultures *(27)*.

References

1. Lowis, G. W., Sheremata, W. A., and Minagar, A. (2002) Epidemiologic features of HTLV-II: serologic and molecular evidence. *Ann. Epidemiol.* **12,** 46–66.
2. Zehender, G., Colasante, C., Santambrogio, S., De Maddalena, C., Massetto, B., Cavalli, B., et al. (2002) Increased risk of developing peripheral neuropathy in patients coinfected with HIV-1 and HTLV-2. *J. Acquir. Immune Defic. Syndr.* **31,** 440–447.

3. Etzel, A., Shibata, G. Y., Rozman, M., Jorge, M. L., Damas, C. D., and Segurado, A. A. (2001) HTLV-1 and HTLV-2 infections in HIV-infected individuals from Santos, Brazil: seroprevalence and risk factors. *J. Acquir. Immune Defic. Syndr.* **26,** 185–190.
4. Khabbaz, R. F., Hartel, D., Lairmore, M., Horsburgh, C. R., Schoenbaum, E. E., Roberts, B., et al. (1991) Human T lymphotropic virus type II (HTLV-II) infection in a cohort of New York intravenous drug users: an old infection? *J. Infect. Dis.* **163,** 252–256.
5. Bangham, C. R. (2000) HTLV-1 infections. *J. Clin. Pathol.* **53,** 581–586.
6. Kalyanaraman, V. S., Sarngadharan, M. G., Robert-Guroff, M., Miyoshi, I., Golde, D., and Gallo, R. C. (1982) A new subtype of human T-cell leukemia virus (HTLV-II) associated with a T-cell variant of hairy cell leukemia. *Science* **218,** 571–573.
7. Murphy, E. L., Grant, R. M., Kropp, J., Oliveira, A., Lee, T. H., and Busch, M. P. (2003) Increased human T-lymphotropic virus type II proviral load following highly active retroviral therapy in HIV-coinfected patients. *J. Acquir. Immune Defic. Syndr.* **33,** 655–656.
8. Silva, E. A., Otsuki, K., Leite, A. C., Alamy, A. H., Sa-Carvalho, D., and Vicente, A. C. (2002) HTLV-II infection associated with a chronic neurodegenerative disease: clinical and molecular analysis. *J. Med. Virol.* **66,** 253–257.
9. Madeleine, M. M., Wiktor, S. Z., Goedert, J. J., Manns, A., Levine, P. H., Biggar, R. J., and Blattner, W. A. (1993) HTLV-I and HTLV-II world-wide distribution—Reanalysis of 4,832 immunoblot results. *Int. J. Cancer* **54,** 255–260.
10. Jacobson, S., Lehky, T., Nishimura, M., Robinson, S., McFarlin, D. E., and Dhib Jalbut, S. (1993) Isolation of HTLV-II from a patient with chronic, progressive neurological disease clinically indistinguishable from HTLV-I-associated myelopathy/tropical spastic paraparesis. *Ann. Neurol.* **33,** 392–396.
11. Lairmore, M. D., Jacobson, S., Gracia, F., De, B., Castillo, L., Larreategui, M., et al. (1990) Isolation of human T-lymphotropic virus type 2 from Guaymi Indians in Panama. *Proc. Natl. Acad. Sci. USA* **87,** 8840–8844.
12. Kelen, G. D., DiGiovanna, T. A., Lofy, L., Junkins, E., Stein, A., Sivertson, K. T., et al. (1990) Human T-lymphotropic virus (HTLV I-II) infection among patients in an inner-city emergency department. *Ann. Intern. Med.* **113,** 368–372.
13. Shindo, N., Alcantara, L. C., Van Dooren, S., Salemi, M., Costa, M. C., Kashima, S., et al. (2002) Human retroviruses (HIV and HTLV) in Brazilian Indians: seroepidemiological study and molecular epidemiology of HTLV type 2 isolates. *AIDS Res. Hum. Retroviruses* **18,** 71–77.
14. Khabbaz, R. F., Douglas, J. M., Judson, F. N., Spiegel, R. A., St. Louis, M. E., Whittington, W., et al. (1990) Seroprevalence of human T-lymphotropic virus type I or II in sexually transmitted disease clinic patients in the USA. *J. Infect. Dis.* **162,** 241–244.
15. Anderson, D. W., Epstein, J. S., Lee, T. H., Lairmore, M. D., Saxinger, C., Kalyanaraman, V. S., et al. (1989) Serological confirmation of human T-lymphotropic virus type I infection in healthy blood and plasma donors. *Blood* **74,** 2585–2591.

16. Williams, A. E., Fang, C. T., Slamon, D. J., Poiesz, B. J., Sandler, G., Darr, W. F., et al. (1988) Seroprevalence and epidemiological correlates of HTLV-I infection in U.S. blood donors. *Science* **240,** 643–646.
17. Rosenblatt, J. and Harrington, W. J., Jr. (2003) Leukemia and myelopathy: the persistent mystery of pathogenesis by HTLV-I/II. *Cancer Invest.* **21,** 323–324.
18. Echeverria de Perez, G., Leon Ponte, M., Noya, O., Botto, C., Gallo, D., and Bianco, N. (1993) First description of endemic HTLV-II infection among Venezuelan Amerindians. *J. Acquir. Immune. Defic. Syndr.* **6,** 1368–1372.
19. Albrecht, B. and Lairmore, M. D. (2002) Critical role of human T-lymphotropic virus type 1 accessory proteins in viral replication and pathogenesis. *Microbiol. Mol. Biol. Rev.* **66,** 396–406.
20. Mortreux, F., Gabet, A. S., and Wattel, E. (2003) Molecular and cellular aspects of HTLV-1 associated leukemogenesis in vivo. *Leukemia* **17**, 26–38.
21. Ye, J., Xie, L., and Green, P. L. (2003) Tax and overlapping rex sequences do not confer the distinct transformation tropisms of human T-cell leukemia virus types 1 and 2. *J. Virol.* **77,** 7728–7735.
22. Ishak, R., Vallinoto, A. C., Azevedo, V. N., Lewis, M., Hall, W. W., and Guimaraes Ishak, M. O. (2001) Molecular evidence of mother-to-child transmission of HTLV-IIc in the Kararao village (Kayapo) in the Amazon region of Brazil. *Rev. Soc. Bras. Med. Trop.* **34,** 519–525.
23. Caterino-De-Araujo, A. and Los Santos-Fortuna, E. (1999) No evidence of vertical transmission of HTLV-I and HTLV-II in children at high risk for HIV-1 infection from Sao Paulo, Brazil. *J. Trop. Pediatr.* **45,** 42–47.
24. Murphy, E. L., Watanabe, K., Nass, C. C., Ownby, H., Williams, A., and Nemo, G. (1999) Evidence among blood donors for a 30-year-old epidemic of human T lymphotropic virus type II infection in the United States. *J. Infect. Dis.* **180,** 1777–1783.
25. Murphy, E. L., Mahieux, R., Dethe, G., Tekaia, F., Ameti, D., Horton, J., and Gessain, A. (1998) Molecular epidemiology of HTLV-II among United States blood donors and intravenous drug users: An age-cohort effect for HTLV-II RFLP type aO. *Virology* **242,** 425–434.
26. Thorstensson, R., Albert, J., and Andersson, S. (2002) Strategies for diagnosis of HTLV-I and -II. *Transfusion* **42,** 780–791.
27. Poiesz, B. J., Dube, S., Choi, D., Esteban, E., Ferrer, J., Leon-Ponte, M., et al. (2000) Comparative performances of an HTLV-I/II EIA and other serologic and PCR assays on samples from persons at risk for HTLV-II infection. *Transfusion* **40,** 924–930.
28. Silva, E. A., Otsuki, K., Leite, A. C., Alamy, A. H., Sa-Carvalho, D., and Vicente, A. C. (2002). HTLV-II infection associated with a chronic neurodegenerative disease: clinical and molecular analysis. *J. Med. Virol* **66,** 253–257.
29. Cimarelli, A., Duclos, C. A., Gessain, A., Casoli, C., and Bertazzoni, U. (1996) Clonal expansion of human T-cell leukemia virus type II in patients with high proviral load. *Virology* **223,** 362–364.

10

Isolation of Foamy Viruses From Peripheral Blood Lymphocytes

Joëlle Tobaly-Tapiero, Patricia Bittoun, and Ali Saïb

Summary

The isolation of a retrovirus from peripheral blood lymphocytes/monocytes can be a difficult task, requiring the fulfillment of three essential parameters. First, this viral agent must infect such cells in vivo. Second, these circulating cells should harbor wild-type proviruses. Finally, the viral agent has to express, at least when these cells are cultured in vitro, the structural proteins necessary for the production of viral particles. Foamy viruses (FVs), also known as spumaviruses, are complex retroviruses whose genomic organization has been known since the cloning of the prototypic primate foamy virus type 1. These retroviruses infect most cell lines in culture, but circulating lymphocytes seem to represent their major reservoir in vivo. FV infection leads to the formation of multinucleated giant cells, resulting from the fusion of adjacent infected cells, which present multiple vacuoles giving the monolayer culture a foam aspect. These two features, combined with electron microscopy studies, have helped investigators in their attempt to isolate new FVs. These viruses were described and isolated from different animal species, mostly in nonhuman primates. Here we present the successive steps leading to the isolation of the equine foamy virus from peripheral blood lymphocytes of infected horses.

Key Words: Spumavirus; PBMC; cytopathic effect; syncytia; λ phage; zoonosis.

1. Introduction

Foamy viruses (FVs), also called spumaviruses, were described for the first time in the early 1950s in cell cultures derived from monkey kidneys *(1)*. Using electron microscopy, they appear as spheres of 100–140 nm diameter, coated with 10/15 nm protuberant spikes (**Fig. 1B**,**C**). Similar to type B/D retroviruses, the viral capsid is preassembled in the cytoplasm of infected cells before budding *(2)*. Studies on the genomic organization of the prototypic PFV-1 (primate foamy virus type 1, also called HFV for human foamy virus) isolate

From: *Methods in Molecular Biology, Vol. 304: Human Retrovirus Protocols: Virology and Molecular Biology*
Edited by: T. Zhu © Humana Press Inc., Totowa, NJ

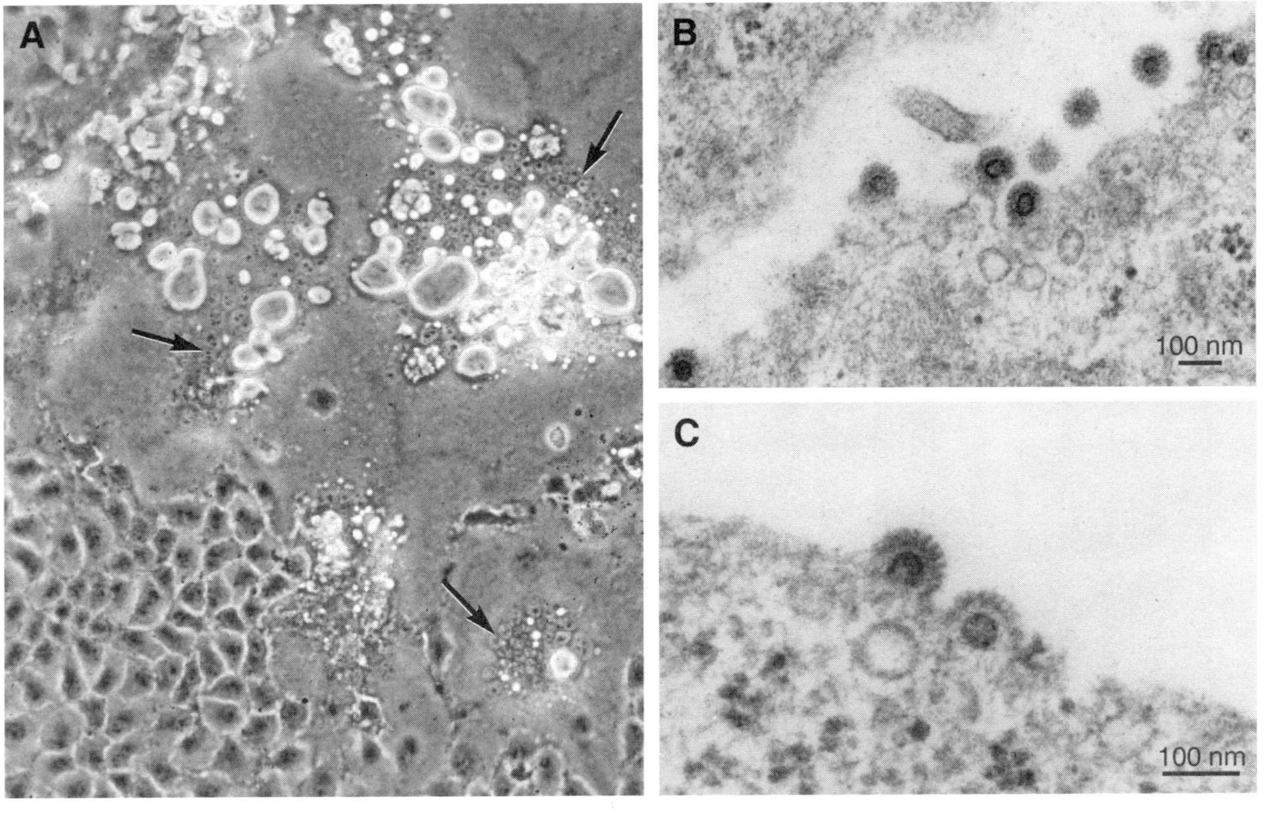

Fig. 1. **(A)** Cytopathic effects observed on co-cultures between horse peripheral blood mononuclear cells and human U373-MG cells. Note the formation of syncytia (arrows) presenting numerous vacuoles. **(B)** and **(C)** Electron microscopy of ultrathin sections from infected cells. The viral particle has the typical foamy virus appearance, enveloped particles surrounded by spikes and a clear central core. Preassembled viral capsids are present in the cytoplasm of infected cells similar to type B/D retroviruses.

revealed that FVs are complex retroviruses because in addition to the structural and enzymatic *gag*, *pol*, and *env* genes, they harbor two additional open reading frames (ORF1 and ORF2) at the 3' end of their genome. At the replication level, FVs share several characteristics with pararetroviruses, exemplified by the hepatitis B virus where there is a production of a specific mRNA for the synthesis of Pol and a late-occurring reverse transcription *(3)*.

In vitro, FV infection induces the formation of multinucleated giant cells, which result from the fusion of adjacent infected cells, which present multiple vacuoles giving the monolayer culture a foam aspect (**Fig. 1A**). These cytopathic effects rapidly lead to cell lysis. Despite these dramatic phenotypic changes in vitro, FVs seem to be harmless in naturally or experimentally infected animals in which they induce a life-long persistent infection *(4)* with circulating lymphocytes that represent their major reservoir *(5)*. Interestingly, these specific cytopathic effects (CPE), together with electron microscopy studies, have helped investigators in their attempt to isolate new FVs.

Foamy viruses are widely represented in nature, mostly among nonhuman primates; although several accidental zoonotic infections have been reported, FVs are not prevalent in humans *(6)*. Nonprimate FVs have also been characterized *(7)*, namely the bovine foamy virus (BFV), the feline foamy virus (FFV), and more recently the equine foamy virus (EFV) *(8)*. We describe in this chapter the main steps leading to the isolation and molecular cloning of EFV from peripheral blood cells of infected horses.

2. Materials

1. FITC-coupled anti-horse immunoglobulin (Ig)G antibodies.
2. Tween-20.
3. Phosphate-buffered saline.
4. Mowiol 4-88 (Calbiochem).
5. Paraformaldehyde.
6. Methanol.
7. RPMI and Dulbecco's modified Eagle medium (DMEM), antibiotics, sodium pyruvate, fetal calf serum (FCS).
8. Phytohemagglutinin P (PHA-P).
9. Heparin-coated tubes.
10. Ficoll-Hypaque preparation (Pharmacia).
11. Oligonucleotide adaptors (Biolabs).
12. Gigapack III Gold cloning kit (Stratagene).
13. T4 DNA ligase, T4 polynucleotide kinase, restriction enzymes.
14. Pancreatic RNase, proteinase K.
15. Glyoxal 30% (w/v) in H_2O (Fluka).
16. Formamide.
17. γ^{32}PATP and α^{32}PdCTP.

18. Prime a Gene labeling kit (Promega).
19. Wizard Lambda Preps DNA Purification System (Promega).
20. Sephacryl S-300 DNA spun columns (Pharmacia).

3. Methods

3.1. Virus Isolation From Whole Blood Samples

3.1.1. Presence of Anti-PFV-1 Antibodies in Blood Plasma From Horses

Interestingly, the distribution of FVs among animal species mirrors that of lentiviruses, except in the case of horses, for which no data were available *(7)*. Thus, the presence of a foamy virus was investigated in domestic horses. Because (1) natural or experimental infection with FVs leads to a strong antibody response and (2) several structural domains are conserved among known FV proteins, the presence of antibodies directed against FV antigens in the plasma of horses was initially determined to assess the presence of a FV in this animal species. Blood samples from 36 horses were collected and serum was separated from whole blood cells by centrifugation at 425*g* for 15 min at 4°C. Sera were tested for the presence of anti-FV antibodies by indirect immunofluorescence (IF) on PFV-1-infected adherent U373-MG cells, 3 d postinfection (*see* **Note 1**). For this purpose:

1. Perform cell fixation with a 4% paraformaldehyde (PFA) solution for 10 min at 4°C and cell permeabilization with methanol for 5 min at 4°C.
2. Following a rapid wash with phosphate-buffered saline (PBS)-Tween 0.1%, dilute horse sera at 1/100 in PBS-Tween 0.1% and incubate with infected cells for 1 h at 37°C.
3. After three washes with PBS-Tween 0.1%, incubate cells with anti-horse IgG, coupled with fluorescein (Biosys), at 1/500 dilution in PBS-Tween 0.1%, for 30 min at 37°C.
4. After three washes with PBS-Tween 0.1%, mount slides with Mowiol.

By confocal microscopy (MRC-600, Bio-Rad), a strong fluorescent staining was detected for 9 sera among 36 tested, whereas mock-infected cells were strictly negative under these settings (**Fig. 2**).

3.1.2. Cell Culture and Virus Isolation

Blood samples from two positive horses were collected in heparin-coated tubes. Peripheral blood cells were isolated by standard density gradient centrifugation. For this purpose:

1. Dilute each blood sample in 3 vol of sterile PBS.
2. Delicately load 30 mL on 20 mL of Ficoll-Hypaque solution in a 50-mL Falcon tube.
3. After a centrifugation step at 840*g* at 20°C for 30 min, collect peripheral blood mononuclear cells (PBMC) from the interphase, wash twice in sterile PBS by

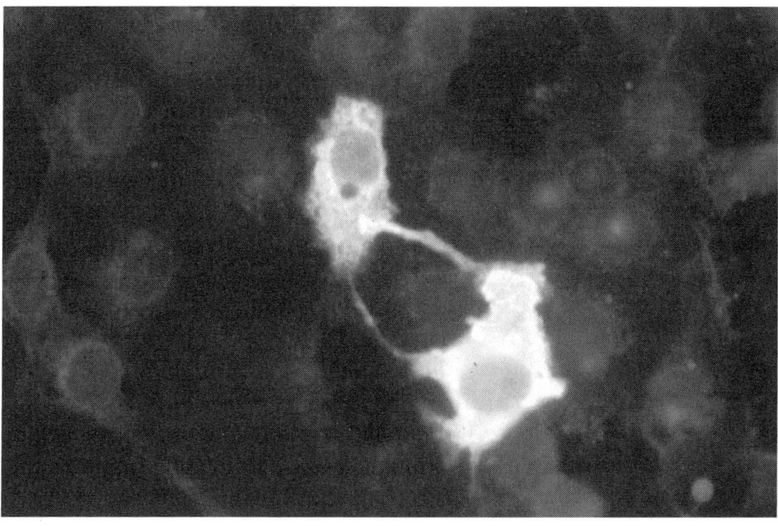

Fig. 2. Reactivity of horse sera on PFV-1 antigens tested by indirect immunofluorescence assay on infected human U373MG cells.

centrifugation (840*g*, 10 min) and harvest in RPMI medium containing 20% FCS, antibiotics (100 µg/mL streptomycin and 100 U/mL penicillin) and the mitogenic lectin PHA-P (3 mg/mL).

4. Two days later, wash the culture in sterile PBS and co-cultivate for 48 h with a highly FV-permissive adherent cell line (either human U373-MG or hamster BHK21 cells [*see* **Note 2**]), in a ratio of 1:1.
5. Remove the nonadherent cells, split the adherent culture twice a week at 1/20, and observe daily for the appearance of cytopathic effects such as syncytia and vacuolization, two features of FV infection.

These culture conditions lead to giant multinucleated cells with numerous vacuoles, which will be observed in infected cell cultures only after 4 wk of cultivation (**Fig. 1A**). These CPE will expand to the entire culture in 2 d, followed by a drastic cell lysis. Meanwhile, cell-free supernatant can be collected, filtered (0.45-µm pore diameter, Nalgene), and used to infect naïve permissive cells for further analysis (*see* **Note 3**).

3.2. Detection of Virus

To directly visualize the presence of a foamy virus in these cultures, ultrathin sections obtained from cells presenting extensive CPE were analyzed by electron microscopy (EM). For that purpose, co-cultures of horse PBMC and human U373MG cells were fixed *in situ* with 1.6% glutaraldehyde (Taab Laboratory Equipment Ltd, Reading, UK) in 0.1 *M* Sörensen phosphate buffer (pH 7.3–7.4) for 1 h at 4°C. Cells were scraped from the plastic substratum and centrifuged at 840*g* for 15 min. The resulting pellet was postfixed with 2% aqueous osmium tetroxide for 1 h at room temperature, dehydrated in ethanol, and embedded in Epon. Ultrathin sections were collected on 200-mesh copper grids coated with Formvar and carbon and stained with uranyl acetate and lead citrate prior to being observed with a Philips 400 transmission electron microscope (80 kV) at a magnification of ×28,000 to ×36,000 ***(9)***.

These sections reveal the presence of viral particles of about 100 nm in diameter, coated with multiple spikes and harboring a clear central core, the ultrastructural features of known FVs (**Fig. 1B,C**). Moreover, assembled viral capsids are present in the cytoplasm, reminiscent of other FVs and type B/D retroviruses. Interestingly, budding images were only observed at the plasma membrane, in contrast to what has been reported for primate FVs, which bud mainly from internal cellular membranes ***(8,10)***.

3.3. Molecular Cloning of the Complete Provirus

After the confirmation of the presence of an FV in cell culture by EM, the next step is to isolate the entire provirus for further genomic analysis. For that purpose, two procedures can be followed. The first one generates a genomic DNA bank from partial fragmentation of total DNA of infected cells with integrated proviruses ***(11)***. An alternative method takes advantage of the presence of multiple copies of unintegrated viral DNA following infection with FVs, as is the case for lentiviruses. In the case of PFV-1, unintegrated viral DNA represents up to 50% of total viral DNA depending on the phases of infection ***(12)***. Therefore, this later approach may be preferred (*see* **Note 4**).

3.3.1. DNA Extraction From Infected Cells

Although this approach could be performed using total DNA, preliminary Hirt extraction of DNA was performed to enrich the preparation with unintegrated viral DNA ***(13)***. For this purpose:

1. Wash the monolayer of co-cultures presenting extensive CPE from a 25-cm^2 flask with PBS.
2. Add 2.5 mL of lysis buffer (10 m*M* Tris-HCl, pH 7.4; 10 m*M* ethylenediamine tetraacetic acid (EDTA), pH 8.0; 0.6% SDS) and gently swirl for a few seconds.
3. Transfer the cell lysate to 2-mL Eppendorf tubes and add NaCl to 1 *M*. After at least 8 h at 4°C, spin the lysate for 30 min at 14,000*g* at 4°C. The supernatant collected is enriched with unintegrated DNA.
4. Treat the supernatant with pancreatic RNAse (20 µg/mL) for 1 h at 37°C, and digest with proteinase K (100 µg/mL) for 3 h at 50°C.
5. Purify the DNA by phenol/chloroform extractions, ethanol precipitation, and centrifugation according to standard methods ***(14)***. Dissolve the pellet in 60 µL of TE (10 m*M* Tris-HCl, pH 8.0; 10 m*M* EDTA, pH 8.0).

3.3.2. Analysis of Viral DNA

In a preliminary experiment, native Hirt supernatant DNA extracted from infected co-cultures was analyzed by Southern blotting following electrophoresis in a 1.1% agarose gel and transfer onto a nitrocellulose membrane following standard procedures ***(14)***. Membrane hybridization was performed at low stringency in standard buffer with 50% formamide at 37°C, and washes were done in a solution containing 0.1% SSC, 0.1% SDS, at room temperature. The full-length PFV-1 genome was used as a probe labeled with α^{32}PdCTP (3000 Ci/mmol) using the Prime a Gene labeling kit. Autoradiography revealed a weak band at 12 kb, whereas no signal was visualized in uninfected U373-MG cells (data not shown).

In all FVs sequenced so far, the primer binding site (*pbs*) is complementary to the 3' end of the cellular tRNA1,2 Lys, which is used as a primer for minus-strand DNA synthesis during reverse transcription (*see* **Note 5**) ***(4)***. One important feature of FVs (shared with lentiviruses) is the dual initiation of plus-strand DNA synthesis being primed at the conventional 3' long terminal repeat (LTR) polypurine tract (PPT) and also at a central PPT sequence located at the 3' end of *pol*. This mode of replication results in the formation of gapped linear DNA duplex intermediates ***(15)***. To determine the structure of the viral genome present in horse PBMC and human U373MG co-cultures, Hirt supernatant DNA was first denatured in 1 *M* glyoxal, 50% (v/v) DMSO, 10 m*M* sodium phosphate, pH 7.0 at 50°C for 1 h (for a final volume of 50 μL) ***(16)***, electrophoretically separated on a 1.1% agarose gel and analyzed by Southern blotting. Following membrane transfer, hybridization was performed with strand-specific oligonucleotide probes of the 18-nt *pbs* sequence, either in a direct (*pbs* +, [5'-TGGCGCCCAACGTGGGGC-3']) or reverse (*pbs* -, [5'-GCCCCACGTTGGGCGCCA-3']) orientation. Radio-labeling of the probes was performed using the T4 polynucleotide kinase in the presence of γ^{32}PATP (3000 Ci/mmol) following standard procedures ***(14)***. For control, glyoxal denatured PFV-1 DNA was run in parallel. As shown in **Fig. 3**, profiles are similar in both samples: the *pbs*(+) probe revealed a single band at 12 kb; with the *pbs*(–) probe, two bands of 7 and 1.4 kb were detected. By analogy to PFV-1, these three bands correspond to the full-length strand, the 5' half part of the viral genome, and the LTR, respectively as shown in the schematic representation of the unintegrated proviral DNA (**Fig. 3**). These results indicate that both viral genomes are comparable in size and that the positive strand of the unintegrated viral genome found in horse PBMC is gapped as previously shown for PFV-1.

Altogether, immunological cross-reactivity with PFV-1, electron microscopy observations, and structural analysis of the viral genome demonstrated that the virus isolated from seropositive horses belongs to the *Spumavirinae* subfamily and was therefore named equine foamy virus, EFV.

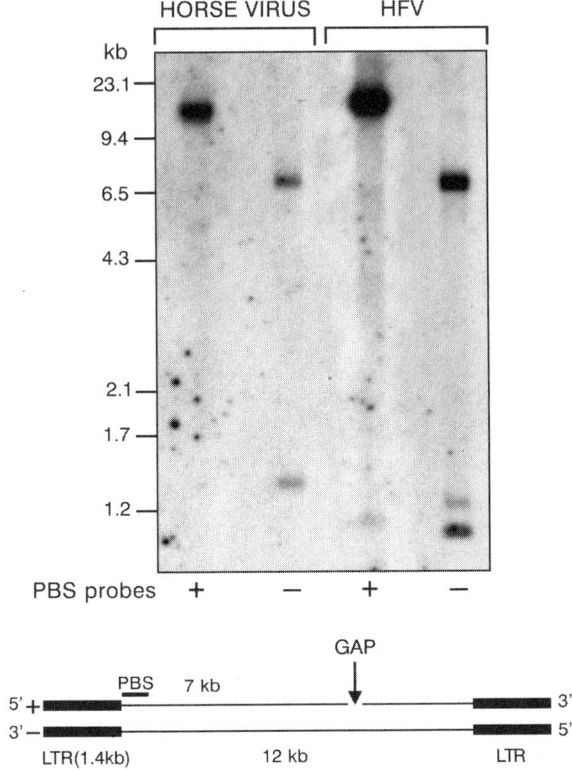

Fig. 3. Analysis of glyoxal denatured viral DNA by Southern blotting. The conserved *pbs* (+) and *pbs* (–) sequences were used as probes and revealed that the viral genome found in horse peripheral blood mononuclear cells is also gapped on the positive strand. Schematic representation of the viral unintegrated double-strand DNA harboring a gap on the positive strand.

3.3.3. Cloning of Viral DNA in λEMBL3

To gain insight into the genomic structure of EFV, viral DNA from Hirt supernatants of acutely infected U373-MG cells was cloned in the λEMBL3 phage after addition of *Bam*HI-*Xmn*I adaptors, using the Gigapack III Gold cloning kit (Stratagene). For this purpose,

1. Anneal nonpalindromic single-stranded adaptors (50 μ*M*), *Xmn*I [d(CGAACCCTTGG)] and *Bam*HI-*Xmn*I [d(GATCCGAAGGGGTTCG)] v/v by heating in a 95°C water bath for 5 min and slowly cooling to room temperature for 1 h.
2. Ligate 5 μL of the annealed product to 30 μL of Hirt supernatant DNA using 3 U of T4 DNA ligase in a 100 μL final volume (adjust with water). After overnight incubation at 12°C, inactivate the enzyme at 65°C for 10 min (*see* **Note 6**).

3. Pass the entire mixture over a Sephacryl 300 spin column (Pharmacia) to remove excess nonligated adaptors as described by the manufacturer. Reduce the effluent elution collected to 3 µL using speed vacuum.
4. Perform the ligation of 3 µL of adapted DNA with 1 µg of λEMBL3 vector arms using 1.5 U of T4 DNA ligase at 4°C for 48–72 h. All the subsequent steps are strictly performed according to the manufacturer's instructions. Use the Gigapack III packaging extract to package the ligation mixture and titer on *E. coli* LE 392 (P2).
5. Plate the library onto large 220-mm agar plates and lift duplicate nitrocellulose filters from each plate.
6. Following the preparation of the membranes for hybridization, perform hybridization, probe preparation and washes for the *pbs+* oligonucleotide probe, then expose the membranes to film. Perform secondary and tertiary screenings also outlined in the standard methodology texts *(14)*.

By this method, 10 positive clones were isolated from a total of approx 10,000 phage plaques. λ-Recombinant DNAs were purified using the Wizard Lambda Preps Kit (Promega).

Digestion with *Bam*HI was performed to excise the insert between the phage arms as advised by the manufacturer. The presence of a 7-kb fragment, which can be detected in Hirt supernantant DNA from EFV-infected cells, attests that at least part of a provirus is cloned. Clones 1 and 2 harbor this band, whereas only clone 1 contains apparently a full-length viral genome since the summation of the fragments generated by the *Bam*HI digestion leads to an insert of approx 12 kb (**Fig. 4**), a size comparable to that of other cloned FV proviruses. Clone 1 was entirely sequenced in both strand directions and critical regions were sequenced on both strands of another full-length clone, leading to the schematic representation of the provirus (**Fig. 5**) *(8)*.

Because sequence homologies were reported between FVs and human endogenous retroviruses *(17)*, a supplementary experiment was performed to verify that the viral products detected were not of endogenous origin. For that purpose, genomic DNA (which can be obtained by different methods using commercial kits) from uninfected horse ED adherent fibroblastic cells and from EFV-infected equine dermal cells digested with several enzymes (*Bam*HI, *Blg*II, *Xmn*I, *Hind*III, *Sal*I) was analyzed by Southern blotting using a probe specific to EFV (for example, the viral LTR generated by PCR *(8)*. Only genomic DNA extracted from infected cells gave rise to positive signals; those sizes are in agreement with sequence analysis of the EFV provirus (**Fig. 6**).

4. Notes

1. Other methods can be used to assess the presence of specific antibodies in horse sera, such as radio immunoprecipitation assay or Western blot, as already described *(8)*.

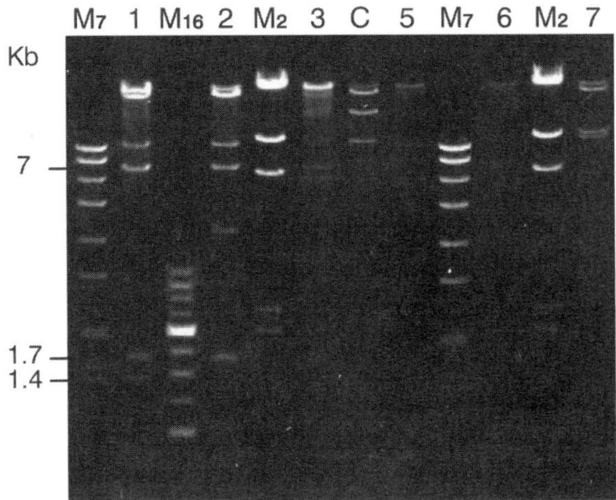

Fig. 4. Ethidium bromide stained 0.7% agarose gel of six recombinant λEMBL3 clones digested with *Bam*HI. The numbers indicate the numerical designation of each clone. Lane C: λ-recombinant clone of a 12-kb test insert provided by the manufacturer and used as internal markers of λEMBL3 arms. Lanes M2, M7, and M16 represent DNA molecular weight markers (Roche). Clones 1 and 2 contain a 7-kb *Bam*HI fragment, while clone 1 alone contains a full-length provirus with the 7-kb band, a doublet at 1.7 kb, and a 1.4-kb fragment. This leads to an insert of approx 12 kb, which is the average size of known foamy virus provirus.

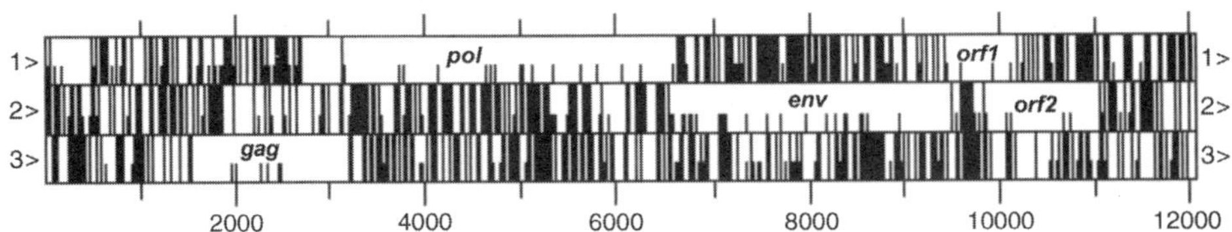

Fig. 5. Schematic representation of predicted open reading frames (ORFs) of the equine foamy virus provirus. *Orf1* encodes the viral transactivator also called Tas.

2. While little is known about the tropism of FVs in their natural hosts, at least circulating lymphocytes harbor the provirus *(5)*. In contrast, in vitro these retroviruses infect a very wide range of cell lines, demonstrating the ubiquity of the receptor. Generally, two permissive cell lines are used to produce viral stocks, the human U373-MG glioblastoma cell line and the hamster BHK21 cells. Avoid the use of HeLa or COS6 cells, which are not good producers of viruses ***(18)***.
3. Most primate FV particles bud from intracellular membranes (although infectious viruses can also be recovered from the supernatant to a lesser extent), which derive either from cell lysis or a few budding events occurring at the plasma membrane. Therefore, viral stocks are produced following three cycles of freez-

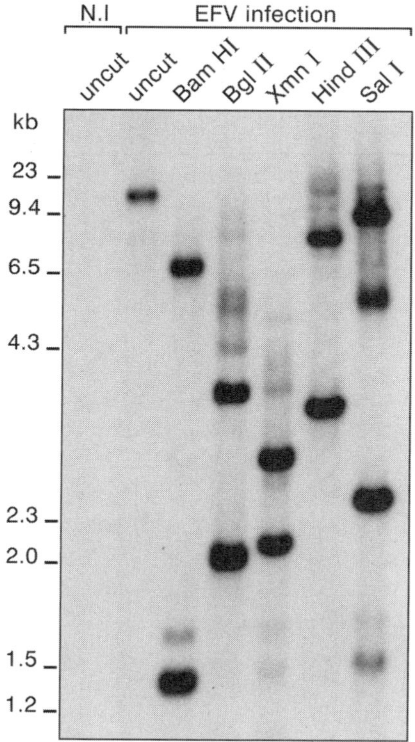

Fig. 6. Southern blot analysis of total DNA or Hirt supernatant DNA from mock-infected (N.I) or equine foamy virus (EFV)-infected equine dermal cells. The EFV long terminal repeat, used as a probe, revealed that the viral genome is absent from uninfected horse cells, thus demonstrating that EFV is an exogenous virus.

ing and thawing of infected cells together with the corresponding supernatant, to free trapped infectious virions from cells. In contrast, in the case of EFV, virus budding mainly occurs at the plasma membrane ***(10)***. This is owing to the lack of a dilysine motif present in the C-terminus of the primate Env glycoprotein, which allows the recycling of this viral protein between the Golgi network and the endoplasmic reticulum and is likely responsible for the budding of primate FVs from intracellular membranes ***(19,20)***. In that case, only the supernatant can be collected to produce virus stocks.

4. As already reported, acute infection by FVs or lentiviruses leads to the production of high amounts of unintegrated viral DNA. A Hirt extraction will allow the enrichment of the DNA preparation with viral DNA ***(12)***. This approach was successfully employed to clone the bovine foamy virus ***(21)***.
5. The detection of viral DNA in these preparations was initially performed using the entire PFV-1 genome as a probe. However, the use of the 18-bp *pbs* probe instead of the entire PFV-1 genome gives strong and specific signals on DNA

from infected cells.

6. Direct cloning of unintegrated viral DNA without a preliminary step of enzymatic blunting suggests that a sizeable proportion of the linear EFV DNA harbors blunt ends as already reported *(21)*. Note that only 9–20 kb insert will give rise to viable phage.

Acknowledgments

We would like to thank the LPH for the photographic work. This work was supported by ARC, *Ensemble Contre le SIDA*, ANRS, and F. Lacoste. We thank Claudine Pique for critical reading of the manuscript.

References

1. Meiering, C. D. and Linial, M. L. (2001) Historical perspective of foamy virus epidemiology and infection. *Clin. Microbiol. Rev.* **14,** 165–176.
2. Fischer, N., Heinkelein, M., Lindemann, D., Enssle, J., Baum, C., Werder, E., et al. (1998) Foamy virus particle formation. *J. Virol.* **72,** 1610–1615.
3. Lecellier, C. H. and Saib, A. (2000) Foamy viruses: between retroviruses and pararetroviruses. *Virology* **271,** 1–8.
4. Saib, A., Peries, J., and de The, H. (1995) Recent insights into the biology of the human foamy virus. *Trends Microbiol.* **3,** 173–178.
5. von Laer, D., Neumann-Haefelin, D., Heeney, J. L., and Schweizer, M. (1996) Lymphocytes are the major reservoir for foamy viruses in peripheral blood. *Virology* **221,** 240–244.
6. Heneine, W., Switzer, W. M., Sandstrom, P., Brown, J., Vedapuri, S., Schable, C. A., et al. (1998) Identification of a human population infected with simian foamy viruses [see comments]. *Nat. Med.* **4,** 403–407.
7. Saïb, A. (2003) Non-primate foamy viruses. *Curr. Top Microbiol. Immunol.* **227,** 197–211.
8. Tobaly-Tapiero, J., Bittoun, P., Neves, M., Guillemin, M. C., Lecellier, C. H., Puvion-Dutilleul, F., et al. (2000) Isolation and characterization of an equine foamy virus. *J. Virol.* **74,** 4064–4073.
9. Saib, A., Puvion-Dutilleul, F., Schmid, M., Peries, J., and de The, H. (1997) Nuclear targeting of incoming human foamy virus Gag proteins involves a centriolar step. *J. Virol.* **71,** 1155–1161.
10. Lecellier, C. H., Neves, M., Giron, M. L., Tobaly-Tapiero, J., and Saib, A. (2002) Further characterization of equine foamy virus reveals unusual features among the foamy viruses. *J. Virol.* **76,** 7220–7227.
11. Saib, A., Peries, J., and de The, H. (1993) A defective human foamy provirus generated by pregenome splicing. *EMBO J* . **12,** 4439–4444.
12. Delelis, O., Saib, A., and Sonigo, P. (2003) Biphasic DNA synthesis in spumavirus. *J. Virol.* **77,** 8141–8146.
13. Hirt, B. (1967) Selective extraction of polyoma DNA from infected mouse cell cultures. *J. Mol. Biol.* **26,** 365–369.

14. Sambrook, J., Fritsch, E. F., and Maniatis, T. (1989) *Molecular Cloning: A Laboratory Manual.* 2nd ed., Cold Spring Harbor Laboratory Press, Cold Spring Harbor, NY.
15. Kupiec, J. J. and Sonigo, P. (1996) Reverse transcriptase jumps and gaps. *J. Gen. Virol.* **77,** 1987–1991.
16. Thomas, P. S. (1983) Hybridization of denatured RNA transferred or dotted to nitrocellulose paper. *Methods Enzymol.* **100B,** 255–266.
17. Cordonnier, A., Casella, J. F., and Heidmann, T. (1995) Isolation of novel human endogenous retrovirus-like elements with foamy virus-related pol sequence. *J. Virol.* **69,** 5890–5897.
18. Hill, C. L., Bieniasz, P. D., and McClure, M. O. (1999) Properties of human foamy virus relevant to its development as a vector for gene therapy. *J. Gen. Virol.* **80,** 2003–2009.
19. Goepfert, P. A., Shaw, K., Wang, G., Bansal, A., Edwards, B. H., and Mulligan, M. J. (1999) An endoplasmic reticulum retrieval signal partitions human foamy virus maturation to intracytoplasmic membranes. *J. Virol.* **73,** 7210–7217.
20. Goepfert, P. A., Shaw, K. L., Ritter, G. D., Jr., and Mulligan, M. J. (1997) A sorting motif localizes the foamy virus glycoprotein to the endoplasmic reticulum. *J. Virol.* **71,** 778–784.
21. Renshaw, R. W., Gonda, M. A., and Casey, J. W. (1991) Structure and transcriptional status of bovine syncytial virus in cytopathic infections. *Gene* **105,** 179–184.

11

Alu-LTR Real-Time Nested PCR Assay for Quantifying Integrated HIV-1 DNA

Audrey Brussel, Olivier Delelis, and Pierre Sonigo

Summary

An improved *Alu*-long terminal repeat (LTR) polymerase chain reaction (PCR) assay is described for the quantification of integrated HIV-1 DNA in infected cells. The method includes generation of an infected cell line containing numerous randomly distributed HIV-1 integrated DNA for the construction of the DNA standard and a two-step real-time PCR assay in which the first-round PCR amplifies the DNA sequence between the HIV-1 LTR and the nearest chromosomal *Alu* element, and the nested PCR specifically amplifies PCR products from the first-round PCR. This assay allows us to quantify proviral DNA with both accuracy and high sensitivity (six proviruses within 50,000 cell equivalents) and exhibits a broad range of quantification spanning 5 $\log_{10}$ provirus copies. This *Alu*-LTR-based real-time nested PCR assay may be particularly useful to quantify integrated HIV-1 DNA in patients. It may also allow for the precise study of integration of HIV-1 DNA or HIV-1 based lentiviral vectors and may be a valuable tool to test future inhibitors of integration.

Key Words: HIV-1; viral DNA integration; provirus; *Alu* element; *Alu*-LTR PCR; real-time PCR.

1. Introduction

Quantitative and sensitive analysis of integrated HIV-1 DNA has long been hampered by the lack of a reliable assay, especially because the sequence of the integrated HIV-1 DNA is indistinguishable from its unintegrated linear form. Previous studies have attempted to develop methods for quantifying integrated viral DNA by exploiting the occurrence of *Alu* sequences that are interspersed and highly repeated within the human genome ***(1,2)***. These methods relied on a two-step polymerase chain reaction (PCR) assay named *Alu*-long

From: *Methods in Molecular Biology, Vol. 304: Human Retrovirus Protocols: Virology and Molecular Biology*
Edited by: T. Zhu © Humana Press Inc., Totowa, NJ

terminal repeat (LTR) PCR *(3–7)*. These *Alu*-LTR PCR assays comprise a first round of PCR in which the DNA sequence between the HIV-1 LTR and the nearest chromosomal *Alu* element is amplified and a second round of PCR consisting of the nested amplification of the LTR region to increase detection sensitivity. However, these *Alu*-LTR PCR assays still had limited sensitivity and were not fully quantitative. In particular, they used DNA standards prepared from infected cell lines in which proviruses are integrated at specific, fixed distances from the *Alu* repeats whereas, in the context of a natural HIV-1 infection, the distance between HIV-1 integration site and *Alu* sequences is variable. In a setting where the development of antiretroviral treatments targeting the HIV-1 proviral latent reservoir in patients is needed, a new quantitative and sensitive integrated HIV-1 DNA assay would be particularly useful. Herein, some improvements of the *Alu*-LTR PCR strategy are described, including (1) the generation of an infected cell line containing numerous randomly distributed HIV-1 proviruses to construct the DNA standard and (2) modifications of the amplification procedures in order to prevent amplification of nonpreamplified DNA during the second round of PCR and to allow the real-time fluorescence-based detection of PCR products using the LightCycler thermal cycler *(8)*. This *Alu*-LTR-based real-time nested PCR assay allows us to quantify integrated HIV-1 DNA with high sensitivity (six proviruses within 50,000 cell equivalents) and displays a wide range of quantification that spans 5 $\log_{10}$ provirus copies.

2. Materials

1. Dulbecco's modified Eagle medium (DMEM) supplemented with 10% fetal calf serum (FCS), 100 U/mL penicillin, and 100 µg/mL streptomycin.
2. Phosphate-buffered saline (PBS).
3. HEK 293 cell line (ATCC number: CRL-1573) *(9)*. Material classified in biosafety level 2 that should be handled under biosafety level 2 guidelines.
4. pR7 Neo Δ*env* vector (*see* **Subheading 3.1.1.**). Material classified in biosafety level 2 that should be handled under biosafety level 2 guidelines.
5. pVSV-G vector (*see* **Subheading 3.1.1.**).
6. 1X HEPES-buffered saline (HBS).
7. 2 *M* $CaCl_2$.
8. 0.45-µm pore-size filter.
9. HeLa cell line (ATCC number: CCL-2) *(10)*. Material classified in biosafety level 2 that should be handled under biosafety level 2 guidelines.
10. Trypsin.
11. G418.
12. Paraformaldehyde (PAF).
13. Giemsa.
14. Aerosol-barrier tips.

15. 20 μ*M* stock solutions of fluorescent oligonucleotide probes (store at 4°C for months or at –20°C for longer storage period; protect from light).
16. 50 μ*M* stock solutions of oligonucleotide primers (store at 4°C for months or at –20°C for longer storage period).
17. LightCycler instrument (Roche Diagnostics, Meylan, France).
18. LightCycler capillaries (Roche Diagnostics).
19. LightCycler cooling block (Roche Diagnostics).
20. LightCycler FastStart DNA hybridization probes (Roche Diagnostics).
21. *Optional:* QIAamp blood DNA minikit (Qiagen, Courtaboeuf, France).
22. *Optional:* Control kit DNA (Roche Diagnostics).

3. Methods

The methods described below outline (1) the construction of the integrated HIV-1 DNA standard and (2) the *Alu*-LTR-based real-time nested PCR assay.

3.1. Construction of Integrated HIV-1 DNA Standard

To generate a proper integrated HIV-1 DNA standard containing integration sites with a wide distribution of distances between the provirus LTR and the nearest *Alu* sequence, we report the preparation of DNA from an infected cell line containing numerous proviruses as a result of a viral infection by a replication-incompetent virus carrying an antibiotic resistance gene.The steps required for the generation of an accurate integrated HIV-1 DNA standard and its characterization are described in **Subheadings 3.1.1.–3.1.6.** (**Fig. 1**). This includes (1) the description of the pR7 Neo Δ*env* and pVSV-G vectors, (2) the production of VSV-G pseudotyped HIV-1 R7 Neo virus stocks by calcium phosphate cotransfection assay, (3) the infection of HeLa cells with the VSV-G pseudotyped HIV-1 R7 Neo virus, (4) the selection of neomycin-resistant cell clones, (5) the preparation of cell DNA from the neomycin-resistant cell line, and (6) the determination of integrated HIV-1 DNA copy number within the DNA standard.

Note that steps described in **Subheadings 3.1.2.–3.1.5.** must be carried out under biosafety level-3 guidelines.

3.1.1. pR7 Neo Δenv and pVSV-G Vectors

The pR7 Neo Δ*env* and pVSV-G vectors were kindly provided by Professor U. Hazan ***(11)***. The pR7 Neo Δ*env* vector is derived from the pR7 Neo vector ***(12)*** and carries an envelope-deleted HIV-1 R7 genome. The pR7 Neo Δ*env* vector also contains a neomycin-resistance gene that replaces the *nef* coding sequence so that the neomycin-resistance gene expression is driven by the viral promoter within the 5' LTR. The pVSV-G vector encodes the G glycoprotein from the vesicular stomatitis virus (VSV-G), whose expression is driven by the cytomegalovirus immediate-early promoter.

Production of VSV-G pseudotyped HIV-1 R7 Neo virus stocks

↓

Infection of HeLa cells with VSV-G pseudotyped HIV-1 R7 Neo virus

↓

Selection of neomycin resistant cell clones

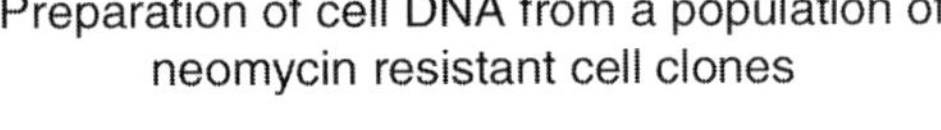

Preparation of cell DNA from a population of neomycin resistant cell clones

Determination of integrated HIV-1 DNA copy number within cell DNA

Fig. 1. Steps required for the generation of an integrated HIV-1 DNA standard.

3.1.2. Production of VSV-G Pseudotyped HIV-1 R7 Neo Virus Stocks by Calcium Phosphate Cotransfection Assay

1. Plate HEK 293 cells in a 10-cm^2 Petri culture dish in DMEM supplemented with 10% FCS, 100 U/mL penicillin, and 100 µg/mL streptomycin. Culture cells at 37°C in an incubator with a humidified atmosphere containing 5% CO_2. Use cells that are 50% confluent on the day of the experiment.
2. Replace the medium approx 8 h prior to the transfection assay using 37°C prewarmed medium.
3. Prepare the transfection mixture. Mix 500 µL of 1X HBS with 10 µL of pVSV-G vector (1 µg/µL) and 10 µL of pR7 Neo Δ*env* vector (1 µg/µL). Add 32 µL of 2 *M* $CaCl_2$ drop by drop while gently vortexing the mixture. Incubate at room temperature for 20–30 min.
4. Add the transfection mixture to HEK 293 cells and swirl the dishes to ensure dispersal. Quickly return the cells to the incubator. Incubate cells overnight.
5. Remove medium and add 5 mL of 37°C prewarmed medium to the cells and return the cells to the incubator for 1 d.
6. Collect supernatant containing VSV-G pseudotyped HIV-1 R7 Neo viruses and filter the supernatant through a 0.45-µm pore-size filter. Freeze viral supernatants at –80°C until use.

3.1.3. Infection of HeLa Cells With VSV-G Pseudotyped HIV-1 R7 Neo Virus

1. Plate HeLa cells into four T-75 flasks in 10 mL DMEM supplemented with 10% fetal calf serum, 100 U/mL penicillin, and 100 µg/mL streptomycin. Culture cells at 37°C in an incubator with a humidified atmosphere containing 5% CO_2. On the day of the experiment, infect cells that are 60% confluent.
2. Thaw VSV-G pseudotyped HIV-1 R7 Neo virus stock in a 37°C water bath.
3. Dilute 25-fold, 125-fold, and 625-fold the VSV-G pseudotyped HIV-1 R7 Neo virus stock in 2 mL of medium (*see* **Note 1**).
4. Add 2 mL of diluted viral supernatant to the cells and return the cells to the incubator for 4 h. Then add 8 mL of medium and return the cells to the incubator for 24 h to allow for G418 resistance gene expression. As a control, perform the same experiment without viral supernatant.

3.1.4. Neomycin-Resistant Cell Clone Selection

1. Trypsinize infected and uninfected HeLa cells and replate them into T-162 flasks in 20 mL medium supplemented with 500 µg/mL G418 to select cells that contain integrated viral DNA.
2. Replace medium every 2 d using fresh medium supplemented with 500 µg/mL G418. Culture the cells as long as viable adherent cells are present within the control flask containing uninfected cells (10 to 14 d are necessary). Trypsinize and split the cells when needed (do not split cells more than one-fifth to preserve cell population heterogeneity).
3. Determine the number of independent cell clones that carry at least one provirus by counting G418 resistant cell clones when G418 selection is achieved. Select a flask that has not been trypsinized since **step 2** and that contains well-separated cell clones. Fix cells with 4% PAF in PBS for 10 min and stain them with 10% Giemsa in PBS for 15 min at room temperature before rinsing in tap water and then perform counting. Determine the number of neomycin-resistant independent cell clones within the HeLa cell population infected by the 25-fold dilution of viral supernatant by multiplying the number of counted clones by the appropriate dilution factor. Neomycin-resistant cell clones within the whole cell population should number at least 1000 clones.
4. Culture the HeLa cell population infected by the 25-fold dilution of viral supernatant again for at least 3 wk in the presence of G418 to dilute away all unintegrated forms of viral DNA. The obtained cell line will be used to generate the integrated HIV-1 DNA standard.

3.1.5. Preparation of Cell DNA From Neomycin-Resistant Cell Line

DNA from 5×10^6 cells of the neomycin-resistant cell line can be prepared using standard isolation techniques (*see* **Note 2**) ***(13)***. Store DNA in aliquots at –20°C until use (avoid repeated thawing and freezing).

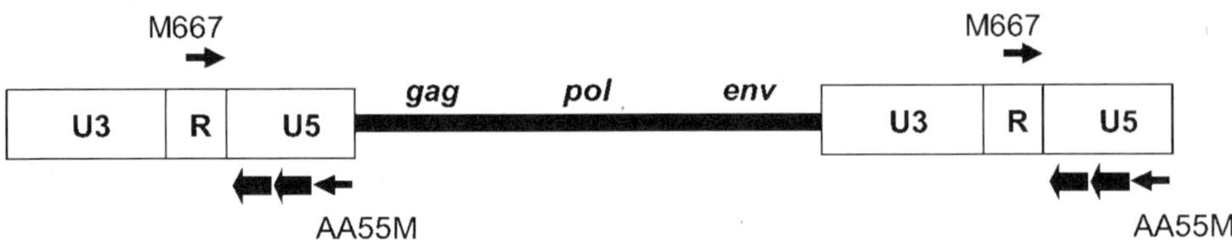

Fig. 2. HIV-1 genome and location of primers and probes to quantify integrated HIV-1 DNA within the DNA standard. Based on total HIV-1 DNA quantification, it is possible to estimate the proviral DNA content of the neomycin-resistant cell line. Total HIV-1 DNA copy number is determined by quantification of viral DNA molecules by real-time polymerase chain reaction (PCR) using primers annealing in the R and in the U5 regions of the long terminal repeat (LTR). Thin black arrows indicate primers and large black arrows indicate probes.

Nevertheless, we recommend the purification of cellular DNA by absorption onto a silica-gel membrane to avoid use of phenol/chloroform that may inhibit subsequent PCR reactions. For instance, prepare cell DNA using the QIAamp DNA blood mini kit (Qiagen) according to the manufacturer's instructions. Elute DNA with 100 µL buffer AE following a 5-min incubation step onto the silica-gel membrane.

3.1.6. Determination of Integrated HIV-1 DNA Copy Number Within Standard DNA

When all unintegrated forms of viral DNA are lost, the neomycin-resistant cell line contains a number of integrated viral DNA that matched with the total HIV-1 DNA copy number. Thus, based on total HIV-1 DNA quantification, it is possible to estimate the proviral DNA content per cell. Total HIV-1 DNA copy number is determined by quantification of viral DNA molecules by real-time PCR using primers annealing in the R and in the U5 regions of the LTR (**Fig. 2**). Herein we present a protocol using the LightCycler instrument and the hybridization probe format ***(14)*** for the real-time detection of PCR products (*see* **Note 3**).

1. Prepare fresh 10-fold serial dilutions of the pR7 Neo Δ*env* vector to obtain final concentrations ranging from 5×10^0 to 5×10^5 copies/mL.
2. Thaw an aliquot of the integrated HIV-1 DNA standard at room temperature.
3. Place 10 LightCycler capillaries in a 4°C precooled LightCycler cooling block.
4. Prepare a quantity of PCR master mix corresponding to 11 reactions (protect the master mix from prolonged exposure to light). Primer and probe sequences are given in **Table 1**. Add items from **Table 2** in the order shown for one 20-µL amplification reaction (*see* **Note 4**).
5. Mix gently.

Table 1
Primer and Probe Sequences

Primer or probe name	Sequence
AA55M	5'-GCT AGA GAT TTT CCA CAC TGA CTA A-3'
M667	5'-GGC TAA CTA GGG AAC CCA CTG-3'
LTR FL*	5'-CAC AAC AGA CGG GCA CAC ACT ACT TGA-3'
LTR LC*	5'-CAC TCA AGG CAA GCT TTA TTG AGG C-3'
M667–L	5'-ATG CCA CGT AAG CGA AAC TCT GGC TAA CTA GGG AAC CCA CTG-3'
Alu 1	5'-TCC CAG CTA CTG GGG AGG CTG AGG-3'
Alu 2	5'-GCC TCC CAA AGT GCT GGG ATT ACA G-3'
Lambda T	5'-ATG CCA CGT AAG CGA AAC T-3'

*Indicates a probe sequence. The long terminal repeat (LTR) FL probe is modified with fluorescein at the 3' end and the LTR LC probe is phosphorylated at the 3' end and modified with LCred640 dye at the 5' end. The lambda phage specific heel sequence of the L-M667 primer is underlined.

Table 2
Add in the Following Order for One 20-μL Amplification Reaction (*see* Note 4)

Volume	Component (initial concentration)	Final concentration
12.96 μL	H_2O sterile, PCR grade*	
2.4 μL	$MgCl_2$*	4 m*M*
0.12 μL	Primer AA55M (50 μ*M*)	300 n*M*
0.12 μL	Primer M667 (50 μ*M*)	300 n*M*
2 μL	Reconstituted LightCycler Fast Start DNA Master Hybridization probes (10X)*	1X
0.2 μL	Fluorogenic hybridization probe LTR FL (20 μ*M*)	200 n*M*
0.2 μL	Fluorogenic hybridization probe LTR LC (20 μ*M*)	200 n*M*
18 μL	Total volume	

*Supplied with the LightCycler Fast Start DNA Master Hybridization probe kit (Roche Diagnostics).

6. Pipet 18 μL of master mix into the precooled LightCycler capillaries.
7. Add 2 μL of each dilution of the pR7 Neo Δ*env* vector into the capillaries.
8. Add 2 μL of the integrated HIV-1 DNA standard in triplicate into the capillaries.
9. Add 2 μL of sterile H_2O as a negative control into one capillary.
10. Seal each capillary with a stopper and place the adapters containing the capillary into a standard benchtop microcentrifuge.
11. Centrifuge at 700*g* for 5 s.

12. Place the capillaries into the rotor of the LightCycler instrument.
13. Program the LightCycler instrument using the parameters given below:
    a. 95°C for 8 min to denature DNA and activate the FastStart DNA polymerase.
    b. Amplify target DNA by performing 50 cycles as follows:
       - 95°C for 10 s to denature DNA.
       - 60°C for 10 s to anneal the primers and probes. **Fluorescence acquisition must be performed at the end of this step.**
       - 72°C for 6 s to extend the annealed primers.
    c. 40°C for 30 s to cool the thermal chamber.
14. Cycle standards and samples.
15. Use the LightCycler quantification software supplied with the LightCycler instrument to display fluorescence signals corresponding to the quotient of the channel F2 over channel F1 (**Fig. 3A**). Generate a linear standard curve by plotting the crossing cycle number versus the logarithm of the initial copy number for each dilution of the pR7 Neo Δ*env* vector (**Fig. 3B**). This standard curve allows us to interpolate the proviral copy number in each integrated DNA standard replicate from its own crossing cycle number. Calculate the actual copy number of integrated HIV-1 DNA in 2 μL of the DNA standard by averaging the copy numbers of the replicates.
16. Calculate cell equivalents in HIV-1 integrated DNA standard to estimate the proviral DNA content per cell. This could be performed by optical density at 260 nm or, to obtain a more reliable quantification, by real-time quantitative PCR using primers annealing in a chromosomal gene of known copy number within the human genome. For instance, cell equivalents can be calculated based on the amplification of the β globin gene (two copies per diploid cell) using the Control kit DNA (Roche Diagnostics) with the LightCycler Instrument according to the manufacturer's instructions.

3.2. Alu-*LTR*-Based Real-Time Nested PCR Assay

The assay for quantifying integrated HIV-1 DNA relies on a two-step PCR protocol (**Fig. 4**). The first-round PCR uses primers annealing in the 3' LTR and in conserved regions of the *Alu* element. To increase assay sensitivity, a nested PCR is then performed to specifically amplify PCR products from the first-round PCR. In contrast to the first-round PCR that generates fragments of varying lengths, the nested amplification results in discrete DNA fragments that are detected using the hybridization probe format. **Subheadings 3.2.1.–3.2.2.** describe the procedures for performing (1) the first-round PCR and (2) the second-round PCR.

3.2.1. First-Round PCR Amplification

1. Prepare DNA from infected cells as described in **Subheading 3.1.5.** Note that no more than six DNA samples can be quantified simultaneously due to limitation in the number of PCR reactions that can be cycled at once by the LightCycler instrument.

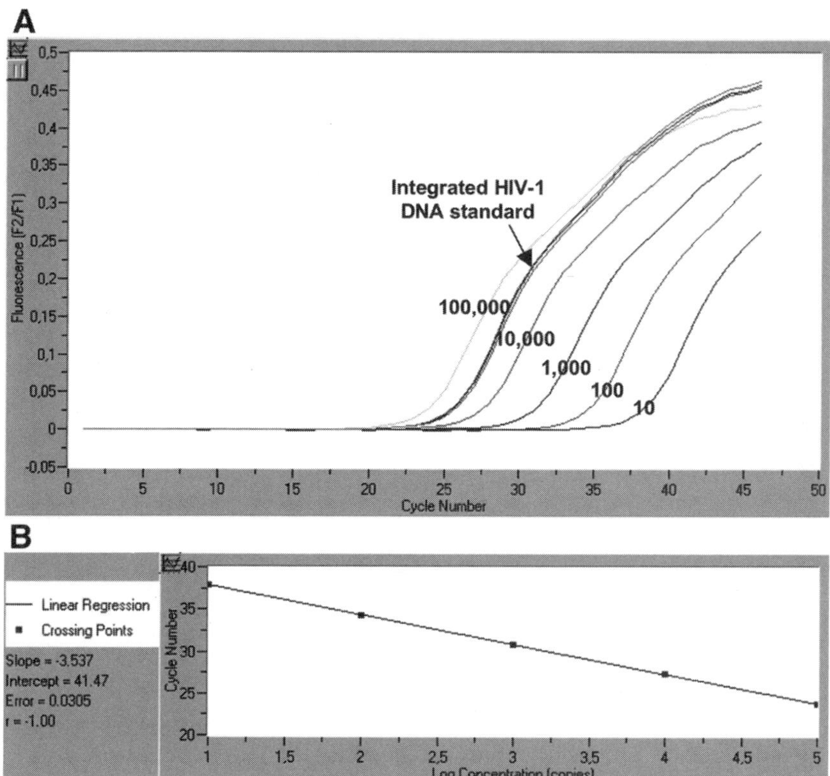

Fig. 3. Generation of a linear standard curve to quantify integrated HIV-1 DNA copy number within the DNA standard. **(A)** Fluorescence curves corresponding to the quotient of the channel F2 over channel F1 generated by amplification of the integrated HIV-1 DNA standard together with serial dilutions of the pR7 Neo Δ*env* vector. The known copy number of each dilution of the pR7 Neo Δ*env* vector is shown over the corresponding fluorescence curve. **(B)** Linear standard curve obtained by plotting the crossing cycle number versus the logarithm of the initial copy number for each dilution of the pR7 Neo Δ*env* vector.

2. Prepare DNA from uninfected cells as described in **Subheading 3.1.5.**
3. Prepare fresh 10-fold serial dilutions of the integrated DNA standard in uninfected cell DNA to obtain final concentrations ranging from 3×10^0 to 3×10^4 integrated DNA copies/µL.
4. Place LightCycler capillaries in a 4°C precooled LightCycler cooling block.
5. Prepare a first PCR master mix containing two *Alu* primers (*see* **Note 5**) and an LTR primer extended at its 5' end with a lambda phage-specific heel sequence (*see* **Note 6**) (**Table 1**). Assemble enough master mix to amplify each dilution of the standard DNA, each DNA sample in duplicate, one negative control plus one extra reaction. Add in the following order for each 20-µL amplification reaction:

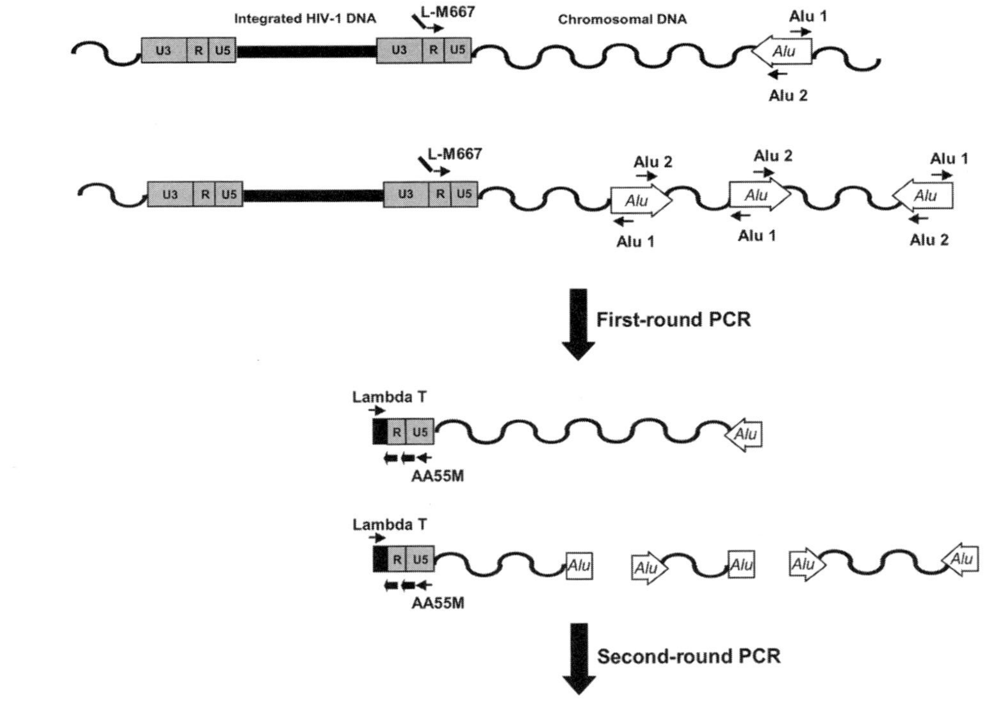

Fig. 4. Assay for quantifying integrated HIV-1 DNA using an *Alu*-long terminal repeat (LTR)-based real-time nested polymerase chain reaction (PCR) strategy. The first-round PCR uses an LTR primer extended at its 5' end with a lambda phage specific heel sequence and two outward-facing *Alu* primers annealing in conserved regions of the *Alu* element. The second-round PCR is carried out with the heel-specific primer and a reverse LTR-primer to prevent amplification of nonpreamplified DNA. In contrast with the first-round PCR that generates fragments of varying lengths, the nested amplification results in discrete DNA fragments that are detected using hybridization probes. Thin black arrows indicate primers and large black arrows indicate probes.

Volume	Component (initial concentration)	Final concentration
13.32 µL	Sterile H_2O, PCR grade[a]	
2.4 µL	$MgCl_2$[a]	4 m*M*
0.04 µL	Primer L-M667 (50 m*M*)	100 n*M*
0.12 µL	Primer *Alu* 1 (50 m*M*)	300 n*M*
0.12 µL	Primer *Alu* 2 (50 m*M*)	300 n*M*
2 µL	Reconstituted LightCycler FastStart DNA Master Hybridization probes (10×)[a]	1X
18 µL	Total volume	

[a]Supplied with the Light-Cycler FastStart DNA Master Hybridization probe kit (Roche Diagnostics).

6. Prepare a second PCR master mix that does not comprise *Alu* primers (*see* **Note 7**). Assemble enough master mix to amplify each DNA sample in duplicate, one negative control plus one extra reaction. Add in the following order for each 20 µL amplification reaction:

Volume	Component (initial concentration)	Final concentration
13.56 µL	Sterile H_2O, PCR grade[a]	
2.4 µL	$MgCl_2$[a]	4 m*M*
0.04 µL	Primer L-M667 (50 µ*M*)	100 n*M*
2 µL	Reconstituted LightCycler FastStart DNA Master Hybridization probes (10×)[a]	1X
18 µL	Total volume	

[a]Supplied with the Light-Cycler FastStart DNA Master Hybridization probe kit (Roche Diagnostics).

7. Mix each master mix gently.
8. Pipet 18 µL of the first master mix into the required number of precooled LightCycler capillaries.
9. Pipet 18 µL of the second master mix into the required number of precooled LightCycler capillaries.
10. Add 2 µL of the serial dilutions of the DNA standard into capillaries containing the first master mix.
11. Add 2 µL of each DNA sample in duplicate into capillaries containing the first master mix.
12. Add 2 µL of each DNA sample in duplicate into capillaries containing the second master mix.
13. Add 2 µL of uninfected cell DNA as negative controls into one capillary containing the first master mix and into one capillary containing the second master mix.
14. Seal each capillary with a stopper and place the adapters containing the capillary into a standard benchtop microcentrifuge.

15. Centrifuge at 700*g* for 5 s.
16. Place the capillaries into the rotor of the LightCycler instrument.
17. Program the LightCycler instrument using the parameters given below:
    a. 95°C for 8 min to denature DNA and activate the FastStart DNA polymerase.
    b. Amplify target DNA by performing 12 cycles as follows (*see* **Note 8**):
    - 95°C for 10 s to denature DNA.
    - 60°C for 10 s to anneal the primers (no fluorescence acquisition is required).
    - 72°C for 170 s to extend the annealed primers.

    c. 40°C for 30 s to cool the thermal chamber.
18. Cycle standards and samples.
19. Remove rotor from the thermal chamber and collect capillaries in an empty LightCycler capillary box (be careful to preserve order).
20. Remove stopper and turn each capillary inside out in a sterile 1.5-mL tube (take care not to cross-contaminate samples with each other; change gloves as often as necessary).
21. Place tubes into a standard benchtop microcentrifuge.
22. Centrifuge at 700*g* for 5 s.
23. Discard capillaries from the 1.5-mL tubes that now contain PCR products.
24. Prepare a 10-fold dilution of PCR products by adding 180 µL of sterile H_2O to each 1.5-µL tube.

3.2.2. Second-Round PCR Amplification

1. Place a number of LightCycler capillaries in a 4°C precooled LightCycler cooling block corresponding to the number of 10-fold diluted first-round PCR products.
2. Prepare a PCR master mix containing the heel specific primer (**Table 1**), the reverse LTR-primer AA55M, and the fluorogenic hybridization probes LTR FL and LTR LC. Assemble an amount of master mix corresponding to the number of 10-fold diluted first-round PCR products plus one extra reaction (protect the master mix from prolonged exposure to light). Add in the following order for each 20-µL PCR reaction:

Volume	Component (initial concentration)	Final concentration
12.96 µL	Sterile H_2O, PCR grade[a]	
2.4 µL	$MgCl_2$[a]	4 m*M*
0.12 µL	Primer AA55M (50 µ*M*)	300 n*M*
0.12 µL	Primer Lambda T (50 µ*M*)	300 n*M*
2 µL	LightCycler FastStart DNA Master Hybridization probes (10×)[a]	1X
0.2 µL	Fluorogenic hybridization probe LTR FL (20 µ*M*)	200 n*M*
0.2 µL	Fluorogenic hybridization probe LTR LC (20 µ*M*)	200 n*M*
18 µL	Total volume	

3. Mix gently.
4. Pipet 18 µL of master mix into the precooled LightCycler capillaries.
5. Add 2 µL of the 10-fold diluted PCR products from the first-round PCR into the capillaries.
6. Seal each capillary with a stopper and place the adapters containing the capillaries into a standard benchtop microcentrifuge.
7. Centrifuge at 700*g* for 5 s.
8. Place the capillaries into the rotor of the LightCycler instrument.
9. Program the LightCycler instrument using the parameters given below:
   a. 95°C for 8 min to denature DNA and activate the FastStart DNA polymerase.
   b. Amplify target DNA by performing 50 cycles as follows:
   - 95°C for 10 s to denature DNA.
   - 60°C for 10 s to anneal primers and probes. **Fluorescence measurement must be performed at the end of this step.**
   - 72°C for 9 s to extend the annealed primers.

   c. 40°C for 30 s to cool the thermal chamber.
10. Cycle PCR reactions.
11. Use the LightCycler quantification software to display fluorescence signals corresponding to the quotient of the channel F2 over channel F1 (**Fig. 5A**). Generate a linear standard curve by plotting the crossing cycle number vs the logarithm of the initial copy number for each dilution of the integrated DNA standard (**Fig. 5B**). This standard curve provides by interpolation a DNA copy number for all PCR reactions from their own crossing cycle number. Average copy numbers of the replicates.
12. Calculate the actual proviral copy number for each sample by subtracting the DNA copy number obtained in the absence of *Alu* primers from that obtained in the presence of *Alu* primers. Note that this correction should always be performed especially at early times following infection ***(11)***.
13. Calculate cell equivalents in sample DNA to estimate the proviral DNA content per cell as described in **Subheading 3.1.6., step 16**.

4. Notes

1. The dilution factors may be adjusted to yield well-separated neomycin-resistant independent clones in **Subheading 3.1.4., step 3**. The dilution factors will depend on the infection efficiency and so mostly on the quantity of infectious viruses within the viral stock.
2. Use aerosol-barrier tips and wear fresh gloves when preparing and handling DNA to minimize contamination from extraneous DNA templates and to prevent sample cross-contamination. Set up cell DNA preparation in a physically separated working place from that used for plasmid DNA isolation.
3. The protocol may be adapted for other fluorimeter-coupled thermal cyclers. Several modifications may be necessary—probe sequence and fluorescence detection format, PCR buffer composition, and cycling parameters—depending on the instrument used. Contact your fluorimeter-coupled thermal cycler supplier for more information.

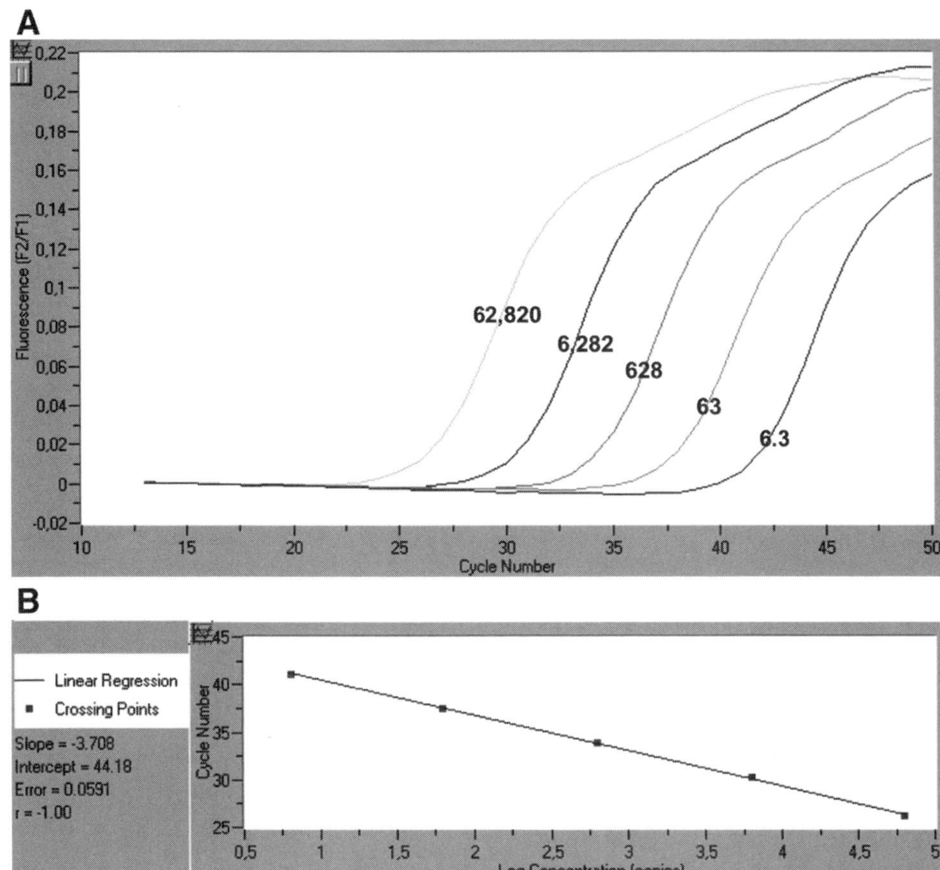

Fig. 5. Generation of a linear standard curve to quantify integrated HIV-1 DNA copy number within samples. **(A)** Fluorescence curves corresponding to the quotient of the channel F2 over channel F1 generated by the two-step amplification of serial dilutions of the integrated HIV-1 DNA standard. The known copy number of each standard dilution is shown over the corresponding fluorescence curve. **(B)** Linear standard curve obtained by plotting the crossing cycle number vs the logarithm of the initial copy number for each dilution of the integrated DNA standard.

4. Contamination of the PCR reaction mixture is the major expected problem when performing the *Alu*-LTR-based real-time nested PCR procedure. Several precautions can be followed to minimize contamination during preparation of the PCR master mix: always wear gloves and change them regularly; use dedicated pipets with aerosol-barrier tips and set up physically separated working areas for DNA and PCR master mix preparation. In a preferred embodiment, PCR mixture is prepared under a fume hood equipped with ultraviolet light.

5. As previously proposed ***(6)***, the use of two outward-facing *Alu* primers should increase the probability to amplify an LTR sequence, as *Alu* elements are present in either orientation relative to the integrated provirus (**Fig. 4**). Given the high number of *Alu* sequences within the human genome (at least 900,000 *Alu* elements per haploid genome ***[2]***), the greatest distance between the 3' LTR and its nearest *Alu* element averages only 4 kb, whatever the *Alu* orientation, which is a suitable length for PCR amplification.
6. The use of an LTR primer extended at its 5' end with a lambda phage-specific heel sequence that does not hybridize with the human genome will allow us to prevent second-round amplification of non-preamplified DNA. In the second round of PCR, using the heel-specific primer and a reverse LTR-primer, only products from the first-round PCR will be amplified (**Fig. 4**).
7. At each cycle during the first-round amplification, the LTR primer primes the formation of single-stranded DNA from all LTR, even coming from unintegrated HIV-1 DNA. To prevent overestimation of the actual integrated HIV-1 DNA copy number in samples, these linear amplifications are quantified in parallel by performing a PCR procedure in the absence of *Alu* primers. The true copy number of HIV-1 integrated DNA would be obtained by subtracting the DNA copy number quantified in the absence of *Alu* primers from that quantified in the presence of *Alu* primers.
8. Because abundant amplifications of inter-*Alu* sequences occur simultaneously with the amplification of *Alu*-LTR sequences, only 12 PCR cycles are performed so that all the dilutions of the integrated HIV-1 DNA standard remain within the exponential phase of amplification. However, PCR cycle number may be modified according to individual requirements. To increase assay sensitivity, a greater PCR cycle number may be carried out; nevertheless, the assay dynamic range will be narrower since the upper quantification limit will decrease. On the other hand, if PCR cycle number is decreased, the upper quantification limit should be higher but the assay sensitivity will be lower.

Acknowledgments

A. Brussel was supported by fellowships from the Agence Nationale de Recherche sur le SIDA and from the Fondation pour la Recherche Médicale. We thank U. Hazan for providing the pR7 Neo Δ*env* and pVSV-G vectors and for constant advice in the development of the integration assay. We acknowledge Olfert Landt (TIB MOLBIOL) for excellent technical assistance with the design of primers and hybridization probes. We would also like to thank C. Petit, S. Pierre, M. Alizon, and B. Canque for helpful discussions.

References

1. Mighell, A. J., Markham, A. F., and Robinson, P. A. (1997) Alu sequences. *FEBS Lett.* **417,** 1–5.
2. Britten, R. J., Baron, W. F., Stout, D. B., and Davidson, E. H. (1988) Sources and evolution of human Alu repeated sequences. *Proc. Natl. Acad. Sci. USA* **85,** 4770–4774.

3. Benkirane, M., Corbeau, P., Housset, V., and Devaux, C. (1993) An antibody that binds the immunoglobulin CDR3-like region of the CD4 molecule inhibits provirus transcription in HIV-infected T cells. *EMBO J.* **12,** 4909–4921.
4. Chun, T. W., Stuyver, L., Mizell, S. B., Ehler, L. A., Mican, J. A., Baseler, M., et al. (1997) Presence of an inducible HIV-1 latent reservoir during highly active antiretroviral therapy. *Proc. Natl. Acad. Sci. USA* **94,** 13,193–13,197.
5. Courcoul, M., Patience, C., Rey, F., Blanc, D., Harmache, A., Sire, J., et al. (1995) Peripheral blood mononuclear cells produce normal amounts of defective Vif- human immunodeficiency virus type 1 particles which are restricted for the preretrotranscription steps. *J. Virol.* **69,** 2068–2074.
6. Sonza, S., Maerz, A., Deacon, N., Meanger, J., Mills, J., and Crowe, S. (1996) Human immunodeficiency virus type 1 replication is blocked prior to reverse transcription and integration in freshly isolated peripheral blood monocytes. *J. Virol.* **70,** 3863–3869.
7. Carteau, S., Hoffmann, C., and Bushman, F. (1998) Chromosome structure and human immunodeficiency virus type 1 cDNA integration: centromeric alphoid repeats are a disfavored target. *J. Virol.* **72,** 4005–4014.
8. Wittwer, C. T., Ririe, K. M., Andrew, R. V., David, D. A., Gundry, R. A., and Balis, U. J. (1997) The LightCycler: a microvolume multisample fluorimeter with rapid temperature control. *Biotechniques* **22,** 176–181.
9. Graham, F. L., Smiley, J., Russell, W. C., and Nairn, R. (1977) Characteristics of a human cell line transformed by DNA from human adenovirus type 5. *J. Gen. Virol.* **36,** 59–74.
10. Scherer, W. F., Syverton, J. T., and Gey, G. O. (1953) Studies on the propagation in vitro of poliomyelitis viruses. IV. Viral multiplication in a stable strain of human malignant epithelial cells (strain HeLa) derived from an epidermoid carcinoma of the cervix. *J. Exp. Med.* **97,** 695–710.
11. Brussel, A. and Sonigo, P. (2003) Analysis of early human immunodeficiency virus type 1 DNA synthesis by use of a new sensitive assay for quantifying integrated provirus. *J. Virol.* **77,** 10,119–10,124.
12. Feinberg, M. B., Baltimore, D., and Frankel, A. D. (1991) The role of Tat in the human immunodeficiency virus life cycle indicates a primary effect on transcriptional elongation. *Proc. Natl. Acad. Sci. USA* **88,** 4045–4049.
13. Sambrook, J., Fritsch, E. F., and Maniatis, T. (1989) *Molecular Cloning: A Laboratory Manual*, 2nd ed., Cold Spring Harbor Laboratory Press, Cold Spring Harbor, NY.
14. Wittwer, C. T., Herrmann, M. G., Moss, A. A., and Rasmussen, R. P. (1997) Continuous fluorescence monitoring of rapid cycle DNA amplification. *Biotechniques* **22,** 130–131, 34–38.

12

Quantification of HFV-Integrated DNA in Human Cells by *Alu*-LTR Real-Time PCR

Olivier Delelis, Audrey Brussel, and Pierre Sonigo

Summary

Integration is described as a key step in viral replication of all retroviruses. A sensitive and quantitative measure of an integrated molecule is a good way to examine the importance of the integration step and to evaluate efficiency of retroviral vectors for gene transfer or anti-integrase drugs. Here, we report a sensitive and quantitative real-time polymerase chain reaction (PCR) technique to measure integrated viral DNA in human cells during a foamy virus (HFV) infection. This technique is based on two steps of PCR. The first round amplifies *Alu*-LTR (long terminal repeat) sequences resulting from viral integration. The second round of PCR is performed to quantify these events of integration. Quantification is monitored by the comparison of the amplification curve of the sample against a standard scale constituted of viral DNA from chronically infected cells. Sensitivity of this technique allows us to detect as few as 25 copies of HFV-integrated DNA in 50,000 cells.

Key Words: Spumavirus; provirus; integration; *Alu*-LTR real-time PCR.

1. Introduction

Viral DNA integration is a key step in retroviruses' life cycle *(1,2)*. During infection, unintegrated viral DNA is synthesized and accumulates into the infected cell. Only a small proportion is integrated randomly into the host genome. Consequently, detection and precise quantification of such a molecule is hampered by its weak representation and the difficulty to distinguish an integrated molecule from an unintegrated one. Three main strategies have been developed to measure viral integration: inverse polymerase chain reaction (PCR) *(3)*, linker-primer PCR *(4)*, and *Alu* PCR *(5–10)*. However, these techniques have limited sensitivity and are not strictly quantitative, as they lack real-time monitoring and polyclonal standards.

From: *Methods in Molecular Biology, Vol. 304: Human Retrovirus Protocols: Virology and Molecular Biology*
Edited by: T. Zhu © Humana Press Inc., Totowa, NJ

In this chapter, we report a real-time PCR method to quantify human foamy virus (HFV) viral-integrated DNA in human cells.

2. Materials

1. pHSRV13 *(11)*.
2. Human U373-MG glioblastoma cell line; BHK-21 (baby hamster kidney) cells.
3. AZT (3'-azido-3'-deoxythymidine) (Sigma-Aldrich, Saint Quentin Fallavier, France).
4. Transfection reagents: 2X HEPES-buffered saline (HBS); 2.5 *M* $CaCl_2$.
5. Phosphate-buffered saline (PBS).
6. Qiamp DNA blood extraction kit (Qiagen, France).
7. NucleoSpin extract (Macherey-Nagel, Düren, Germany).
8. Agarose gel and Southern blot equipment.
9. Nona primer kit (Stratagene).
10. (α-^{32}P) dCTP (Amersham Biosciences, Saclay, France).
11. Nick column (Pharmacia Biotech).
12. Dulbecco's modified Eagle medium (DMEM); fetal calf serum (FCS); penicillin; streptomycin.
13. Oligonucleotide primers and probes (TIB MOLBIOL, Berlin, Germany).
14. PCR hood.
15. Filtered tips.
16. LightCycler instrument (Roche Diagnostics, Meylan, Germany).
17. LightCycler capillaries (Roche Diagnostics).
18. LightCycler cooling block (Roche Diagnostics).
19. FastStart DNA hybridization probes (Roche Diagnostics).
20. FastStart DNA Syber green (Roche Diagnostics).
21. LightCycler control kit DNA (Roche Diagnostics).

3. Methods

The method described below outlines (1) the construction of the standard for quantification of human foamy virus (HFV)-integrated DNA in human cells and (2) a description of a real-time PCR protocol.

3.1. Construction of Standard Scale for Calibration of HFV-Integrated DNA Quantification

To quantify HFV-integrated DNA a polyclonal standard, constituted of viral DNA integrated randomly into the cell genome that mimics a wild-type infection, must be constructed. In this section, we describe construction of human chronically infected cells for calibration of the technique.

3.1.1. Transfection of pHSRV13

Plasmid pHSRV13 was transfected in BHK-21 (baby hamster kidney cells) by standard protocol using a calcium phosphate method *(12)*:

1. The day before transfection, plate BHK-21 cells in a T-75 flask (*see* **Note 1**) and incubate at 37°C, 5% CO_2. When cells are approx 60% confluent, mix 10 µg of pHSRV13 with 50 µL of 2.5 *M* $CaCl_2$ in a final volume of 500 µL; and add drop by drop to 500 µL of 2X HBS.
2. Incubate 30 min at room temperature.
3. Put the mix drop by drop on BHK-21 cells.
4. 16 h posttransfection, wash the cells abundantly with PBS.
5. Add prewarmed medium to transfected cells.
6. Collect the supernatant 48 h posttransfection.
7. Filter with a 0.45-µm filter and make aliquots frozen at –80°C.

3.1.2 Infection of U373-MG Cells

1. Use one aliquot to infect 10^5 U373-MG cells (*see* **Note 1**).
   Several weeks postinfection, after the occurrence of intense cytopathic effect and cell lysis owing to the spread of infection, some cells resistant to infection can be obtained. Integration can occur up to 20 provirus copies per cell in chronically infected cells ***(13,14)***. These cells, coming from different clones, can be maintained in culture in the same conditions as uninfected cells.
2. To eliminate unintegrated viral DNA ***(1,15)***, add and maintain AZT into the medium at a final concentration of 100 µ*M* during 1 wk ***(16)***. Amplify chronically infected cells under AZT pressure.
3. Trypsinize and count cells. Make aliquots by centrifugation of 5×10^6 cells (1500*g*) for 5 min and freeze at –80°C until DNA extraction. These aliquots are further used to quantify HFV-integrated DNA.

3.1.3. Verification of Viral DNA Integration

3.1.3.1. DNA Extraction

DNA was extracted according to the manufacturer's protocol (QIAamp DNA blood mini kit, Qiagen) (*see* **Note 2**).

1. Perform as described by the manufacturer. Don't forget to proceed with optional steps (second centrifugation step at full speed after washing the silica-gel membrane with AW2) to prevent any ethanol contamination, which would inhibit subsequent reactions.
2. Elute DNA with 200 µL of buffer AE following incubation for 2 min at room temperature.
3. Make aliquots of eluted DNA; store at –20°C.

3.1.3.2. Southern Blotting

Genomic DNA from chronically infected cells (20 mg), treated or not with AZT, were loaded on 0.8% agarose gel.

1. Transfer DNA by capillarity in a 20X SSC buffer on a Hybond-N+ membrane (Amersham Biosciences) and fix DNA with UV.

2. Prehybridize the Hybond-N+ membrane for 4 h at 65°C in a 15-mL prehybridization buffer (5X SSPE, 5X Denhardt's, 0.5% SDS, 100 μg/mL denatured salmon sperm).
3. Construct a DNA probe with the full-length pHSRV13 and radiolabel with a Prime-It II kit (Stratagene) in the presence of α^{32}PdCTP.
4. Denature DNA probe at 95°C for 5 min, chill immediately at 4°C, and add the probe to the prehybridization buffer.
5. Wash the membrane four times with 2X SSPE/0.1% SDS for 5 min at room temperature and two times with 0.1X SSPE/0.01% SDS for 15 min at 50°C. Put an x-ray film on the radiolabeled membrane for 24 h. Results are represented in **Fig. 1**. In the presence of AZT in the medium, autoradiography revealed that viral DNA is present only in an integrated form as shown by the arrow.

3.1.4. Quantification of Proviral DNA in Infected U3MG Cells

After DNA extraction from chronically infected cells, viral integrated DNA was quantified by real-time PCR using Syber Green procedure and primers annealing in the *env* gene.

3.1.4.1. Quantification of Viral DNA (*see* **Note 3**)

Briefly centrifuge one vial 1a (colorless cap) containing the LightCycler FastStart Enzyme and one vial 1a (green cap) containing the LightCycler FastStart Reaction Mix Syber Green I. One vial 1a contains enzyme concentrate for three vials PCR master mix.

1. Pipet a total volume of 10 μL from vial 1a into vial 1b. Mix gently by pipetting up and down but do not vortex. The resulting "Hot Start" master mix (10X) in one vial is sufficient for 32 reactions. When the master mix is reconstituted, **protect from light and avoid repeated freezing and thawing**. After the first thawing, store master mix at 2–8°C for a maximum of 1 wk.
2. Depending on the total number of reactions, place LightCycler capillaries in precooled centrifuge adapters. Mix is prepared by multiplying the number of reactions by the quantity of each component necessary for one reaction.
3. In a 1.5-mL reaction tube, add the following components in the order mentioned below:

			Final concentration
$3H_2O$, PCR grade	:	11.6 mL	
$MgCl_2$	:	2.4 mL	4 m*M*
SpuIn F (10 m*M*)	:	1 mL	0.5 m*M*
Spu R (10 m*M*)	:	1 mL	0.5 m*M*
LightCycler FastStart DNA master Syber green	:	2 mL	1X
Total volume	:	18 mL	

A

Primer or Probe	Sequence	
Spu INF :	(5'-GGA CCT GTA ATA GAC TGG AA-3')	**Total Viral DNA**
Spu R :	(5'-ATT TGC AGG TCT AAT ACT CTC C-3')	
ALU 1:	(5'-TCC CAG CTA CTG GGG AGG CTG AGG-3')	**Integrated viral DNA**
ALU 2 :	(5'-GCC TCC CAA AGT GCT GGG ATT ACA G-3')	**(first round PCR)**
LambdaSpA :	(5'-**ATG CCA CGT AAG CGA AAC T** TA GTA TAA TCA TTT CCG CTT TCG-3')	
Lambda :	(5'-ATG CCA CGT AAG CGA AAC T-3')	**Integrated viral DNA**
NestedR :	(5'-GAA ACT AGG GAA AAC TAG G-3')	**(second round PCR)**
SP FL* :	(5'-GAG AGA CAC AAG GTT CTT AAA TTG TCC TCA TTC GC-3'a)	
SP LC* :	(5'-ACT CCC TCT GAC ATC CAA CGC TGG GCT AC-3'b)	

* : Probe sequence.
A: modified probe with a 3' end fluorescein.
B: modified probe with LC red 640 dye at the 5' end and phosphorylated at the 3' end.
Primers and probes were purchased from TIB MOLBIOL (Berlin, Germany).

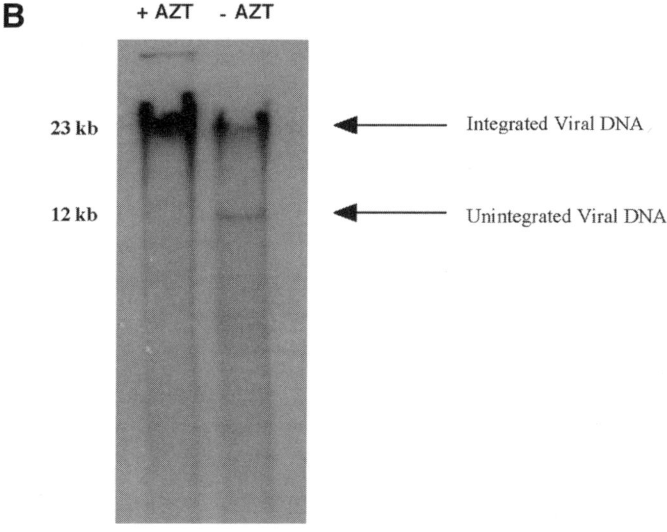

Fig. 1. **(A)** Primers and probe sequences used for quantification of viral DNA. **(B)** Southern blot of chronically infected U373-MG cells with or without 100 μ*M* AZT. Linear unintegrated and integrated viral DNA are represented by an arrow.

For the primer stocks (*see* **Note 4**), $MgCl_2$ was supplied by the manufacturer.

4. Pipet 18 μL master mix into the precooled LightCycler capillary on a cooling block under a PCR hood to prevent any DNA contamination. Note that DNA must not be introduced into the PCR hood.

5. Transfer the cooling block to a culture hood. A standard scale is made by fresh 10-fold dilution series of plasmid pHSRV13. Aliquots of pHSRV13 plasmid are made at 5×10^9 copies/µL and stored at –20°C. For each viral DNA quantification, one aliquot is thawed and dilutions down to 5×10^0 copies/µL are prepared.
6. Complete the first capillary with 2 µL of sterile water and seal immediately with a stopper. This is the negative control of the PCR reaction.
7. Complete the other capillaries with 2 µL of DNA sample. After each deposit, each capillary is sealed with a stopper.
8. Perform the PCR with 2 µL of each plasmid dilution of the standard scale from 1×10^1 copies to 1×10^6 copies. To prevent plasmid contamination, add dilutions of weak concentration first and continue with dilutions of higher concentration.
9. Seal each capillary with a stopper and place the adapters, containing the capillary, into a standard benchtop microcentrifuge.
10. Centrifuge at 700*g* for 5 s and place the capillaries in the rotor of the LightCycler instrument.

All real-time PCR protocols are composed of three programs:

Program 1: Preincubation and denaturation of the template DNA.
Program 2: Amplification of the target DNA.
Program 3: Cooling the rotor and thermal chamber.

Set the values for the preincubation and denaturation of the template DNA as follows:

Cycle program data	Value
Cycles	1
Target temperature (°C)	95
Incubation time (s)	480
Temperature transition rate (°C/s)	20

Set the values for amplifying the target DNA as follows:

Cycle program data	Value
Cycles	45
Analysis mode	Quantification

Temperature targets	segment 1	segment 2	segment 3
Target temperature (°C)	95	58	72
Incubation time (s)	10	10	20
Temperature fusion rate (°C/s)	20	20	20
Acquisition mode	none	none	single

Set the values for cooling the rotor and thermal chamber at the end of the protocol as follows:

Cycle program data	Value
Cycles	1
Analysis mode	none

Temperature targets	segment 1
Target temperature (°C)	40
Incubation time (s)	30
Temperature fusion rate (°C/s)	20
Acquisition mode	none

Quantification are performed with the LightCycler software Version 3.5 according to manufacturer's instructions (*see* **Note 5**).

3.1.4.2. Quantification of Human β-Globin

The quanitification of human β-globin DNA in 2 µL of DNA from U3MG cells is performed using the commercially available material (control kit DNA; Roche Diagnostics combined with the LightCycler FastStart DNA Master Hybridization Probes) following the manufacturer's instructions. Amplification of β-globin DNA is monitored using hybridization probes labeled with LC Red 640 that hybridize to an internal sequence of the amplified fragment.

1. In a 1.5-mL reaction tube, add the following components in the order mentioned below:

			Final concentration
H_2O, PCR grade	:	3.8 µL	
$MgCl_2$	:	1.2 µL	4 m*M*
β-globin primer mix	:	1 µL	0.5 µ*M*
β-globin hybridization probe mix, LC-red 640 labeled	:	1 µL	probe 1: 0.2 µ*M* probe 2: 0.4 µ*M*
LightCycler FastStart DNA master hybridization probes mix	:	1 µL	1X
Total volume	:	8 µL	

2. Proceed with the same precautions described previously for viral quantification.

Two μL of sample DNA are quantified by real-time PCR.

3. The standard scale is made by fresh dilutions of human genomic DNA (purple cap) (15 ng/μL). Two mL of these dilutions are used to quantify DNA sample. Thirty ng of genomic DNA represents 10,000 genome equivalents.

As described previously, the β-globin protocol consists of three programs:

- Denaturation program (same as described for viral DNA).
- Amplification program.

Cycle program data	Value
Cycles	45
Analysis mode	Quantification

Temperature targets	segment 1	segment 2	segment 3
Target temperature (°C)	95	55	72
Incubation time (s)	0	10	5
Temperature fusion rate (°C/s)	20	20	20
Acquisition mode	none	none	single

- Cooling program (same as described for viral DNA).

Quantifications are performed with the LightCycler software Version 3.5 according to manufacturer's instructions (*see* **Note 5**). If the U373-MG cells are diploid, then the number of cells in a 2-μL sample is equal to half human β-globin gene quantification of the analyzed sample. *Make aliquots of this DNA; it represents the standard scale for integrated viral DNA quantification.*

3.2. Principle of Integrated HFV DNA Quantification (see Note 6)

To measure integrated DNA, a two-round PCR protocol was developed. Viral DNA integration occurs randomly into the host cell genome. Consequently, the distance between a determined sequence and viral genome is different for all integration sites. To measure all integration events, we focus on *Alu* elements, which are the most numerous repetitive elements in primate genomic DNA, with more than 1 million copies per diploid genome *(17,18)*. *Alu* elements are randomly distributed, roughly 5000 bp apart, and are randomly oriented. Thus, in a statistical manner, one event of viral DNA integration is 5000 bp from an *Alu* sequence.

3.2.1. First Round of PCR: Pre-Amplification

As shown in **Fig. 2**, the first round of PCR is performed with two *Alu* primers and an anchored primer named LambdaSpA:

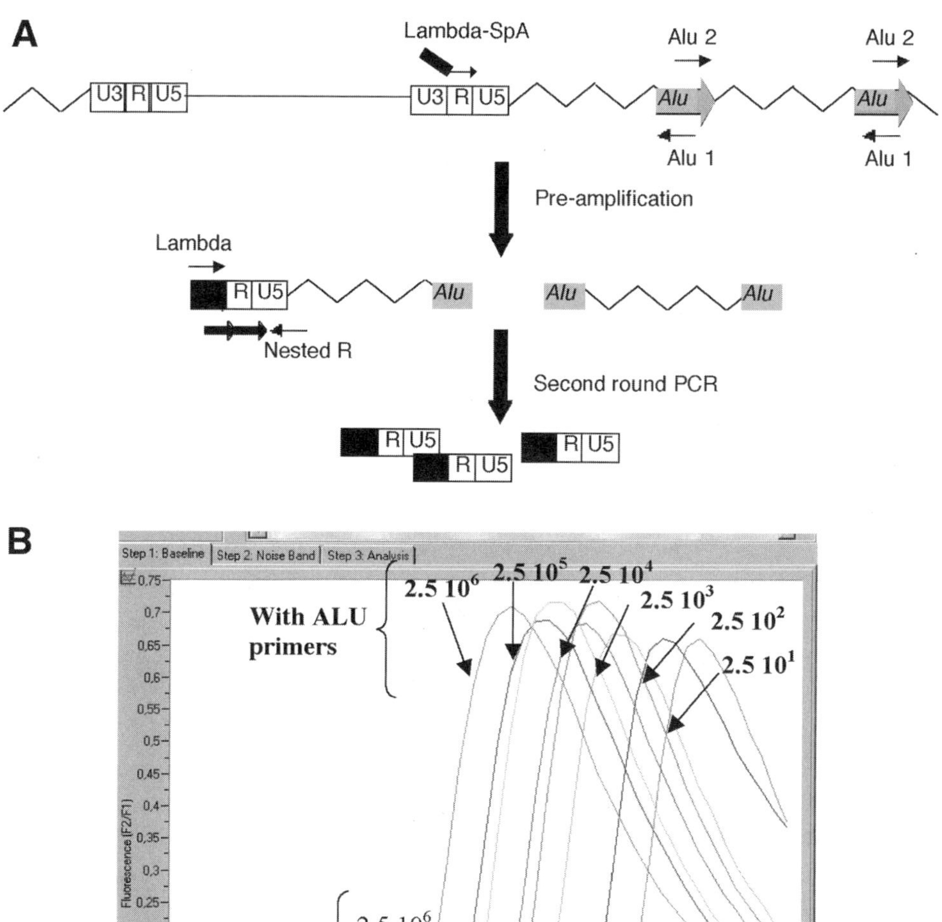

Fig. 2. **(A)** Principle of integrated DNA measure. The first round of polymerase chain reaction (PCR) consists of a preamplification of *Alu*-long termimal repeat (LTR) sequences of human foamy virus (HFV)-integrated DNA. The second round of PCR allows to quantify viral integrated DNA by comparison to a standard scale. **(B)** Amplification of a standard scale. Serial dilutions of U373-MG chronically infected cells from 25 copies to 2.5 million copies of integrated DNA were amplified with (bold) or without *Alu* primers.

ALU1: (5'-TCC CAG CTA CTG GGG AGG CTG AGG-3')
ALU2: (5'-GCC TCC CAA AGT GCT GGG ATT ACA G-3')
LambdaSpA: (5'-**ATG CCA CGT AAG CGA AAC T**TA GTA TAA TCA TTT CCG CTT TCG-3')

The sequence in bold type does not hybridize on viral LTR and is absent from all mammalian databanks. Amplification is performed with 500 n*M* of the two *Alu* primers and 300 n*M* of the anchored primer in a 20-µL reaction volume containing 1X LightCycler FastStart DNA Master Probe technique mixture (Roche Diagnostics) and 4 m*M* $MgCl_2$. Heterogeneous amplicons of variable length are produced with the Lambda sequence at one extremity when integrated DNA is present (*see* **Note 7**).

3.2.2. First Round of PCR

1. Prepare two mixtures:

MIX 1: with *Alu* primers

			Final concentration
H_2O, PCR grade	:	11 µL	
$MgCl_2$	:	2.4 µL	4 m*M*
ALU1 (10 µ*M*)	:	1 µL	0.5 µ*M*
ALU2 (10 µ*M*)	:	1 µL	0.5 µ*M*
LambdaSpA	:	0.6 µL	0.3 µ*M*
LightCycler FastStart DNA master hybridization probes	:	2 µL	1X
Total volume	:	18 µL	

MIX 2: without *Alu* primers

			Final concentration
H_2O, PCR grade	:	13 µL	
$MgCl_2$	:	2.4 µL	4 m*M*
ALU1 (10 µ*M*)	:	0 µL	0 µ*M*
ALU2 (10 µ*M*)	:	0 µL	0 µ*M*
LambdaSpA	:	0.6 µL	0.3 µ*M*
LightCycler FastStart DNA master hybridization probes	:	2 µL	1X
Total volume	:	18 µL	

2. The second mix is then identical to the first one but does not contain ALU1 and ALU2 primers. Two microliters of DNA sample is then tested in duplicate with the first mixture and in simplicate with the second one.
3. For the first round of PCR, set the values of PCR protocol:
   - Denaturation program (as described previously).
   - Set the values for amplifying the target DNA as follows:

Cycle program data	Value
Cycles	12
Analysis mode	Quantification

Temperature targets	segment 1	segment 2	segment 3
Target temperature (°C)	95	60	72
Incubation time (s)	10	10	170
Temperature fusion rate (°C/s)	20	20	20
Acquisition mode	none	none	none

   - Then a cooling protocol is executed as described previously.
4. Pipet 2 μL of each DNA sample in two capillaries containing the first mixture (with *Alu* primers).
5. Pipet 2 μL of each DNA sample in one capillary containing the second mixture (with no *Alu* primers). **Use at least four serial dilutions of the standard scale (DNA extracted for U373-MG cells) for quantification.**
6. Pipet 2 μL of each dilution of standard scale into capillaries containing mixture with *Alu* primers.
7. After the PCR, recover all capillaries, put on the stopper, and put the top of each capillary into a 1.5-mL Eppendorf tube.
8. Centrifuge at 700*g* for 5 s.
9. Recover the 20 μL content of each capillary and add 180 μL of water under a culture hood.

These dilutions constitute the template of the second round of PCR. The second round of PCR is performed to detect and quantify in a specific manner amplicons of the first PCR, testifying to integration events.

3.2.3. Second Round of PCR

The second-round real-time PCR was performed using 2 μL from the first PCR.

1. In a 1.5-mL reaction tube, add the following components in the order mentioned below:

			Final concentration
H_2O, PCR grade	:	11.2 µL	
$MgCl_2$	:	2.4 µL	4 m*M*
Lambda (10 µ*M*)	:	1 µL	0.5 µ*M*
Nested R (10 µ*M*)	:	1 µL	0.5 µ*M*
Sp FL* (20 µ*M*)	:	0.2 µL	0.2 µ*M*
Sp LC* (20 µ*M*)	:	0.2 µL	0.2 µ*M*
LightCycler FastStart DNA master hybridization probe	:	2 µL	1X
Total volume	:	18 µL	

Sequences of primers and probes are given in **Fig. 1**.

2. Pipet 18 µL of the mixture under a PCR hood.
3. Add 2 µL of diluted DNA under a culture hood and seal each capillary with a stopper immediately after each addition.
4. Centrifuge at 700*g* for 5 s.
5. Set the parameters of the PCR protocol.
   - Denaturation program (as described previously)
   - Amplifying program:

Cycle program data	Value
Cycles	45
Analysis mode	Quantification

Temperature targets	segment 1	segment 2	segment 3
Target temperature (°C)	95	60	72
Incubation time (s)	0	10	8
Temperature fusion rate (°C/s)	20	20	20
Acquisition mode	none	none	single

Cool the thermal chamber at 40°C for 30 s.

Values of integrated DNA of each sample are monitored by the LightCycler software Version 3.5 according to manufacturer's instructions (*see* **Note 5**).

Even when the mixture does not contain *Alu* primers, nonintegrated DNA can be amplified linearly by lambdaSpA primer during the first PCR. To control for such linear amplification, a measure without *Alu* primers is realized in parallel. As a correction, the background signal contributed from unintegrated DNA (i.e., without *Alu* primers) is then subtracted from the value obtained with *Alu* primers (*see* **Note 8**).

4. Notes

1. All cells: BHK-21 and U373-MG cells were grown in DMEM with 10% fetal bovine serum (FBS) and antibiotics (penicillin, streptavidine) at 37°C with 5% CO_2.
2. With the rapid advancement in molecular biology techniques, multiple methods can be used for DNA extraction. However, take care that when ethanol is used preferentially in wash buffer, it is eliminated. When some drops are still present in the eluate, they can inhibit subsequent reactions as real-time PCRs.
3. Alternative methods of quantification can be used with the LightCycler instrument. To quantify viral DNA in U3MG chronically infected cells, Syber green technique can be used, but hybridization probes methods may also be used. All primers and probes used in these techniques were designed by Olfert Landt (TIB MOLBIOL, Germany)
4. All primers and probes are resuspended in PCR-grade H_2O and stored at 4°C. Primers are stocked at a final concentration of 100 µ*M* and probes at 20 µ*M*. For better conservation, protect DNA probes from light.
5. Sigmoidal amplification can be observed between 10 and 1 million HFV copies per reaction. The regression line, calculated with LightCycler Software Version 3.5, usually has a *y* intercept of 40 cycles and a negative slope near –3.4 per 10-fold increase of a standard curve concentration. For DNA quantification, check the baseline adjustment "Arithmetic" and channel F1 when Syber green is used. Check adjustment "Proportional" and channel F2 over F1 when hybridization probes are used. Then move the noise band up or down to define which data points will be excluded in the analysis. All data points that fall below the noise band will be excluded. Once background fluorescence has been defined, the LCDA software uses the intersection between each fluorescence curve and a crossing line, set to the level of the noise band to give a value for a crossing point. The crossing points values for a set of standards, expressed as fractional cycle number can be plotted against log concentration to give a standard curve and to determine values of the unknown samples.
6. A general consideration: PCR parameters (the number of cycles and more particularly the number of cycles of the first integrated PCR, temperature hybridization, and so on) have been determined empirically by trial and error.
   a. Set up physically separated working places for template preparation and setting up the PCR reactions.
   b. To prevent any contamination, always prepare PCR mix under a fume hood with filter tips and dedicated (PCR use only) pipets, microcentrifuges, and disposable gloves. Wear a "PCR coat" for preparation of PCR mixtures. Always use fresh gloves and change them regularly.
   c. Have your own set of PCR reagents and solutions and use them only for PCR reactions.

A

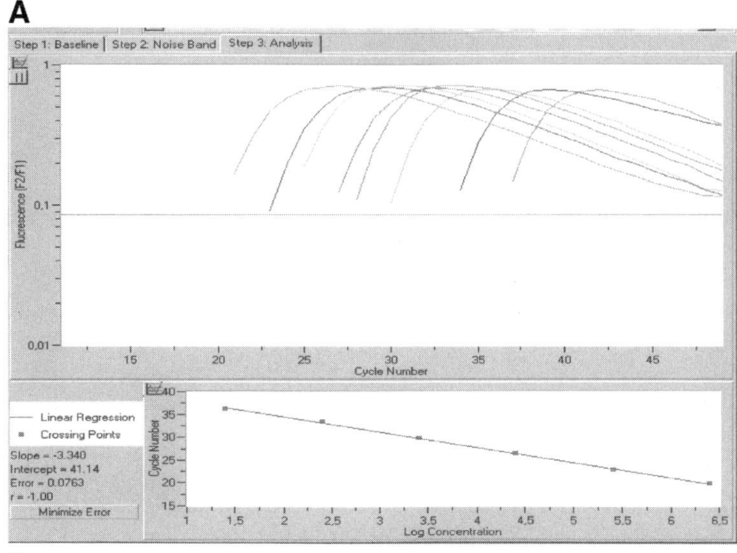

B

Total viral DNA in U373-MG chronically infected cells	Quantification with ALU primers	Quantification with NO ALU primers	Quantification with NO ALU primers / with ALU primers
$2.5\ 10^6$	$2.37\ 10^6$	$0.16\ 10^6$	6.7 %
$2.5\ 10^5$	$2.83\ 10^5$	10^4	3.7 %
$2.5\ 10^4$	$2.47\ 10^4$	692	2.8 %
$2.5\ 10^3$	$2.54\ 10^3$	46	1.8 %
$2.5\ 10^2$	$1.97\ 10^2$	4	2 %
$2.5\ 10^1$	$2.9\ 10^1$	0	0 %

Fig. 3. **(A)** Parameters of the second polymerase chain reaction (PCR). After removing the background noise, log-lines are calculated for each curve and extrapolated to the threshold line. Slope of this PCR: –3.340 and intercept, 41.14, are determined by LightCycler Software Version 3.5. **(B)** Quantification of the standard scale with and without *Alu* primers. Results demonstrate that values with no *Alu* primers represent a maximum of 6.7% of integrated viral DNA with *Alu* primers.

d. Store these reagents in aliquots.
e. **Do not introduce any DNA under the PCR hood.** Samples must be prepared under a culture hood.
f. If negative control samples of any PCR show positive amplification curves, exchange all critical solutions.

7. Even without viral integrated DNA in the sample, primer LambdaSpA can hybridize on LTR of unintegrated DNA and amplify viral DNA in a linear manner.

To quantify this phenomenon, which might overestimate viral integrated measure, a second mixture is realized without *Alu* primers.

8. For standard curve, amplification without *Alu* primers represent only 6% of the value determined with *Alu* primers (**Fig. 3B**). However, it can reach 25% of the value during acute infection when integrated DNA represent a minor form. In this latter case, correction is required.

Acknowledgments

We are grateful to Olfert Landt (TIB MOLBIOL) for technical assistance with the design of primers. We thank Ali Saïb, Caroline Petit, and Thierry Leste-Lasserre for helpful discussions.

References

1. Englund, G., Theodore, T. S., Freed, E. O., Engleman, A., and Martin, M. A. (1995) Integration is required for productive infection of monocyte-derived macrophages by human immunodeficiency virus type 1. *J. Virol.* **69,** 3216–3219.
2. Quinn, T. P. and Grandgenett, D. P. (1988) Genetic evidence that the avian retrovirus DNA endonuclease domain of pol is necessary for viral integration. *J. Virol.* **62,** 2307–2312.
3. Chun, T. W., Finzi, D., Margolick, J., Chadwick, K., Schwartz, D., and Siliciano, R. F. (1995) In vivo fate of HIV-1-infected T cells: quantitative analysis of the transition of stable latency. *Nat. Med.* **1,** 1284–1290.
4. Vandegraaff, N., Kumar, R., Burrell, C. J., Li, P., Bell, P., Montaner, L. J., and Maul, G. G. (2001) Kinetics of human immunodeficiency virus type 1 (HIV) DNA integration in acutely infected cells as determined using a novel assay for detection to integrated HIV DNA. *J. Virol.* **75,** 11,253–11,260.
5. Benkirane, M., Corbeau, P., Housset, V., and Devaux, C. (1993) An antibody that binds the immmunoglobulin CDR3-like region of the CD4 molecule inhibits provirus transciption of HIV-infected T cells. *EMBO J.* **12,** 4909–4921.
6. Bouyac-Bertoia, M., Dvorin, J. D., Fouchier, R. A., Jenkins, Y., Meyer, B. E., Wu, L. I., et al. (2001) HIV-1 infection requires a functional integrase NLS. *Mol. Cell* **7,** 1025–1035.
7. Butler, S. L., Hansen, M. S., and Bushman, F. D. (2001) A quantitative assay for HIV DNA integration in vivo. *Nat. Med.* **7,** 631–634.
8. Courcoul, M., Patience, C., Rey, F., Blanc, D., Harmache, A., Sire, J., et al. (1995) Peripheral blood mononuclear cells produce normal amounts of defective Vif-human immunodeficiency virus type 1 particles which are restriced for the preretrotranscription steps. *J. Virol.* **69,** 2068–2074.
9. Sonza, S., Maerz, A., Deacon, N., Meanger, J., Mills, J., and Crowe, S. (1996) Human immunodeficiency virus type 1 replication is blocked prior to reverse transcription and integration in freshly isolated peripheral blood monocytes *J. Virol.* **70,** 3863–3869.
10. O'Doherty, U., Swiggard, W. J., Jeyakumar, D., McGain, D., and Malim, M. H. (2002) A sensitive, quantitative assay for human immunodeficiency virus type 1 integration. *J. Virol.* **76,** 10,942–10,950.

11. Lochelt, M., Zentgraf, H., Flugel, R. M., Bell, P., Montaner, L. J., and Maul, G. G. (1991) Construction of an infectious DNA clone of the full-length human spumaretrovirus genome and mutagenesis of the bel 1 gene. *Virology* **184,** 43–54.
12. Graham, F. L., van der Eb, A. J., O'Doherty, U., Swiggard, W. J., Jeyakumar, D., McGain, D., and Malim, M. H. (1973) A new technique for the assay of infectivity of human adenovirus 5 DNA. *Virology* **52,** 456–467.
13. Meiering, C. D., Comstock, K. E., and Linial, M. L. (2000) Multiple integrations of human foamy virus in persistently infected human erythroleukemia cells. *J. Virol.* **74,** 1718–1726.
14. Saib, A., Koken, M. H., van der Spek, P., Peries, J., and de The, H. (1995) Involvement of a spliced and defective human foamy virus in the establishment of chronic infection. *J. Virol.* **69,** 5261–5268.
15. Bell, P., Montaner, L. J., and Maul, G. G. (2001) Accumulation and intranuclear distribution of unintegrated human immunodeficiency virus type 1 DNA. *J. Virol.* **75,** 7683–7691.
16. Linial, M. L. (1999) Foamy viruses are unconventional retroviruses. *J. Virol.* **73,** 1747–1755.
17. Jelinek, W. R., Schmid, C. W., Lochelt, M., Zentgraf, H., Flugel, R. M., Bell, P., et al. (1982) Repetitive sequences in eukaryotic DNA and their expression. *Annu. Rev. Biochem.* **51,** 813–844.
18. Mighell, A. J., Markham, A. F., Robinson, P. A., Lochelt, M., Zentgraf, H., Flugel, R. M., et al. (1997) Alu sequences. *FEBS Lett.* **417,** 1–5.

13

Detection of HIV-1 Provirus and RNA by *In Situ* Amplification

Alcina Nicol and Gerard J. Nuovo

Summary

The detection of the HIV-1 provirus that can integrate into a host cell nucleus and remain latent for years is problematic. The threshold of *in situ* hybridization, which is about 10 copies per cell, is too high to detect one integrated copy of the provirus. Although polymerase chain reaction (PCR) can detect 1 provirus per 100,000 cells, it cannot determine the specific cellular localization of the virus. These problems can be resolved with PCR *in situ* hybridization. Adapting this method to RNA detection (reverse transcriptase [RT] *in situ* PCR) allows one to determine whether viral infection is latent or productive as well as to detect the host response in the form of cytokine mRNA expression. These methodologies have demonstrated that (1) there is massive infection of CD4 cells by HIV-1 prior to AIDS-defining symptomatology, (2) progression of AIDS is marked by the progressive destruction of CD4 cells, as evidenced by an increased ratio of productively to latently infected cells, (3) the primary target of the virus in the uterine cervix, lung, central nervous system, and skeletal muscle is the macrophage and its derivatives, and (4) AIDS-related diseases such as AIDS dementia are marked by both many viral-infected cells and upregulation of a wide variety of cytokines, primarily in the neighboring noninfected cells. This chapter will describe the methodologies for detecting HIV-1 DNA and RNA in paraffin-embedded tissue sections as well as the colabeling experiments needed to define the host response to the viral invasion.

Key Words: *In situ* hybridization; PCR; *in situ* PCR; reverse transcriptase; HIV-1; TNF-α.

1. Introduction

Polymerase chain reaction (PCR) has become the most important and commonly used technique in such diverse fields as basic research, forensic pathology, and the study of infectious disease. However, unless one does

From: *Methods in Molecular Biology, Vol. 304: Human Retrovirus Protocols: Virology and Molecular Biology*
Edited by: T. Zhu © Humana Press Inc., Totowa, NJ

microdissection experiments, it cannot determine what specific cell type(s) contains the DNA or RNA sequence of interest. Even with tissue microdissection, one loses important information such as the molecular interactions between neighboring cells and the histologic distribution of a given target. There are many experimental questions where this information would be of obvious interest ***(1–14)***. This chapter will focus on how this information can provide much insight into HIV-1 pathogenesis.

Although useful for targets with high copy numbers (such as productive DNA viral infections), standard *in situ* hybridization requires that at least 10 target sequences be present in an intact cell in order for one to see the signal ***(1)***. In many instances only one or a few targets may be present in a given cell. HIV-1 is a classic example, where one copy of the DNA provirus of HIV-1 can integrate into the nucleus of the host cell, and remain latent for many years. By combining the cell-localizing ability of *in situ* hybridization with the high sensitivity of PCR, one can readily and routinely detect one to a few target sequences in a cell and—by either using its cytologic features or colocalizing experiments—determine the exact nature of that cell. Thus, PCR *in situ* hybridization (for DNA) and reverse transcriptase (RT) *in situ* PCR (for RNA) are the methods of choice for detecting low copy sequences in intact tissue samples.

2. Materials

2.1. Pretreatment of Tissue or Cell Preparations Prior to PCR Amplification of DNA or cDNA

1. Fresh xylene.
2. Fresh 100% ethanol.
3. Protease (pepsin—DAKO, cat. no. S3002) (10 mL of pepsin prepared by adding 10 mg of pepsin to 9.5 mL of RNAse-free water and 0.5 mL of 2 *N* HCl) (*see* **Note 1**).
4. RNAse-free DNAse (Applied Biosystems, cat. no. 776 785).
5. Silane-coated glass slides.

2.2. PCR Amplification of DNA or cDNA

1. EZ buffer (buffer, nucleotides, and rTth are part of the EZ RT-PCR kit from Applied Biosystems, cat. no. N808-0179).
2. dATP, dCTP, dGTP, dTTP.
3. Bovine serum albumin (BSA).
4. Digoxigenin dUTP (Enzo Biochemicals, 0.5 m*M* cat. no. 42822).
5. Primers (20 m*M* each; *see* **Note 2**) (Sigma Genosys).
6. RNase inhibitor (Applied Biosystems, cat. no. 799 017).
7. rTth polymerase (Applied Biosystems, cat. no. N808-0097).

2.3. In Situ *Detection of Digoxigenin-Labeled Amplicons*

1. High-stringency wash (0.2X SSC, 2% BSA).
2. Antidigoxigenin-alkaline phosphatase conjugate (*see* **Note 3**) (Boehringer Mannheim cat. no. 1093274).
3. Nitroblue tetrazolium and bromochloroindolyl phosphate (NBT/BCIP) (*see* **Note 4**).
4. Nuclear fast red counterstain.
5. Xylene, 100% ethanol, permount, and glass cover slips.

3. Methods

The RT *in situ* PCR assay requires (1) the pretreatment of the tissue with protease and DNase, (2) the RT and PCR step with direct incorporation of the reporter nucleotide into the amplicon, and (3) detection of the reporter nucleotide by a colorimetric method, followed by routine cover slipping of the slide.

3.1. Pretreatment of Tissue With Protease and DNAse

It is important to stress that coated slides must be used for either standard *in situ* hybridization or any method in which cDNA/DNA is being amplified *in situ*. Coating the slides with silane will allow the tissue to remain firmly adhered to the slide during the multiple steps of the *in situ* procedure. Silane-coated slides are readily available from many commercial sources and have a variety of trade names, such as PLUS coated slides (Fisher Scientific). Another useful tip to remember is to place two (for large tissue sections) or three (for surgical biopsy material) tissue sections (of the usual thickness, typically 4 µm) on a given silane-coated glass slide. This allows one to perform the essential controls (*see* **Note 5**) on the same glass slide as the test reaction. Finally, recall that formalin fixation (10% buffered formalin is the standard in laboratories in the United States) is the optimal fixative for any *in situ* assay, with or without *in situ* amplification.

The most important variable for success with RT *in situ* PCR is the protease digestion step. To understand why, one may recall that formalin fixation crosslinks the DNA and RNA in the cell with the abundant cellular proteins. Thus, for the reagents used during the RT *in situ* PCR process to access the target RNA, the protein-RNA crosslinks must be much reduced. More importantly, the high temperatures invariably used during paraffin-embedding of formalin-fixed tissue induces nicks in the genomic DNA. These nicks can induce DNA synthesis in a process that is analogous to the generation of labeled probes using nick translation. The areas of DNA with nicks are protected from DNAse digestion by the protein-DNA crosslinks formed during formalin fixation. Unless these crosslinks are removed, the DNAse cannot degrade the genomic DNA to prevent nonspecific DNA repair/synthesis during the PCR step of RT *in situ* PCR. This is the key given that a reporter nucleotide (digoxigenin dUTP) will be incorporated into any newly made DNA during RT *in situ* PCR, be it the target cDNA or nonspecific DNA repair-based synthesis (*see* **Note 5**). Thus, after one removes the paraffin using a wash in fresh xylene (2–5 min), 100% ethanol (2–5 min), then air-drying the slide, the tissue sections must be incubated in a protease for the optimal digestion time. As described in **Note 5**, the optimal protease digestion time is defined as a strong primer-independent nuclear-based signal in at least 50% of the cells, which is totally eliminated by overnight pretreatment with DNase digestion (**Fig. 1**).

After optimal protease digestion, one needs to treat the tissue in DNase solution overnight at 37°C. It is important to stress that, with three tissue sections per slide, one can leave one tissue without DNase (this will be the positive PCR control—at least 50% of the nuclei should turn dark blue after the assay is done), DNase another tissue followed by RT and PCR with an irrelevant primer set (this is the negative PCR control; the signal noted with the positive control should be eliminated), and DNAse another tissue followed by RT and PCR with the primer set of interest (test run). Given that one is using glass slides, it is very simple to remove the protease and DNase: wash the slides for 1 min in RNase-free water followed by a 1-min rinse in 100% ethanol, which facilitates the air-drying of the slide.

3.2. RT and PCR Step With Direct Incorporation of Reporter Nucleotide Into the Amplicon

After the slide is air-dried, one is ready to add the RT and PCR reagents to the tissues. Prepare the following solution (most are part of the EZ rTth RNA PCR kit, Applied Biosystems cat. no. N808-0179):

- 10 µL of rTth buffer
- 1.6 µL each of dATP, dCTP, dGTP, dTTP
- 1.6 µL 2% BSA
- 2 µL of rTth
- 12.4 µL of 10 m*M* Mn acetate
- 0.6 mL of 1 m*M* digoxigenin dUTP
- 15.0 µL of sterile RNase-free water
- 2.0 µL of primers (stock solution, 200 µ*M*; *see* **Note 2**).

Add enough of the solution to cover the tissue. There are several alternatives for preventing evaporation of the amplifying solution. The simplest way is to use polypropylene plastic cover slips cut to size. They can be anchored with a small drop of glue or nail polish. Another alternative is Ampliclips and Amplicovers from Applied Biosystems (cat. nos. N804-0501 and N804-0600). Finally, a reagent called SelfSeal (MJ Research) can be added to the amplifying solution to prevent evaporation.

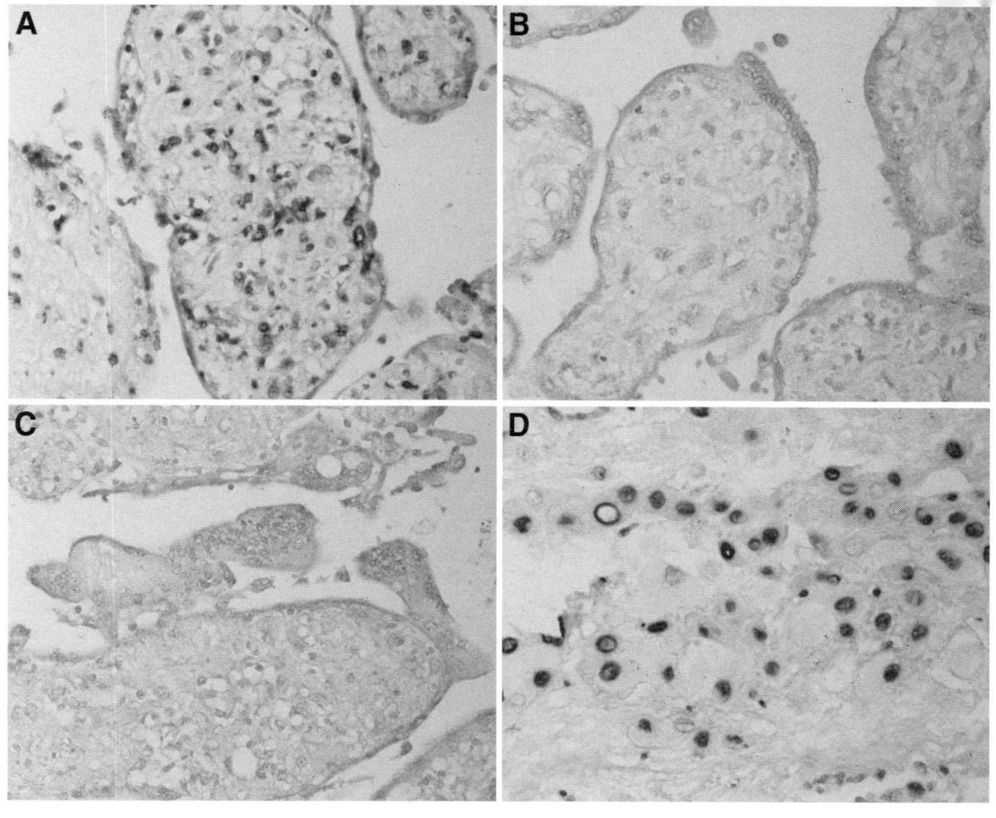

Fig. 1. Determining optimal protease digestion with the start-up protocol. This section of placenta tissue was digested in pepsin for 60 min **(A)** and 10 min **(C)** and treated per the start-up protocol. Note the lack of detectable DNA synthesis in **C** and the strong nuclear-based signal in **A**, denoting that 60 min of digestion is the optimal time. With the optimal protease digestion, the signal is completely lost with overnight digestion with RNase-free DNase (negative control, **B**). However, as expected, background is still evident after DNase digestion if 10 min of protease digestion is used **(D)**.

The RT step is done for 30 min at 65°C. This is followed by PCR amplification of the cDNA, which, after an initial denaturation of 95°C for 3 min, is achieved by cycling at 60°C for 1 min, 30 s, followed by denaturation at 95°C for 45 s, for 20 cycles. If one uses the polypropylene cover slip/mineral oil overlay method, then the slides need to be washed in fresh xylene for 3–5 min, followed by 100% ethanol for 3–5 min, followed by air-drying.

3.3. Hot-Start PCR Step Followed by In Situ Hybridization for Detection of HIV-1 Proviral DNA

There are two fundamental differences between PCR *in situ* hybridization for DNA and RT *in situ* PCR for RNA. First, one does not use direct incorporation of a reporter nucleotide with paraffin-embedded tissue because the nonspecific DNA repair pathway would render the results useless; thus, a hybridization step is needed. Second, one must use the hot-start maneuver, which inhibits mispriming and primer dimerization (although the latter does not appear to occur in the milieu of the cell) since, of course, one cannot use DNase digestion to eliminate these unwanted pathways.

One may use the same reagents with PCR *in situ* hybridization as with RT *in situ* PCR with the importance exception that the labeled reporter nucleotide (digoxigenin dUTP) is omitted. Also, it is recommended to use the hot-start maneuver. One has several options. There is manual hot-start, where the polymerase is withheld until the temperature of the cycler reaches 50°C. Alternatively, one can use AmpliGold (with the appropriate buffer and magnesium concentrations, Applied Biosystems, cat. no. N808-0241).

After the PCR step, perform standard *in situ* hybridization. It is strongly recommended that one use as large a probe as possible (80–120 base pairs [bp]), even if it includes the region of the primers, as primer oligomerization does not appear to occur with *in situ* PCR amplification. With these larger probes, one can use the same stringent wash and detection system as with RT *in situ* PCR, as described in **Subheading 3.4.**

3.4. Detection of Reporter Nucleotide by Colorimetric Method

The key step here is the stringency wash. It is important to remember that the amplifying solution most likely will contained labeled DNA that is generated from primer-dimerization and, perhaps (with too strong a protease step), target-specific amplicons that diffused out of the cells. This leads to a situation analogous to standard *in situ* hybridization, where the labeled probe in the hybridization cocktail can nonspecifically bind to cellular proteins and nucleic acids. These complexes are weakly held, and easily removed by a stringent wash. Thus, incubate the slides at 60°C for 10 min in a solution of 0.1X SSC that contains 2% BSA. Then remove the slide from the stringent wash, dry the back of the slide, and place over the tissue the antidigoxigenin-alkaline phosphatase conjugate and incubate at 37°C for 30 min. The slides can then be placed directly in the NBT/BCIP solution (*see* **Note 4**). The slides should be monitored under the microscope to determine the amount of time for which the reaction should proceed. After the signal is determined to be optimal and, of course, background is minimal, the slides should be washed in tap water for 1 min, counterstained in nuclear fast red for 3 min, and cover slipped with Permount.

3.5. Examples of Use of In Situ *Amplification for Study of HIV-1-Related Pathogenesis*

Cytokines are soluble molecules, secreted by white blood cells and other cells in response to certain types of stimuli. The major producers of cytokines are monocytes/macrophages and T-helper cells, but most immune cells are capable of some cytokine secretion following stimulation ***(15)***. A cytokine and its cytokine receptor are complementary in shape and, when bound together, initiate an intracellular biochemical cascade that leads to the expression of new genes by the cell ***(16)***.

HIV-1 infection can affect the function of the immune system. There is both immunodeficiency due to the loss of $CD4^+$ T-helper cells and hyperactivity as a result of B-cell activation; therefore, both increases and decreases are seen in the production and/or activity of cytokines. Several cytokines are responsible for the progression of HIV-associated disease, although there is disagreement about the role of some cytokines known as regulatory molecules such as Th1 (interleukin [IL]-2, interferon [IFN]-γ) or Th2 (IL-4, IL-10, IL-6) as prognostic to AIDS evolution during HIV-1 infection ***(17)***. The patterns of innate cytokine production have been postulated to switch from TH1- to TH2-type cytokines with the progression of HIV-associated disease owing to the cellular activation and consequently the increase of HIV-1 viral replication. A recent study that utilized RT *in situ* PCR and immunohistochemistry ***(18)*** showed in vitro that epithelial cells and keratinocytes containing human papillomavirus (HPV) were clearly more potent positive regulators for HIV activation than HPV-negative cells in human macrophages. This upregulation is mainly the result of induction of pro-inflammatory cytokines, such as IL-6 and tumor necrosis factor (TNF)-α. In vitro studies also suggest that IL-6 may contribute to HIV burden and to immunological abnormalities in HIV-infected patients ***(19)***. We also demonstrated that cervical biopsies from HIV/HPV coinfected woman showing high-grade cervical intraepithelial neoplasia had high statistical significance and correlation with IL-4, IL-10, IL-8, and IFN-γ expression. Further, there was a significant correlation between IL-4 and IL-10 production in the HIV-positive patients, suggesting a switch to a TH2 cytokine profile even in the presence of significant levels of IFN-γ (Th1 response). The HIV-negative patients with HPV infection showed high expression of IL-6 and TNF-α, compared with HIV/HPV coinfected women.

There are mechanisms involved to controlling the regulation of TH1 or TH2 responses. Recently, a novel class of suppressors belonging to the cytokine signaling (SOCS/SSI) family of proteins have been implicated in the negative feedback regulation of cytokine receptor signaling and cytoplasmic signaling adaptor molecules. We have recently shown (Nicol and Nuovo, unpublished data) using RT *in situ* PCR for SOCS and HIV-1 gag RNA in serial sections, that there was a strong inverse correlation between HIV-1 RNA expression and SOCS expression, suggesting that the former may be modulating SOCS expression, which, of course, could dramatically impact cytokine expression (**Fig. 2**).

Clearly, cytokine expression in HIV-1 patients is a complex, multifactorial process. RT *in situ* PCR done on serial sections and with colabeling will aid greatly in better defining the complex interplay between viral infection and the host response.

4. Notes

1. There are a wide variety of proteases that one can use with either *in situ* hybridization or RT *in situ* PCR. The two most commonly used proteases are proteinase K and pepsin. Pepsin has the advantage that it rarely overdigests tissue and, thus, tends to give excellent histologic detail. Also, it is easy to inactivate by simply flooding the slide with water. We prefer pepsin and use it at a high concentration (2 to 10 mg/mL). For most tissues, 30–60 min of digestion in this protease will give good results. However, for tissues that have been fixed for long periods of time in 10% buffered formalin, such as autopsy material, proteinase K digestion may be necessary. Proteinase K gives excellent digestion but it is more likely to overdigest the tissue, which, of course, defeats the purpose of any *in situ* methodology. For proteinase K, it is recommended that one use a concentration of 250 μg/mL. For most tissues, 10–30 min of digestion will give good results.
2. It has been our experience that primers that are acceptable for solution phase RT PCR will also be yield good results with RT *in situ* PCR. The size of the amplicon is typically 100–300 bp. We have used amplified cDNAs as long as 1000 bp with RT *in situ* PCR with good results.
3. We recommend the antidigoxigenin-alkaline phosphatase conjugate. However, it is important to realize that there are many choices that are also acceptable. For example, we at times use the antidigoxigenin-peroxidase conjugate with DAB as the chromogen. Alternatively, one may use biotin as the reporter nucleotide with the streptavidin-alkaline phosphatase as the conjugate. If one uses biotin, it is important to realize that the ratio of unlabeled (dTTP) to labeled (biotin dUTP) needs to be about 1:1; the window of optimal labeled to unlabeled is narrow. However, with digoxigenin, there is a broad window of unlabeled (dTTP) to labeled (digoxigenin dUTP); the range is from 1:50 to 1:1.

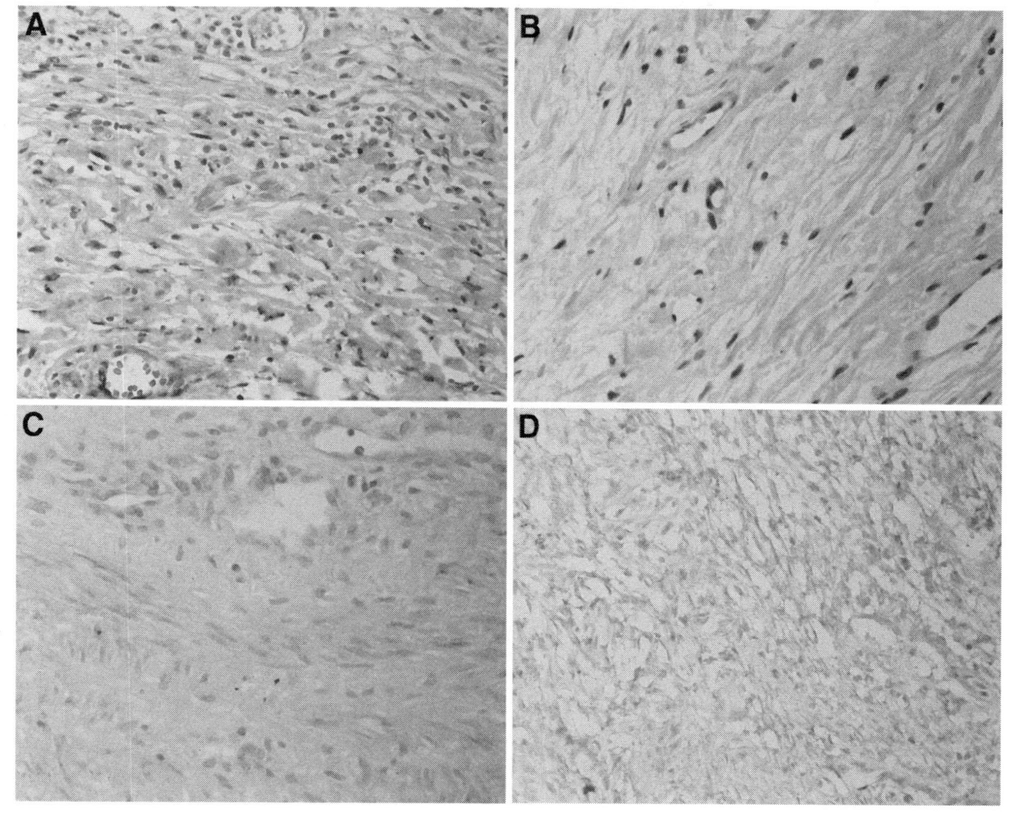

Fig. 2. Correlation of HIV-1 RNA detection with tumor necrosis factor (TNF)-α and SOCS mRNA in cervical biopsies from women with AIDS. **A** represents a section of cervical tissue near the transformation zone in a woman with AIDS; note the many HIV-1 infected cells (gag transcript detected by reverse transcriptase [RT] *in situ* polymerase chain reaction [PCR]) in cells with the cytologic features of macrophages. Analysis for TNF-α expression showed an equivalent histologic pattern **(B)**; however, SOCS mRNA was not detected in this area **(C)**. However, in another area where neither HIV-1 RNA or TNF-α mRNA was evident, many cells expressed SOCS mRNA as determined by RT *in situ* PCR **(D)**.

4. There are many sources of NBT/BCIP commercially available. We have noted that the concentration of the reagents is critical; too much can yield a blue-green background signal that is primarily cytoplasmic. As a starting point, it is recommended that one use 45 µL of each to 15 mL of the pH 9.5 Tris-HCl solution (Biogenex, cat. no. HK097-5K). If the signal-to-background ratio is not optimal, consider increasing the amounts of NBT and BCIP to 65 µL per 15 mL of the Tris-HCl buffer.
5. It cannot be overstressed that one cannot interpret RT *in situ* PCR without the proper controls. The negative control section (which has been treated with DNase) gets the RT *in situ* PCR solution either without the primers or (as we prefer) with irrelevant primers. The definition of irrelevant primers is that the target they represent could not possibly be present in the tissue being analyzed. Viral-specific primers work well in this regard. We routinely use either rabies or hantavirus specific primers, as—obviously—these viruses are not going to be present in most samples. We prefer to use irrelevant primers as this also serves as a check for nonspecific DNA synthesis from mispriming, though adequate DNase digestion will not allow this to occur. The positive control section has not been treated in DNase and, thus, there will be robust nonspecific DNA synthesis regardless of the primers. Thus, one can use the RT *in situ* PCR solution with the specific primers of interest (realizing that the latter are not necessary but we use them for the convenience of making the same solution for the positive control and for the test slide). The test control section has been treated with DNase and is incubated with the RT *in situ* PCR solution that contains the target specific primers of interest. The importance of having a basic knowledge of surgical pathology for the interpretation of RT *in situ* PCR data cannot be stressed enough. Of course, this is just as true for the interpretation of immunohistochemistry, *in situ* hybridization, and *in situ* PCR. For simplicity's sake, let us describe target-specific signal using the word "signal" and the word "background" for any blue coloration owing to any nonspecific pathway. Background is just as common with *in situ* hybridization and immunohistochemistry as it is with RT *in situ* PCR. The key to recognizing involves using two different tools:

   Correct interpretation of the controls. This is simple if one does the controls recommended above on the same glass slide as the test was done. A successful RT *in situ* PCR run is *defined by no background with the negative control* (DNAse with irrelevant primers) and an *intense nuclear-based signal in at least 50% of the cells with the positive control* (no DNAse) *in different cell types*. It follows that the definition of an unsuccessful RT *in situ* PCR run is also simple: *background, nuclear-based colorization is evident with the negative control* (DNAse with irrelevant primers). When this occurs, one typically sees either weak or no "signal" with the positive control (no DNase) but, for ease of interpretation, it is simpler to focus one's attention on the negative control (DNase with irrelevant primers). When this occurs, one must increase the time of protease digestion, as there is inadequate protease digestion to remove the DNA-protein crosslinks and allow the DNase to digest the template DNA. Of course, another possible result is no signal with the negative and positive controls. If this occurs, and the tissue morphology is poor, the obvious solution is to reduce the protease digestion time. If one sees no signal, and the morphology is good, then either the antidigoxigenin-alkaline phosphatase conjugate has expired (or was left in the freezer), the polymerase did not work, or the NBT/BCIP was no longer active. This is very unlikely; if no signal is evident, we recommend that one test the detection reagents with standard *in situ* hybridization and a high copy target, such as HPV in genital warts. The absence of a signal would document that one of the reagents is not effective.

Correct interpretation of the histologic pattern of the signal. A good surgical pathologist will appreciate the value of this tool. If one is looking for a lymphocyte marker and does immunohistochemistry for such, and then sees a strong "signal" in lymphocytes and squamous cells, then by definition there is background, and the experiment must be redone. This is particularly useful when studying viruses, as their specific cell tropism is usually well known. Of course, there will be instances where the specific histologic distribution of a given target will not be known and one must rely more on the correct interpretation of the controls as detailed above. Still, one should always be suspect when a "signal" is seen in many diverse cell types with a given probe, primer set, or antibody, and suspect background.

Acknowledgments

The authors greatly appreciate the financial support from the Lewis Foundation; also supported by NIH grant NCI (RO1 HL-00-012) (GJN).

References

1. Nuovo, G. J. (1997) *PCR in situ hybridization*; 3rd ed. Lippincott-Raven Press, NY.
2. Nuovo, G. J., Gallery, F., MacConnell, P., Becker, J., and Bloch, W. (1991) An improved technique for the detection of DNA by *in situ* hybridization after PCR-amplification. *Am. J. Pathol.* **139,** 1239.
3. Chou, Q., Russell, M., Birch, D. E., Raymond, J., and Bloch, W. (1992) Prevention of pre-PCR mispriming and primer dimerization improves low copy amplification. *Nucl. Acid Res.* **20,** 1717.
4. Nuovo, G. J., Gallery, F., Hom, R., MacConnell, P., and Bloch, W. (1993) Importance of different variables for optimizing *in situ* detection of PCR-amplified DNA. *PCR Method Applic.* **2,** 305.
5. Nuovo, G. J., Becker, J., Margiotta, M., Burke, M., Fuhrer, J., and Steigbigel, R. (1994) *In situ* detection of PCR-amplified HIV-1 nucleic acids in lymph nodes and peripheral blood in asymptomatic infection and advanced stage AIDS. *J. Acquired Immun. Def.* **7,** 916.
6. Embretson, J., Zupancic, M., Ribas, J. L., Racz, P., Haase, A. T. (1993) Massive covert infection of helper T lymphocytes and macrophages by HIV during the incubation period of AIDS. *Nature* **362,** 359.

7. Bagasra, O., Lischer, H. W., Sachs, M., and Pomerantz, R. (1992) Detection of HIV-1 provirus in mononuclear cells by *in situ* PCR. *N. Engl. J. Med.* **326,** 1385.
8. Nuovo, G. J., Forde, A., MacConnell, P., and Fahrenwald, R. (1993) *In situ* detection of PCR-amplified HIV-1 nucleic acids and tumor necrosis factor cDNA in cervical tissues. *Am. J. Pathol.* **143,** 40.
9. Nuovo, G. J., Becker, J., Simsir, A., Margiotta, M., and Shevchuck, M. (1994) *In situ* localization of PCR-amplified HIV-1 nucleic acids in the male genital tract. *Am. J. Pathol.* **144,** 1142.
10. Nuovo, G. J., Gallery, F., MacConnell, P., and Braun, A. (1994) *In situ* detection of PCR-amplified HIV-1 nucleic acids and tumor necrosis factor RNA in the central nervous system. *Am. J. Pathol.* **144,** 659.
11. Dubrovsky, L., Ulrich, P., Manogue, K. R., Cerami, A., and Burkinsky, M. (1995) Nuclear localization signal of HIV-1 as a novel target for therapeutic intervention. *Mol. Med.* **1,** 217.
12. Schmidtmayerova, H., Notte, H. S., Nuovo, G. J., Raabe, T., Flanagan, C. R., Dubrovsky, L., et al. (1996) HIV-1 infection alters chemokine β peptide expression in human monocytes: implications for recruitment of leukocytes into brain and lymph nodes. *Proc. Natl. Acad. Sci. USA* **93,** 700.
13. Euscher, E., Davis, J., Holtzman, I., and Nuovo, G. J. (2001) Coxsackie virus infection of the placenta associated with neurodevelopmental delays in the newborn. *Obstet. Gynecol.* **98,** 1019.
14. Cioc, A. and Nuovo, G. J. (2002) Correlation of viral detection with histology in cardiac tissue from patients with sudden, unexpected death. *Mod. Pathol.* **15,** 914–922.
15. Parham, P. (2000) *The Immune System.* Garland Publishing, NY.
16. Goldsby, R. A, Kindt, T. J., Kuby, J., and Osborn, B. A. (2000) *Immunology*, 4th ed. New York: W. H. Freeman and Co.
17. Manetti, R., Annunziato, F., Giannò, V., Tomasévic, L., Beloni, L., Mavilia, C., and Maggi, E. (1996) Th1 and Th2 cells in HIV infection. In: *Th1 and Th2 in Health and Disease* (Romagnani, S. ed.) *Chem. Immunol.* Basel, Karger, **63,** 138–157.
18. Breen, E. C. (2002) Pro-and anti-inflammatory cytokines in human immunodeficiency infection and acquired immunodeficiency syndrome. *Pharmacol. Ther.* **95,** 295–304.
19. Gage, J. R., Sandhu, A. K., Nihira, M., Bonecini-Almeida, M. G., Cristoforoni, P., Kishimoto, T., et al. (2000) Cervical cancer cell lines and human papillomavirus (HPV)—immortalized keratinocytes induce HIV-1 in the U1 monocyte line. *J. Obst. Gynecol.* **96,** 879–885.

14

Detection of HTLV-1 Gene on Cytologic Smear Slides

Kenji Kashima, Tsutomu Daa, and Shigeo Yokoyama

Summary

In this chapter we describe a method for the detection of human T-cell leukemia virus type 1 (HTLV-1) genes in cytologic smears by polymerase chain reaction (PCR). First, already-stained and covered slides should be immersed in xylene for removal of cover slips. After passage through a descending ethanol series, slides are ready for DNA extraction. If the neoplastic cells on slides are mixed with nonneoplastic lymphocytes, cells of interest are isolated by microdissection. Two easy methods to dissect the samples using hydrophobic and hydrophilic mounting media are detailed. Second, microdissected cells are collected in microtubes and digested with proteinase K. The cells that did not undergo the microdissection are digested and dissolve in the proteinase K solution on the slides. Last, the template DNA is extracted from the solution and provided to PCR. We use two sets of primers for detection of HTLV-1 genes, and the products of amplification by PCR that correspond to the pX and tax regions are expected to be 127 and 159 base pairs long, respectively. Although this method does not provide proof of the monoclonal integration of HTLV-1 genes, it can be applied when adult T-cell leukemia/lymphoma is suspected cytologically but fresh samples for Southern blotting are unavailable.

Key Words: Cytologic samples; HTLV-1 gene; pX region; tax region; polymerase chain reaction; microdissection.

1. Introduction

Adult T-cell leukemia/lymphoma (ATLL) is a very aggressive neoplasm that is caused by human T-cell leukemia virus type-1 (HTLV-1). Clinically, ATLL is classified into four types: acute, chronic, smoldering, and lymphoma. Cytologic smears that include neoplastic cells are usually obtained from the peripheral blood of patients with acute ATLL and from the pleural effusions or cerebrospinal fluid of patients with advanced acute or lymphoma-type ATLL. Cytologic examination reveals that neoplastic cells have hyperlobulated nu-

From: *Methods in Molecular Biology, Vol. 304: Human Retrovirus Protocols: Virology and Molecular Biology*
Edited by: T. Zhu © Humana Press Inc., Totowa, NJ

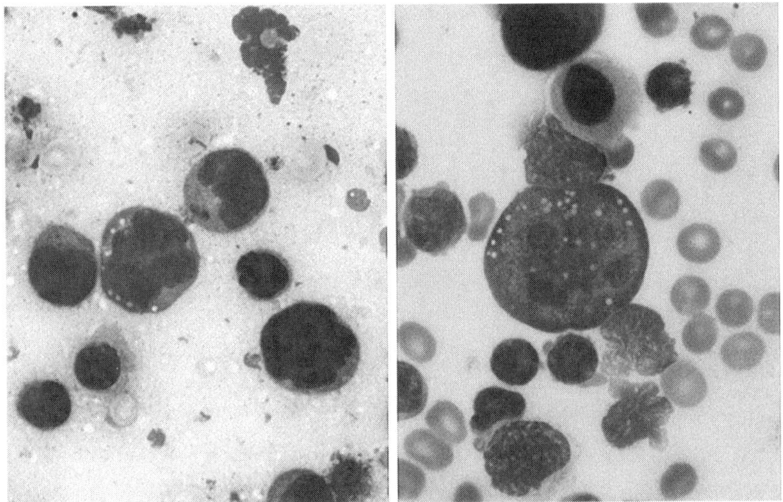

Fig. 1. Atypical lymphocytes with hyperlobulated nuclei from a patient with adult T-cell leukemia/lymphoma.

clei; cells with this peculiar morphologic feature are known as "cloverleaf cells" (**Fig. 1**). ATLL is defined as a peripheral T-cell neoplasm resulting from infection by human T-cell leukemia virus type 1 (HTLV-1). Thus, by themselves, the morphologic features of the neoplastic cells are insufficient for an unequivocal diagnosis of ATLL. An accurate diagnosis of ATLL requires proof of the monoclonal integration of HTLV-1 into the neoplastic cells, which is usually obtained by Southern blotting analysis of fresh samples, such as lymphoma tissue or peripheral blood.

We describe here a method for the detection of HTLV-1 genes in cytologic smears by polymerase chain reaction (PCR). This method can be applied when ATLL is suspected cytologically but fresh samples for Southern blotting are unavailable (*see* **Note 1**).

2. Materials

1. Proteinase K.
2. Tris-buffered saline (pH 8.0): 10 m*M* Tris-HCl buffer, pH 8.0, 150 m*M* NaCl.
3. TE buffer: 10 m*M* Tris-HCl buffer, pH 8.0, 10 m*M* ethylenediamine tetraacetic acid (EDTA), 150 m*M* NaCl, and 0.1% sodium dodecyl sulfate (SDS).
4. Humid chamber designed for immunohistochemistry.
5. PCR buffer: 50 m*M* KCl, 10 m*M* Tris-HCl buffer, pH 8.0, 0.2 m*M* dNTPs, and 1 m*M* $MgCl_2$ in sterile H_2O.
6. *Taq* polymerase (Perkin Elmer, Norwalk, CT).

7. Oligonucleotide primers (*see* **Subheading 3.3.**) for HTLV-1 pX and tax genes (TLV-1019/1020; Maxim Biotech Inc., South San Francisco, CA).
8. PCR thermal cycler (GeneAmp™ PCR System 2400, Perkin-Elmer).
9. Equipment for polyacrylamide gel electrophoresis (PhastSystem™; Pharmacia Biotech, Uppsala, Sweden).
10. Polyacrylamide gel (Homogeneous 12.5™; Pharmacia Biotech).
11. DNA silver-staining kit for polyacrylamide gels (Pharmacia Biotech).
12. Equipment for agarose gel electrophoresis, agarose and ethidium bromide as substitutes for **items 9–11**.

3. Methods

The method can be divided into four steps: (1) pretreatment of cytologic smears; (2) extraction of DNA from cytologic smears; (3) preparation of primers for amplification of HTLV-1 genes; and (4) amplification by PCR.

3.1. Pretreatment of Cytologic Smears

3.1.1. Removal of Cover Slips

Both ethanol-fixed/Papanicolaou-stained slides and air-dried/May-Giemsa-stained slides can be used *(1)*. Covered slides should be immersed in xylene for removal of cover slips. If cover slips cannot be removed easily, slides can be incubated in xylene at 37°C for several days. Slides without the cover slips are rinsed several times in fresh xylene, and in 100% ethanol for 5 min to remove xylene. After passage through a descending ethanol series (95%, 90%, and 80%; 5 min each), slides are washed in sterile H_2O. The cells on slides are now ready for the extraction of DNA (*see* **Subheading 3.2.1.**)

3.1.2. Microdissection

Omit this step if it is not necessary, and proceed to the next step, namely, extraction of DNA.

If the neoplastic cells on slides are mixed with nonneoplastic lymphocytes, cells of interest are isolated by microdissection. Laser-capture microdissection equipment allows microdissection at the single-cell level, but we recommend two easier and less expensive methods *(2,3)*.

3.1.2.1. Transfer of Cells Using Hydrophobic Mounting Medium (*see* **Note 2**)

After removal of cover slips with xylene and rinsing of slides in fresh xylene, hydrophobic mounting medium (Mount-Quick™; Daido Sangyo Co., Tokyo, Japan) is dropped on the target areas on the slides, which are then heated at 60°C for 2 h until the medium hardens. The slides with hardened media are dipped in water at 37°C, and the medium becomes soft enough to be peeled off.

The blade of a scalpel is used to remove the medium plus target cells from each slide. Unwanted cells can be removed by cutting the solidified medium with a knife. The trimmed solidified medium with the target cells is placed in a microtube and treated with xylene, a graded ethanol series, and H_2O, as described above.

3.1.2.2. Transfer of Cells Using Hydrophilic Mounting Medium (*see* **Note 3**)

After removal of cover slips with xylene and passage through the descending ethanol series and sterile H_2O, hydrophilic mounting medium (Crystal Mount™; Biomeda, Foster City, CA) is dropped on the slides. After 20–60 min at 37°C, the medium becomes hard enough to be cut with a knife. Under the light microscope, a small piece of solidified medium, containing target cells, can be isolated with a surgical knife. The small piece of solidified medium is transferred to a microtube with forceps and washed several times with sterile H_2O to remove the medium.

3.2. Extraction of DNA

3.2.1. Digestion of Cells on Slides or in Tubes

The re-hydrated smears on slides are washed in 10 m*M* Tris-buffered saline, pH 8.0, and placed in a humid chamber for immunocytochemical staining. Proteinase K, diluted at 200 µg/mL in TE buffer, is applied to the glass slides and then incubated at 37°C for 24 h. The cells are digested and dissolve in the solution on the slides. The solution on each slide is collected with a micropipet and transferred to a microtube (*see* **Note 4**).

Microdissected cells in microtubes, as described in **Subheading 3.1.2.**, are also digested with proteinase K under the same conditions but, in this case a more concentrated solution of proteinase K (400 µg/mL) is added to an equal volume of cell suspension.

3.2.2. Extraction of DNA

The samples, after digestion on slides or in tubes, are boiled for 5 min to inactivate proteinase K. Then an equal volume of phenol is added to each sample, with a brief mixing on a vortex mixer. After centrifugation at 10,000*g* for 30 min, 3.3 µL of 3 *M* sodium acetate, 1 µL of Ethacinmate (Nippon Gene, Toyama, Japan), and 250 µL of 100% ethanol are added to 100 µL of each supernatant. The mixture is mixed thoroughly on a vortex mixer and centrifuged at 10,000*g* for 5 min. The pellet is washed with 70% ethanol, air-dried, and resuspended in 50 µL of sterile H_2O.

3.3. Primers for Amplification of HTLV-1 Genes

We use two sets of primers for detection of HTLV-1 genes: one for detection of the pX region, designed by Hall et al. *(4)*, and one for detection of the tax region, which was designed by and can be obtained from Maxim Biotech. The sequences of primers for the pX region are as follows: forward, CCA ATC ACT CAT ACA ACC CCC A; and reverse, CTG GAA AAG ACA GGG TTG GGA G. The products of amplification by PCR that correspond to pX and tax are expected to be 127 base pairs (bp) and 159 bp long, respectively.

3.4. Amplification by PCR (see *Note 5*)

The reaction mixture (total volume, 50 µL) contains 20 pmol of each primer, 2.5 U of *Taq* polymerase, 1 µL of the solution of template DNA, and 35 µL of PCR buffer (50 m*M* KCl, 10 m*M* Tris-HCl [pH 8.0], 0.2 m*M* dNTPs, and 1 m*M* $MgCl_2$). The conditions for amplification are as follows: for the pX region, denaturation at 94°C for 1 min, annealing at 56°C for 1.5 min, and primer extension at 72°C for 2 min; and for the tax region, denaturation at 94°C for 1 min, annealing at 58°C for 1 min, and primer extension at 72°C for 1 min. The final extension is allowed to proceed for an additional 10 min. Amplification is allowed to proceed for 30 cycles in a thermal cycler (GeneAmp PCR System 2400). The products of PCR are analyzed by electrophoresis on a polyacrylamide gel (Homogeneous 12.5) and staining with a DNA silver-staining kit (PhastSystem) or on a 1% agarose gel supplemented with 0.1% ethidium bromide and a subsequent examination on an ultraviolet (UV) illuminator.

Products of approx 130 bp and 160 bp are obtained for pX and tax, respectively (**Fig. 2**).

4. Notes

1. This method does not provide proof of the monoclonal integration of HTLV-1 genes. Furthermore, contaminating nonneoplastic T-cells that have been infected by HTLV-1 can interfere with the results. To solve this problem, we need a method that demonstrates the existence of the virus at the single-cell level. In our hands, nonradioactive *in situ* hybridization using biotin-labeled probes did not work well, probably because of the low number of copies of HTLV-1 genes in infected cells. It has been reported that a technique involving PCR and *in situ* hybridization allows confirmation of the presence of HTLV-1 genes at the single-cell level with a high degree of sensitivity *(5)*.
2. The cell-transfer method using hydrophobic mounting medium can be used for the transfer of cytologic samples from one slide to another. Samples that have been peeled off slides with hardened medium can be divided with scissors or a knife and can be transferred to several separate slides for immunocytochemical analysis with multiple antibodies.
3. The most critical issue when hydrophilic mounting medium is used is the duration of heating. When the medium is heated too long and becomes too hard, it is difficult to remove from slides and to cut. Optimal conditions vary from case to case, depending on the amount of residual water on each slide, and it is recommended that a shorter heating period (20–30 min) be tested initially.

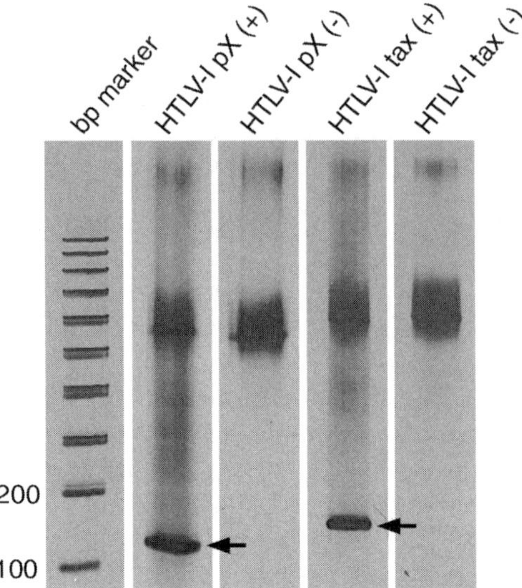

Fig. 2. Products of amplification of the pX and tax region of human T-cell leukemia virus type-1 on a polyacrylamide gel after staining with silver.

4. Contamination of one sample by others must be avoided after cover slips have been removed. The inside of the humid chamber must be cleaned for each trial. The microdissection methods described in this chapter should help to prevent contamination by other samples.
5. When the results of PCR are negative, the integrity of the extracted DNA should be examined. We routinely amplify the gene for nucleophosmin or exon 3 of the gene for β-catenin by PCR since the products of amplification of these sequences are longer than the products expected after the amplification of the pX or tax region. The respective primers and conditions for PCR have been described by Wood et al. *(6)* and Kitaeva et al. *(7)*, respectively. The product of amplification of the gene for nucleophosmin is 185 bp long and that of exon 3 of the gene for β-catenin is 228 bp long.

Acknowledgments

The authors thank Drs. H. Kikuchi, E. Ohtsuka, Y. Saburi, K. Tsuji, S. Urabe, K. Ono, A. Gamachi, and T. Nakayama for their cooperation in preparing cytologic samples from patients with ATLL.

References

1. Kashima, K., Nagahama, J., Sato, K., Tanamachi, H., Gamachi, A., Daa, T., et al. (2002) Detection of the HTLV-I gene on cytologic smear slides. *Acta Cytol.* **46,** 709–712.

2. Sherman, M. E., Jimenez-Joseph, D., Gangi, M. D., and Rojas-Corona, R. R. (1994) Immunostaining of small cytologic specimens: facilitation with cell transfer. *Acta Cytol.* **38,** 18–22.
3. Honda, K., Kashima, K., Daa, T., Yokoyama, S., and Nakayama, I. (2000) Clonal analysis of the epithelial component of Warthin's tumor. *Hum. Pathol.* **31,** 1377–1380.
4. Hall, W. W., Liu, C. R., Shneewind, O., Takahashi, H., Kaplan, M. H., Röupe, G., and Vahlne, A. (1991) Deleted HTLV-I provirus in blood and cutaneous lesions of patients with mycosis fungoides. *Science* **253,** 317–320.
5. Setoyama, M., Kerdel, F. A., Elgart, G., Kanzaki, T., and Byrnes, J. J. (1998) Detection of HTLV-1 by polymerase chain reaction *in situ* hydridization in adult T-cell leukemia/lymphoma. *Am. J. Pathol.* **152,** 683–689.
6. Wood, G. S., Schaffer, J. M., Boni, R., Dummer, R., Burg, G., Takeshita, M., and Kikuchi, M. (1997) No evidence of HTLV-I proviral integration in lymphoproliferative disorders associated with cutaneous T-cell lymphoma. *Am. J. Pathol.* **150,** 667–673.
7. Kitaeva, M. N., Grogan, L., Williams, J. P., Dimond, E., Nakahara, K., Hausner, P., et al. (1997) Mutations in β-catenin are uncommon in colorectal cancer occurring in occasional replication error-positive tumors. *Cancer Res.* **57,** 4478–4481.

15

Detection of HIV-2 by PCR

Feng Gao

Summary

Genomes of human immunodeficiency virus type 2 (HIV-2), like those of HIV-1 or other retroviruses, are highly variable. These genetic variants have been classified into seven genetic subtypes ***(1–4)***. The average genetic divergence between different subtypes is about 20% in the *gag* gene, which is higher than those among HIV-1 group M subtypes ***(1)***. The current serological tests cannot distinguish the different subtypes from one another. To understand genetic variation, evolution, and subtype distribution of HIV-2, polymerase chain reaction (PCR) technology has been widely used. The PCR products that are amplified from highly conserved regions among all subtypes can be either cloned into plasmid vectors for sequence analysis or directly sequenced without a cloning step. Phylogenetic analysis of newly obtained sequences with reference sequences can determine the subtype classification or identify new subtypes if the sequences do not belong to any known subtypes.

Key Words: HIV-2; PCR; detection; subtype; genome; infection; amplification.

1. Introduction

Polymerase chain reaction (PCR) has become a powerful tool for detection of viral infection. DNA or RNA molecules can be extracted from blood or tissue samples and used as templates for PCR amplification. Although the high level of genetic variation among HIV-2 subtypes makes it difficult to detect all genetic variants, some regions within viral genes are conserved possibly owing to biological function constraints. By comparing known HIV-2 sequences and designing primers at these highly conserved regions, it has been possible to amplify multiple HIV-2 subtypes with broadly reactive PCR primer sets ***(1–4)***.

From: *Methods in Molecular Biology, Vol. 304: Human Retrovirus Protocols: Virology and Molecular Biology*
Edited by: T. Zhu © Humana Press Inc., Totowa, NJ

2. Materials

1. Genomic DNA extraction kit (Qiagen, Valencia, CA).
2. Viral RNA extraction kit (Qiagen).
3. Agarose.
4. DNA marker: φX174 DNA/*Hae* III markers (Promega, Madison, WI).
5. Platinum *Taq* High Fidelity polymerase (Invitrogen, Carlsbad, CA).
6. dNTP mix: 10 m*M* each dATP, dGTP, dCTP, and dTTP.
7. TBE buffer: 90 m*M* Tris-borate and 2 m*M* EDTA.
8. Oligonucleotide primers.
9. Ethanol (96–100%).
10. Microcentrifuge tubes (1.5 mL).
11. RNase-free pipet tips with aerosol barrier.
12. Random hexamers.
13. Dithiothreitol (DTT).
14. SuperScript III RT (Invitrogen).
15. Diethylpyrocarbonate (DEPC)-treated water.
16. $MgCl_2$.

3. Methods

3.1. PCR Amplification of DNA Templates

Although many tissues from HIV-2-infected patients can be used to detect integrated HIV-2 genomes, blood samples can be obtained easily for DNA extraction and used as a template for PCR amplification. With development of new methods for DNA extraction from whole blood or peripheral blood mononuclear cells (PBMC), only small amounts of samples are needed for sample preparation and HIV-2 proviral DNA can be routinely obtained at high quality for PCR amplification ***(1–4)***.

3.1.1. DNA Extraction

Many methods can be used for DNA extraction from HIV-2-infected cells. They include the traditional phenol:chloroform method and a number of commercial DNA extraction kits. Because the phenol:chloroform method requires separation of white blood cells from patient whole blood samples and the chemicals are toxic to human health, it has not been used as frequently as before. Instead, the commercial kits can be used to extract DNA from whole blood samples or PBMC without any pretreatment. This has made DNA sample preparation much easier. Among these kits, the QIAamp DNA Blood Mini Kit has been widely used and is used as an example for DNA extraction (*see* **Note 1**). The detailed protocol can be obtained from the manufacturer (Qiagen). A brief protocol for the QIAampSpin Column is described below (*see* **Note 2**):

1. Add 20 μL Qiagen Protease to the bottom of a 1.5-mL tube.
2. Add 200 μL whole blood or 5×10^6 PBMC (or cultured cells) in 200 μL PBS (*see* **Note 3**).

3. Add 200 µL buffer AL to the sample and mix by pulse-vortexing for 15 s.
4. Incubate at 56°C for 10 min.
5. Briefly centrifuge the 1.5-mL tube to remove drops from the inside of the lid.
6. Add 200 µL ethanol (96–100%) to the sample, and mix again by pulse-vortexing for 15 s. After mixing, briefly centrifuge the 1.5-mL tube to remove drops from the inside of the lid.
7. Carefully apply the mixture from **step 6** to the QIAamp Spin Column (in a 2-mL collection tube) without wetting the rim, close the cap, and centrifuge at 6000*g* for 1 min. Place the QIAamp Spin Column in a clean 2-mL collection tube, and discard the tube containing the filtrate.
8. Carefully open the QIAamp Spin Column and add 500 µL buffer AW1 without wetting the rim. Close the cap and centrifuge at 6000*g* for 1 min. Place the QIAamp Spin Column in a clean 2-mL collection tube, and discard the tube containing the filtrate.
9. Carefully open the QIAamp Spin Column and add 500 µL buffer AW2 without wetting the rim. Close the cap and centrifuge at 20,000*g* for 3 min.
10. Place the QIAamp Spin Column in a clean 1.5-mL tube and discard the collection tube containing the filtrate. Carefully open the QIAamp Spin column and add 200 µL distilled water. Incubate at room temperature (15–25°C) for 1 min, and then centrifuge at 6000*g* for 1 min.
11. Determination of DNA concentration: Add 20 µL of DNA samples from **step 10** to 80 µL water (1:5 dilution) and measure the absorbance at 260 nm. The concentration of the DNA samples (µg/mL) equals OD value × 50 × 5 (dilution factor). Normally, the preparation will yield a total of 3–18 µg genomic DNA (*see* **Note 4**).
12. The DNA samples are ready to be used for PCR amplification or stored at –20°C for later use.

3.1.2. PCR Amplification

HIV-2 genomes are widely variable; however, some regions are highly conserved possibly owing to functional constraints. Therefore, these regions become attractive targets for designing broadly reactive primers to amplify diverse HIV-2 genetic variants. Two highly cross-reactive and sensitive primers will be used to amplify partial HIV-2 genomes in this protocol. Because the number of proviral copies in infected patient cells is generally low, one round of PCR amplification usually will not yield enough products to be detected by the ethidium bromide (EB) staining method. A second round of PCR (nested PCR) needs to be carried out to obtain enough PCR products for detection. Because nested PCR can be sensitive enough to detect as low as one copy template, a clean working environment is required to avoid false-positive results resulting from sample contamination.

3.1.2.1. Amplification of the *gag* Gene

1. Prepare the master PCR reaction mix in an autoclaved tube at ambient temperature or on ice. The following recipe is for one reaction. Always prepare one extra reaction to have enough master mix for all tested samples.

Component	Volume	Final concentration
10X high fidelity buffer	5 µL	1X
10 m*M* dNTP mixture	1 µL	0.2 m*M* each
50 m*M* $MgCl_2$	2 µL	2 m*M*
Primer gagA	1 µL	20 pmol
Primer gagB	1 µL	20 pmol
Platinum *Taq* High Fidelity	0.2 µL	1 U (*see* **Note 5**)
Autoclaved distilled water	to 45 µL (*see* **Note 6**)	

2. Vortex the mixture and transfer 45 µL master mix to each 200-µL thin-wall PCR tube.
3. Add 5 µL DNA sample (0.1–1 µg) to each tube.
4. Cap the tubes, mix, and centrifuge briefly to collect the contents to the bottom of the tubes.
5. Put tubes into the heat blocks in the thermocycler (with the heating block on the top).
6. Denature the templates for 2 min at 94°C. Then perform 30 cycles of PCR amplification as follows:
   Denature: 94°C for 15 s
   Anneal: 40°C for 30 s (*see* **Note 7**)
   Extend: 68°C for 2 min
7. Extend for additional 5 min at 68°C after the last cycle of PCR amplification.
8. Maintain the reaction at 4°C after cycling. Samples can be stored at –20°C for later use.
9. Prepare master PCR reaction mix as in **step 1** with nested primers gagC and gagF (20 pmol).
10. Follow **steps 2–8** for second-round PCR amplification using 5 µL of first-round PCR products as templates (*see* **Note 8**).
11. Load 5 µL PCR products on a 0.7% agarose gel with 0.5 µg/mL ethidium bromide. In a separate well, load DNA marker for estimation of the size of the PCR products. The expected size of PCR product is 839 bp.
12. Take photos or digital images for record (*see* **Note 9**).
13. Primer sequences:

    First-round PCR primers:
    gagA: 5'-AGGTTACGGCCCGGCGGAAAGAAAA-3' (nt 603–627 according to the positions in the HIV-2_{ROD} genome)
    gagB: 5'-CCTACTCCCTGACAGGCCGTCAGCATTTCTTC-3' (nt 1581–1612)
    The length of PCR products is 1010 bp.
    Second-round PCR primers:
    gagC: 5'-AGTACATGTTAAAACATGTAGTATGGGC-3' (nt 628–655)
    gagF: 5'-CCTTAAGCTTTTGTAGAATCTATCTACATA-3' (nt 1437–1466)
    The length of PCR products is 839 bp.

3.1.2.2. AMPLIFICATION OF PBS REGION

As a result of the high level of genetic variation, some mutations may even occur in the highly conserved regions. This may cause PCR amplification to fail even with the PCR primers that are designed at these regions. Therefore, if the above PCR amplification fails, HIV-2 genomes of multiple subtypes can be amplified using another set of PCR primers that are designed based on a highly conserved nontranslation region in the HIV-2 genome. The following are the primers for the PCR amplification of the nontranslation region:

First-round PCR primers:
PBS2A: 5'-GAGGTTCTCTCCAGCACTAGCAGGT-3' (nt 38–62)
PBS2B: 5'-TTTTCTAATTCATCTGCTTTTTTCCCT-3' (nt 572–598)
The length of PCR products is 563 bp.
Second-round PCR primers:
PBS2C: 5'-TCCCATCTCTCCTAGTCGCCGCCTGGT-3' (nt 188–214)
PBS2D: 5'-CAAGACGGAGTTTCTCGCGCCCAT-3' (nt 546–569)
The length of PCR products is 383 bp.

The same DNA templates and PCR conditions for the amplification of the *gag* gene can be used for the amplification of the nontranslation region in HIV-2 genomes. Since the PCR fragments are smaller with this set of primers, the extension time can be reduced to 1 min.

3.1.3. Amplification of HIV-2 Genomes From Cultured Cells

If HIV-2 isolates have been successfully adapted to grow in either normal donor PBMC or T-cell lines, the viral genome copy numbers in the infected cells are much higher than those in uncultured patient PBMC samples. Therefore, it often needs only a single round of PCR amplification to obtain enough PCR products to be detected on an agarose gel with EB staining. The same DNA templates, PCR conditions, and PCR primers for the amplification of the *gag* gene and the nontranslation region can be used for the PCR amplification of HIV-2 genomes in the infected cell culture.

3.2. PCR Amplification of RNA Templates

HIV-2 genomes can also be amplified from patient plasma samples using the reverse transcriptase (RT)-PCR method. Plasma samples are more easily collected, transported, and stored than patient PBMC samples. Patient plasmas are occasionally the only materials for genetic detection and subtype classification because patient PBMC or virus isolates are not available for DNA extraction. It is important to note that plasma samples can also be used to determine the viral load in patient blood for predicting disease outcomes and to monitor efficacy of antiretroviral therapy.

3.2.1. RNA Extraction

RNA is extremely sensitive to RNases and should be prepared with great care. RNases are very active enzymes and difficult to inactivate. Only minimal amounts of RNase are sufficient to destroy RNA. The most common sources of RNase contamination are from hands and dust particles. Therefore, extra caution is needed during preparation and storage of RNA samples. The tubes, tips, and water used for RNA preparation and storage should be RNase-free. Gloves should be worn at all times when working with RNA. RNA samples should be stored at –80°C; avoid freezing and thawing cycles. Many commercial kits have been used to extract viral RNA from plasma samples. Among these kits, the QIAamp Viral RNA Mini Kit has been widely used and is used as an example for RNA extraction in this protocol (*see* **Note 10**) The detailed protocol can be obtained from the manufacturer. A brief protocol for QIAamp Spin Procedure is described as follows:

1. Pipet 560 µL of buffer AVL containing carrier RNA into a 1.5-mL tube.
2. Add 140 µL plasma to the buffer AVL/carrier RNA and mix by pulse-vortexing for 15 s (*see* **Note 3**).
3. Incubate at room temperature (15–25°C) for 10 min.
4. Briefly centrifuge the 1.5-mL tube to remove drops from the inside of the lid.
5. Add 560 µL ethanol (96–100%) to the sample, and mix again by pulse-vortexing for 15 s. After mixing, briefly centrifuge the 1.5-mL tube to remove drops from the inside of the lid.
6. Carefully apply the mixture from **step 5** to the QIAamp Spin Column (in a 2-mL collection tube) without wetting the rim, close the cap, and centrifuge at 6000*g* for 1 min. Place the QIAamp Spin Column in a clean 2-mL collection tube and discard the tube containing the filtrate.
7. Carefully open the QIAamp Spin Column and add 500 µL buffer AW1. Close the cap and centrifuge at 6000*g* for 1 min. Place the QIAamp Spin Column in a clean 2-mL collection tube and discard the tube containing the filtrate.
8. Carefully open the QIAamp Spin Column and add 500 µL buffer AW2. Close the cap and centrifuge at 22,000*g* for 3 min.
9. Place the QIAamp Spin Column in a new 2-mL collection tube and discard the tube containing the filtrate. Centrifuge at full speed for 1 min.
10. Place the QIAamp Spin Column in a clean 1.5-mL RNase-free tube. Discard the old collection tube containing the filtrate. Carefully open the QIAamp Spin Column and add 60 µL of buffer AVE equilibrated to room temperature. Close the cap, and incubate at room temperature for 1 min. Centrifuge at 6000*g* for 1 min.
11. The RNA samples are ready to be used for reverse transcription or stored at –80°C for later use.

3.2.2. Reverse Transcription and cDNA Synthesis

1. Combine the following in a 0.5-mL tube:

Component	Amount
RNA template	6 µL
Primers	1 µL
Random hexamers (50 ng/µL) or	
gagB (10 pmol) or	
PBS2B (10 pmol)	
10 m*M* dNTP mix	1 µL
DEPC-treated water	2 µL

2. Incubate at 65°C for 5 min, then place on ice for at least 1 min.
3. Prepare the master cDNA synthesis mix in an RNase-free tube at ambient temperature. The following formula is for one reaction. Always prepare one extra reaction to have enough master mix for all tested samples.

Component	Amount
10X RT buffer	2 µL
25 m*M* $MgCl_2$	4 µL
0.1 *M* DTT	2 µL
RNaseOUT (40 U/ µL)	1 µL
SuperScript III RT (200 U/µL)	1 µL

4. Add 10 µL of cDNA synthesis mix to each RNA/primer mix (**step 2**), mix gently, and collect by brief centrifugation. Incubate as follows:
   At 50°C for 50 min for gagB or PBS2B primers.
   At 25°C for 10 min and then at 50°C for 50 min for random hexamers.
5. Terminate the reaction at 85°C for 5 min.
6. The cDNA synthesis reaction can be stored at –20°C or used for PCR amplification immediately.

3.2.3. Nested PCR Amplification

Four microliters of cDNA synthesis reaction is used for each PCR reaction using HIV-2 *gag* or PBS primers sets as described in **Subheading 3.1.2.** (For troubleshooting and precautions of PCR amplification, *see* **Notes 11** and **12**).

3.2.4. Reaction Setup

Because many steps are involved in the RT-PCR procedure, it is critical to include a number of controls to ensure that the final PCR amplification is specific for the target genes. For each RT-PCR experiment, the following controls should be included:

	Sample	No RT control	No sample control	Negative PCR control	Positive control
RNA template	+	+	–	–	known RNA
RT	+	–	+	–	+
PCR	+	+	+	+	+

4. Notes

1. Although the DNA extraction kit made by Qiagen has been widely used, the following similar kits from other manufacturers can also be used to obtain the same quality of DNA samples: GenElute Mammalian Genomic DNA Miniprep kit from Sigma, St. Louis, MO; Easy-DNA™ Kit from Invitrogen; and Wizard® Genomic DNA Purification Kit from Promega, Madison, WI.
2. The buffers used in this protocol come with the kits. There is no need to prepare them.
3. All infectious materials (whole blood, plasma, or infected cells) that contain live HIVs should be prepared under biosafety level 2 (BSL-2) laminar flow hoods. Extreme cautions should be taken to avoid any direct contacts to the infectious materials.
4. When more DNA samples are needed, DNA extraction kits for larger amounts of samples are available from the manufacturer.
5. The supplied Platinum *Taq* High Fidelity enzyme binds to the specific Platinum *Taq* antibody and is inactive. After heating at 94°C for 2 min, the polymerase activity will be restored. This will provide an automatic hot-start and increase specificity, sensitivity, and yield. Up to 2.5 U can be used for each reaction. Buffers and reagents needed for the PCR amplification come with the kit. Other *Taq* polymerases can also be used.
6. The total volume of the PCR reaction is 50 μL. Five microliters of each sample were subtracted from the master mix.
7. Annealing temperature is critical for successful PCR amplification. Since the genetic variation of HIV-2 genomes is very high, it is difficult to design primers that can perfectly match the templates of all subtypes. The primers generally have some mismatches with the template. Therefore, the annealing temperature is relatively low to ensure the binding between primers and templates. The annealing temperature should be adjusted for any particular PCR reaction for optimal amplification.

8. In general, 5 µL of the first-round PCR reaction is a good starting point for the second round PCR amplification. However, depending on the DNA sample quality and the copy numbers of target molecules, more or less of the first round PCR reaction may be needed for the optimal second round PCR reaction.
9. The PCR products can also be detected by other more sensitive methods if the final PCR products do not need to be visualized or recovered for direct sequencing or cloning. The most common methods will be the use of radioactive compound-labeled PCR products, hybridization, or enzyme-linked assays.
10. Although the viral RNA extraction kit made by Qiagen has been widely used, the following similar kits from other manufactures can also be used to obtain RNA templates: Tri Reagent RNA Isolation Reagent from Sigma and Micro-to-Midi Total RNA Purification System from Invitrogen.
11. Troubleshooting for PCR amplification:

Problem	Solution
No band	Lower annealing temperature
	Increase amounts of DNA and/or RNA templates
	Increase $MgCl_2$ concentration
	Increase cycle number
	RNA degraded; use new RNA samples
Too many bands	Increase annealing temperature
	Decrease $MgCl_2$ concentration
	Decrease amount of template
	Decrease primer concentration
Wrong size bands	Increase annealing temperature
Primer-dimer	Decrease primer concentration
	Increase annealing temperature
Band in negative control	Contamination; prepare new reagents, buffers, and water

12. Precautions:
    a. Never use PCR products or work with molecular clones in PCR reaction preparation areas or rooms.
    b. Always use the following controls:
        - Low copy positive control.
        - Negative control (normal genomic DNA).
    c. Always use aerosol-resistant tips.
    d. Always wear new gloves.

References

1. Gao, F., Yue, L., Robertson, D. L., Hill, S. C., Hui, H., Biggar, R. J., et al. (1994) Genetic diversity of human immunodeficiency virus type 2: evidence for distinct sequence subtypes with differences in virus biology. *J. Virol.* **68,** 7433–7447.

2. Chen, Z., Luckay, A., Sodora, D. L., Telfer, P., Reed, P., Gettie, A., et al. (1997) Human immunodeficiency virus type 2 (HIV-2) seroprevalence and characterization of a distinct HIV-2 genetic subtype from the natural range of simian immunodeficiency virus-infected sooty mangabeys. *J. Virol.* **71,** 3953–3960.
3. Gao, F., Yue, L., White, A. T., Pappas, P. G., Barchue, J., Hanson, A. P., et al. (1992) Human infection by genetically diverse SIVSM-related HIV-2 in west Africa. *Nature* **358,** 495–499.
4. Yamaguchi, J., Devare, S. G., and Brennan, C. A. (2000) Identification of a new HIV-2 subtype based on phylogenetic analysis of full-length genomic sequence. *AIDS Res. Hum. Retroviruses* **16,** 925–930.

16

Quantitation of HIV-1 Viral RNA in Blood Plasma and Genital Secretions

Susan A. Fiscus

Summary

Quantitation of HIV RNA in blood is commonly used to monitor progression of the disease and to assess the effect of antiretroviral therapy in individuals. Although not approved in the US for diagnosis of HIV infection, the finding of a positive HIV RNA with a negative HIV enzyme immunoassay and Western blot (or evolving Western blot) is an indication of primary HIV infection and should be followed up closely. Large clinical trials and cohort studies have demonstrated the importance of HIV RNA as an indicator of drug efficacy and as a factor in HIV transmission.

Sexual intercourse is the most common method of transmission of HIV-1. Several studies have demonstrated that blood plasma viral load is significantly correlated with the risk of sexual HIV transmission. Additional investigations have found a significant correlation between the viral load in blood plasma and in the genital tract. This chapter describes methods of collection, processing, and testing in blood, plasma, and male and female genital secretions for quantifying HIV RNA.

Key Words: HIV RNA; viral load; seminal plasma; cervicovaginal secretions; genital tract viral load.

1. Introduction

The ability to measure the concentration of HIV RNA in blood and genital fluids accurately and sensitively has provided insights into the transmission and pathogenesis of HIV infection. For instance, Ho et al. *(1)* and Wei et al. *(2)* have demonstrated the staggering dynamics of HIV virion production daily in infected individuals. Several studies have demonstrated that blood plasma viral load is significantly correlated with the risk of sexual and perinatal HIV transmission *(3–7)*.

From: *Methods in Molecular Biology, Vol. 304: Human Retrovirus Protocols: Virology and Molecular Biology*
Edited by: T. Zhu © Humana Press Inc., Totowa, NJ

Additional investigations have found a significant correlation between blood plasma viral load and the viral burden in the genital tract *(8–12)*. This has led to speculation that one could reduce HIV transmission by treating infected individuals with highly active antiretroviral therapy (HAART). Certainly this has been demonstrated in perinatal HIV, where transmission has been reduced from approx 25% in the absence of therapy to less than 2% in women who receive combination antiretroviral treatment *(13)*.

Cross-sectional studies have demonstrated that approx 60–70% of HIV-infected men and women shed HIV in the genital tract *(8–12)*. Longitudinal studies have shown that HIV RNA in the genital tract increases with time in individuals who progress to AIDS and decreases with effective antiretroviral therapy *(14–17)*.

Quantitating HIV RNA in the male genital tract is relatively straightforward. There is, after all, only one major male genital secretion—semen—and we have developed ways to measure the viral load in it. One impediment to determining viral load in seminal plasma was the frequent occurrence of inhibition of PCR *(18)*. This has been obviated by use of the Boom silica extraction method *(19)* as described by Dyer et al. *(18)* and Coombs et al. *(9)*.

Female genital secretions are somewhat more complicated. First, there is the question of whether to sample the vagina or the cervix. Second, if one uses lavage to collect the specimen, one is left with the issue of how to correct for dilution. Third, the menstrual cycle complicates collection and possibly interpretation of the data. Some of these questions have been studied in detail *(11,20–22)*.

We have chosen to use commercially available viral load assays for several reasons: they are readily available and there are considerable published data both for measuring HIV in the genital tract *(9,11,12,14–18)* as well as a wealth of papers comparing the two most commonly used assays, Roche Amplicor Monitor and the bioMerieux NucliSens assays *(23–27)*. Commercial kits have a good record of quality control of all kit components, both of these assays contain internal standards that are particularly important when dealing with specimens that may contain inhibitors such as seminal plasma, and they allow cross-study comparisons.

2. Materials

2.1. Collection, Processing, and Testing of Blood (see *Notes 1 and 2*)

2.1.1. Collection of Blood

Plasma for HIV RNA quantitation should be collected in ethylenediamine tetraacetic acid (EDTA) or acid-citrate-dextrose (ACD) anticoagulated blood collection tubes, not heparin-containing tubes (*see* **Note 3**).

2.1.2. Processing of Blood

1. 2.0-mL cryovials.
2. Pipets.

2.1.3. Quantitation of HIV RNA in Blood Plasma

One of the three US Food and Drug Administration (FDA)-approved HIV RNA licensed kits—Roche Amplicor Monitor, bioMerieux NucliSens, or Bayer Versant assay (*see* **Note 4**).

2.2. Collection, Processing, and Testing of Seminal Plasma

2.2.1. Collection of Semen

1. Sterile urine container.
2. Antiseptic towelette.

2.2.2. Processing of Semen

1. 15-mL conical centrifuge tube.
2. 2.0-mL cryovials.
3. Pipets.

2.2.3. Quantitation of HIV RNA in Seminal Plasma

Seminal plasma contains inhibitors of RT-PCR that can be removed during RNA isolation using the Boom extraction technique *(19)*. After RNA isolation either the bioMerieux NucliSens HIV QT or the Roche Amplicor Monitor assays can be used for HIV RNA quantitation.

2.2.3.1. bioMerieux Nucli-Sens HIV QT Assay

Use bioMerieux's NucliSens Assay following instructions found in the package insert. The Boom silica bead nucleic acid purification method is part of the NucliSens assay so the seminal plasma can be tested directly with the kit with no need to make additional reagents.

2.2.3.2. Roche Amplicor Monitor Assay (Boom Silica Bead Extraction Required)

1. Reagents
   a. Diethylpyrocarbonate (DEPC) distilled water.
   b. Trizma base.
   c. Concentrated HCl.
   d. Guanidinium thiocyanate.
   e. Triton-X 100.
   f. Silicon dioxide (Sigma).
   g. 70% Ethanol (Prepare fresh: 11 mL 95% ETOH plus 4 mL dH_2O for six specimens).

   h. Acetone (reagent grade >99% pure).
   i. 0.2 *M* EDTA, pH 8.0.
   j. Roche Monitor Kit.
2. Equipment and supplies
   a. 2.0-mL Sarstedt tubes.
   b. Vortex mixer.
   c. Microcentrifuge (12,000*g*).
   d. Transfer pipets with thin tips.
   e. Dry heating block.
   f. Water bath.
   g. Aspiration device or transfer pipets.
   h. pH meter.
   i. P1000 and P200 pipettors.
   j. Aerosol barrier pipet tips.

2.3. Collection, Processing, and Testing of Cervical Fluids

2.3.1. Collection of Cervical Fluid

Various genital fluids can be used to measure HIV RNA in the female genital tract. Among the most sensitive are cervical fluid collected in Sno-Strip™ wicks (Akorn, Decatur, IL). However, these are sometimes difficult to obtain. As an alternative, a Dacron swab can be used to collect the specimen (*see* **Note 5**). Both of these methods, however, require placing a woman in stirrups for the collection. Webber et al., have described a methods of self-collection using a tampon ***(28,29)***.

1. Vaginal speculum.
2. Sno-Strips (Akorn).
3. Cotton swabs for removing mucus clot, if necessary.
4. Dacron swabs if Sno-Strips are not available.
5. Ring or sponge forceps.
6. NASBA lysis buffer (bioMerieux, Durham, NC) or 4 *M* guanidine isothiocyanate.
7. 1.5-mL cryovials.

2.3.2. Processing of Cervicovaginal Fluid

No reagents or equipment necessary.

2.3.3. Quantitation of HIV-1 RNA in Cervicovaginal Fluid

Either the NucliSens or the Roche Monitor assays can be used following the package inserts. Cervical fluids do not appear to have the inhibitors commonly found in seminal plasma ***(30–32)***.

3. Methods

3.1. Collection, Processing, and Testing of Blood

3.1.1. Collection of Blood

Anticoagulated blood should be collected in EDTA or ACD tubes, not in heparin (*see* **Notes 1** and **3**).

3.1.2. Processing of Blood

Blood tubes should be centrifuged at 800–1600*g* for 20 min or more at room temperature. Alternatively, the EDTA tubes should be centrifuged at 800*g* for 10 min at room temperature. Plasma should be carefully removed from each tube, placed in a sterile polypropylene conical centrifuge tube and centrifuged again (at 800*g*) at room temperature (21–26°C) for 10 min to completely remove platelets and cell debris (*see* **Note 6**).

Aliquots of 1 mL should be stored at –70°C until needed (*see* **Note 7**).

3.1.3. Quantitation of HIV RNA in Blood Plasma

Follow the directions found in the manufacturer's package insert.

3.2. Collection, Processing, and Testing of Seminal Plasma

3.2.1. Collection of Semen

1. The subject should refrain from sexual activity for at least 48 h prior to donation.
2. The subject should wash his hands and penis and then use an antiseptic towelette to wipe the head of the penis including the opening. If the subject is uncircumcised, the foreskin should be pulled back before cleaning the head and opening.
3. The subject should masturbate and collect the specimen in a sterile container, such as a sterile urine collection cup.
4. The container should be placed in a zip-lock bag and then in an appropriate transport carrier. The mode of transport will determine the type of carrier required. For example, specimens transported by cab must comply with Department of Transportation regulations for infectious substances. Efforts should be made to keep the specimen container upright.
5. The specimen should be rapidly transported at room temperature to the clinic or directly to the lab. The laboratory should receive the specimen within 2 h if possible.

3.2.2. Processing of Seminal Plasma

1. Allow liquefaction of the semen to occur. This typically occurs within 20–45 min of specimen collection.
2. Transfer the sample to a conical centrifuge tube using a pipet, then measure and record the volume of semen.

3. Centrifuge at 600–800*g* for 10 min.
4. Remove supernatant, divide into 0.25-mL aliquots, and freeze at –70°C (*see* **Note 7**). Seminal cells may be used for HIV-1 culture (*see* Chapters 5 and 6).

3.2.3. Quantitation of HIV-1 RNA in Seminal Plasma

One of the following two assay options may be used:

3.2.3.1. bioMerieux NucliSens HIV QT Assay

Use bioMerieux's NucliSens HIV QT assay following instructions found in the package insert.

3.2.3.2. Roche Amplicor Monitor Assay (Boom Silica Bead Extraction Required)

1. 0.2 *M* EDTA, pH 8.0: Add 9 g EDTA to 121 mL DEPC distilled water and check pH.
2. L2 buffer (2 L).
   a. Dissolve 24.22 g Trizma base in 1600 mL DEPC distilled water.
   b. Adjust pH to 6.4 with concentrated HCl (approx 15.5 mL).
   c. Cool to room temperature.
   d. Check pH and readjust to pH 6.4.
   e. Add DEPC water to make a total volume of 2000 mL.
   f. Store at room temperature in the dark up to 6 months.
3. Washing buffer (approx 2 L).
   a. Dissolve 1200 g guanidinium thiocyanate (GuSCN) in 1000 mL L2 buffer. Heating to 60–65°C with shaking facilitates this process.
   b. Store at room temperature in the dark for up to 6 mo.
4. Lysis buffer (about 800 mL).
   a. Dissolve 660 g GuSCN in 550 mL L2 buffer (*not* washing buffer). Heating to 60–65°C with shaking facilitates the process.
   b. Add 121 mL 0.2 *M* EDTA, pH 8.0.
   c. Add 14.3 g Triton-X 100.
   d. Mix and store at room temperature in the dark up to 6 mo.
5. Silica reagent
   a. Place 60 g silicon dioxide in a 500-mL glass cylinder.
   b. Add dH_2O to 500 mL and allow to stand overnight at room temperature.
   c. Remove supernatant (approx 430 mL) by suction.
   d. Add dH_2O to 500 mL and shake vigorously to resuspend silica.
   e. Let stand 5 h at room temperature.
   f. Remove supernatant (approx 440 mL) by suction.
   g. Adjust pH to 2.0 with approx 400 µL of concentrated HCl.
   h. Dispense in glass containers and autoclave for 15 min.
   i. Store at room temperature in the dark for up to 1 yr.
6. Roche monitor HIV RNA assay.
   a. Add 100 µL of Roche QS to 12 mL of lysis buffer. Use the QS from the kit

you intend to use to assay the specimens. Do not mix QS lots. If the specimen is already in lysis buffer, add 6.9 µL of the Roche QS to each sample in lysis buffer.

b. Mix well by vortexing for 5 s and tilting the tube several times.
c. Aliquot 900 µL into each labeled 2.0-mL Sarstedt tube.
d. Add 200 µL of seminal plasma or kit controls. Roche Monitor kit controls should be processed per Roche package insert instructions (50 µL control mixed with 200 µL negative human plasma).
e. Cap and invert tube three to five times, then microfuge briefly (quick spin-down).
f. Resuspend silica solution by vigorous mixing.
g. Add 50 µL to each tube of Lysis buffer.
h. Cap the tubes and vortex immediately until solution is homogeneous (5–10 s).
i. Incubate at room temperature for 10 min.
j. Vortex for 5 s.
k. Centrifuge for 15 s at 12,000*g*. Aspirate supernatant with fine-tipped transfer pipet and discard.
l. Add 1 mL of washing buffer and vortex until pellet is completely resuspended.
m. Centrifuge for 15 s at 12,000*g*.
n. Use a fine-tipped transfer pipet to aspirate supernatant.
o. Repeat **steps l–n**.
p. Add 1.0 mL 70% ethanol and vortex until pellet is completely resuspended. This may be somewhat difficult—shaking the vial may help.
q. Centrifuge for 15 s at 12,000*g*.
r. Use fine-tipped transfer pipet to aspirate supernatant.
s. Repeat **steps p–r**.
t. Add 1.0 mL acetone and vortex thoroughly.
u. Centrifuge for 15 s at 12,000*g*.
v. Use fine-tipped transfer pipet to aspirate supernatant. Pellet may be slick. Recentrifuge if pellet is accidentally aspirated.
w. With a 200-µL ART tip, aspirate any remaining acetone.
x. Evaporate acetone by incubating open vials in 56°C heating.block for 10–15 min. Pellet must be dry.
y. Add 400 µL of Roche elution buffer.
z. Recap vials and vortex until pellet is resuspended.
aa. Incubate 10 min at 56°C, vortexing once after 5 min.
bb. Centrifuge at 12,000*g* for 2 min to pellet silica. The supernatant contains the RNA.
cc. Amplify immediately or store frozen at –20°C until ready to proceed with the HIV Monitor Assay for up to 1 wk. If samples are frozen prior to amplification, thaw them, vortex to resuspend the silica, heat to 56°C for 10 min and centrifuge at 12,000*g* for 2 min before adding 50 µL of the supernatant containing RNA to the Roche PCR tubes.
dd. Proceed with the Roche Monitor Assay following the package insert.

3.3. Collection, Processing, and Testing of Cervical Fluid

3.3.1. Collection of Cervical Fluid

Endocervical wicking (Sno-strips) for HIV-1 RNA quanitition: Sno-strips can be used as wicks to collect cell-free virions from the endocervix. If excess mucus or menses clot has accumulated near the cervix, a large-tipped cotton swab may be used to gently remove this material before inserting the Sno-strips.

1. Align three Sno-strips. Using forceps (ring or sponge forceps work well), gently insert the three Sno-strips into the vagina, place through the cervical os into the distal endocervical canal, and hold to adsorb sample. (Each Sno-strip adsorbs approx 8 µL of specimen. Adsorption usually takes approx 1 min, but may take longer.)
2. Hold the narrow end of the three strips over and slightly inside one labeled *plastic* transport tube (1.5-mL cryovial) containing 500 µL of 4 *M* guanidinium solution or 500 µL NASBA lysis buffer. Cut the strips at the junction of the "shoulder and neck" with scissors, allowing the narrow end to fall into the tube. Send the sealed transport vial to the local laboratory for microcentrifugation.

Alternatively, an endocervical swab can be used for HIV RNA quantitation.

1. The swabs for HIV RNA quantitation collection should be made from 100% synthetic materials. Puritan Sterile Dacron Polyester Tip Applicator 25-806-1PD is recommended.
2. Gently insert a Dacron-tipped swab with a plastic shaft into the vagina until it reaches the cervix. Rotate the swab at least 360° on the os, and, if possible, rotate 720°. A smaller swab may also be used if a woman's cervical anatomy or personal comfort level warrants it (Harwood Products Co., cat. no. 25-800D or 25-801D).
3. Place the swab for HIV RNA PCR immediately in a plastic vial containing 500 µL of 4 *M* guanidine isothiocyanate or 500 µL NASBA lysis buffer. Seal swabs in the appropriate vials before drying occurs. Swabs can be stored at 4°C for up to 72 h. Long-term storage should be at or below –70°C.

3.3.2. Processing of Cervicovaginal Fluid

The specimens are ready for testing and require no more processing.

3.3.3. Quantitation of HIV-1 RNA in Cervicovaginal Fluid

HIV RNA can be quantitated from the Sno-Strips using either the Roche Amplicor assay following the procedure outlined above in **Subheading 3.2.3.2.** for seminal plasma or the bioMerieux NucliSens HIV QT assay following the manufacturer's package insert.

4. Notes

1. Universal precautions should be observed when handling blood or other potentially infectious specimens, including genital secretions. Essential precautions include adequate hand-washing facilities and appropriate work practice policies, as well as the proper use of warning labels and protective equipment and clothing, such as gloves and lab coats, the routine decontamination of work areas, and the proper disposal of waste material. All workers should be trained on an annual basis regarding blood-borne pathogens.
2. As with all nucleic acid amplification techniques, contamination and subsequent false-positive results must be avoided. Because of the exquisite sensitivity of the amplification techniques, meticulous care must be taken to avoid contamination. Separate rooms, hoods, or designated lab areas should be used for each of the following steps: sample processing, preamplification, amplification, and detection. Dedicated supplies, reagents, and pipettors should be used before and after amplification and should never be used interchangeably. Reagents should be aliquoted and used only once. Positive displacement pipets or aerosol-resistant pipet tips should be used. Aerosols should be carefully avoided and gloves changed frequently.
3. EDTA is the preferred anticoagulant for several reasons. Heparin has been shown to inhibit PCR reactions and can be used only if one performs the Boom silica RNA isolation procedure. Acid citrate dextrose dilutes the plasma to a certain degree depending on how full the tube of blood was and can, hence, lead to variable results. EDTA has none of these problems and has the additional benefit of HIV RNA stability up to at least 30 h prior to centrifugation ***(33)***.
4. There are three FDA-approved assays for the quantitation of HIV RNA in blood plasma: the Roche Amplicor Monitor, bioMerieux's NucliSens QT, and Bayer's Versant assay. The Roche assay appears to be somewhat more sensitive than the other two assays (Brambilla and Bremer, personal communication). The Roche (version 1.5) and Versant assays are better at detecting HIV RNA in nonsubtype B specimens ***(34,35)***. The NucliSens assay works best with specimens that may have inhibitors such as seminal plasma and saliva ***(18,24,30)***. If following a patient it is advisable to use the same assay throughout, although one may safely use one assay for blood and another assay for the genital tract.
5. Data from the WIHS study suggest that cervical wicking with Sno-Strips is a much more sensitive method of obtaining viral loads from the female genital tract than cervical swabs or cervicovaginal lavages ***(11)***.
6. Cells and platelets can interfere with accurate and reproducible testing and must be removed by adequate centrifugation. Cellular proviral DNA or mRNA might be co-isolated and amplified if not removed from the preparation. Platelets can bind HIV RNA and confound quantitation.
7. Data from several studies have demonstrated that HIV RNA is stable after up to three to four freeze–thaw cycles ***(33,34)***.

References

1. Ho, D. D., Neumann, A. U., Perelson, A. S., Chen, W., Leonard, J. M., and Markowitz, M. (1995) Rapid turnover of plasma virions and CD4 lymphocytes in HIV-1 infection. *Nature* **373,** 123–126.
2. Wei, X., Ghosh, S. K., Taylor, M. E., Johnson, V. A., Emini, E. A., Deutsch, P., et al. (1995) Viral dynamics in human immunodeficiency virus type 1 infection. *Nature* **373,** 117–122.
3. Quinn, T. C., Wawer, M. J., Sewankambo, N., Serwadda, D., Li, C., Wabwire-Mangen, F., et al. (2000) Viral load and heterosexual transmission of human immunodeficiency virus type 1. Rakai Project Study Group. *N. Engl. J. Med.* **342,** 921–929.
4. Gray, R. H., Wawer, M. J., Brookmeyer, R., Sewankambo, N. K., Serwadda, D., Wabwire-Mangen, F., et al., Rakai Project Team. (2001) Probability of HIV-1 transmission per coital act in monogamous, heterosexual, HIV-1-discordant couples in Rakai, Uganda. *Lancet* **357,** 1149–1153.
5. Mofenson, L. M., Lambert, J. S., Stiehm, E. R., Bethel, J., Meyer, W. A., 3rd, Whitehouse, J., et al. (1999) Risk factors for perinatal transmission of human immunodeficiency virus type 1 in women treated with zidovudine. Pediatric AIDS Clinical Trials Group Study 185 Team. *N. Engl. J. Med.* **341,** 385–393.
6. Garcia, P. M., Kalish, L. A., Pitt, J. , Minkoff, H., Quinn, T. C., Burchett, S. K., et al. (1999) Maternal levels of plasma human immunodeficiency virus type 1 RNA and the risk of perinatal transmission. Women and Infants Transmission Study Group. *N. Engl. J. Med.* **341,** 394–402.
7. Ioannidis, J. P., Abrams, E. J., Ammann, A., Bulterys, M., Goedert, J. J., Gray, L., et al. (2001) Perinatal transmission of human immunodeficiency virus type 1 by pregnant women with RNA virus loads <1000 copies/mL. *J. Infect. Dis.* **183,** 539–545.
8. Vernazza, P. L., Gilliam, B. L., Dyer, J., Fiscus, S. A., Eron, J. J., Frank, A. C., and Cohen, M. S. (1997) Quantification of HIV in semen: correlation with antiviral treatment and immune status. *AIDS* **11,** 987–993.
9. Coombs, R. W., Speck, C. E., Hughes, J. P., Lee, W., Sampoleo, R., Ross, S. O., et al. (1998) Association between culturable human immunodeficiency virus type 1 (HIV-1) in semen and HIV-1 RNA levels in semen and blood: evidence for compartmentalization of HIV-1 between semen and blood. *J. Infect. Dis.* **177,** 320–330.
10. Hart, C. E., Lennox, J. L., Pratt-Palmore, M., Wright, T. C., Schinazi, R. F., Evans-Strickfaden, T., et al. (1999) Correlation of human immunodeficiency virus type 1 RNA levels in blood and the female genital tract. *J. Infect. Dis.* **179,** 871–882.
11. Coombs, R. W., Wright, D. J., Reichelderfer, P. S., Burns, D. N., Cohn, J., Cu-Uvin, S., et al., Women's Health Study 001 Team. (2001) Variation of human immunodeficiency virus type 1 viral RNA levels in the female genital tract: implications for applying measurements to individual women. *J. Infect. Dis.* **184,** 1187–1191.

12. Kovacs, A., Wasserman, S. S., Burns, D., Wright, D. J., Cohn, J., Landay, A., et al., DATRI Study Group; WIHS Study Group. (2001) Determinants of HIV-1 shedding in the genital tract of women. *Lancet* **358,** 1593–1601.
13. Fiscus, S. A., Adimora, A. A., Schoenbach, V. J., McKinney, R., Lim, W., Rupar, D., et al. (1999) Trends in HIV counseling, testing, and anti-retroviral treatment of HIV-infected women and perinatal transmission in North Carolina. *J. Infect. Dis.* **180,** 99–105.
14. Vernazza, P. L., Gilliam, B. L., Flepp, M., Dyer, J. R., Frank, A. C., Fiscus, S. A., et al. (1997) Effect of antiviral treatment on the shedding of HIV-1 in semen. *AIDS* **11,** 1249–1254.
15. Vernazza, P. L., Troiani, L., Flepp, M. J., Cone, R. W., Schock, J., Roth, F., et al. (2000) Potent antiretroviral treatment of HIV-infection results in suppression of the seminal shedding of HIV. The Swiss HIV Cohort Study. *AIDS* **14,** 117–121.
16. Gilliam, B. L., Dyer, J. R., Fiscus, S. A., Marcus, C., Zhou, S., Wathen, L., et al. (1997) Effects of reverse transcriptase inhibitor therapy on the HIV-1 viral burden in semen. *J. Acquir. Immune Defic. Syndr. Hum. Retro.* **15,** 54–60.
17. Gupta, P., Mellors, J., Kingsley, L., Riddler, S., Singh, M. K., Schreiber, S., et al. (1997) High viral load in semen of human immunodeficiency virus type 1-infected men at all stages of disease and its reduction by therapy with protease and nonnucleoside reverse transcriptase inhibitors. *J. Virol.* **71,** 6271–6275.
18. Dyer, J. R., Gilliam, B. L., Eron, J. J., Jr., Grosso, L., Cohen, M. S., and Fiscus, S. A. 1996. Quantitation of human immunodeficiency virus type 1 RNA in cell free seminal plasma: comparison of NASBA with Amplicor reverse transcription-PCR amplification and correlation with quantitative culture. *J. Virol. Methods* **60,** 161–170.
19. Boom, R., Sol, C. J., Salimans, M. M., Jansen, C. L., Wertheim-van Dillen, P. M., and van der Noordaa, J. (1990) Rapid and simple method for purification of nucleic acids. *J. Clin. Microbiol.* **28,** 495–503.
20. Reichelderfer, P. S., Combs, R. W., Wright, D. J., Cohn, J., Burns, D. N., Cu-Uvin, S., et al. (2000) Effect of menstrual cycle on HIV-1 levels in the peripheral blood and genital tract. *AIDS* **14,** 2101–2107.
21. Mohammed, A. S., Becquart, P., Hocini, H., Metais, P., Kazatchkine, M., and Belec, L. (1997) Dilution assessment of cervicovaginal secretions collected by vaginal washing to evaluate mucosal shedding of free HIV. *Clin. Diag. Lab. Immunol.* **4,** 624–626.
22. Min, S. S., Corbett, A. H. Rezk, N., Fiscus, S. A. Cohen, M. S., and Kashuba, A. D. M. Differential penetration of protease inhibitors (PI), non-nucleoside reverse transcriptase inhibitors (NNRTI) into female genital tract (GT). XIV International AIDS Conference, Barcelona Spain, July 7–12, 2002.
23. Dyer, J. R., Pilcher, C. D., Shepard, R., Schock, J., Eron, J. J., and Fiscus, S. A. (1999) Comparison of NucliSens and Roche Monitor assays for plasma HIV-1 RNA quantitation. *J. Clin. Microbiol.* **37,** 447–449.

24. Fiscus, S. A., Brambilla, D., Coombs, R. W., Yen-Lieberman, B., Bremer, J., Kovacs, A., et al. (2000) Multi-center evaluation of methods to quantitate HIV type 1 RNA in seminal plasma. *J. Clin. Microbiol.* **38,** 2438–2353.
25. Murphy, D. G., Cote, L., Fauvel, M., Rene, P., and Vincelette, J. (2000) Multicenter comparison of Roche COBAS AMPLICOR MONITOR version 1.5, Organon Teknika NucliSens QT with Extractor, and Bayer Quantiplex version 3.0 for quantification of human immunodeficiency virus type 1 RNA in plasma. *J. Clin. Microbiol.* **38,** 4034–4041.
26. Skidmore, S. J., Zuckerman, M., and Parry, J. V. (2000) Accuracy of plasma HIV RNA quantification: a multicentre study of variability. *J. Med. Virol.* **61,** 417–422.
27. Holguin, A., de Mendoza, C., and Soriano, V. (1999) Comparison of three different commercial methods for measuring plasma viraemia in patients infected with non-B HIV-1 subtypes. *Eur. J. Clin. Microbiol. Infect. Dis.* **18,** 256–259.
28. O'Shea, S., de Ruiter, A., Mullen, J., Corbett, K., Chrystie, I., Newell, M. L., and Banatvala, J. E. (1997) Quantification of HIV-1 RNA in cervicovaginal secretions: an improved method of sample collection. *AIDS* **11,** 1056–1058.
29. Webber, M. P., Schoenbaum, E. E., Farzadegan, H., and Klein, R. S. (2001) Tampons as a self-administered collection method for the detection and quantification of genital HIV-1. *AIDS* **15,** 1417–1420.
30. Shepard, R. N., Schock J., Robertson, K., Shugars, D. C., Dyer, J., Vernazza, P., et al. (2000) Quantitation of HIV-1 RNA in different biological compartments. *J. Clin. Microbiol.* **38,** 1414–1418.
31. Holodniy, M., Anderson, D., Wright, D., Sharma, O., Cohn, J., Alexander, N., et al. (1998) HIV quantitation in spiked vaginocervical secretions: lack of non-specific inhibitory factors. DATRI 005 Study Team. Division of AIDS Treatment Research Initiative. *J. Virol. Methods* **72,** 185–195.
32. Bremer, J., Nowicki, M., Beckner, S., Brambilla, D., Cronin, M., Herman, S., et al. (2000) Comparison of two amplification technologies for detection and quantitation of human immunodeficiency virus type 1 RNA in the female genital tract. Division of AIDS Treatment Research Initiative 009 Study Team. *J. Clin. Microbiol.* **38,** 2665–2669.
33. Holodniy, M., Mole, L., Yen-Lieberman, B., Margolis, D., Starkey, C., Carroll, R., et al. (1995) Comparative stabilities of quantitative human immunodeficiency virus RNA in plasma from samples collected in VACUTAINER CPT, VACUTAINER PPT, and standard VACUTAINER tubes. *J. Clin. Microbiol.* **33,** 1562–1566.
34. Burgisser, P., Vernazza, P., Flepp, M., Boni, J., Tomasik, Z., Hummel, U., et al. (2000) Performance of five different assays for the quantification of viral load in persons infected with various subtypes of HIV-1. Swiss HIV Cohort Study. *J. Acquir. Immune Defic. Syndr.* **23,** 138–144.
35. Antunes, R., Figueiredo, S., Bartolo, I., Pinheiro, M., Rosado, L., Soares, I., et al. (2003) Evaluation of the clinical sensitivities of three viral load assays with plasma samples from a pediatric population predominantly infected with human immunodeficiency virus type 1 subtype G and BG recombinant forms. *J. Clin. Microbiol.* **41,** 3361–3367.

36. Lew, J., Reichelderfer, P., Fowler, M., Bremer, J., Carrol, R., Cassol, S., et al. (1998) Determinations of levels of human immunodeficiency virus type 1 RNA in plasma: reassessment of parameters affecting assay outcome. TUBE Meeting Workshop Attendees. Technology Utilization for HIV-1 Blood Evaluation and Standardization in Pediatrics. *J. Clin. Microbiol.* **36,** 1471–1479.
37. Griffith, B. P., Rigsby, M. O., Garner, R. B., Gordon, M. M., and Chacko, T. M. (1997) Comparison of the Amplicor HIV-1 monitor test and the nucleic acid sequence-based amplification assay for quantitation of human immunodeficiency virus RNA in plasma, serum, and plasma subjected to freeze-thaw cycles. *J. Clin. Microbiol.* **35,** 3288–3291.

17

Quantification of Proviral DNA Load of Human Immunodeficiency Virus Type 2 Subtypes A and B Using Real-Time PCR

Marie Gueudin, Florence Damond, and François Simon

Summary

HIV-2 infection is confined mostly to West Africa. Seven HIV-2 subtypes have so far been described; only HIV-2 subtypes A and B are prevalent, the others being considered self-limiting infections at the epidemiological level. The main limitation for the HIV-2 DNA proviral quantification is the lack of HIV-2 DNA standard. We designed and tested a new HIV-2 primer couple that amplifies both the HIV-2 ROD strain and HIV-1 LAV/BRU strain. These HIV-2 primers were used to quantified an HIV-2 standard comparatively to a standard widely used in proviral DNA HIV-1 quantification, i.e., the 8E5 cell line transfected by a single defective integrated provirus of HIV-1 BRU/LAV by cell. The primers and probe used to quantify HIV-2 DNA are located in a long terminal repeat (LTR) region with low variability. These primers amplify both HIV-2 subtypes A and B. The relevance of the follow-up of the infected patients by the quantification of the proviral DNA HIV-2 is currently studied.

Key Words: HIV-2; proviral DNA; real-time PCR; quantification.

1. Introduction

Human immunodeficiency virus type 2 (HIV-2) was first isolated in 1986 from peripheral blood mononuclear cells (PBMC) from patients in the Cape Verde Islands and Guinea-Bissau *(1)*. Seven HIV-2 subtypes have so far been described *(2–4)*. Only HIV-2 subtypes A and B are prevalent, the others being considered self-limiting infections at the epidemiological level. HIV-2 infection is confined mostly to West Africa *(5)*.

For HIV-1 infection, it has been demonstrated that proviral load correlates with disease progression *(6)*. For HIV-2, certain studies that assessed the quan-

From: *Methods in Molecular Biology, Vol. 304: Human Retrovirus Protocols: Virology and Molecular Biology*
Edited by: T. Zhu © Humana Press Inc., Totowa, NJ

titative proviral load have proposed, as for HIV-1, an inverse correlation with the CD4+ cell count and clinical outcome *(7–9)*. However, other reports have suggested that HIV-2 proviral load does not correlate with the CD4+ cell count *(10,11)*.

HIV-2 is much less pathogenic than HIV-1 *(12,13)*. HIV-2 infection is associated with plasma viral loads that are significantly lower than those found in HIV-1 infection *(14,15)*. Proviral load does not correlate with plasma HIV-2 RNA load *(16)*.

The main limitation for the HIV-2 DNA proviral quantification is the lack of an HIV-2 DNA standard. Here, we designed and tested a new HIV-2 primer couple that amplifies both the HIV-2 ROD and the HIV-1 LAV/BRU strains. These HIV-2 primers were used to quantify and compare an HIV-2 standard with a standard widely used in proviral DNA HIV-1 quantification, i.e., the 8E5 cells line transfected by a single defective integrated provirus of HIV-1 BRU/LAV by cell.

2. Materials

1. QIAamp DNA Blood Mini Kit (Qiagen GmbH, Germany).
2. Purified oligonucleotide primers and probe (Eurogentec, Seraing, Belgium).
3. 8E5 cell line (ATCC, Manassas, VA).
4. Ficoll-Hypaque.
5. LC—Faststart DNA master hybridization probes (Roche Diagnostics, Basel, Switzerland).
6. LC—FastStart DNA Master SYBR Green I (Roche Diagnostics).

3. Methods

DNA was extracted (*see* **Subheading 3.1.**) from the 8E5 cells line transfected with HIV-1 and from a pellet of cells of the HIV-2 ROD strain to determine the value of the standard. The cellular extract of HIV-2 ROD strain was quantified by report of a range of DNA coming from 8E5 cells (*see* **Subheading 3.2.**). DNA was extracted similarly from patients' peripheral blood mononuclear cells (PBMC). The DNA concentration in these patients' extracts was determined in order to standardize the DNA amount used in each vial for PCR HIV-2. The HIV-2 standard was serially diluted to establish the reference curve for the quantification of the HIV-2 proviral load by real-time PCR (*see* **Subheading 3.3.**).

3.1. DNA Extraction

To quantify the standard, two extracts were necessary: one from a pellet of the 8E5 cell line and one from a pellet of cells obtained starting from a culture of the HIV-2 ROD strain. Samples of blood were collected onto ethylenediamine tetraacetic acid (EDTA) or acid-citrate-dextrose (ACD). The PBMC were isolated by Ficoll-Hypaque density gradient centrifugation and were stored at –80°C ***(17)***. DNA was extracted from 5 × 10^6 cells (or less, with at least 1 × 10^6) with the QIAamp DNA Blood Mini Kit according to the manufacturer's recommendations. The elution volume was 50 µL when one to 1–3 × 10^6 cells were extracted and 100 µL when 3–5 × 10^6 cells were extracted (*see* **Note 1**). The DNA concentration of each extract was determined by measurement of the optic density at 260 nm (*see* **Note 2**).

3.2. Standard

3.2.1. Quantification of Standard

The HIV-2 DNA standard was a cellular extract of the HIV-2 ROD strain (Genbank accession no. X05291) quantified using primers defined specifically to amplify both HIV-2 ROD and HIV-1 BRU strains (Genbank accession no. K02013). The cell line 8E5 contains a single provirus of HIV-1 per cell. With these primers, we quantified the extracted DNA of HIV-2 ROD with a range of DNA extract from the 8E5 cell line (*see* **Note 3**). The DNA extracted from the 8E5 cell line was quantified four times spectrophotometrically and the average of the measurements was selected. The number of copies was calculated using the equivalence 1 µg of DNA = 150,000 cells = 150,000 copies.

3.2.2. Primers and Amplification

The primers used for the quantification of the cellular extract of the HIV-2 ROD strain were:

BRU ROD U: 5' AATGAGGAAGCWGCAGAAT 3' at position 951 on HIV-1 BRU/LAV
BRU ROD L: 5' GCTATGTCASWTCCCCTTG 3' at position 1027

The primers generate a product of 95 bp with a melting temperature (Tm) of 84.7°C. The LightCycler instrument (LC, Roche Diagnostics) was used to amplify and quantify the PCR product at each PCR cycle. PCR was performed with LC—FastStart DNA Master SYBR Green I (Roche Diagnostics). The LC master mix (2 µL) was mixed with 3 m*M* $MgCl_2$, 0.5 µ*M* each primer. The volume of 10 µL was reached by supplementing with water. A range of the 8E5 extract was diluted from 10 to 10,000 copies/reaction in nuclease-free water, the HIV-2 ROD extract was diluted at 1/10, and 10 µL of each extract was added to the mix. The amplification was carried out as follows:

- 95°C for 10 min (one cycle).
- Denaturation at 95°C for 10 s, annealing at 62°C for 15 s (45 cycles).
- Melting curve 95°C—10 s, 65°C—30 s, and 95°C—0 s with a slope of temperature of 0.1°C/s.

Table 1
Primer and Probe Positions Defined From GenBank

Primers and probe	Sequence from 5'→3'	Position on HIV-2 BEN (Accession no. NC001722)
HIV-2 U	TAG TCG CCG CCT GGT CA	756
HIV-2 L	TTC CTG CCG CCC TTA CT	920
Probe HIV-2	TGG TCT GTT AGG ACC CTT CTT GCT TTG	801

The melting curve analysis allowed us to confirm that the expected PCR product presents the right Tm. The sensitivity and the range of standard linearity were determined. The quantification threshold was considered to be 5 copies/reaction and the PCR reaction is linear between 10,000 and 5 copies/reaction.

3.3. Quantification of Proviral DNA HIV-2

3.3.1. Primers and Probe

The primers and probe (**Table 1**) were synthesized by Eurogentec (Seraing, Belgium). These primers generate a PCR product of 181 bp in an LTR region with low variability. These genomic regions are perfectly conserved between the HIV-2 subtypes and these primers amplify both HIV-2 subtypes A and B. A highly HIV-2-specific probe was also selected and synthesized with a reporter fluorescent dye (FAM) attached to the 5' end and the TAMRA linked to the 3' end for detection (*see* **Note 4**).

3.3.2. Real-Time PCR HIV-2

PCR was performed with LC—FastStart DNA master hybridization probes. The LC master mix (2 μL) was mixed with 4 m*M* $MgCl_2$, 0.25 μ*M* each primer, and 0.25 μ*M* probe. A variable volume of extract, containing 500 ng of DNA, was added. The final volume of 20 μL was reached by supplementing with water.

The amplification was carried out as follows: 95°C for 10 min (one cycle), followed by denaturation at 95°C for 10 s, and annealing at 60°C for 30 s (45 cycles).

The sensitivity and the range of standard for which the reaction is linear were determined. The quantification threshold is five copies/reaction and the PCR reaction is linear between 100,000 and five copies/reaction.

4. Notes

1. An automation of the extraction procedure has been tested. DNA was extracted using the automat Magna Pure LC® (Roche Diagnostics) with the DNA Isolation Kit Large Volume according to the manufacturer's recommendations but with an

external lysis. A pellet of 1–5 × 10^6 cells was suspended in 200 µL of PBS. 300 µL of lysis/binding buffer were added. Fifteen min later, the 500 µL were placed in the sample cartridge. In the reagent rack the lysis/binding buffer was replaced by PBS; the protocol used was the LV Blood 300–500 µL with an elution volume of 100 µL. We obtained good results with this kit but other extraction procedures may be used.

2. To quantify the DNA concentration in the patients' extracts, a spectrophotometer could be used. Another way is to quantify the albumin gene in a range of human genomic DNA *(18)*.
3. The use of an extract of 8E5 cells is usual in the techniques of quantification of the proviral DNA of HIV-1. Our approach makes it possible to obtain comparable results between HIV-1 and HIV-2. Another way is to synthesize a plasmid containing the amplified sequence. This type of standard is completely usable with our technique of quantification.
4. The probe is particularly unstable. It is necessary to avoid freeze–thaw cycles. The probe is reconstituted in nuclease-free water at 100 µ*M* and must be aliquoted.

Acknowledgments

We thank all the technicians of the Virology Department at the University Hospital Charles Nicolle in Rouen, France. This work was supported by ANRS (Agence Nationale de Recherche sur le SIDA).

References

1. Clavel, F., Guetard, D., Brun-Vezinet, F., Chamaret, S., Rey, M. A., Santos-Ferreira, M. O., et al. (1986) Isolation of a new human retrovirus from West African patients with AIDS. *Science* **233,** 343–346.
2. Chen, Z., Luckay, A., Sodora, D. L., Telfer, P., Reed, P., Gettie, A., et al. (1997) Human immunodeficiency virus type 2 (HIV-2) seroprevalence and characterization of a distinct HIV-2 genetic subtype from the natural range of simian immunodeficiency virus-infected sooty mangabeys. *J. Virol.* **71,** 3953–3960.
3. Gao, F., Yue, L., Robertson, D. L., Hill, S. C., Hui, H., Biggar, R. J., et al. (1994) Genetic diversity of human immunodeficiency virus type 2: evidence for distinct sequence subtypes with differences in virus biology. *J. Virol.* **68,** 7433–7447.
4. Yamaguchi, J., Devare, S. G., and Brennan, C. A. (2000) Identification of a new HIV-2 subtype based on phylogenetic analysis of full-length genomic sequence. *AIDS Res. Hum. Retroviruses* **16,** 925–930.
5. De Cock, K. M. and Brun-Vezinet, F. (1989) Epidemiology of HIV-2 infection. *AIDS* **3(Suppl 1),** S89–S95.
6. Schechter, M. T., Neumann, P. W., Weaver, M. S., Montaner, J. S., Cassol, S. A., Le, T. N., et al. (1991) Low HIV-1 proviral DNA burden detected by negative polymerase chain reaction in seropositive individuals correlates with slower disease progression. *AIDS* **5,** 373–379.

7. Berry, N., Ariyoshi, K., Jobe, O., Ngum, P. T., Corrah, T., Wilkins, A., et al. (1994) HIV type 2 proviral load measured by quantitative polymerase chain reaction correlates with CD4+ lymphopenia in HIV type 2-infected individuals. *AIDS Res. Hum. Retroviruses* **10,** 1031–1037.
8. Norrgren, H., Marquina, S., Leitner, T., Aaby, P., Melbye, M., Poulsen, A. G., et al. (1997) HIV-2 genetic variation and DNA load in asymptomatic carriers and AIDS cases in Guinea-Bissau. *J. Acquir. Immune Defic. Syndr. Hum. Retrovirol* **16,** 31–38.
9. Sarr, A. D., Popper, S., Thior, I., Hamel, D. J., Sankale, J. L., Siby, T., et al. (1999) Relation between HIV-2 proviral load and CD4+ lymphocyte count differs in monotypic and dual HIV infections. *J. Hum. Virol.* **2,** 45–51.
10. Gomes, P., Taveira, N. C., Pereira, J. M., Antunes, F., Ferreira, M. O., and Lourenco, M. H. (1999) Quantitation of human immunodeficiency virus type 2 DNA in peripheral blood mononuclear cells by using a quantitative-competitive PCR assay. *J. Clin. Microbiol.* **37,** 453–456.
11. Popper, S. J., Sarr, A. D., Gueye-Ndiaye, A., Mboup, S., Essex, M. E., and Kanki, P. J. (2000) Low plasma human immunodeficiency virus type 2 viral load is independent of proviral load: low virus production in vivo. *J. Virol.* **74,** 1554–1557.
12. Kanki, P. J., Travers, K. U., Mboup, S., Hsieh, C. C., Marlink, R. G., Gueye, N. A., et al. (1994) Slower heterosexual spread of HIV-2 than HIV-1. *Lancet* **343,** 943–946.
13. Marlink, R. G., Ricard, D., M'Boup, S., Kanki, P. J., Romet-Lemonne, J. L., N'Doye, I., et al. (1988) Clinical, hematologic, and immunologic cross-sectional evaluation of individuals exposed to human immunodeficiency virus type-2 (HIV-2). *AIDS Res. Hum. Retroviruses* **4,** 137–148.
14. Damond, F., Gueudin, M., Pueyo, S., Farfara, I., Robertson, D. L., Descamps, D., et al. (2002) Plasma RNA viral load in human immunodeficiency virus type 2 subtype A and subtype B infections. *J. Clin. Microbiol.* **40,** 3654–3659.
15. Popper, S. J., Sarr, A. D., Travers, K. U., Gueye-Ndiaye, A., Mboup, S., Essex, M. E., and Kanki, P. J. (1999) Lower human immunodeficiency virus (HIV) type 2 viral load reflects the difference in pathogenicity of HIV-1 and HIV-2. *J. Infect. Dis.* **180,** 1116–1121.
16. Damond, F., Descamps, D., Farfara, I., Telles, J. N., Puyeo, S., Campa, P., et al. (2001) Quantification of proviral load of human immunodeficiency virus type 2 subtypes A and B using real-time PCR. *J. Clin. Microbiol.* **39,** 4264–4268.
17. Simon, F., Matheron, S., Tamalet, C., Loussert-Ajaka, I., Bartczak, S., Pepin, J. M., et al. (1993) Cellular and plasma viral load in patients infected with HIV-2. *AIDS* **7,** 1411–1417.
18. Laurendeau, I., Bahuau, M., Vodovar, N., Larramendy, C., Olivi, M., Bieche, I., et al. (1999) TaqMan PCR-based gene dosage assay for predictive testing in individuals from a cancer family with INK4 locus haploinsufficiency. *Clin. Chem.* **45,** 982–986.

18

Plasma RNA Viral Load in HIV-1 Group O Infection by Real-Time PCR

Marie Gueudin and François Simon

Summary

HIV-1 group O strains are highly divergent, and are found mainly in central Africa. The clinical course of group O infection is identical to that of HIV-1 group M infection, with rapid onset of immunodeficiency. The important divergence of the HIV-1 group O strains lead to high limitations of the commercial tests. We describe here a method based on real-time polymerase chain reaction (PCR) to quantify plasma HIV-1 group O RNA. Primers amplify both HIV-1 group O and HIV-1 group M strains. Conversely, the probe is HIV-1 group O-specific. The standard used to quantify the clinical samples is an RNA solution resulting from the transcription of a plasmid including the amplified fragment of PCR. Our technique is capable of amplifying a wide range of HIV-1 group O strains belonging to the three current clades. This technique can be used to monitor HIV-1 group O viral load, which has previously been difficult.

Key Words: HIV-1 group O; real-time PCR; RNA viral load; quantification.

1. Introduction

In the early 1990s, HIV-1 group O strains (O for outlier) were identified *(1–4)*. This new group of HIV corresponds to highly divergent strains, found mainly in central Africa with relatively low prevalence (**5**). Sporadic cases have been reported in Europe and the United States *(1,6–9)*, but Cameroon remains the epicenter where most documented cases have so far occurred *(5,10–14)*. The clinical course of group O infection is identical to that of HIV-1 group M infection, with rapid onset of immunodeficiency *(15)*. Cases of vertical transmission have also been reported *(16)*.

The important divergence of the HIV-1 group O strains leads to high limitations of the commercial tests, which either do not detect or underquantify these strains *(17–20)*. Recently, using the LCx® HIV RNA QT kit (Abbott, Chicago,

From: *Methods in Molecular Biology, Vol. 304: Human Retrovirus Protocols: Virology and Molecular Biology*
Edited by: T. Zhu © Humana Press Inc., Totowa, NJ

IL), satisfactory results were obtained with a limited number of samples ***(21,22)***, but the group O major variability remains a limitation for a standardized group O quantitative assay ***(23)***. This approach based on manufactured assays requires costly reagents, making it unsuitable for use in most developing countries.

Real-time PCR is less costly and can encompass a wide range of genome diversity. It may thus represent an alternative for the management of patients infected by rare variant strains. We have developed such a method to quantify plasma HIV-1 group O RNA. The assay is sensitive and specific, and overcomes the problems posed by the high genomic variability of group O strains ***(24)***.

2. Materials

1. QIAamp Viral RNA minikit (Qiagen GmbH, Germany).
2. Purified oligonucleotidic primers and probe (Eurogentec, Seraing, Belgium).
3. Omniscript™ reverse transcriptase (RT) kit (Qiagen).
4. LC–FastStart DNA master hybridization probes (Roche Diagnostics, Basel, Switzerland).
5. TOPO TA® cloning kit (Invitrogen, Carlsbad, CA).
6. Restriction enzymes *Hind*III and *Spe*I.
7. QIAprep Spin Miniprep® (Qiagen).
8. Riboprobe® in vitro transcription system (Promega, Madison, WI).
9. DNA-free™ system (Ambion Inc., Austin, TX).
10. Diethylpyrocarbonate-treated water.

3. Methods

With the primer and probe selection (*see* **Subheading 3.1.**), the main steps of this technique are the RNA extraction procedure, the reverse transcription, and the amplification (*see* **Subheading 3.2.**). For the external standard used to quantify the clinical samples, a plasmid including the fragment of HIV-1 group O genome amplified in the polymerase chain reaction (PCR) was constructed, linearized, and transcripted (*see* **Subheading 3.3.**).

3.1. Primers and Probe

Primers were selected to amplify both HIV-1 group O and HIV-1 group M strains (*see* **Note 1**). Conversely the probe is HIV-1 group O-specific. The primers and probe were synthesized by Eurogentec:

- P1: 5' CTC AAT AAA GCT TGC CTT GA 3' at position 524 on HXB2 K03455
- P2: 5' CGC CAC TGC TAG AGA TTT T 3' at position 622

These primers generate a product of 112 bp in a long terminal repeat (LTR) region with lower variability.

A highly specific probe for group O identification was also selected at position 551 of HXB2/position 584 of Ant 70: 5' AAG CAG TGT GTG CTC ATC TGT TG 3'.

The HIV-1 group O probe was synthesized with a reporter fluorescein dye (FAM) attached to the 5' end and the Eclipse® Dark Quencher (Eurogentec) linked to the 3' end for detection. Eclipse® Dark Quencher is a nonfluorescent molecule quenching effective fluorescence over a broad wavelength range from about 400 nm to 650 nm (*see* **Note 2**).

The specificity of this technique was evaluated. A limited number of HIV-1 group M could be amplified ***(24)*** and the diagnosis of group O must not be determined with the results of this PCR. Sequence of the small fragment of PCR and the analysis of the result by Blast is sufficient to determine if the virus is belongs to HIV-1 group O or not.

3.2. Construction of an HIV-1 Group O Clone

3.2.1. Cloning

The plasmid (**Fig. 1**) is carried out by using the TOPO TA Cloning kit by Invitrogen according to the manufacturer's recommendations. The plasmid vector (pCR®2.1-TOPO®) supplied is linearized.

1. Generate the DNA fragment of interest with the primers used for real-time PCR from a positive sample.
2. Mix the PCR product together with the vector. The fragment is inserted downstream of the T7 RNA polymerase promoter.
3. Do the transformation step with cells of *Escherichia coli* (strain TOP10) provided competent.
4. Put the transformed bacteria in culture on Luria-Bartani (LB) plates containing ampicillin as the antibiotic of selection. A solution of X-Gal should be spread out over these LB plates beforehand. Only the bacteria having integrated the plasmid will be able to multiply. Only the white colonies contain the plasmid in which the fragment of PCR is integrated.
5. Analyze the clones by PCR by using primers M13R and M13 F-20 of the kit, which generate a fragment of 313 bp. The direction of insertion of the fragment is checked by enzymatic restriction of this PCR product by the *Hind*III enzyme. When the fragment is inserted in the good direction the cut by *Hind*III generates three fragments of 30,118, and 165 bp.
6. Put the clone selected in culture in LB medium + ampicillin and the following day extract the plasmid DNA with the QIAprep Spin Miniprep® kit.
7. Determine the concentration of the plasmid by a measurement on spectrophotometer at 260 nm. The selected clone of *E. coli* is frozen with sterile glycerol for long-term storage.

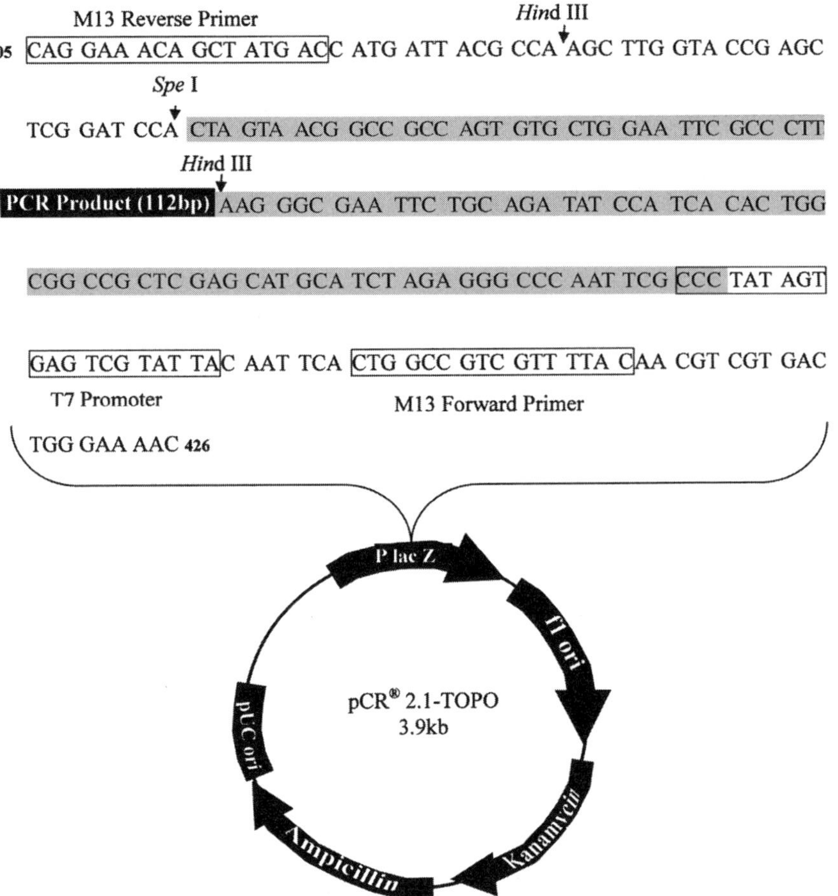

Fig. 1. Schematic drawing of pCR 2.1 TOPO plasmid adapted from Invitrogen. The transcripted sequence is drawn in gray.

3.2.2. Linearization and Transcription of Plasmid

In order to obtain transcripts of equal length, 5 µg of plasmid is cut, downstream of the introduced fragment, using the restriction enzyme *Spe*I. We then transcribed 5 µL of the linearized plasmid using T7 RNA polymerase from the Riboprobe in vitro transcription system according to the manufacturer's instructions. The transcript corresponds to a 219-bp length.

3.2.3. DNAse Treatment

A treatment by DNase was performed to eliminate DNA and the sample tested to be DNA-free.

1. Treat the RNA collected after the transcription step with the DNA-free system (Ambion).
2. Remove the DNase with the provided system that does not require phenol/chloroform extraction or heating.
3. Check if RNA is still contaminated with DNA by performing PCR without the RT step on RNA. No amplification must be observed. If any amplification occurs, a new DNase treatment must be performed (*see* **Note 3**).
4. Calculate the mean of four measurements of RNA concentration assessed by spectrophotometry.
5. Aliquot and store the standard at a known concentration at –80°C. Thawed aliquots must never be re-frozen. For each quantification run, an aliquot of the standard is serially diluted 10-fold, in duplicate, in diethylpyrocarbonate-treated water (*see* **Note 4**).

3.3. RT-PCR HIV-1 Group O LTR

3.3.1. RNA Extraction

The extraction is performed on 500 µL of plasma obtained after centrifugation at 3000*g* of a sample of blood collected onto ethylenediamine tetraacetic acid (EDTA) or acid-citrate-dextrose (ACD). The 500 µL of plasma is centrifuged at high speed for 1 h at 24,000*g*. 360 µL of supernatant are eliminated and the extraction is performed on 140 µL. The QIAamp Viral RNA minikit is used according to the manufacturer's recommendations (*see* **Note 5**).

- The optional step of centrifugation, proposed by the manufacturer between the washing step AW2 and the step of elution, is performed.
- The volume of elution is 40 µL.
- Perform a second elution: the first eluate is passed one time again on the column.

The standard is not extracted. It is simply serially diluted 10-fold, in duplicates, from 2×10^6 to 200 copies/mL.

3.3.2. Reverse Transcription

Five µL of the serially diluted standard and 5 µL of the clinical samples are reverse-transcribed in parallel. The Omniscript reverse transcriptase kit is used as recommended by the manufacturer at 37°C for 1 h and 93°C for 5 min. Primer P2 (1 µ*M*) is then added to the RT mix.

3.3.3. Measurement of HIV-1 Group O cDNA by Real-Time PCR

The LightCycler instrument (Roche Diagnostics) was used to amplify and quantify the PCR product after each cycle. PCR was performed with LC–FastStart DNA master hybridization probes. The LC master mix (2 µL) was mixed with 4 m*M* $MgCl_2$, 0.5 µ*M* of each primer, and 0.5 µ*M* probe. cDNA (5 µL) was added to 15 µL of this mixture and amplification was carried out as follows: 95°C for 10 min (one cycle), followed by denaturation at 95°C for 10 s, and annealing at 58°C for 30 s (45 cycles).

4. Notes

1. Primers and probe are delivered lyophilized. They must be reconstituted with nuclease-free water. Bidistilled water is not recommended.
2. The probe is particularly unstable. It is necessary to avoid cycles of freezing and thawing. The probe is reconstituted at 100 μ*M* and must be aliquoted.
3. It is often difficult to completely eliminate the DNA after the transcription. However, if the control PCR is negative on the dilution of RNA used as standard, an amplification in the solution of transcript not diluted can be tolerated.
4. It is possible to use a supernatant of a culture as a standard. In this case, an aliquot of supernatant diluted in negative human plasma is extracted like the other patients' and this extract is serially diluted in water to obtain a range of standard.
5. An automation of the extraction procedure has been tested. RNA was extracted using the automat Magna Pure LC® (Roche Diagnostics) with the total nucleic acid isolation kit according to the manufacturer's recommendations. The initial volume of plasma was 200 μL and the elution volume was 50 μL. We obtained reproducible results with this kit but other extraction procedures may probably be used.

Acknowledgments

We thank all the technicians of the Virology Department of the University Hospital Charles Nicolle in Rouen, France, and J. C. Plantier for helpful discussions. This work was supported by ANRS (Agence Nationale de Recherche sur le SIDA).

References

1. Loussert-Ajaka, I., Chaix, M. L., Korber, B., Letourneur, F., Gomas, E., Allen, E., et al. (1995) Variability of human immunodeficiency virus type 1 group O strains isolated from Cameroonian patients living in France. *J. Virol.* **69,** 5640–5649.
2. Charneau, P., Borman, A. M., Quillent, C., Guetard, D., Chamaret, S., Cohen, J., et al. (1994) Isolation and envelope sequence of a highly divergent HIV-1 isolate: definition of a new HIV-1 group. *Virology* **205,** 247–253.
3. Gurtler, L. G., Hauser, P. H., Eberle, J., von Brunn, A., Knapp, S., Zekeng, L., et al. (1994) A new subtype of human immunodeficiency virus type 1 (MVP-5180) from Cameroon. *J. Virol.* **68,** 1581–1585.
4. Vanden Haesevelde, M., Decourt, J. L., De Leys, R. J., Vanderborght, B., van der Groen, G., van Heuverswijn, H., and Saman, E. (1994) Genomic cloning and complete sequence analysis of a highly divergent African human immunodeficiency virus isolate. *J. Virol.* **68,** 1586–1596.
5. Mauclere, P., Loussert-Ajaka, I., Damond, F., Fagot, P., Souquieres, S., Monny Lobe, M., et al. (1997) Serological and virological characterization of HIV-1 group O infection in Cameroon. *AIDS* **11,** 445–453.

6. Hampl, H., Sawitzky, D., Stoffler-Meilicke, M., Groh, A., Schmitt, M., Eberle, J., and Gurtler, L. (1995) First case of HIV-1 subtype O infection in Germany. *Infection* **23,** 369–370.
7. Rayfield, M. A., Sullivan, P., Bandea, C. I., Britvan, L., Otten, R. A., Pau, C. P., et al. (1996) HIV-1 group O virus identified for the first time in the United States. *Emerg. Infect. Dis.* **2,** 209–212.
8. Soriano, V., Gutierrez, M., Garcia-Lerma, G., Aguilera, O., Mas, A., Bravo, R., et al. (1996) First case of HIV-1 group O infection in Spain. *Vox Sang.* **71,** 66.
9. Sullivan, P. S., Do, A. N., Ellenberger, D., Pau, C. P., Paul, S., Robbins, K., et al. (2000) Human immunodeficiency virus (HIV) subtype surveillance of African-born persons at risk for group O and group N HIV infections in the United States. *J. Infect. Dis.* **181,** 463–469.
10. Ayouba, A., Mauclere, P., Martin, P. M., Cunin, P., Mfoupouendoun, J., Njinku, B., et al. (2001) HIV-1 group O infection in Cameroon, 1986 to 1998. *Emerg. Infect. Dis.* **7,** 466–467.
11. Peeters, M., Gueye, A., Mboup, S., Bibollet-Ruche, F., Ekaza, E., Mulanga, C., et al. (1997) Geographical distribution of HIV-1 group O viruses in Africa. *AIDS* **11,** 493–498.
12. Zekeng, L., Obiang Sima, J., Hampl, H., Ndemesogo, J. M., Ntutumu, J., Sima, V., et al. (1997) Update on HIV-1 group O infection in Equatorial Guinea, Central Africa. *AIDS* **11,** 1410–1412.
13. Yamaguchi, J., Vallari, A. S., Swanson, P., Bodelle, P., Kaptue, L., Ngansop, C., et al. (2002) Evaluation of HIV type 1 group O isolates: identification of five phylogenetic clusters. *AIDS Res. Hum. Retroviruses* **18,** 269–282.
14. Roques, P., Robertson, D. L., Souquiere, S., Damond, F., Ayouba, A., Farfara, I., et al. (2002) Phylogenetic analysis of 49 newly derived HIV-1 group O strains: high viral diversity but no group M-like subtype structure. *Virology* **302,** 259–273.
15. Nkengasong, J. N., Fransen, K., Willems, B., Karita, E., Vingerhoets, J., Kestens, L., et al. (1997) Virologic, immunologic, and clinical follow-up of a couple infected by the human immunodeficiency virus type one, group O. *J. Med. Virol.* **51,** 202–209.
16. Chaix-Baudier, M. L., Chappey, C., Burgard, M., Letourneur, F., Igual, J., Saragosti, S., and Rouzioux, C. (1998) First case of mother-to-infant HIV type 1 group O transmission and evolution of C2V3 sequences in the infected child. French HIV Pediatric Cohort Study Group. *AIDS Res. Hum. Retroviruses* **14,** 15–23.
17. Coste, J., Montes, B., Reynes, J., Peeters, M., Segarra, C., Delaporte, E., and Segondy, M. (1997) Effect of HIV-1 genetic diversity on HIV-1 RNA quantification in plasma: comparative evaluation of three commercial assays. *J. Acquir. Immune Defic. Syndr. Hum. Retrovirol.* **15,** 174–175.
18. Debyser, Z., Van Wijngaerden, E., Van Laethem, K., Beuselinck, K., Reynders, M., De Clercq, E., et al. (1998) Failure to quantify viral load with two of the three commercial methods in a pregnant woman harboring an HIV type 1 subtype G strain. *AIDS Res. Hum. Retroviruses* **14,** 453–459.

19. Gobbers, E., Fransen, K., Oosterlaken, T., Janssens, W., Heyndrickx, L., Ivens, T., et al. (1997) Reactivity and amplification efficiency of the NASBA HIV-1 RNA amplification system with regard to different HIV-1 subtypes. *J. Virol. Methods* **66,** 293–301.
20. Alaeus, A., Lidman, K., Sonnerborg, A., and Albert, J. (1997) Subtype-specific problems with quantification of plasma HIV-1 RNA. *AIDS* **11,** 859–865
21. de Mendoza, C., Alcami, J., Sainz, M., Folgueira, D., and Soriano, V. (2002) Evaluation of the Abbott LCx quantitative assay for measurement of human immunodeficiency virus RNA in plasma. *J. Clin. Microbiol.* **40,** 1518–1521.
22. Swanson, P., Harris, B. J., Holzmayer, V., Devare, S. G., Schochetman, G., and Hackett, J., Jr. (2000) Quantification of HIV-1 group M (subtypes A-G) and group O by the LCx HIV RNA quantitative assay. *J. Virol. Methods* **89,** 97–108.
23. Plantier, J. C., Gueudin, M., Ogel, P., Damond, F., Mauclere, P., and Simon, F. (2003) Plasma RNA Quantification and HIV-1 divergent strains. *J. AIDS* **33,** 1–7.
24. Gueudin, M., Plantier, J. C., Damond, F., Roques, P., Mauclere, P., and Simon, F. (2003) Plasma viral RNA assay in HIV-1 group O infection by real-time PCR. *J. Virol. Methods* **113,** 43–49.

19

A New Combined HIV p24 Antigen and Anti-HIV-1/2/O Screening Assay

Susanne Polywka, Hedwig Duttmann, Frank Lübben, Rainer Laufs, and Jürgen Feldner

Summary

It is important to shorten the window period after acute HIV infection in which infected individuals are still antibody-negative, especially in blood donors. Newly developed fourth-generation assays detect antibodies to HIV-1, including subtype O, and to HIV-2 and, simultaneously, p24 antigen of HIV-1. To evaluate this assay for daily routine work we compared it with different third-generation assays using sera from uninfected patients and patients with known HIV infection. The most interesting sera are those drawn during seroconversion from freshly infected patients. Whenever we encounter such a patient with acute HIV infection we store the serum in aliquots at –20°C. Thus, we were able to establish our own seroconversion panel and use it in our laboratory for evaluation of new assays. The new test was shown to be able to detect all chronically HIV-infected individuals and four of six patients during seroconversion although in two of these patients conventional assays for HIV antibodies were still negative. The rate of unspecific reactivities was slightly higher as compared with third-generation assays.

Key Words: HIV; combined antigen/antibody test; early detection of HIV infection; seroconversion; p24.

1. Introduction

The most efficient way of acquiring infection with the human immunodeficiency virus type 1 (HIV-1) or type 2 (HIV-2) is by transfusion of contaminated blood or blood products. The inclusion of serologic screening for antibodies to HIV-1 or HIV-2 in Western countries has led to a dramatic decrease of this risk. However, in the early phase of HIV infection, the virus is already replicating while specific antibodies are still lacking. Newly infected individuals would not be shown to be infectious during this window period of

From: *Methods in Molecular Biology, Vol. 304: Human Retrovirus Protocols: Virology and Molecular Biology*
Edited by: T. Zhu © Humana Press Inc., Totowa, NJ

6 to 12 wk after infection while their blood donations would be highly infectious. Thus, by testing only for HIV antibodies the estimated risk of acquiring HIV by blood donations in the United States was 1/493,000/U *(1)*.

To minimize this antibody-negative phase, attempts were made to improve both specificity and sensitivity of the assays. This was hampered by the fact that different species and subtypes of the virus exist containing antigens that lead to only weakly cross-reactive antibodies. The first available enzyme-linked immunosorbent assays (ELISAs) for detection of HIV-specific antibodies used HIV-1 lysate from infected cell lines as an antigen. The introduction of recombinant HIV proteins and synthetic peptides in ELISAs of the second generation considerably improved the specificity of reactions. Tests of the third generation were additionally able to detect antibodies to HIV-1 group O. They used HIV-1 and HIV-2 antigens both on the solid phase and in the conjugate; thus, antibodies of the immunoglobulin (Ig)G as well as of the IgM type could be detected. This development shortened the seronegative window period by 1 to 2 wk *(2)*, but HIV infected individuals were still false-negative for 4 to 6 wk after infection. To shorten this period and to exclude seronegative viremic individuals from blood donations, testing for HIV RNA was introduced in transfusion institutes, leading to an earlier detection of infection by approx 1 wk *(3)*. To reduce costs for blood products, most transfusion centers pool sera from many donors for the reverse transcriptase polymerase chain reaction (RT-PCR) procedure; therefore, false-negative results are observed in patients with only a few copies of HIV-1 RNA *(4)*. The risk of acquiring HIV by blood transfusion in Germany has been reduced to approx 1 in 18 million *(5)*.

Another approach to reduce the time after infection during which an infected individual appears to be seronegative is to test for HIV-specific antigens. In newly infected individuals, high levels of p24 antigen can be detected before the onset of production of specific antibodies. The testing of blood samples for both HIV-1 and HIV-2 antibodies and for p24 antigen would allow for an earlier detection of HIV-infected individuals but would increase the costs. The introduction of assays capable of detecting both antibodies and p24 antigen would have the advantage of being less expensive than PCR testing of single donations or conduction of several serological assays. Previous studies evaluating these fourth-generation assays showed promising results *(2,6–11)*. However, to establish a new assay in routine laboratory work, this assay must be carefully evaluated for its sensitivity, specificity, and reproducibility. We describe here how we achieved this for a new fourth-generation assay.

2. Materials

2.1. Combined HIV p24 Antigen and Anti-HIV-1/2/O Screening Assay

A fourth-generation enzyme-linked immunosorbent assay based on the determination of HIV p24 antigen and antibodies to HIV-1 and HIV-2 including

HIV-1 group O (Enzygnost® HIV Integral, Dade Behring Marburg GmbH, Marburg, Germany).

2.2. Conventional Assays for Detection of Antibodies to HIV-1 and/or HIV-2

Two assays for sole detection of HIV-1 antibodies including group O and HIV-2, as well as two assays for the separate detection of antibodies to HIV-1 and HIV-2, respectively; two Western blots as confirmatory assays for HIV-1 and HIV-2, respectively:

1. Enzygnost Anti-HIV-1/2 Plus (Dade Behring Marburg GmbH).
2. AxSYM® HIV-1/2 gO (Abbott GmbH Diagnostika, Wiesbaden-Delkenheim, Germany).
3. ELAVIA® Ac-Ab-Ak I (Sanofi Diagnostics Pasteur, Marnes-la-Coquette, France).
4. ELAVIA® Ac-Ab-Ak II (Sanofi Diagnostics Pasteur).
5. NovaPath™ HIV-1 Immunoblot (Bio-Rad Laboratories, München, Germany).
6. New LAV Blot II Ac-Ab-Ak (Sanofi Diagnostics Pasteur).

2.3. Assay for Detection and Quantitation of p24

HIVAG™-1 Monoclonal (Abbott GmbH Diagnostika).

2.4. Detection of HIV-1 RNA

Amplicor HIV-1 Monitor™ Test 1.5 (Roche Diagnostics GmbH, Mannheim, Germany).

2.5. Sera Used for Evaluation of Fourth-Generation Assay

A variety of stored sera are used to evaluate the sensitivity and specificity of a new assay:

1. Sera drawn from HIV-negative patients (*see* **Note 1**).
2. Sera drawn from HIV-infected patients (*see* **Note 2**).
3. Stored sera taken during seroconversion (*see* **Notes 3** and **4**).

3. Methods

3.1. Fourth-Generation Assay

The fourth-generation assay uses microtiter plates coated with recombinant proteins (HIV-1 gp41, HIV-1 group O gp41, and HIV-2 gp36) expressed in *Escherichia coli* and synthetic peptides (HIV-1 gp41) as well as a polyclonal p24 antibody.

For the Enzygnost HIV Integral assay, introduce 25 μL of sample buffer into each well. Add 100 μL of control sera or sample. Cover the test plate with adhesive foil and incubate at 37°C for 30 min. During this step HIV antibodies in the test samples or control react with the recombinant proteins or synthetic peptides that coat the wells of the microtiter plate. Simultaneously, any p24 antigens present in the test sample bind to the polyclonal p24 antibodies attached to the wells. After this incubation step, wash the wells four times with 300 µL washing solution per well to remove unbound constituents of the samples or controls. Then add 100 µL conjugate 1 to each well containing six different biotin-conjugated antigens or antibodies (one recombinant HIV-1 group M gp41 antigen; three different synthetic peptides: HIV-1 group M gp41, HIV-1 group O gp 41, HIV-2 gp36; two monoclonal anti-p24 antibodies) and incubate the microtiter plate for 30 min at 37°C. During this step, the labeled antigens and antibodies react with the specifically bound antibodies or p24 antigens, respectively. Wash each well four times with 300 µL washing solution to remove unbound biotin conjugates. After this, add 100 µL of conjugate 2 consisting of a streptavidin-peroxidase complex to each well. During the incubation for 30 min at 37°C, this conjugate binds to the biotin-labeled antigen or antibodies fixed to the solid phase. Remove the excess conjugate by washing the plate as mentioned before. Add 75 µL of the chromogen tetramethylbenzidine dihydrochloride (TMB) to each well to determine the solid-phase-bound enzyme activity and incubate at 15–25°C for 30 min. The peroxide bound to the solid phase reacts with TMB and the chromogen turns blue. Stop the enzymatic reaction of this step by adding 75 µL of sulfuric acid to each well; this leads to conversion of the blue color of the chromogen to yellow. The color intensity, which is roughly proportional to the concentration of HIV antibodies and/or p24 antigen in the sample, is measured at 450 nm against 650 nm. Samples are considered reactive if their absorbances exceed those of the mean values of three negative controls plus 0.450.

3.2. Third-Generation Assays

Perform conventional assays according to the manufacturers' instructions. The Enzygnost Anti-HIV-1/2 Plus assay uses microtiter plates coated with a mixture of recombinant HIV proteins (HIV-1 gp41, HIV-1 group O gp41, and HIV-2 gp36) expressed in *E. coli*. HIV-specific Ig contained in the test sample or control bind to the recombinant HIV-1, HIV-2, or HIV-1 group O proteins bound to the surface of the microtiter plate. Unbound components are removed by washing. Conjugate consisting of peroxidase-labeled recombinant HIV proteins and synthetic peptides similar to those coated on the surface is added and binds to the HIV-specific antibodies on the microtiter plate. The excess conjugate is removed by washing and TMB is added. The peroxidase of the bound conjugate turns the TMB blue. The reaction is stopped by 0.5 *N* sulfuric acid and the plate is measured at a wavelength of 450 nm and 650 nm as reference. The color intensity is proportional to the concentration of antibody present in the sample or control.

The AxSYM HIV-1/2gO assay based on the microparticle enzyme immunoassay (MEIA) technique utilizes recombinant HIV proteins corresponding to four viral proteins (HIV-1 envelope, HIV-1 group O envelope, HIV-1 core, and HIV-2 envelope) expressed in *E. coli* or *Bacillus megaterium* and two synthetic peptides corresponding to HIV-1 and HIV-2 envelope. These proteins and peptides are coated on a solid phase to capture antibodies to HIV-1 or HIV-2 from the test sample or control. A portion of the reaction mixture is transferred to the matrix cell. Microparticles bind irreversibly to the glass fiber matrix. The matrix cell is washed to remove materials not bound to the microparticles. Biotinylated recombinant antigens and synthetic peptides are dispensed onto the matrix cell, forming an antigen–antibody–antigen complex. Unbound biotinylated antigens are removed by washing. The antibiotin:alkaline phosphatase conjugate is dispensed onto the matrix cell and binds with the antigen–antibody–antigen complex. Unbound conjugate is removed by washing. The substrate, 4-methylumbelliferyl phosphate (MUP), is added. The alkaline phosphatase-labeled conjugate catalyzes the removal of a phosphate group from the substrate, yielding the fluorescent product, 4-methylumbelliferone, which is measured by the MEIA optical assembly.

3.3. Tests for Separate Detection of Antibodies to HIV-1 or HIV-2

The ELAVIA Ac-Ab-Ak I and the ELAVIA AV-Ab-Ak II have identical procedures and use cell-culture-derived inactivated virus and uninfected cells as controls. The test is based on the use of two solid supports; the first one is coated with purified and inactivated virus antigens of HIV-1 and HIV-2, respectively (Ag-positive), and the other one is coated with cellular and serum antigens (Ag-negative). Samples and controls are added to both the Ag-positive and the Ag-negative wells. Antibodies to HIV-1 or HIV-2 bind to the immobilized virus antigens. The wells are washed to remove unbound parts of the sample or control. Peroxidase-labeled anti-human IgG antibody is added and binds to the solid-phase-retained HIV antibodies. The microtiter plate is washed to remove excess conjugate and the substrate o-phenylenediamine containing hydrogen peroxide is added. The peroxidase of the bound conjugate reacts with hydrogen peroxide group of the substrate, leading to a yellow-orange color. The reaction is stopped by the addition of 4 *N* sulfuric acid. The absorbance is read by a spectrophotometer at 492 nm to 620 nm. The observed absorbance difference for the same sample (Ag-positive minus Ag-negative) indicates the presence or absence of antibodies to HIV-1 and HIV-2, respectively.

3.4. Confirmatory Assays

Every positive result of an HIV screening assay was confirmed by immunoblot. Both blots for the confirmation of reactivities in HIV antibody screening assays, New LAV Blot I and New LAV Blot II, have the identical principle. They use the indirect ELISA technique on a nitrocellulose strip containing all constitutive proteins of the HIV-1 or HIV-2 virus and an internal anti-IgG control. The latter is located at the bottom of the strip below the p17 (HIV-1) and p16 (HIV-2) protein and is used to validate the addition of sample and reagents as well as the correct execution of the test protocol. Inactivated proteins of HIV-1 or HIV-2 are separated according to their molecular weights by polyacrylamide gel electrophoresis in dissociating and reducing medium and subsequently transferred onto a nitrocellulose membrane sheet. The strips are rehydrated in washing solution. Afterward, sample or control is added. If anti-HIV-1 or anti-HIV-2 antibodies are present, they will bind to the viral proteins on the strip. After washing, the alkaline phosphatase-labeled anti-human IgG conjugate is added and incubated to allow the binding to the anti-HIV antibodies captured on the solid phase. Excess conjugate is removed by washing the substrate, 5-bromo-4-chloro-3-indolyl phosphate (BCIP), is added. If the conjugate is bound to an HIV-specific antibody, the substrate's color changes to gray-blue. To confirm the reactivity of a screening assay at least one band of the env-region (gp41, gp120, gp160) or two bands of the pol- (p31, p66) and gag-region (p17, p24, p55) of HIV must be positive. Indeterminate reactivities are those restricted to one or more bands of only one region of HIV-1.

3.5. Assay for Detection of p24 Antigen

In the evaluation of the fourth-generation assay's sensitivity during seroconversion, an ELISA is used for the separate detection of p24 antigen. In this assay HIV-1 virions are disrupted by the addition of Triton X-100 containing specimen diluent. Beads coated with monoclonal antibody to HIV-1 p24 are incubated with the pretreated sample and HIV-1 p24 present in it binds to the monoclonal antibody on the bead. Unbound material is removed by washing. The bead is then incubated with rabbit antibody to HIV-1, which binds to the p24 antigen on the bead. Excess material is again removed by washing. A conjugate of goat antibody to rabbit IgG and horseradish peroxidase is added. During the next incubation the conjugate binds rabbit IgG on the bead. Unbound conjugate is removed by another washing step and the beads are subsequently incubated with *o*-phenylenediamine (OPD) containing hydrogen peroxide. The reaction of the substrate with peroxidase yields a yellow-orange color. The intensity of the color is proportional to the amount of uncomplexed HIV-1 p24 antigen present in the sample. The reaction is stopped by the addition of 1 *N* sulfuric acid and the intensity of color is read using a spectrophotometer set at 492 nm. Repeatedly reactive samples are tested again with HIVAG-1 monoclonal blocking antibody. This procedure uses a specific antibody-neutralizing step followed by testing for p24 antigen as described above.

By adding the neutralizing antibody, the amount of free binding sites on the p24 antigen is reduced so that the rabbit antibody to p24 cannot bind to it. This leads to a reduction of the optical density measured. To confirm a positive result as specific the OD signal has to be reduced by at least 50% compared to the sample without blocking antibodies.

3.6. Detection of HIV-1 RNA by PCR

In all confirmed reactive sera, the viral load is determinded by RT-PCR. The AMPLICOR HIV-1 Monitor test v1.5 is based on five major processes: specimen preparation; reverse transcription of target RNA to generate complementary DNA (cDNA); PCR amplification of target cDNA using HIV-1-specific complementary primers; hybridization of the amplified DNA to oligonucleotide probes specific to the target; and detection of the probe-bound amplified DNA by colorimetric determination. HIV-1 RNA is isolated either directly from plasma by lysis of virus particles with a chaotropic agent followed by precipitation with alcohol, or, in the case of low viremia, by concentrating viral particles in plasma by high-speed centrifugation followed by lysis and alcohol precipitation. A known number of quantitation-standard RNA molecules is introduced into each specimen with the lysis buffer and is used for the quantitation of HIV-1 RNA in the sample. The reverse transcription and PCR amplification reactions are performed with the thermostable recombinant enzyme *Thermus thermophilus* DNA polymerase (r*Tth* pol). In the presence of manganese, r*Tth* pol has both reverse transcriptase and DNA polymerase activity. This allows both reverse transcription and PCR amplification to occur in the same reaction mixture. Processed specimens are added to the amplification mixture. The downstream or antisense primer and the upstream or sense primer are biotinylated at the 5' ends. The reaction mixture is heated to allow the downstream primer to anneal specifically to the HIV-1 target RNA. In the presence of Mn^{2+} and excess deoxynucleoside triphosphates (dNTPs), r*Tth* pol extends the annealed primer to a DNA strand (cDNA) complementary to the RNA target. The reaction mixture is heated to denature the RNA:cDNA hybrid and expose the primer target sequences. As the mixture cools, the upstream primer anneals specifically to the cDNA strand, the r*Tth* pol extends the primer, and a second DNA strand is synthesized. This completes the first cycle, yielding a double-stranded copy of the target region. To separate the double-stranded DNA and to expose the primer target sequences, the mixture is heated again. During the next cycle, both sense and antisense primers anneal to the target DNA and are extended along the target templates to produce a 155-base-pair double-stranded DNA molecule termed the amplicon. This process is repeated for a designated number of cycles, with each cycle effectively doubling the amount of amplicon DNA. After the last cycle, the amplicon is chemically denatured to form single-stranded DNA by the addition of denaturation solution. Aliquots of denatured amplicon are added to separate wells of a microwell plate coated with HIV-1-specific oligonucleotide probes. The amplicon is bound to the wells by hybridization to the oligonucleotide probes on the solid phase. To achieve quantitative results over a large dynamic range, serial dilutions of the denatured amplicon are analyzed in the microwell plate. After hybridization, the plate is washed to remove any unbound material and avidin-horseradish peroxidase conjugate is added to each well. The conjugate binds to the biotin-labeled amplicon hybridized to the target-specific oligonucleotide probes on the solid phase. The microwell plate is washed again to remove unbound conjugate and a substrate solution containing hydrogen peroxide and 3,3' 5,5'-tetramethylbenzidine (TMB) is added to the wells. In the presence of hydrogen peroxide the horseradish peroxidase catalyzes the oxidation of TMB to form a colored complex. The reaction is stopped by the addition of a weak acid and the optical density is measured at 450 nm. The probe-binding region of the HIV-1 quantitation standard amplicon has been modified to differentiate it from the sample amplicon. The amount of HIV-1 RNA in each specimen is calculated from the ratio of the total HIV-1 OD to the total HIV-1 quantitation standard OD and the input number of RNA molecules contained in the latter.

3.7. Algorithm for Retests

If discrepant results with the fourth-generation and any conventional assay appear, repeat the test twice. To consider any reactivity as specific, use confirmatory assays according to the algorithm shown in **Fig. 1**.

4. Notes

1. To evaluate the fourth-generation assay's specificity, use a number of sera drawn from HIV-negative patients. These can be sera drawn from patients who have been previously tested and shown to be negative or you can use sera that are currently tested in your laboratory and have negative results in your reference assay. In our study we included 546 sera from noninfected patients. Of these, 537 sera were negative in all three screening assays (that is, the fourth-generation assays, the Enzygnost Anti-HIV-1/2 Plus, and the AxSYM HIV-1/2 gO assay). Nine patients had positive results in at least one assay, but due to retests and additional tests such as the Western blot they could be classified as uninfected. Overall, the rate of unspecific reactivities was slightly higher in the fourth generation assay (1.1%) as compared to conventional screening assays (Enzygnost Anti-HIV-1/2 Plus: 0.4%; AxSYM HIV-1/2 gO: 0.4%). The results of the HIV-negative patients are shown in **Table 1**. Note that in one of them an unspecific p24 band was found in the Western blot, and the ELAVIA Ac-Ab-Ak I was also unspecifically reactive.

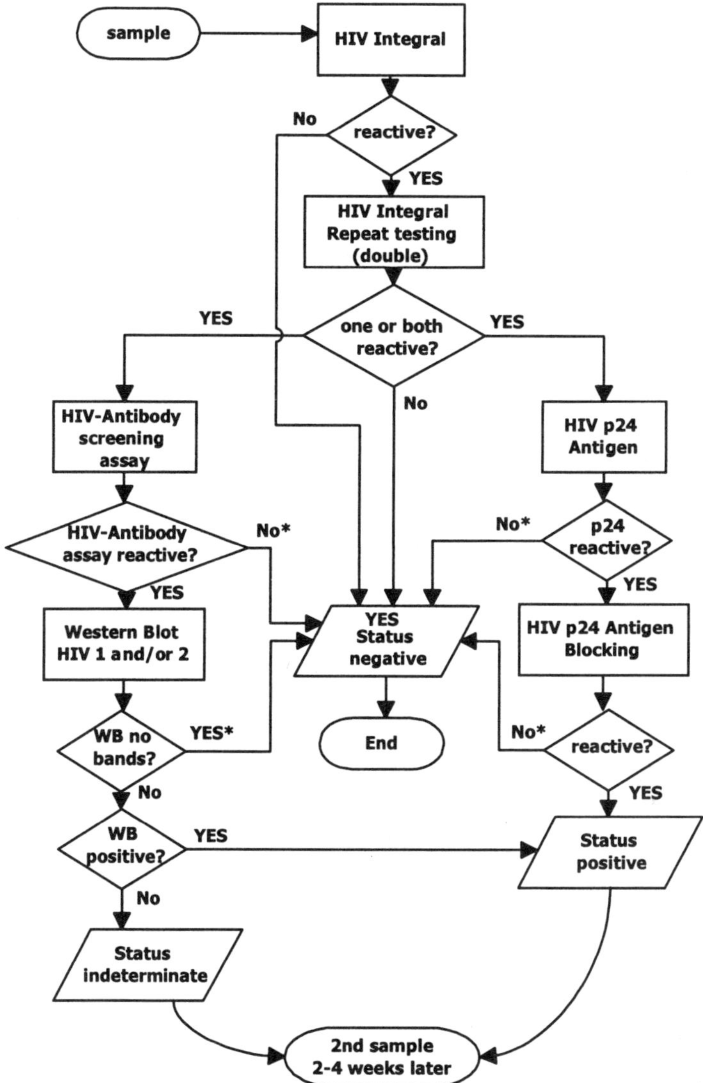

Fig. 1. Retest algorithm for Enzygnost® HIV Integral. *High-risk persons with suspected HIV infection should give a second sample within the next 2–4 wk.

2. Sensitivity of the test is evaluated by a number of sera drawn from patients known to be HIV-infected. Because the fourth-generation assay should detect HIV-1 infections as well as HIV-2 infections, it would be best to use sera from both HIV-1- and HIV-2-infected patients. The results of the new assay and the reference assay(s) should not only be reported as "positive" or "negative" but the intensity of the OD values should also be considered.

Table 1
Results of 546 HIV-Negative Patients

N sera	Enzygnost HIV Integral	Enzygnost anti-HIV-1/2 Plus	AxSYM HIV-1/2 gO (MEIA)	ELAVIA Ac-Ab-Ak I	Western blot	HIVAG (p24 antigen)
537	negative	negative	n.d.	n.d.	n.d.	n.d.
3	3 × reactive	3 × negative	2 × negative	negative	n.d.	negative
1	3 × reactive	3 × negative	2 × negative	HIV-1 reactive	p24.000	negative
1	1 × borderline, 2 × negative	negative	negative	n.d.	n.d.	n.d.
1	1 × reactive, 1 × borderline, 1 × negative	3 × negative	2 × negative	n.d.	n.d.	n.d.
1	3 × negative	3 × reactive	2 × reactive	negative	negative	negative
1	3 × negative	1 × reactive, 2 × negative	negative	n.d.	n.d.	n.d.
1	3 × negative	3 × negative	2 × reactive, 1 × borderline	n.d.	n.d.	negative

n.d., not done.

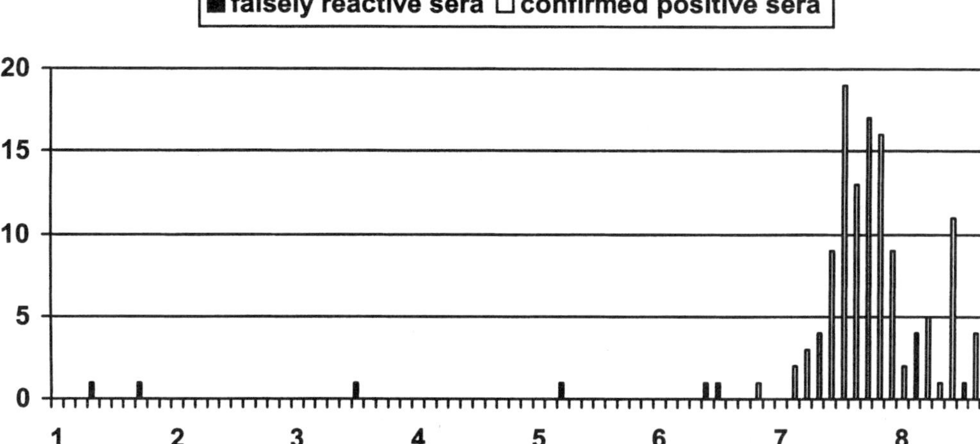

Fig. 2. Distribution of sample-to-cutoff ratio (S/CO) values of falsely reactive (n = 6) and confirmed HIV-1-positive (n = 121) sera in the Enzygnost® HIV Integral assay.

The prevalence of HIV-2 in Germany is very low and largely restricted to patients coming from endemic areas. Thus, in our study we did not include any serum from HIV-2-infected individuals. In total we used sera from 121 patients with documented HIV-1 infections who had already been tested in our laboratory before. All these sera were reactive in the fourth generation, the reference assays (Enzygnost Anti-HIV-1/2 Plus and AxSYM HIV-1/2 gO), the ELAVIA Ac-Ab-Ak I, and the Western blot. We found no difference in the reactivities in the fourth-generation assay between sera positive or negative for p24 antigen. Overall, 7 of the 121 patients had antigenemia (13–504 pg/mL; median, 19 pg/mL). Thirty-eight patients were also positive by PCR with viral loads varying between 54 copies/mL and >750,000 copies/mL (median, 1000 copies/mL). In the combined antigen-antibody assay, all sample to cut-off (S/CO) ratios of confirmed HIV-1 positive sera exceeded those of the six falsely reactive sera shown in **Table 1**. Thus, in most cases it is possible to distinguish between specific and unspecific reactivities by evaluating the intensity of the reactivity, but in every reactive serum however weakly positive it may be, confirmatory assays are mandatory. **Figure 2** compares the S/CO values of falsely reactive and confirmed positive sera.

3. The most interesting point in testing the sensitivity of a combined assay for the detection of HIV antibodies and p24 antigen is to use it in patients with acute HIV infections. This is especially important since a test like this is declared to shorten the period in HIV seroconversion. There are commercially available

seroconversion kits but these are expensive. Freshly infected patients are encountered only occasionally in our laboratory, but when they are found, we store these sera apart from normal routine sera in aliquots at –20°C to prevent them from repeated freezing and thawing. Over the years we have established our own seroconversion panel to use for evaluations like this. Of these sera, we have protocols with the results of every assay ever run.

We tested sera from six freshly infected patients; for one of them three sera were available, drawn 6 and 21 d after the first sample. In the other five patients, HIV seroconversion was confirmed 7–17 d later (data not shown). In two patients (patient 2 and patient 5; *see* **Table 2**) the fourth-generation assay was already reactive while conventional screening assays were not, confirming the higher sensitivity during seroconversion. These patients had high amounts of p24 antigen (976 pg/dL and 410 pg/dL, respectively). In patient 2, the immunoblot also confirmed a positive antibody reactivity, but only after prolonged incubation for 16 h instead of 30 min. In sera obtained 6 and 21 d later, the conventional ELISAs also became positive. In contrast to this, the immunoblot of patient 5 revealed only an indeterminate reactivity restricted to antibodies against proteins of the gag region. In patient 3, all antibody assays were reactive, including the Western blot. In patient 4 only the fourth-generation and the conventional ELISA of the same manufacturer were positive. The ELISA for the sole detection of HIV-1 antibodies failed to react and the immunoblot was also indeterminate with visible bands only with gag proteins after prolonged incubation overnight. In two other patients with low amounts of p24 antigen (patient 1 and patient 6), the combined antibody/antigen assay was negative; in the latter case, the conventional ELISA from the same manufacturer was weakly reactive. In both cases, the immunoblot assay showed bands only after prolonged incubation.

4. Sera of our own seroconversion panel were also tested for p24 antigen when they were first sent to us so we could also evaluate the sensitivity of the new assay for the detection of p24 antigen. The lower detection limit of p24 antigen with the Enzygnost HIV Integral assay is approximately 110 pg/mL *(12)*. Our study confirmed that the assay failed to detect patients during seroconversion with antigen concentrations of 38 and 88 pg/mL (patients 1 and 6). In patient 4 the antigen concentration was even lower (14 pg/mL) but the fourth generation nevertheless was weakly reactive. In this case, part of the p24 antigen might have been complexed by specific antibodies. The fact that even the conventional Enzygnost Anti-HIV-1/2 Plus assay as well as the blot after prolonged incubation were positive supports the assumption that low amounts of antibodies were already present. However, it remains unclear why in patient 6 only the conventional Anti-HIV-1/2 Plus assay was reactive while the fourth-generation assay was not. In patient 2, from whom three sera were available, the S/CO value declined from 8.17 to 5.64 and afterward rose again to a value of 7.31. This course probably reflects the decrease in antigen concentration while only small amounts of antibodies were present. Interestingly, in another study the authors described a second diagnostic window with fourth-generation assays during which these tests become negative again as a result of the absence of HIV antibodies and the decrease of antigen

Table 2
Comparison of Reactivities in Six Patients During Seroconversion

Patient No.[a]	Enzygnost® Integral (S/CO)	Enzygnost Anti-HIV-1/2 Plus (S/CO)	ELAVIA I (S/CO)	Western blot (reactive bands in kD)	PCR (copies/mL)	p24 antigen (pg/mL)
1	0.49	0.30	0.29	17, 24, 31, 55, 66[b]	n.d.	38
2 d 1	8.17	0.04	0.03	17, 24, 31, 41, 160[b]	n.d.	976
d 7	5.64	1.37	8.33	17, 24, 31, 41, 120, 160[b]	n.d.	176
d 22	7.31	4.71	2.21	24, 55, 160	positive	negative
3	7.82	8.43	4.55	17, 24, 31, 55, 66, 160	1000	36
4	1.67	1.12	0.61	17, 24, 55[b,c]	100	14
5	7.38	0.19	0.67	17, 24, 55[c]	10,000	410
6	0.82	1.75	0.48	17, 24, 55, 66[b]	100	88

[a]In patient 2, the data of the three tested samples are shown (d 1, d 7, and d 22).
[b]Visible bands only after modified incubation overnight.
[c]Pattern considered as indeterminate.
S/CO, sample to cutoff ratio; positive results with the individual assays are underlined; PCR, polymerase chain reaction; n.d., not done.

concentration below the detection limit *(13)*. Therefore, it is important to further enhance the sensitivity of fourth-generation assays for p24 antigen. This can be achieved, for example, by modifying the arrangement of antigens and antibodies on the solid phase to enlarge the number of binding sites. Also, using a monoclonal antibody instead of a polyclonal to p24 antigen might improve sensitivity as well as specificity of the assay. However, even when an assay has been optimized, it must be re-evaluated carefully on a regular basis to avoid failure of new mutants of the virus imported from endemic areas.

References

1. Aubuchon, J. P., Birkmeyer, J. D., and Busch, M. P. (1997) Cost-effectiveness of expanded human immunodeficiency virus-testing protocols for donated blood. *Transfusion* **45,** 45–51.
2. Brust, S., Duttmann, H., Feldner, J., Gürtler, L., Thorstensson, R., and Simon, F. (2000) Shortening of the diagnostic window with a new combined HIV p24 antigen and anti-HIV-1/2/O screening assay. *J. Virol. Methods* **90,** 153–165.
3. Dodd, R. Y., Notari, E. P., and Stramer, S. L. (2002) Current prevalence and incidence of infectious disease markers and estimated window-period risk in the American Red Cross blood donor population. *Transfusion* **42,** 975–979.
4. Barlow, K. L., Tosswill, J. H. C., Parry, J. V., and Clewly, J. P. (1997) Performance of the Amplicor human immunodeficiency virus type 1 PCR and analysis of specimens with false negative results. *J. Clin. Microbiol.* **35,** 2846–2853.
5. Roth, W. (2003) Virussicherheit von Blutprodukten: Erfolge und Lücken. Arbeitstagung der Deutschen Vereinigung zur Bekämpfung der Virusinfektionen (DVV) und der Gesellschaft für Virologie (GfV), Gießen.
6. Weber, B., Fall, E. H. M., Berger, A., and Doerr, H. W. (1998) Reduction of diagnostic window by new fourth-generation human immunodeficiency virus screening assays. *J. Clin. Microbiol.* **36,** 2235–2239.
7. Martínez-Martínez, P., Martín del Barrio, E., De Benito, and J., Landinez, R. (1999) New Lineal immunoenzymatic assay for simultaneous detection of p24 antigen and HIV antibodies. *Eur. J. Clin. Microbiol. Infect. Dis.* **18,** 591–594.
8. Gürtler, L., Muhlbacher, A., Michl, U., Hofmann, H., Paggi, G., Bossi, V., et al. (1998) Reduction of the diagnostic window with a new combined p24 antigen and human immunodeficiency virus antibody screening assay. *J. Virol. Methods* **75,** 27–38.
9. Van Binsbergen, J., Siebelink, A., Jacobs, A., Keur, W., Bruynis, F., van de Graaf, M., et al. (1999) Improved performance of seroconversion with a 4th generation HIV antigen/antibody assay. *J. Virol. Methods* **82,** 77–84.
10. Hashida, S., Hashinaka, K., Nishikata I, Saito, A., Takamizawa, A., Shinagawa, H., and Ishikawa, E. (1996) Earlier diagnosis of HIV-1 infection by simultaneous detection of p24 antigen and antibody IgGs and reverse transcriptase in serum with enzyme immunoassay. *J. Clin. Lab. Anal.* **10,** 213–219.
11. Yerly, S., Simon, F., and Perrin, L. (1999) Early diagnosis of primary HIV infections: using a combined screening test (p24 antigen and anti-HIV antibodies). *Schweiz. Med. Wochenschr.* **129,** 319–322.

12. le Group de Travail Rétrovirus de la Société Francaise de Transfusion Sanguine, Couroucé, A. (1999) Tests de dépistage combiné des anticorps anti-VIH et de l'antigène p24. *Gazette Trans.* **155,** 4–18.
13. Meier, T., Knoll, E., Henkes, M., Enders, G., and Braun, R. (2001) Evidence for a diagnostic window in fourth generation assays for HIV. *J. Clin. Virol.* **23,** 113–116.

20

A New Automated Fourth-Generation HIV Screening Assay With Sensitive Antigen Detection Module and High Specificity

Bernard Weber

Summary

New screening enzyme immunoassays, which permit the simultaneous detection of HIV antigens reduce the diagnostic window period between the time of immunodeficiency virus (HIV) infection and seroconversion. The VIDAS HIV DUO Ultra is an enzyme-linked fluorescent assay (ELFA) for the screening of HIV infection. It is performed with the fully automated VIDAS or mini-VIDAS instruments, which are so-called walk away systems. The detection limit is 3 pg of HIV-1 p24 Ag/mL serum. HIV antibody is detected with the same sensitivity as stand-alone third-generation antibody tests. The total incubation time is about 2 h. Results are calculated, interpreted, and printed by the VIDAS instrument. Usually, fourth-generation assays demand a special algorithm for the analysis of reactive samples. For the anti-HIV part of the assay, confirmation of reactivity should be done with an assay that lacks the p24-antigen detection module and when reactivity persists subsequently by immunoblot. For the p24-antigen part, confirmation of reactivity should be analyzed in an assay that lacks the anti-HIV detection part.

Key Words: Primary infection; enzyme immunoassay (EIA); automated assay; HIV screening.

1. Introduction

In order to reduce the diagnostic window period between the time of human immunodeficiency virus (HIV) infection and seroconversion, new screening enzyme-linked immunosorbent assays (ELISA), which permit the simultaneous detection of HIV antigen and antibody, have been available on the international market since 1997 ***(1–4)***. On average, the diagnostic window is 4 to 5 d less for these new, so-called fourth-generation HIV assays compared with an-

From: *Methods in Molecular Biology, Vol. 304: Human Retrovirus Protocols: Virology and Molecular Biology*
Edited by: T. Zhu © Humana Press Inc., Totowa, NJ

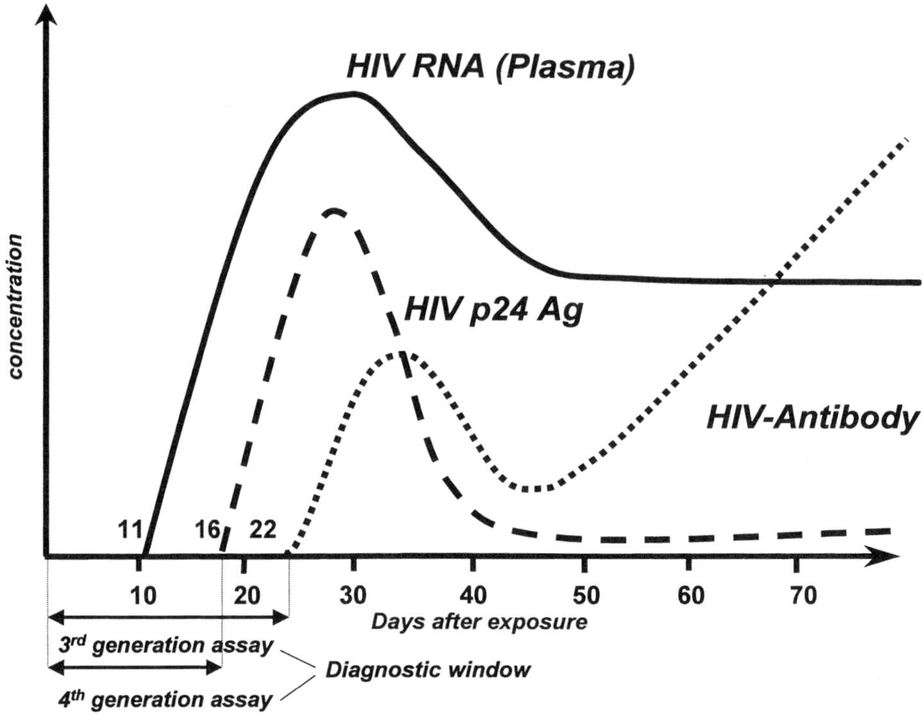

Fig. 1. Kinetics of HIV markers during primary infection and diagnostic window.

tibody ELISAs (third-generation) *(1–4)* (**Fig. 1**). The most recently launched commercial fourth-generation test kits have a detection threshold of 3–10 pg of p24 Ag/mL, which is within the range of single HIV p24 Ag assays *(3,4)*. HIV-1 non-B subtypes are detected with a comparable sensitivity, in dilutions of virus lysates and in HIV-1 subtype E (CRFO1_AE) primary infection, to p24 Ag assays *(5)*. An important issue is the role that the new combination assays should play in blood donor screening. Combination assays are of particular interest for lower-income nations that may not be able to afford nucleic acid amplification technology (NAT) since these assays are easy to use, show a high stability and cost-effectiveness (list price in the same range than antibody-alone assays), and have low false-positive rates (comparable to that of third-generation assays). In high-income countries, NAT will probably replace the need for HIV-1 Ag testing in blood donor screening *(6)*. In Europe, fouth-generation assays are used for screening and diagnosis of HIV infection in the clinical laboratory and have proven their efficiency for the detection of primary infection in risk populations in urban centers *(7)*.

VIDAS HIV DUO Ultra is an enzyme-linked fluorescent assay (ELFA), which permits the simultaneous detection of the p24 antigen and antibodies against HIV-1 (including group O) and HIV-2. The upper part of the solid phase receptacle (SPR) coated with three different anti-p24 monoclonal antibodies is used for p24 Ag detection (**Fig. 2**). The lower part of the SPR serves for immunoglobulin (Ig)G, IgM, and IgA antibody detection and is sensitized with gp160 of HIV-1 and peptides representing immunodominant epitopes of gp41 from HIV-1 group O and gp36 from HIV-2.

During the first incubation step, the sample and biotinylated rabbit anti-p24 antibody are aspirated into the SPR. p24 antigen is released through virus lysis and binds to the monoclonal antibodies on the SRP and also to the biotinylated anti-p24 antibody. Simultaneously, anti-HIV-1 and anti-HIV-2 antibodies bind to gp160 and/or gp36/gp41 peptides on the lower part of the SPR. The second incubation with biotinylated antigens (identical to those coated to the solid phase) is performed only on the lower part of the SPR.

The third reaction step is performed with the entire SPR. Alkaline phosphatase labeled streptavidine is added and binds to biotinylated anti-p24 antibody on the upper part of the SPR, if present, and to biotinylated antigens in the lower part, if present. The substrate (4-methylumbelliferyl phosphate) first reacts with the lower part of the SPR and fluorescence is measured at a wavelength of 450 nm. The intensity of fluorescence is proportional to anti-HIV antibody activity. Afterward, the substrate is incubated with the entire SPR and the fluorescence is measured a second time. The intensity of the reaction is proportional to the p24 Ag concentration.

At the end of the assay, results are automatically calculated by VIDAS in relation to standards and printed. A separate test value for p24 antigen and HIV antibody is calculated by dividing the patient relative fluorescence value (RFV) by the RFV of the respective standard.

2. Materials

1. **Specimen**
   Serum or plasma (EDTA, citrate, thrombin, lithium heparinate, lithium heparinate gel) (*see* **Note 1**). Decanted EDTA plasma must be centrifuged at 1000 to 1300*g* for 10 to 15 min in order to avoid false-positive results. Samples containing particles must be centrifuged. Samples can be stored at 2–8°C for a maximum of 48 h; if longer storage is required, they should be frozen. Specimens should not be inactivated.
2. **Reagents (supplied)**
   Indicate the first day of use of the kit on the test package or in the dedicated laboratory files.
   Store reagents at 2–8°C. Shelf life is approx 9 mo (*see* **Note 2**).

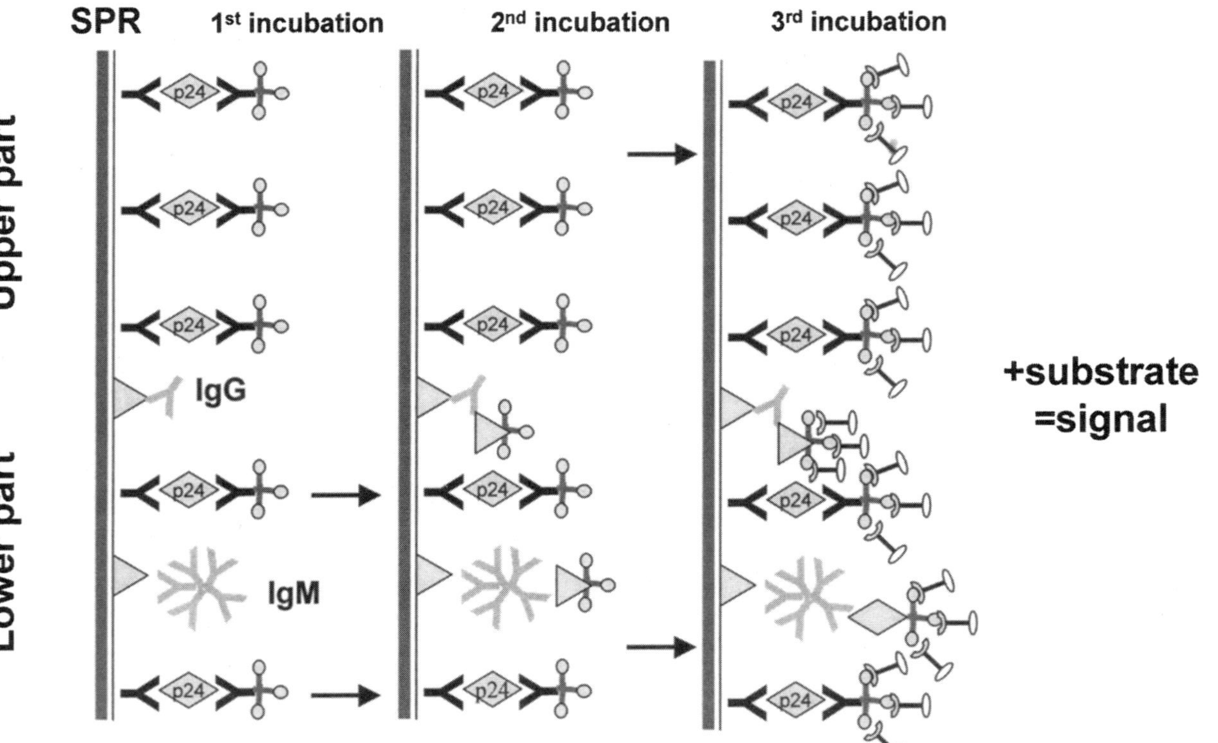

Fig. 2. Test principle of the VIDAS HIV DUO Ultra assay.

Table 1
Description of Reagent Strip

Well	Reagents
1	Sample well
2	HEPES buffer (20 mmol/L, pH 6.6) + biotin-labeled rabbit anti-24 antibody + Triton X-100 (2%) + goat serum 10% + protein and chemical stabilizers + gentamicin sulfate 0.2 g/L + sodium azide 0.9 g/L (300 μL). Conjugate 1 (alkaline phosphatase labeled mono clonal mouse anti-human IgG antibody)
3–5–6–8–9	Washing buffer (Tris-HCl [0.2 mol/L, pH 7.8], protein and chemical stabilizers, sodium azide 0.9 g/L [600 μL])
4	HEPES buffer (20 mmol/L, pH 7.2) + gp160 protein and gp41 and gp36 peptides biotin labeled + defatted milk protein and chemical stabilizers + gentamicin sulfate 0.2 g/L + sodium azide 0.9 g/L (300 μL)
7	Tracer (alkaline phosphatase labelled streptavidin, Tris-HCl (0.1 mol/L, pH 7.4), defatted milk + bovine albumin 2.5 g/L + Triton X100 (0.25%) + protein and chemical stabilizers, sodium azide 0.9 g/L (400 μL)
10	Substrate 4-methyl-umbelliferyl phosphate (0.6 mmol/L), diethanolamine (DEA) (0.62 mol/L, pH 9.2), sodium azide 1 g/L (300 μL)

a. Antibody-positive control (C1) (inactivated anti-HIV-1 IgG positive human serum, Tris-HCl [0.1 mol/L, pH 7.4], protein and chemical stabilizers and preservative); for reconstitution *see* **Subheading 3.3.1.**
b. Negative control (C2) (human HIV negative serum with sodium azide as preservative), ready to use.
c. Antigen-positive control (C3) (human serum with inactivated HIV-1 viral lysate, Tris-HCl [0.1 mol/L, pH 7.4], protein and chemical stabilizers and preservative); for reconstitution *see* **Subheading 3.3.1.**
d. Antibody standard (S1) (inactivated anti-HIV-1 IgG positive human serum, Tris-HCl [0.1 mol/L, pH 7.4], protein and chemical stabilizers and preservative); for reconstitution *see* **Subheading 3.3.1.**
e. Antigen standard (S2) (human serum with inactivated HIV-1 viral lysate, Tris-HCl [0.1 mol/L, pH 7.4], protein and chemical stabilizers and preservative); for reconstitution *see* **Subheading 3.3.1.**
f. Polypropylene reagent strips consisting of 10 wells (1–10) covered with a labeled foil seal (**Table 1**). The foil of the first sample well is perforated to facilitate the introduction of the sample. The last well of each strip is a cuvet in which the fluorometric reading is performed. The eight wells in the center section of the strip contain the various reagents required for the assay.

3. **Supplies**
   a. Pipet with disposable tip calibrated to dispense 200 µL.
   b. Powderless, disposable latex gloves.
4. **Equipment**
   a. VIDAS or mini-VIDAS analyzer (*see* **Note 4**).
   b. Vortex mixer.

3. Methods

3.1. Master Lot Data Entry

Before each new lot of reagents is used, specifications (or factory master calibration curve data) must be entered into the instrument (VIDAS or mini-VIDAS) using the master lot entry (MLE) card (specifications sheet) included in each kit. If this operation is not performed before initiating tests, the protocol will not run. The master lot data need to be entered only once for each lot.

It is possible to enter data automatically, using the MLE card, or manually. For complete instructions, see the VIDAS or mini-VIDAS operator's manual. The instrument will be able to check the control value only if the positive controls are identified by C1 and C3, the negative control by C2, and the standards by S1 and S2.

3.2. Calibration

Calibration, using the standard provided in the kit, must be performed upon reception of a new lot of reagents after the master lot data have been entered. Recalibration should be performed every 14 d. This operation provides instrument-specific calibration curves and compensates for possible minor variations in assay signal throughout the shelf life of the kit.

The standards identified by S1 and S2 must be tested in duplicate (see the VIDAS user manual) and their values must be within the set RFV range or the mean value will not be stored in the memory. If this is not the case, recalibrate.

3.3. Test Procedure

3.3.1. Before Testing

1. Remove the test kit from the refrigerator and allow it to come to room temperature for at least 30 min (*see* **Notes 4–7**).
2. Place powder-free disposable gloves on both hands (*see* **Note 8**).
3. Reconstitution of reagents C1, C3, S1, and S2: Add 2 mL of distilled water (*see* **Note 9**). Wait 5–10 min and homogenize with the vortex. The reconstituted standards and controls are stable for 2 mo. Write the date of reconstitution on the control and standard recipient labels.
4. Remove one HIV5 strip and one HIV5 SPR for each sample, control, or standard to be tested. Make sure the storage pouch has been resealed after the required

SPRs have been removed.

5. Place the HIV5 strip and HIV5 SPR on the VIDAS preparation/loading tray.
6. Enter the appropriate assay and patient data using the keyboard to create a work list (code HIV5). Type "HIV5" to enter the code, and indicate the number of tests to be run. If calibrators need to be tested, enter "S1"and "S2" for the sample identification. If the positive controls are to be tested, they will be identified by "C1" and "C3." If the negative controls need to be tested, they should be identified by "C2."
7. If necessary, centrifuge the samples prior to testing.
8. Mix the sample standards and/or controls using a vortex.

3.3.2. Testing

1. Pipet exactly 200 μL of sample standards or controls into the sample well (*see* **Note 9**).
2. Insert the VIDAS SPRs and strips into the positions indicated on the screen.
3. Initiate the analysis as directed in the VIDAS User Manual. All the assay steps are automatically controlled by the instrument. Results are obtained within approx 120 min.
4. After the assay is completed, dispose of the used SPRs and strips into an appropriate receptacle (*see* **Notes 10** and **11**).

3.3.3. Results

Once the assay is completed, results are analyzed automatically by the computer. Fluorescence is measured twice in each reagent strip's reading cuvet for each sample tested. The first reading is a background reading of the cuvet and substrate before the SPR is introduced into the substrate. The second reading is taken for HIV antibody detection after incubation of the substrate with the enzyme present on the lower part of the SPR. The third reading is performed for the detection of HIV p24 antigen after incubation of the substrate with the enzyme present on the total surface of the SPR. The RFVs are printed on the result report. The tests results are calculated for each antibody response (AB on the result report) and antigen (AG on the result report) by the VIDAS instrument as follows:

$$\text{Test value} = \text{specific patient RFV/specific standard RFV}$$

The interpretation for each sample (HIV5 on the result report) is calculated according to the value of the antigen and antibody test (**Table 2**). High RFV values for antibody detection may mask the antigen response or vice versa. For the masked response, test value is not indicated, and the message N.D. for "Not Determinable" appears instead. In these cases, the interpretation is valid. The message "incorrect" for the sample interpretation invalidates all the results for that sample, which needs to be retested. Samples with a negative interpretation

Table 2
Interpretation of Test Results

Test value	Interpretation
<0.25 (for antigen and antibody detection)	Negative
≥0.25 (for antigen or antibody detection)	Positive

are considered to be negative, within the performance limitations of the reagent. In cases of suspected primary infection, values ≅ 0.25 must be interpreted with caution.

Samples with a positive interpretation must be retested in duplicate.

- Nonrepeatable initially positive samples (two negative reactions for three tests) are considered to be negative, within the performance limitations of the reagent.
- Repeatedly positive samples (at least two of three positive tests) must be confirmed with complementary tests, in accordance with the algorithm shown in **Fig. 3**.

A positive result for one of the two tests (antibody or antigen) may help to orientate the choice of the complementary assays:

- If the value for the antibody detection is ≥0.25, the complementary (and confirmatory) assay can be the Western blot and/or another antibody screening assay.
- If the value for the antigen detection is ≤0.25, the complementary assay can be the HIV p24 stand-alone test or HIV-1 RNA NAT (*see* **Notes 12** and **13**).

The interpretation of test results should consider the clinical background and eventually the results of other tests. **Figure 3** shows the algorithm for the interpretation of fourth-generation test results.

3.3.4. Quality Control

One antibody-positive control (C1), one antigen-positive control (C3), and one negative control (C2) are included in each VIDAS HIV DUO Ultra kit. These controls must be performed immediately after opening a new kit to ensure that reagent performance has not been altered. Each recalibration must also be checked using these controls. Controls must be identified with C1, C2, and C3 in order to permit a verification of their values by the instrument. Results cannot be validated if the control values deviate from the expected values (*see* **Note 14**).

3.3.5. Limitations of Procedure

VIDAS HIV DUO Ultra is a screening assay for HIV infection and should not be used as a specific HIV-1 p24 antigen test.

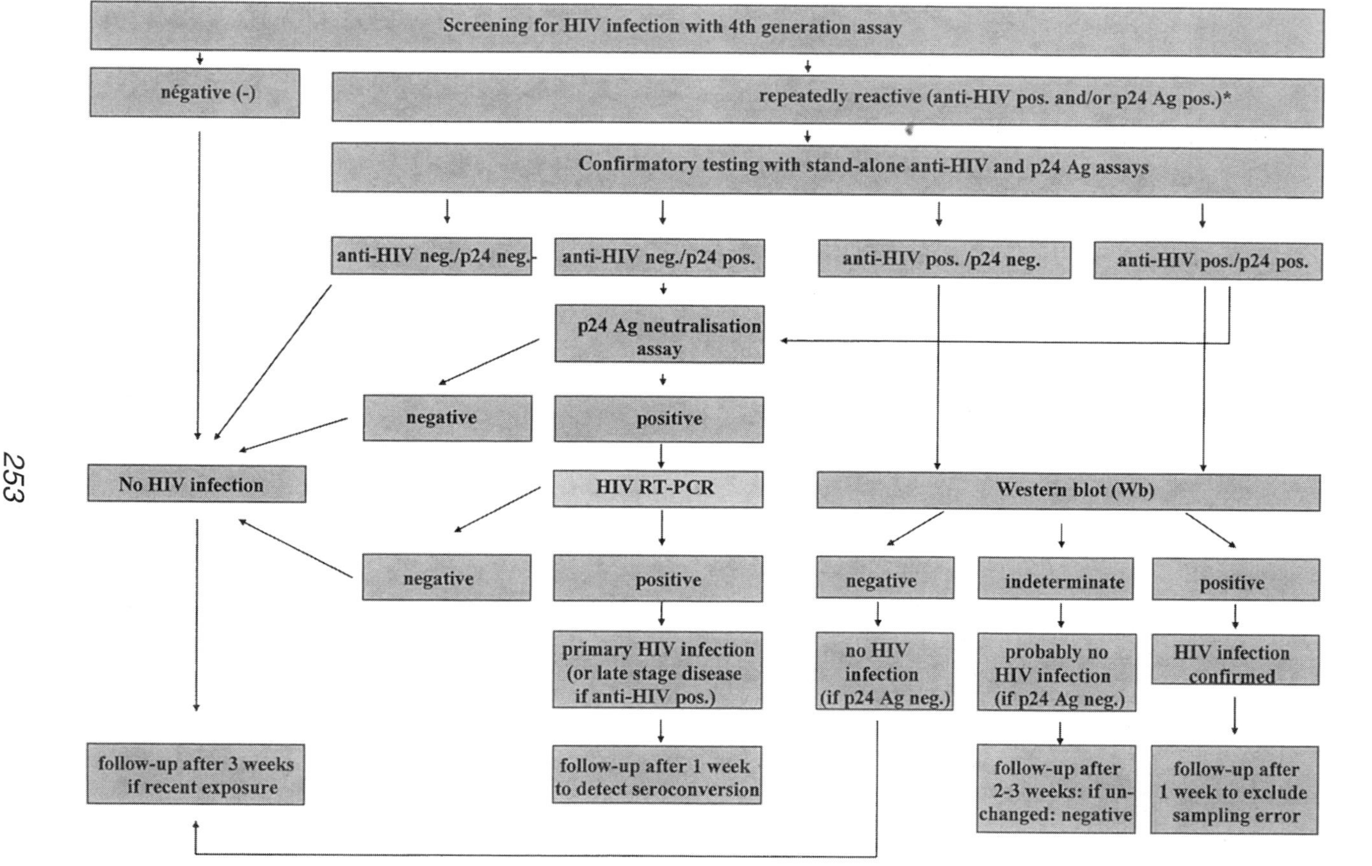

Fig. 3. Algorithm for the interpretation of test results. *With VIDAS HIV DUO Ultra, mostly only one of the parts (HIV Ag or HIV antibody) needs to be verified depending on which of the two is repeatedly reactive.

The assay has been validated on serum and plasma samples. The test has not been validated for other biological fluids including urine and cerebrospinal fluid. Pooling of samples is proscribed.

Interferences may be observed with certain serum samples, which contain antibody against kit components (*see* **Note 15**). Results should therefore be interpreted in consideration of the clinical context and eventually the results of other assays.

4. Notes

1. Universal precautions should be used when collecting and manipulating any blood specimen, and all blood products should be considered potentially infectious in this testing procedure.
2. Do not freeze reagents. Do not use reagents after their expiration date. Do not mix reagents and disposables from different kit lots. The kit contains human blood products and products of animal origin. No known analysis method can guarantee the absence of transmissible pathogenic agents. It is therefore recommended that these products be treated as potentially infectious and handled observing the usual safety precautions *(8)*.
3. Kit reagents contain sodium azide, which can react with lead or copper sink plumbing to form explosive metal azides. If any liquid containing sodium azide is disposed of in the plumbing system, drains should be flushed with water to avoid buildup.
4. Room temperature needs to be under 25°C in order to avoid instrument failure. It is recommended to place the instrument on a 10-cm-high support, which permits appropriate airflow since the cooling fans are on the bottom of the analyzer.
5. After opening the kit, check that the SPR pouch is correctly sealed and undamaged. If not, do not use the SPRs.
6. Carefully reseal the pouch with the desiccant inside after use to maintain stability of the SPRs and return the complete kit to 2–8°C.
7. If stored according to the recommended conditions, all components are stable until the expiration date indicated on the label.
8. Use powderless gloves, as powder has been reported to cause false results for certain enzyme immunoassay tests.
9. Spills should be wiped thoroughly after treatment with liquid detergent and a solution of household bleach containing at least 0.5% sodium hypochlorite. See the VIDAS user manual for cleaning spills on or in the VIDAS instrument. Do not autoclave solutions containing bleach.
10. When the analysis is completed, remove the used SPRs and strips, and dispose of them appropriately (i.e., autoclaving). All other contaminated material, such as disposable gloves and pipet tips, should be disposed of in a similar manner.
11. The VIDAS instrument should be regularly cleaned and decontaminated (see VIDAS user manual).

12. NAT do not amplify HIV-2 RNA; the VIDAS HIV DUO Ultra, however, can detect HIV-2 antigen *(3,5)*. In case of a confirmed HIV Ag positive result but negative NAT, a follow-up sample after 1 wk needs to be tested.
13. National mandatory regulations should be considered, if available.
14. The user is responsible for the implementation of quality control in accordance with the local legislation.
15. The specificity of VIDAS DUO Ultra is 99.5 in 2660 low- and high-risk individuals when compared to Food and Drug Administration licensed tests *(9)*.

References

1. Weber, B., Fall, E. M. B., Berger, A., and Doerr, H. W. (1998) Reduction of diagnostic window by new fourth-generation human immunodeficiency virus screening assays. *J. Clin. Microbiol.* **36,** 2235–2239.
2. Gürtler, L., Mühlbacher, A., Michl, U., Hofmann, H., Paggi, G., Bossi, V., et al. (1998) Reduction of the diagnostic window with a new combined p24 antigen and human immunodeficiency virus antibody screening assay. *J. Virol. Meth.* **75,** 27–38.
3. Weber, B., Berger, A., Rabenau, H., and Doerr, H. W. (2002) Evaluation of a new combined antigen and antibody human immunodeficiency virus screening assay, VIDAS HIV DUO Ultra. *J. Clin. Microbiol.* **40,** 1420–1426.
4. Weber, B., Gurtler, L., Thorstensson, R., Michl, U., Muhlbacher, A., Burgisser, P., et al. (2002) Multicenter evaluation of a new automated fourth-generation human immunodeficiency virus screening assay with a sensitive antigen detection module and high specificity. *J. Clin. Microbiol.* **40,** 1938–1946.
5. Weber, B., Thorstensson, R., Tanprasert, S., Schmitt, U., and Melchior, W. (2003) Reduction of the diagnostic window in three cases of human immunodeficiency – 1 subtype E primary infection with fourth-generation HIV screening assays. *Vox Sanguinis* **85,** 73–79.
6. Weber, B. (2003) HIV seroconversion: performance of combined antigen/antibody assays. *AIDS* **17,** 931–945.
7. Ly, T. D., Edlinger, C., and Vabret, A. (2000) Contribution of combined detection assays of p24 antigen and anti-human immunodeficiency virus (HIV) antibodies in diagnosis of primary HIV infection by routine testing. *J. Clin. Microbiol.* **38,** 2459–2461.
8. Laboratory Bio Safety Manual, 2nd ed., 1993, WHO, Geneva, Switzerland.
9. Saville, R. D., Constantine, N. T., Cleghorn, F. R., Jack, N., Bartholomew, C., Edwards, J., et al. (2001) Fourth-generation enzyme linked immunosorbent assay for the simultaneous detection of human immunodeficiency virus antigen and antibody. *J. Clin. Microbiol.* **39,** 2518–2524.

21

Detection of Antiviral Antibodies Using Enzyme-Linked Immunosorbent Assay

Ingrid G. Winkler

Summary

Following retroviral infection, specific antibodies against viral proteins may be detected in the blood, urine, milk, or saliva of an infected animal. Enzyme-linked immunosorbent assay (ELISA) is the most commonly used method to screen for antiviral antibodies, as the technique is sensitive, reproducible, relatively inexpensive, and easily adapted to large-scale screening. This chapter describes two independent ELISA techniques. The first uses a recombinant biotinylated viral protein antigen expressed in *E. coli* that is purified directly on streptavidin-coated plastic wells. The second uses fixed "virus-infected" cultured cells and can be used either as a second confirmatory assay or initially in a small-scale screening to identify potentially infected individuals for further study.

Key Words: ELISA; anti-viral antibodies; foamy virus; immunodeficiency virus; diagnostic; screening assay.

1. Introduction

Following retroviral infection, specific antibodies against viral proteins may be detected in an infected animal. Enzyme-linked immunosorbent assay (ELISA) is the most commonly used method to screen for antiviral antibodies, as the technique is sensitive, reproducible, relatively inexpensive, and easily adapted to large-scale screening. A positive ELISA result (if confirmed using a second independent test method) can be used as a presumptive diagnosis of viral infection.

ELISA involves plastic wells coated with either "fixed whole" virus or purified viral proteins. Samples (such as serum, saliva, milk, or urine) from infected animals are added and any antibodies in these samples will bind to viral

From: *Methods in Molecular Biology, Vol. 304: Human Retrovirus Protocols: Virology and Molecular Biology*
Edited by: T. Zhu © Humana Press Inc., Totowa, NJ

antigens attached to the plastic. Following washing, another reagent labeled with enzyme is added. This second reagent is specific to immunoglobulin (Ig)G or IgM and binds to the first antibody already bound to the viral peptide. After a wash step, the presence of enzyme is detected by addition of a substrate that changes color.

Critical to the development of any ELISA is the selection of a "good" purified viral antigen to use as target. Ideally, viral proteins that elicit strong and specific antibody responses are used. In practice, most ELISA methods rely on either preparations from whole virus or (especially if the antibody response has been well characterized) selected viral antigens produced as recombinant proteins or synthetic peptides. Considerations involved in the choice of target viral antigens in ELISA are sensitivity (whether enough viral epitopes are present for the detection of immune antibodies in the majority of infected animals) and specificity (are these viral epitopes also detected by antibodies that occur in noninfected animals).

Low specificity may occur if animals have previously been infected with related viruses and have antibodies that cross-react. For this reason viral proteins that are highly homologous between related viruses (such as most of the retroviral Pol protein) are best avoided. For diagnosis of foamy virus (spumaretrovirus) infection, recombinant viral peptides based on regions within the viral Gag (particularly the C-terminal domain) generally provide the best combination of sensitivity and specificity.

Below are described two independent ELISA techniques. The first uses a recombinant biotinylated viral protein antigen expressed in *E. coli* and purified directly on streptavidin-coated plastic wells. The second uses fixed "virus-infected" cultured cells and can be used either as a second confirmatory assay or initially in a small-scale screening to identify potentially infected individuals for further study. To set up and optimize these ELISAs one will require (1) a known positive serum sample and (2) viral DNA (either in infected cells or cloned into a plasmid).

2. Materials

2.1. Culture of Virus-Infected Cells

1. Cell culture media. Cells are cultured in Dulbecco's modified Eagle's medium (DMEM) supplemented with 5% fetal bovine serum and antibiotics (60 μg/mL penicillin, 40 μg/mL gentamicin). May also include 10 μg/mL fungazone.
2. 0.25% trypsin and 1 m*M* ethylenediamine tetraacetic acid (EDTA) in phosphate-buffered saline (PBS) is used to detach cells.
3. PBS: For 10X PBS stock solution, mix 80 g NaCl, 2 g KCl, 11.5 g $Na_2HPO_4 \cdot 7H_2O$, 2 g KH_2PO_4 in 1 L water. Autoclave and store at room temperature, dilute to 1X for use.
4. Virus: "stock virus" stored at –70°C (*see* **Note 1**) or an "infectious" (complete) retroviral genome cloned into a plasmid (*see* **Note 2**).

2.2. Viral DNA Extraction (Salting-Out Method)

1. DNA lysis buffer: 10 m*M* Tris-HCl, pH 8.0; 2 m*M* EDTA; 400 m*M* NaCl. Store at room temperature.
2. Additional reagents required include: 10% (w/v) sodium dodecyl sulfate (SDS); 6 *M* (saturated) NaCl; 100% ethanol; 70% ethanol; and 20 mg/mL proteinase K. All reagents are stored at room temperature with the exception of the proteinase K, which is stored at –20°C.

2.3. Polymerase Chain Reaction Amplification of Viral Gene

1. Polymerase chain reaction (PCR) primers: For the detection of FeFV infection, use the sense primer (FUV2662s) 5'-ACC TCC TCG TGG AAG TGG 3' and the antisense primer (FUV3065a) 5'-TTG CTG CCT AAC AGG TTC TTC TCC 3', which amplify the nucleocapsid domain of the *gag* gene (*see* **Note 3**).
2. Reagents for PCR reaction: pooled dNTPs at 25 m*M* each; *Taq* DNA polymerase and reaction buffer provided by the manufacturer; 25 m*M* $MgCl_2$. Store frozen.
3. Agarose suitable for DNA gels and 5 mg/mL ethidium bromide stock. Store at room temperature protected from light.
4. 5X TAE buffer: 200 m*M* Tris-acetate (24.2 g Tris base, 5.7 mL glacial acetic acid in 1 L water) with 10 m*M* Na_2EDTA (3.7 g/L). Store at room temperature; dilute to 1X concentration for use.
5. 10X DNA loading solution: 60% glycerol, 0.2% SDS in 1X TAE. Add a small crystal of bromophenol blue to give a dark blue color.
6. DNA markers (such as "Kb ladder" 250–12 kB, Stratagene cat. no. 201115, La Jolla, CA), which can be used both to determine the size of PCR product to estimate DNA concentration.

2.4. T-Vector Cloning of PCR Product

1. PinPoint Xa-1 T-Vector System plus competent cells (Promega cat. no. V2850, Madison, WI). Contains vector, T4 DNA ligase buffer, and host strain *E. coli* (JM109) with a competency of $>10^8$ cfu/μg DNA.
2. SOC media: 2% tryptone, 0.5% yeast extract, 10 m*M* NaCl, 2.5 m*M* KCl, 10 m*M* $MgCl_2$, 10 m*M* $MgSO_4$, 20 m*M* glucose. Autoclave, then add the (filter-sterilized) glucose later.
3. Luria broth (LB amp) agar plates with ampicillin: 2% agar, 1% tryptone, 0.5% yeast extract, 20 m*M* NaCl. Autoclave. While still liquid at 50°C, add 100 μg/mL ampicillin mix and pour 10 mL per Petri dish. Allow to solidify.

2.5. Identification of Recombinant Clones

1. Luria broth (LB amp): 1% tryptone, 0.5% yeast extract, 20 m*M* NaCl. Autoclave. Before use add 100 μg/mL ampicillin.
2. Plasmid preparation buffer: 0.1 *M* NaOH; 0.5% SDS; 10 m*M* Tris-HCl, pH 8.0; 1 m*M* EDTA. Make fresh from stock solutions on day of use. Also require stocks of 3 *M* NaAc, pH 5.2; precooled 100% ethanol; and room-temperature 70% ethanol.
3. Restriction enzymes to check plasmid identity. EcoRI and BglII with reaction buffer. Also 1 mg/mL RNAse stock. Store all reagents at –20°C.

4. PinPoint sequencing primer: 5'-CGT GAC GCG GTG CAG GGC G (*see* **Note 4**).

2.6. Production of Recombinant Biotinylated Fusion Protein

1. 4 m*M* D-Biotin stock in PBS and 100 m*M* isopropylthio-β-D-galactoside stock (1.2 g IPTG in 50 mL water). Sterilize by filtration; store at 4°C.
2. Protein buffer: 50 m*M* Tris-HCl, pH 7.5; 200 m*M* NaCl; 1 m*M* EDTA; 5% glycerol. Optional is the addition of protease inhibitors such as phenylmethylsulfonylfluoride (PMSF); use at 1:1000 dilution of a 200 μ*M* stock (stored at –20°C).
3. Tip sonicator (such as 13-mm tip on a Vibra Cell 600W, Sonics and Materials Inc., Danbury, CT).

2.7. Immunoblotting

1. Necessary only if confirmation of recombinant protein expression is required.
2. Two 12% SDS-PAGE mini-gels.
3. Minigel apparatus with transfer system (such as Hoefer miniVE with blotter; Amersham, Little Chalfont, UK).
4. 2X protein loading buffer: 0.1 *M* Tris-HCl, pH 6.8; 20% glycerol; 4% SDS; add small crystal of bromophenol blue to give dark blue color. Store at –20°C.
5. Running buffer: 20 m*M* Tris-HCl; 150 m*M* glycine; 0.1% (w/v) SDS.
6. Transfer buffer: 20 m*M* Tris-HCl; 150 m*M* glycine; 0.1% (w/v) SDS; 20% (v/v) methanol.
7. Membrane to bind proteins, such as Hybond-C super membrane (Amersham).
8. PBST: to 1X PBS (described in **Subheading 2.1.**) add 0.05% Tween-20 detergent.
9. Blocking buffer: PBST with 5% dried skim milk.
10. Horseradish peroxidase (HRP)-conjugated Protein A/G (Pierce, Rockford, IL). Dissolve powder at 0.5 mg/mL in PBS with 50% glycerol. Store at –20°C (*see* **Note 5**).
11. *Optional*: HRP-conjugated streptavidin (such as from Vector, Burlingame, CA). Dissolve powder at 0.5 mg/mL in PBS with 50% glycerol. Store at –20°C.
12. Westen blotting detection system (such as ECL, Amersham).
13. X-ray medical film (such as SuperRX, Fujifilm, Dusseldorf, Germany), X-ray film cassette, and film processor.

2.8. Streptavidin Capture ELISA

1. 96-well flat-bottom plates for ELISA. Nunc polysorb cat. no. 4442404, Roskilde, Denmark (made from polystyrene plastic) (*see* **Note 6**).
2. Streptavidin (purified, such as cat. no. S4762, Sigma Chemicals, St. Louis, MO). Resuspend powder at 2 mg/mL in water and store frozen in 100-μL aliquots.
3. PBST: to 1X PBS (described in **Subheading 2.1.**) add 0.05% Tween-20 detergent.
4. Dried skim milk for blocking (or other blocking agent if preferred; *see* **Note 7**).
5. Horseradish peroxidase (HRP)-conjugated protein A/G (described in **Subheading 2.7.** and *see* **Note 5**).

6. K Blue chromogenic substrate (a stable mixture of 3,3',5,5' tetramethylbenzidine [TMB] and H_2O_2, made by ELISA Technologies, Neogen, Lexington, KY). Protect from air and light. Store at 4°C.
7. 1 *M* H_2SO_4 stock (approx 1:4 dilution of concentrated H_2SO_4).
8. ELISA 96-well plate reader able to read at 450 nm (yellow) wavelength and optionally at 650 nm (blue) wavelength (such as Titretek, Multiskan, Eflab Oy, Helsinki, Finland) with correct filters.

2.9. ELISA Based on Fixed Virus-Infected Cells for Detection of Antibodies to Feline Immunodeficiency Virus and Feline Foamy Virus

1. Tissue culture flat-bottom 96-well plates.
2. Precooled ethanol:acetone (95:5) for fixation of cells.
3. *Optional*: solution containing 0.4% sulfurhodamine B dye solubilized in 1% acetic acid.

3. Methods

3.1. Culture of Virus-Infected Cells

1. Passage Crandell feline kidney (CRFK) cells when approaching confluence. Wash cells with PBS then add trypsin/EDTA to detach cells (usually requires a twice-weekly 1:4 cell split).
2. CRFK cells can be infected by stock virus from primary cat tissues (*see* **Note 1**) or using infectious retroviral clones (*see* **Note 2**). Cells containing a lytic retroviral infection are recognized by the formation of multinucleated syncytia (*see* **Fig. 1**).

3.2. Viral DNA Extraction (Salting-Out Method)

1. Viral DNA can be extracted from virus-infected cells or directly from an infected animal (from blood leukocytes or from the cell pellet of an oropharyngeal swab). To approx 5×10^6 cells pelleted in a microfuge tube, add 360 μL DNA lysis buffer, 17 μL SDS, and 7 μL protease K. Mix and incubate at 55°C from 2 to 16 h for cell lysis.
2. Add 125 μL saturated NaCl solution, mix, and centrifuge 15 min at 10,000*g* to pellet proteins.
3. Transfer supernatant by tipping quickly into a new microfuge tube (do not transfer any white protein precipitate). To precipitate DNA, add 1000 μL absolute ethanol, invert tube five times, and centrifuge 30 s at 10,000*g*.
4. Wash the DNA pellet once with 1 mL room-temperature 70% ethanol to remove excess salt.
5. Dry DNA pellet for 5 min at 55°C, then resuspend in 25 μL water. Use 1–5 μL per PCR reaction.

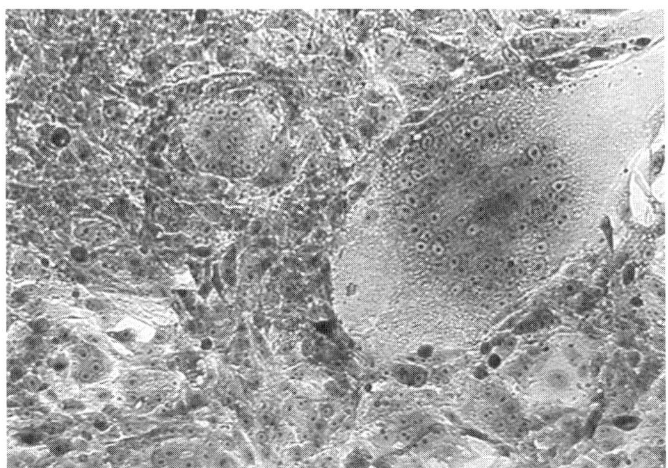

Fig. 1. CRFK cells were incubated with the supernatant derived from an oropharyngeal swab taken from a cat (as described in **Note 1**). The formation of multinucleated syncytia (indicative of feline foamy virus infection) were visible 5 d postinfection. Syncytia form as a result of the retroviral envelope protein "fusing" adjacent cells. Syncytia are typically defined as a single enlarged cell containing three or more nuclei.

3.3. PCR Amplification of Viral Gene

1. Use 1–5 µL of DNA (approx 0.1–1 µg) in a standard 50-µL PCR reaction mix containing 250 µ*M* dNTPs, 1.5 m*M* $MgCl_2$, 140 p*M* of sense and antisense primers (*see* **Note 3**), and 1 unit of DNA polymerase with reaction buffer provided by the manufacturer. Cycle 35 times at 94°C for 30 s, 52°C for 30 s, and 72°C for 60 s with a final extension of 72°C for 5 min.
2. Microwave 1.5% agarose gel in 1X TAE buffer until it boils (to dissolve the agarose). Allow to cool to 50°C, then add ethidium bromide at 1:5000 dilution. Mix well, pour into an agarose gel holder, add combs, and allow to cool.
3. Mix 10 µL of neat PCR product with 1 µL of DNA loading solution and pipet into the agarose wells. Repeat with a 1:10 and 1:100 dilution of the PCR product and load beside 0.5 µg of the DNA marker (this is to estimate DNA concentration). Fill the tank surrounding the agarose with 1X TAE and apply 100 V (should run at <200 mA). The PCR product will move toward the cathode. Stop the voltage when the blue line has moved halfway along the agarose.
4. Visualize bands using ultraviolet light. Wear goggles to protect eyes.
5. The PCR product should be a single band at 404 bp in size when compared to the DNA marker. Estimate concentration of the PCR product by comparison with the 500-bp band of the Kb ladder DNA marker (14 ng of DNA).

3.4. T-Vector Cloning of PCR Product

1. In a microfuge tube aliquot 1 µL (50 ng) of the PinPoint Xa1 vector, 1 µL ligase, 1 µL ligation buffer and 150–200 ng of PCR product (do not use more than 5µL).

Add water to make a final volume of 10 μL. Also prepare a second ligation mix where no PCR product has been added (vector control). Mix and incubate the tubes overnight at 14°C for ligation to occur.

2. For tranformation of *E.coli*, thaw a vial of competent JM109 cells on ice. Transfer 50 μL to sterile microfuge tubes and add 2 μL of the ligation mix. Mix gently and leave on ice for 30 min.
3. Heat-shock cells by immersing tubes in a prewarmed water bath at exactly 42°C for 45 s, then return cells to ice for 5 min.
4. Add 950 μL SOC medium, and shake the tubes for 1 h at 37°C.
5. Spread 100 μL of transfected cells evenly over LB amp agar plates and incubate plates overnight at 37°C. As a backup, store the remaining transformation mix in SOC in the refrigerator overnight.
6. Next morning expect about 10–100 colonies per LB amp plate and 10-fold less in the vector control. Select a dozen of these recombinant clones for further characterization.

3.5. Identification of Recombinant Clones by Plasmid Purification and PCR

1. To determine both the identity and orientation of the insert (the PCR product in T-vector clones can be in both sense and antisense orientation), transfer a small amount of selected colonies into 2 mL of LB containing 100 μg/mL ampicillin and incubate at 37°C overnight in 10-mL tubes with shaking (for aeration).
2. Transfer 1 mL to a microfuge tube and pellet bacteria with 15 s centrifugation at 10,000*g*.
3. Remove and discard supernatant by inverting tubes and resuspend bacterial pellet completely by vortexing (ensure that no clumps remain).
4. Add 300 μL plasmid preparation buffer (to lyse bacteria); vortex for 1 s.
5. Add 150 μL NaAc (to precipitate bacterial genomic DNA); vortex for 1 s only.
6. Centrifuge at 10,000*g* for 5 min. Quickly tip supernatant into a new microfuge tube (do not transfer any white precipitate).
7. Add 900 μL of precooled ethanol, vortex 2 s, and centrifuge again at 10,000*g* for 2 min.
8. Remove all supernatant, then wash the plasmid pellet with 1 mL room-temperature 70% ethanol.
9. Centrifuge again for 2 min, remove supernatant, and dry the pellet at 55°C (about 5 min).
10. Resuspend the plasmid in 50 μL water.
11. Set up a restriction digestion using 10 μL of plasmid preparation, 1 μL EcoRI, 1 μL Bgl II, 2 μL EcoRI 10X buffer, 0.1 μL RNAse. Make up to 20 μL with water. Incubate at 37ºC for 2 h, then run on an agarose gel (as described in **Subheading 3.3.**). Clones with the correct insert will contain an 880-bp band (404 bp of PCR product and 476 bp of vector sequence). Select only clones with an insert of the correct size for further confirmation by PCR.
12. In a PCR reaction (similar to that described in **Subheading 3.3.**) use 0.1 μL of plasmid and a combination of the original antisense primer (FUV3065a) with the

PinPoint sequencing primer (located within the vector 5' of the cloned insert). This will amplify only recombinant inserts that are in the sense orientation.

13. Also perform the converse PCR, that is using the original sense primer (FUV2662s) together with the PinPoint sequencing primer. This will amplify inserts that are in the antisense orientation.
14. Select several bacterial clones that contain the insert in the sense orientation, as well as one that contains the insert in the antisense orientation to be used as a negative control. To store selected clones *see* **Note 8**.

3.6. Production of Recombinant Biotinylated Fusion Protein

1. Culture sense and antisense clones at 37°C overnight in 1 mL LB amp with shaking.
2. Dilute the overnight colonies 1:100 in a large conical flask containing 50 mL fresh LB amp and 50 μL D-biotin.
3. Shake for 3 h at room temperature (20–25°C), then induce the expression of the recombinant protein by adding 50 μL IPTG and reincubate at room temperature for a further 17 h with shaking.
4. Pellet bacteria by centrifuging at 8000*g* for 10 min at 4°C in preweighed tubes and remove all supernatant.
5. Resuspend the bacterial pellet at 10% (w/v) in protein buffer (expect approx 0.5 g bacterial pellet resuspended in 5 mL protein buffer). Mix well until no clumps remain.
6. To lyse bacteria, place the tube in a beaker of ice and immerse the step tip in the protein buffer. Sonicate at high intensity four times, 15 s each, with pauses to allow samples to cool. Try not to "foam" samples (this will denature the proteins). Wear ear protection.
7. Bacterial debris is pelleted by centrifugation at 10,000*g* for 20 min at 4°C. Apply the supernatant fraction either directly to streptavidin-coated ELISA plates (10 μL each well), in wells of SDS-PAGE for immunoblotting, or store frozen, after mixing with an equal volume of glycerol (bacterial pellet may also be used; *see* **Note 9**).

3.7. Immunoblotting

Immunoblotting can be used to check that the recombinant protein is correctly expressed and is recognized by known positive antisera. Immunoblotting may also be used subsequently to confirm the "true" antibody status of sera that react nonspecifically in ELISA.

1. Mix 10 μL of recombinant biotinylated fusion protein with 10 μL of protein-loading buffer, boil tubes for 3 min, then load 10 μL onto two identical 12% SDS-PAGE mini-gels along with molecular weight markers. Assemble mini-gel running apparatus, fill tank with running buffer, and run samples at 60 V until the blue band is at the bottom of the gel.
2. Transfer mini-gels to blotting apparatus, place each mini-gel against (Hybond C) membrane, assemble blotting apparatus, and fill with transfer buffer. Fill the surrounding tank with cold water. Electrotransfer at 28 V for 2 h.

3. Following transfer, block membranes by incubating with blocking buffer at room temperature for 2 h.
4. For the first membrane (to determine if a known positive antiserum will detect the recombinant viral protein): Incubate with blocking buffer containing the antiserum diluted 1:100 for 2 h. Wash the membrane three times with PBST over 10 min. Then add HRP-conjugated Protein A/G at 1:10,000 dilution in blocking buffer for another 1 h incubation at room temperature before a final series of washes in PBST.
5. For the second membrane (to see all biotinylated proteins present in the bacterial lysate): Incubate with blocking buffer containing 1:5000 HRP-conjugated streptavidin (SA) for 1 h, then wash in PBST for 1–3 h.
6. Detect HRP activity on both membranes using ECL (mix 500 μL each of reagent A and B together, and wet membranes with this mixture for 5 min). Place membranes in an X-ray cassette and expose to film for 2–15 min. Develop film.

Expected results: *Second membrane*: The SA-HRP will detect a 22.5-kDa band in all lanes. This is a biotinylated protein that is native to all *E. coli*. Recombinant clones with insert in the sense orientation will have an additional band at 31 kDa, corresponding to the predicted molecular weight of the cloned viral protein and N-terminal vector sequence.

First membrane: The specific antisera should recognize the same recombinant viral protein band at 31 kDa in the bacterial lysates derived from sense clones but not from antisense clones. If no bands are detected *see* **Note 10.**

3.8. Streptavidin Capture ELISA

Each sample to be tested by ELISA will require duplicate "plus" and "minus" antigen wells (for more details on controls *see* **Note 11**).

1. The ELISA plate wells are first coated with streptavidin. For each 96-well plate, dilute 25 μL of streptavidin stock in 5 mL water and add 50 μL per well. Leave the plates in a dry incubator at 37°C overnight to allow complete evaporation. Dried plates are sealed and stored at 4°C for up to 1 mo. Before use in ELISA, wash the streptavidin-coated wells three times for 10 min each with PBST (*see* **Note 12**).
2. On the day of the ELISA, dilute recombinant biotinylated fusion protein (made in **Subheading 3.6.** from either supernatant or the bacterial pellet) 1:20 in PBST with 0.5% dried skim milk and add 200 μL to each well (i.e., 10 μL final of the bacterial extract per well). Incubate for 45 min at room temperature with occasional shaking.
3. Remove unbound proteins by washing the ELISA plate thoroughly four times with PBST.
4. Dilute test sera 1:100 in PBST with 0.5% dried skim milk and add 100 μL to each of the duplicate wells (for optimal dilutions of test samples *see* **Note 13**).
5. Cover ELISA plates (with a lid or with sticky tape) and incubate at room temperature with shaking for 60 min to allow antibody binding before washing four times in PBST.

6. Dilute HRP-conjugated protein A/G 1:50,000 in PBST with 0.5% dried skim milk and add 100 µL to each well (*see* **Note 14**). Cover wells and incubate an additional 45 min at room temperature to allow binding of conjugate.
7. Finally, wash wells six times with PBST.
8. Add 100 µL K Blue chromogenic substrate to wells (note the time). Positive wells will change color from clear to sky blue. Stop the reaction when the negative control wells (minus antigen) remain clear and the positive control wells (plus antigen) have changed color. Optimal development time should be 10 min and largely depends on the dilution of HRP-protein A/G used. All wells will slowly turn blue with longer incubation (*see* **Note 15**).
9. Stop the reaction by addition of 50 µL 1 *M* H_2SO_4 per well (the positive wells will now turn yellow).
10. Record the absorbance of the wells at 450 nm (yellow) in an ELISA plate reader.

Analysis of results:

1. Examine the duplicate wells. If the duplicates are similar, average the absorbance.
2. Subtract the absorbance of the "background" control well from the average absorbance of the test wells.
3. Eliminate nonspecific (false-positive) reactions. These are identified by high absorbance in both the wells containing extract from antisense clones (minus antigen) as well as the wells containing extract from sense clones (plus antigen). The antiviral status of these sera cannot be determined using this method (*see* **Note 16**).
4. For each test serum, divide the average absorbance for the "plus" wells (containing recombinant viral antigen expressed from the sense clone) by the absorbance of the "minus" wells (with no viral antigen).
   a. A **positive** ELISA result is when this value is greater than 2.5.
   b. A **negative** ELISA result is when this value is less than 1.5.
   c. An **intermediate** result (values between 1.5 and 2.5) could be due to either a low titer of antiviral antibodies or nonspecific binding. The true antiviral antibody status of these sera can be resolved by (1) repeating with a different ELISA technique, such as the fixed virus-infected cell method described below, or (2) by immunoblotting.

3.9. ELISA Based on Fixed Virus-Infected Cells for Detection of Antibodies to Feline Immunodeficiency Virus and Feline Foamy Virus

1. For each serum sample to be analyzed, set up duplicate wells of virus-infected cells as well as duplicate "mock-infected" cell wells (minus antigen). Grow virus such as FIV or FeFV in Crandell feline kidney (CRFK) cells (*see* **Notes 1** and **2**). When lytic infection (with multinucleated syncytia) begins (**Fig. 1**), detach cells with trypsin/EDTA, resuspend at 100,000 cells/mL in fresh cell culture medium and transfer 100 µL to the wells of a tissue-culture flat-bottom 96-well plate. Repeat in parallel wells with mock-infected CRFK cells (as negative control).

2. Incubate the 96-well plate at 37°C in 5% CO_2 for 3–5 d, observing daily, until the cell layer is approaching confluency and numerous multinucleated syncytia are observed in the virus-infected wells.
3. Wash wells once with PBS and block with 100 μL PBS containing 0.5% dried skim milk per well for 30 min. Remove the contents of the wells and "fix" the cells with precooled ethanol:acetone (95:5) for 20 min at –20°C. After fixation, remove the ethanol. Plates can be sealed and stored at –70°C for up to 1 mo, or alternatively wash wells three times with PBST and proceed directly to ELISA.
4. Add diluted serum to the wells and perform the ELISA as described above, starting at **Subheading 3.8.4.**, with the following exception: incubation time with the diluted serum is extended to 2–4 h at room temperature (or overnight at 4°C) before washing and proceeding to the HRP-conjugate step.

Analysis of results: Similar to the previous ELISA above, however, sensitivity may be reduced. The absorbance in the "plus" (virus-infected) wells from a positive serum may be only twice that of the "minus" control well.

For this second ELISA method, it is important that each well be initially coated with a similar number of cells (especially difficult if the virus-infected cells grow poorly compared with noninfected cells). If large differences in cell confluency between the "plus" and "minus" wells is observed, the absorbance values taken with this ELISA may need to be normalized against cell number (*see* **Note 17**).

4. Notes

1. More than 50% of mature cats are infected with feline foamy virus (FeFV) and most secrete virus in the saliva *(1)*. Feline immunodeficiency virus (FIV) infection is less common (about 10% of mature male cats are infected; *[2]*) and the virus is shed in the saliva only for a short time after initial infection. Use the following method to isolate and culture virus. Oropharyngeal swabs are collected from cats and vortexed in 2 mL cell cuture media containing 10 μg/mL fungazone to inhibit fungal growth. After centrifugation (8000*g* for 10 min), replace the media on a flask of 50% confluent CRFK cells with the supernatant and incubate 2 h at 37°C in 5% CO_2 for virus absorption. Replace with fresh cell culture media containing 10 μg/mL fungazone and passage cells as required (usually twice weekly). Cells with a lytic viral infection exhibit multinucleated syncytia within 2 wk (**Fig. 1**). To collect virus, freeze-thaw the flask, centrifuge the contents (400*g* for 5 min), and store the supernatant in aliquots at –70°C as stock virus (**caution**: this "stock" virus has not been clonally selected, so may contain other viruses as well).
2. Retroviral genomes can be cloned into bacterial plasmids to create "infectious clones," for example, F31 feline immunodeficiency virus$_{petaluma}$ (AIDs Research Reagent Program, NIH, USA) or pHSRV13 human spumaretrovirus *(3)*. These infectious clones are much easier to transport between laboratories, are well-

characterized, and will not be contaminated with other unwanted viruses.

There are several methods of introducing the purified plasmid of an infectious retroviral clone into cells. The easiest is lipofection techniques, such as using Fugene (Roche, Indianapolis, IN). Mix 3 µL Fugene with 100 µL DMEM (without serum); add 1 µg viral plasmid. Mix gently, incubate 60 min at room temperature, then add dropwise to 50% confluent CRFK cells in a 35-mm dish. Multinucleated syncytia (indicative of lytic viral infection; *see* **Fig. 1**) should be evident within 2 wk.

3. Proteins expressed from any region of the viral gene can be cloned and used as the basis for this ELISA. Design PCR primers to encompass the portion of the gene required. As the expression vector adds an N-terminal consensus biotinylation site, make sure that the 5' nucleotide of the sense primer encodes a protein that is in the same reading frame as the PinPoint Xa-1 vector sequence.
4. Identification of cloned inserts should be confirmed by DNA sequencing using the PinPoint sequencing primer that is located in the vector 5' of the insert. This is to ensure that the cloned insert is in the correct reading frame to produce a recombinant protein, that it is in the sense orientation, and to identify any errors that may have been incorporated in the sequence due to the PCR amplification.
5. A recombinant protein containing the Fc-binding domains of both protein A and protein G, conjugated to HRP, is used in the two ELISA methods and immunoblot. This reagent works well in ELISA and detects most IgG isotypes from most animal species. If desired, HRP-conjugated antifeline antibodies may be substituted for the proteinA/G-HRP.
6. The amount of protein that will absorb to plastic wells depends on the type of plastic used and the surface charge. Generally polystyrene plates absorb proteins well. We have also found that polyvinyl plates (Flat-bottom cat. no. 2595; Costar, Cambridge, MA) give good results. Sensitivity will be decreased if proteins do not absorb well to the plastic; however, too much absorption results in ELISA reactions with high background.
7. The purpose of a blocking agent is to prevent antibodies and other reagents from sticking nonspecifically to the plastic of the test wells. Blocking agents are added to the PBST and used in all incubation steps of the ELISA. Generally 0.5% dried skim milk is most effective (several brands of dried skim milk may need to be tested; avoid baby formula milk). Alternative blocking agents that can be used include 0.5% casein (a purified milk protein), 0.2% gelatin (first dissolve by boiling), and 1.0% bovine serum albumin (BSA). Note that the detergent, Tween-20, which is in the PBST, will also limit proteins from sticking nonspecifically to plastic wells.
8. Bacteria containing the selected clones can be stored for many years at –70°C in 50–80% glycerol. Incubate the *E. coli* containing the clone in LB with ampicillin at 37°C overnight with shaking until turbid. Mix with 1.5 vol of (sterile) glycerol (make sure the mixture is homogenous) and store at –70°C. The plasmid preparations can also be stored frozen.

9. Some recombinant proteins are predominantely expressed in an insoluble form and are found in the bacterial pellet fraction (this can be minimized by using low incubation temperatures, such as 25°C, during expression). Attempt to resolubilize these proteins by resuspending the bacterial pellet at 10% (w/v) in 50 m*M* Tris-HCl, pH 7.5, 6 *M* urea, 1 m*M* EDTA, and gently agitate for 2 h at 4°C. They can now be applied directly to streptavidin-coated ELISA plates (use 10 μL per well), to SDS-PAGE wells for immunoblotting, or alternatively stored frozen.
10. If the 31-kDa band is not detected, do not proceed with ELISA. Either:
    a. Incorrect sequence was amplified (incorrect primer sequences or PCR amplification errors or the expressed protein is not in correct reading frame with vector; check by DNA sequencing).
    b. Insert is correct but the recombinant protein is not being expressed (either no induction of the *tac* promoter with IPTG [try more IPTG or change host strain of *E. coli*], recombinant protein is expressed but is being degraded [sonicate more gently, add more protease inhibitors to protein buffer, try protease-deficient *E. coli* host strain such as BL21], or protein is being expressed in an insoluble form [check bacterial pellet]).
    c. The antiserum does not react with the cloned portion of the virus. If the sequencing is correct and a recombinant biotinylated protein of the correct molecular weight can be seen in the second blot with the streptavidin-HRP, then perhaps the known positive antiserum does not work (from improper storage, frequent freeze-thawing, or bacterial contamination). Alternatively an immune animal simply does not make antibodies to this portion of the virus. Before setting up the ELISA, it is essential to have a biotinylated recombinant protein that can be detected by the sera of an infected animal.
11. "Plus" antigen wells contain recombinant proteins from sense clones, while "minus" antigen wells contain recombinant protein from antisense clones (and should be negative after ELISA reactions). An additional control required for each ELISA plate is duplicate wells that receive no test serum. These will be used to measure "background absorbance."
12. Two different manual methods are commonly used to wash 96-well ELISA plates. Either immerse the upright plate completely in a lunchbox-type container full of PBST (which can be reused for several washes). Once the wells are full, hold the plate over a sink and flick the PBST out of the wells. Gently tap the upside-down plate against clean paper toweling. Work fast enough to ensure that the wells do not have time to completely dry out (drying will denature the antigen and may lead to nonspecific reactions). An alternative washing method utilizes a squeeze bottle of PBST. Simply squirt the PBST along the rows of wells, ensuring that each well is full. Then flick the PBST from the wells as described above.
13. To detect antibodies by ELISA, sera or plasma must be diluted >1:25 in PBST with blocking agent before applying to wells; a 1:100 dilution generally works well. With less dilution, nonspecific binding to the wells may occur, leading to false-positive ELISA results. Urine and milk samples give less nonspecific binding and can be used at a 1:2 dilution or less.

The antiviral titer of a serum is the highest dilution still able to give a positive result in ELISA. The titer of a serum can be measured using serial twofold dilutions until the ELISA reaction becomes negative.

14. HRP-conjugate will steadily deteriorate with age even if stored at –20°C in 50% glycerol. Typically, a fresh HRP-conjugate may need to be diluted 1:100,000 for ELISA, whereas older batches may require 1:10,000 dilution. It is simplest to start with a 1:50,000 dilution and vary this depending on the time it takes for positive control samples to change color (if reactions are developing too fast, a greater dilution of HRP-protein A/G should be used in the future).
15. To rescue a reaction that has developed too rapidly (all wells turn blue within seconds), wash wells again in PBST and repeat with K blue reagent diluted 1:4 in water. Alternatively, if no color develops after 10 min, reincubate the plate in the dark and let the reaction proceed for up to 1 h. Absorbance of wells can be taken at 650 nm (blue) while the color is still developing; however, the difference in absorbance between wells will not be as large as that taken using the yellow (450 nm) filter after the reaction has been stopped with acid.
16. "Minus" antigen wells will detect nonspecific binding of test sample to the plastic plate (known as false-positive reactions). Up to 5% of samples may give false-positive reactions in ELISA. Repeated freeze-thawing of serum/plasma, hemolysis, bacterial contamination, old or otherwise poorly stored samples are more likely to give nonspecific reactions. If a result for these samples is absolutely required, try using either a different ELISA (such as the virus-infected cell ELISA, *see* **Subheading 3.9.**) or immunoblotting (*see* **Subheading 3.7.**).
17. To normalize ELISA absorbance values against number of cells in each well, after the final absorbance readings of the ELISA:
    a. Wash wells again with PBS.
    b. Add 100 µL diluted sulfurhodamine B dye to each well (will stain cells pink). Shake for 30 min at room temperature.
    c. Wash wells twice with water (to remove unbound dye).
    d. Add 100 µL 1% acetic acid per well. Shake for 30 min at room temperature to resolubilize dye from cells.
    e. Read absorbance at 550 nm on an ELISA plate reader. Higher absorbances indicate higher cell densities in that well.

    Interpretation of results:

    a. Average absorbances (at 550 nm) for the duplicate wells.
    b. Choose the highest absorbance for each plate and divide this number by the average absorbances of the other wells.
    c. Multiply these values with the average absorbances obtained during the ELISA (at 450 nm) for the identical wells. This will give a value that has been normalized to cell number.
    d. Now divide the normalized absorbance values of the "plus" wells by the "minus" wells as described previously. A value greater than 2 is considered positive.

References

1. Winkler, I. G., Löchelt, M., Levesque, J-P., Bodem, J., Flügel, R. M., and Flower, R. L. P. (1997) A rapid streptavidin-capture ELISA specific for the detection of antibodies to feline foamy virus. *J. Immunol. Meth.* **207,** 69–77.
2. Winkler, I. G., Löchelt, M., Flower, and R. L. P. (1999) Epidemiology of feline foamy virus and feline immunodeficiency virus infections in domestic and feral cats: a seroepidemiological study. *J. Clin. Microbiol.* **37,** 2848–2851.
3. Löchelt, M., Zentgraf, H., and Flügel, R. M. (1991) Construction of an infectious DNA clone of the full-length human spuma retrovirus genome and mutagenesis of the bel 1 gene. *Virology* **184,** 43–54.

22

Plaque-Reduction Assays for Human and Simian Immunodeficiency Virus Neutralization

Anna Nordqvist and Eva Maria Fenyö

Summary

Research on HIV vaccines, as well as studies on HIV pathogenesis in human and SIV in the macaque model, require the availability of simple and standardized assays for quantification of neutralizing antibodies to primary virus isolates. We have recently developed and standardized assays using human cell lines engineered to express CD4 and co-receptors for HIV and SIV entry. One cell line originated from a glioma (U87) and the other from an osteosarcoma (HOS). Both cell lines and their derivatives form monolayer cultures, a prerequisite for counting plaques. HIV-infected U87.CD4-CCR5 or -CXCR4 cells form syncytia, that is, plaques that can be stained with hematoxylin and enumerated by light microscopy. In addition to CD4 and co-receptors (most often used CCR5 and CXCR6 by SIV), GHOST(3) cells have been engineered to express the green fluorescent protein following virus infection. Infected cells show green fluorescence and can be enumerated by fluorescence microscopy. Neutralization is determined by the ability of a serum to reduce the number of plaque-forming units (PFU) relative to controls exposed to medium or negative serum. Both assays are run in microtiter format and neutralization is evaluated after 3 d. Intra-assay variation has been used for estimation of the cutoff for neutralization. Testing 15 serum-virus combinations in the U87.CD4 assay and four serum-virus combinations in the GHOST(3) assay revealed that standard deviation of differences ranged from 9.1% to 9.9% in the two assays. This allowed the use of a cutoff >3 SD; that is, 30% neutralization. Virus titration experiments showed that neutralization results were dependent on virus dose and therefore the neutralization assays should be performed with a virus dose of 10–100 PFU/well. The assays have high specificity and reproducibility, and are simple and sensitive high-throughput assays.

Key Words: HIV; SIV; plaque assays; neutralization assays; co-receptor use; engineered cell lines.

1. Introduction

In general, viral infections elicit neutralizing antibody responses and persistence of such antibodies has been interpreted as a sign of protection. Thus,

From: *Methods in Molecular Biology, Vol. 304: Human Retrovirus Protocols: Virology and Molecular Biology*
Edited by: T. Zhu © Humana Press Inc., Totowa, NJ

when testing for vaccine efficacy, large seroepidemiological studies are important tools. A major conceptual problem in HIV vaccine development is the lack of information on correlates of immune protection against infection or disease, and HIV neutralization has long been a controversial issue. Primary HIV-1 isolates seem to be more resistant to the effect of neutralizing antibodies than T-cell line adapted strains. Also, not more than 50% of sera from HIV-1 infected individuals show any neutralizing activity, and only half of these have neutralizing titers higher than 1:80 *(1)*. Emergence of virus variants resistant to neutralization by autologous sera is a main feature of progressive HIV disease (reviewed in **ref.** ***2***). The reason for evasion of antibody-mediated neutralization by HIV-1 seems to occur through conformational masking of receptor binding *(3)*. It has been shown that, indeed, conformational masking of neutralizing epitopes may occur through mutations at glycosylation sites *(4)*. This would explain why most antibodies in the HIV-1-infected individual have no access to the receptor-binding site on the virus envelope and cannot inhibit infection of new cells. In contrast, long-term nonprogression has been associated with a broadly cross-reactive, high-titer serum-neutralizing activity *(5)*, suggesting that neutralizing antibodies may have an important role in pathogenesis and their protective role has to be considered in vaccine studies. Similarly, the low-pathogenic HIV-2 infection in humans or experimentally infected macaques seem to elicit broadly cross-reactive neutralizing antibodies *(6)*. In the monkey model, SIVsm infection causes immunodeficiency that is associated with emergence of neutralization resistant viruses, similar to HIV-1 infection in humans *(2,7)*.

One way to learn more about protective immunity and try to resolve the gap in our knowledge is through vaccine trials. There is, however, a need for a strategy to elicit broadly cross-reactive immune responses in vaccinees. Since the form of immunogen affects the type of immune response, to stimulate humoral immunity, antigens in native form must be available outside of cells to the surface immunoglobulin receptor of B-cells. Evaluation of vaccine efficacy will be necessary in large seroepidemiological studies. For this purpose it is feasible to use neutralization assays, since measurement of neutralizing antibody responses is much less difficult than measurement of cytotoxic T-lymphocyte (CTL), where epitope recognition is further complicated by human lymphocyte antigen (HLA) type. Assuming that neutralization is an indicator for protection (from infection or disease), neutralizing antibodies may provide important clues for vaccine efficacy.

The assays for HIV neutralization to be described in this chapter have been developed with the aim to provide simple and standardized methods to test primary HIV and SIV isolates with improved sensitivity and reproducibility. The recognition that HIV-1 requires CD4 and a co-receptor to enter cells and,

Table 1
HIV Phenotypes

Replication capacity in PBMC	Syncytium induction in PBMC/MT-2	Co-receptor use	Phenotype
Rapid/high	SI	CXCR4	X4
		CCR5,CXCR4	R5X4
		CCR3,CCR5,CXCR4	R3R5X4
Slow/low	NSI	CCR5	R5
		CCR3,CCR5	R3R5

furthermore, with the identification of the co-receptors as seven-transmembrane G protein-coupled chemokine receptors, opened up new avenues, not only in studies of pathogenic processes, but also in quantitative measurement of HIV and SIV ***(8–10)***. The two most well-defined HIV-1 co-receptors are CXCR4 and CCR5, members of the CXC (α) and CC (β) chemokine receptor subfamilies, respectively. CXCR4 was the first HIV-1 co-receptor to be characterized, and was shown to be required for the fusion of T-cell line-adapted viruses with nonhuman cells expressing human CD4 ***(8)***. It is expressed on many cell types including transformed T-cells, fibroblasts, primary T-cells, and macrophages. Subsequently, CCR5 was shown to be the principal co-receptor for primary HIV-1 isolates with the nonsyncytium-inducing (NSI) phenotype ***(11,12)***, whereas the syncytium-inducing (SI) phenotype was associated with the use of CXCR4 alone or in combination with CCR5 ***(13)***. Other members of the CC chemokine receptor family, such as CCR2b and CCR3, may also function as co-receptors for HIV entry, although generally in a less efficient manner than CCR5, as assayed in tissue culture. The co-receptor usage patterns of HIV laid the ground for a new classification system (*see* **Table 1**). Thus, viruses previously termed NSI or slow/low viruses are defined by their use of members of the CC-chemokine receptor family, principally CCR5, and are termed R5. Viruses previously termed SI or rapid/high are defined by their use of the CXC-chemokine receptor CXCR4 and are termed X4. Some of the SI viruses use both CXCR4 and CCR5 (and/or CCR3) receptors; these are termed R5X4 (R3R5X4 or R3X4) viruses ***(9)***.

2. Materials

1. U87.CD4 cells with (CCR1, CCR2B, CCR3, CCR5, or CXCR4-expressing cells available) or without co-receptor expression (U87.CD4 parental cells) ***(12,13)***.
2. GHOST(3) cells with (CCR3, CCR5, CXCR4, CXCR6, or BOB-expressing cells available) or without co-receptor expression (GHOST[3] parental cells).

GHOST(3) cells also carry the HIV-2 LTR-driven green fluorescent protein (GFP) marker, which becomes activated upon infection with HIV-1 (HIV-2 or SIV) ***(14,15)***. The cell lines are available from the repositories in the United States and United Kingdom ***(16,17)***.
3. Dulbecco's modified Eagle's medium (DMEM) supplemented with 10% or 7.5% (for U87.CD4 or GHOST[3] cells, respectively) fetal calf serum (FCS, FDA-approved) and antibiotics (final concentration: penicillin 50 U/mL and streptomycin 50 µg/mL).
4. DMEM (high-glucose) supplemented as in **Subheading 3.**
5. 5 m*M* EDTA, pH 8.0.
6. Polybrene, 2 µg/mL final concentration.
7. Phosphate-buffered saline (PBS).
8. Hematoxylin.
9. 48-well or 96-well (for U87.CD4 or GHOST[3] cells, respectively) flat-bottom microtiter plates with lid.
10. Light microscope, inverted, for observation of monolayer cultures in microtiter plates.
11. Fluorescence microscope, inverted.
12. Virus stocks of HIV or SIV isolates produced in peripheral blood mononuclear cells (PBMC) ***(18)***.
13. Sera, plasma, or the IgG derivative, and/or monoclonal antibodies.

3. Methods

The methods described herein outline (1) two plaque assays for quantitative measurement of HIV or SIV infectivity and (2) plaque-reduction assays by antibodies. Both assays utilize human cell lines engineered to stably express CD4 and co-receptors for HIV and SIV entry ***(12,14)***. Since the indicator cell lines are derived from human glioma (U87) or osteosarcoma (HOS) cells they form monolayers and can be easily observed by the microscope. Plaques are composed of virus-infected cells and are scored as syncytia by light (U87 series) or fluorescence microscope (GHOST[3] series).

*3.1. Maintenance of Indicator Cell Lines (*see also *ref.* 18*)*

3.1.1. Culturing of U87.CD4 Cells

1. Thaw cryotube rapidly in lukewarm water.
2. Transfer cells to a 10-mL centrifuge tube. Add 8 mL DMEM 10% (high glucose), prewarmed to 18–22°C.
3. Centrifuge at 156*g* for 10 min.
4. Resuspend cells in 10 mL 10% DMEM and seed cells in a 25-cm^2 flask. Place flasks horizontally in a humidified 5% CO_2 atmosphere at 37°C.
5. Observe cultures; change medium or subculture as necessary (approximately twice weekly). If cultures grow very slowly, use medium containing L-alanyl-L-glutamine (Glutamax) rather than L-glutamine (less stable).

6. Subculture:
   a. Remove medium and rinse the monolayer with 5 mL sterile PBS.
   b. Add 0.5 mL 5 m*M* EDTA, pH 8.0, and place flasks horizontally (preferably at 37°C) for 5 min.
   c. When cells detach add 5 mL 10% DMEM, resuspend, and split cultures 1:3 (1:5).
   d. Add 10 mL 10% DMEM and return flask to the incubator (humidified 5% CO_2 atmosphere at 37°C).

3.1.2. Culturing of GHOST(3) Cells

1. Thaw cryotube rapidly in lukewarm water.
2. Transfer cells to a 10-mL centrifuge tube. Add 8 mL 7.5% DMEM, prewarmed to 18–22°C.
3. Centrifuge at 156*g* for 10 min.
4. Resuspend cells in 8 mL 7.5% DMEM and seed cells in a 25-cm^2 flask. Place flasks horizontally in a humidified 5% CO_2 atmosphere at 37°C.
5. Observe cultures; change medium and subculture twice weekly.
6. Subculture:
   a. Remove medium and rinse the monolayer with 5 mL sterile PBS.
   b. Add 0.5 mL 5 m*M* EDTA, pH 8.0, and place flasks horizontally (preferably at 37°C) for 2–3 min.
   c. When cells become detached add 5 mL DMEM 7.5%, resuspend cells, and split cultures 1:10–1:20, as necessary.
   d. Add 8 mL 7.5% DMEM and return flask to the incubator (humidified 5% CO_2 atmosphere at 37°C).

It is essential that these indicator cell lines not be carried in culture for longer than 2 mo, because they may lose expression of different receptors. The faster a cell line grows the higher the risk for the loss of the markers. For example, GHOST(3) cells, a derivative of HOS cells carrying CD4, grow faster and are more prone to change than U87.CD4 cells. It is important that even during the 2-mo periods in culture, the cell lines be subjected to regular control of the different markers. Details of control and reselection of the different markers are described in **Note 1**.

3.2. Plaque Assays for Quantifying HIV/SIV Infectivity

With both cell lines, the assays are single-cycle assays and measure input virus. By adding protease inhibitors (PIs) simultaneously with virus (HIV-1) to the U87.CD4 cells it has been proven that virus spread does not influence the measurements of plaques ***(19)***. PIs (we used nelfinavir and indinavir in these experiments) act at a late step of the HIV-1 life cycle and do not affect synthesis of envelope protein. Accordingly, addition of PIs to the cultures had no effect on the number of plaques within the limited time frame of the assay (3–4 d).

Readout is by light microscopy (U87.CD4 cells) or fluorescence microscopy (GHOST[3] cells). In both cases, distinct groupings of syncytial cells are counted as plaques and infectious virus titers expressed as plaque-forming units (PFU). Syncytial cells have been verified as virus-infected cells in HIV-1-infected U87.CD4-CCR5 cultures by staining with a mouse anti-HIV-1 p24 monoclonal antibody in the blue cell assay (*see* **Note 2**).

*3.2.1. Plaque Assay on U87.CD4 Cells (*see also *ref.* **19***)*

Seed cells in 48-well plates at a concentration of 1×10^5 cells per well in 1 mL DMEM (10% FCS and antibiotics). Incubate plates at 37°C with 5% CO_2 in a humidified incubator, until cultures reach half confluence (usually 24 h).

Day 1:

1. Virus stocks (the supernatant culture fluids from primary or expanded PBMC cultures) are stored below –70°C. Select the virus to be used for infection and allow it to thaw in the working hood or, more quickly, by placing tubes in cold water.
2. Prepare serial virus dilutions in 10% DMEM (containing 2 μg polybrene/mL) using sterile capped tubes or a 24-well plate. Make six fivefold dilutions in the ranges 1/5 to 1/15,625 (range sufficient for primary isolates) in the following way:
   a. Add 800 μL of medium to six wells.
   b. Add 200 μL of virus stock to the first well (1/5 dilution).
   c. Change tip, mix, and transfer 200 μL to the adjacent well containing 800 μL of medium (a 1/25 dilution), and so on. Ensure thorough and consistent mixing of each well and replace tips between dilutions.
3. Prior to adding virus, rinse the U87.CD4 monolayer cells once with PBS and add fivefold virus dilutions in 200 μL vol/well to triplicate wells. Care must be taken that the U87.CD4 monolayers do not dry out during this procedure. Remove medium from one row at the time and add virus immediately. Alternatively, immediately replace old medium with 50 μL fresh medium (containing 2 μg polybrene/mL) and add virus dilutions in 150 μL vol/well.
4. Incubate overnight at 37°C with 5% CO_2 in a humidified incubator.

Day 2:

1. Wash plates by rinsing two times with PBS (500 μL/well).
2. Add 1 mL of fresh 10% DMEM and incubate plates as above.

Day 4 (or 5):

1. Wash plates by rinsing two times with PBS (500 μL/well).
2. Fix with methanol:acetone (1:1) for 3–5 min.
3. Stain with hematoxylin for 2–3 min, wash with tap water, and air-dry. Hematoxylin stains cell nuclei, which turn dark blue after contact with tap water.
4. Count syncytial cells under a light microscope. The number of plaques (distinct grouping of syncytial cells) is counted in wells infected with the lowest virus dilution that allows identification of individual plaques, that is, in a dilution that produces 20–30 plaques per well. Virus titers are calculated as PFU per milliliter =

$$\frac{\text{average number of plaques in triplicate wells} \times \text{virus dilution}}{\text{volume in the well*}}$$

*Refers to the volume of the virus added, either 200 µL or 150 µL (*see* **day 1 step** above).

3.2.2. Plaque Assay on GHOST(3) Cells

Seed cells in 96-well plates at a concentration of 1 × 10^4 cells/well (or 5 × 10^3 cells/well) in 200 µL DMEM (7.5% FCS and antibiotics). Incubate plates at 37°C with 5% CO_2 in a humidified incubator, until cultures reach half confluence (usually 24 h).

Day 1:

1. Thaw virus stocks (*see* **Subheading 3.2.1.** for details).
2. Prepare serial virus dilutions in 7.5% DMEM, as described in **Subheading 3.2.1.**
3. Prior to adding virus, rinse the monolayer of GHOST(3) cells once with PBS and add the fivefold virus dilutions in 200 µL vol/well to triplicate wells. Care must be taken that the GHOST(3) monolayers do not dry out during this procedure. Remove medium from one row at the time and add virus immediately. Alternatively, replace old medium immediately with 50 µL fresh medium (containing 2 µg polybrene/mL) and add virus dilutions in 150 µL vol/well.
4. Incubate overnight at 37°C with 5% CO_2 in a humidified incubator.

Day 2:

1. Wash plates by rinsing two times with PBS (200 µL/well).
2. Add 200 µL of fresh 7.5% DMEM and incubate plates as above.

Day 4:

Count fluorescent cells in a fluorescence microscope. Fluorescence can be observed 48 h after infection, peaks at 72 h, and decays thereafter. The number of plaques (distinct grouping of syncytial cells or single cells that show green fluorescence) is counted in wells infected with the lowest virus dilution that allows identification of individual plaques—that is, in a dilution that produces 20–30 plaques per well. For calculation of PFUs per milliliter (see above). For quantifying fluorescence by flow cytometry *see* **Note 3**.

3.3. Plaque-Reduction Assays for HIV/SIV Neutralization

Virus must be titrated prior to performing the assay, as neutralization is dependent on the dose of input virus ***(19)***. A high virus dose can mask neutralization of weakly neutralizing sera, as exemplified in **Fig. 1B**; a 1:20 dilution of the HIV-1-positive patient serum neutralized all doses of the isolate TZ97002, but the same serum at a 1:160 dilution neutralized only the lowest virus dose (88 PFU/well). Based on these data it is recommended that the neutralization assay be performed with a virus dose of 10–100 PFU/well. A final dilution of 1/10 (or 1/20) is appropriate for most HIV-1 primary isolates.

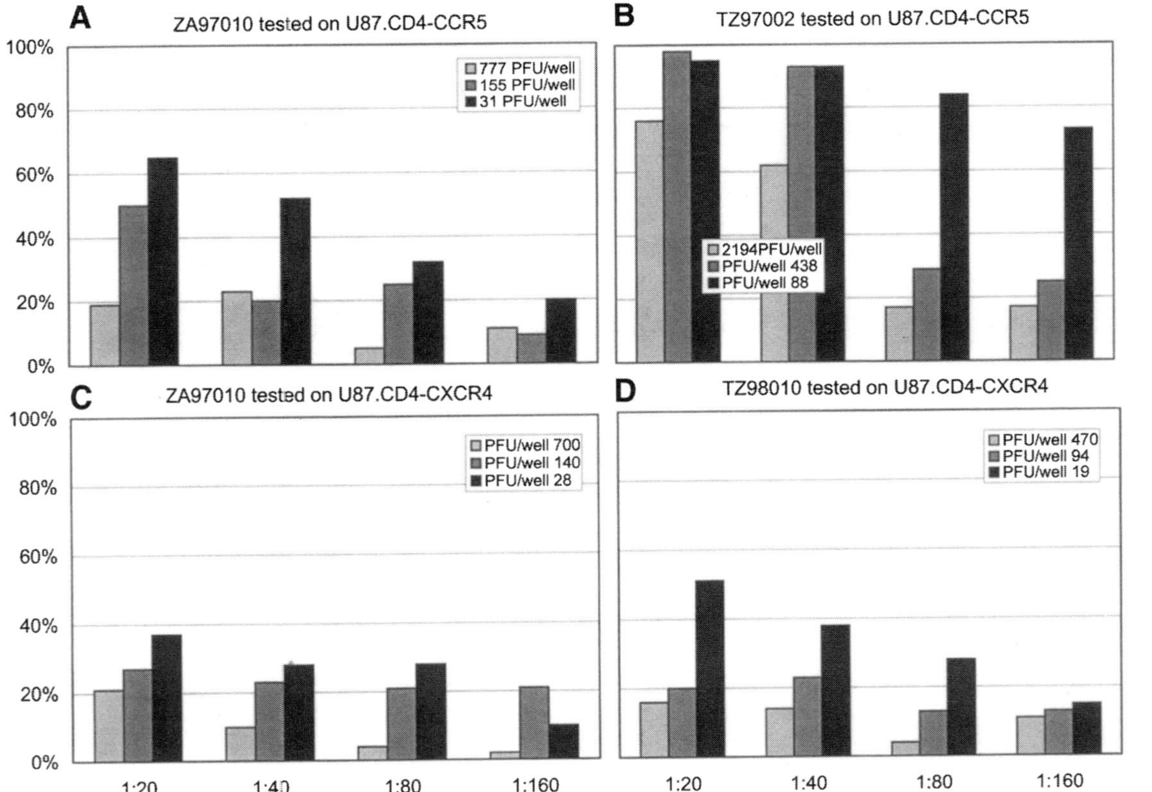

Fig 1. Influence of virus titer and serum dilution on the result of neutralization. The neutralizing capacity of different dilutions of serum S652 (from an HIV-1-infected individual) was tested against three different doses of viruses ZA97010 and TZ97002 on U87.CD4-CCR5 cells and three different doses of viruses ZA97010 and TZ98010 on U87.CD4-CXCR4 cells. Reproduced with permission of AIDS Research and Human Retroviruses ***(19)***.

Table 2
Overview of Methodology of Two Assays

Cells	U87.CD4 and co-receptors	GHOST(3) and co-receptors
Format	48-well plate[a]	96-well plate
Number of cells/well	1 × 105	4 × 103
Readout	Plaques: Syncytial cells Counted by light microscopy after staining with hematoxylin	Plaques: Green cells Counted by UV microscopy (fluorescence can be quantified by flow cytometry)

[a]The wells of 48-well plates provide a larger area than 96-well plates and allow convenient scoring for syncytial cells in the microscope.

The neutralization assay is similar in the two indicator cell lines and therefore it will be described once. It must be remembered, however, that there are differences in the handling of U87.CD4 and GHOST(3) cells. These differences are highlighted in **Table 2**.

Day 0:
Seed cells as described in **Subheading 3.2.1.** or **3.2.2.**

Day 1:

1. Mix serum or plasma (45 µL) and virus stock (usually 90 µL) in a sterile capped tube or 24-well (alternatively, 48-well) microtiter plate and dilute with DMEM to a total volume of 900 µL. This results in a final dilution factor of 1/20 for the serum and 1/10 for the virus.
2. Incubate the serum-virus mixture at 37°C for 1 h.
3. Dilute further in two or three fivefold dilution steps, as described in **Subheading 3.2.1.**
4. Distribute 200 µL of each dilution into triplicate wells containing the cells. Alternatively, to prevent drying out of monolayers, immediately replace old medium with 50 µL fresh medium (containing 2 µg polybrene/mL) and add dilutions of serum-virus mixtures in 150 µL vol/well.
5. Control wells in triplicate:
   a. Positive virus controls consist of wells with cells and virus but no serum and/or HIV/SIV negative serum.
   b. Cell controls consist of wells with cells only.
   c. Positive serum control consists of wells with cells + virus + serum with known neutralizing activity.

Day 2:

1. Wash plates by rinsing two times with PBS (500 µL /well for 48-well plates and 200 µL /well for 96-well plates).

2. Add 500 µL or 200 µL, respectively, of fresh DMEM and incubate plates further.

Day 4:

Readout as detailed above for U87.CD4 and GHOST(3) cells, in **Subheadings 3.2.1.** and **3.2.2.**, respectively. Calculate the neutralizing capacity of the serum by the formula 1– (PFU with serum/PFU without serum) × 100. The neutralizing capacity of a serum is thus expressed as the degree of reduction in plaque-forming units in the presence of serum relative to wells with no serum.

3.4. Calculation of Cutoff Value for Neutralization (see also *ref.* 19)

A major concern was the reproducibility of the virus titer and its implication for the cutoff value of neutralization. Because both neutralization assays are based on plaque reduction in the presence of serum, the intraassay variation is important. To establish the accuracy of the intraassay variation of virus titer determinations, three assays were performed on the same day. Fifteen serum-virus combinations were tested on U87.CD4-CCR5 cells and four serum-virus combinations on GHOST(3)-CCR5 cells. We calculated the percent difference for each individual determination relative to the mean of the three repeat determinations. The range of differences was –19 to +27% on the U87.CD4-CCR5 cells, giving an SD ± 9.1%. On GHOST(3)-CCR5 the range of differences was –12.3 to +11% for the negative sera or +48.8 to +73.4% for the positive sera, giving SD ± 9.66% or SD ± 9.89%, respectively. On the basis of these data we chose a cutoff for neutralization (i.e., plaque reduction) of 30%, which represents 3.3 or 3.1 standard deviations in assays performed in the two different indicator cell systems. By using this cutoff the risk of falsely calling a serum neutralizing should be less than 1%.

4. Notes

1. U87.CD4 and GHOST(3) cell lines have been engineered to stably express CD4, co-receptors for HIV and, in the case of GHOST(3) cells, also the green fluorescent protein (GFP) ***(12,14)***. The cell lines are available from the repositories in the United States and United Kingdom ***(16,17)***. Stability of the different markers is, however, highly variable. It is therefore mandatory that upon receipt of the indicator cell lines, each laboratory freezes a large stock of cells (at least 10 tubes each). Continuous passage of any of the U87.CD4 cell series should not exceed 2 mo. After this time the cultures have to be discarded and a new tube from the stock thawed out. GHOST(3) cells grow faster than U87.CD4 cells and may be even more prone to lose markers. It is therefore advisable to thaw out the cells shortly (2–3 d) before use for experiments. The cell lines may then be used for a maximum of 8 wk.

   It is important to include control viruses with defined co-receptor usage in each experiment. Using the same virus stock, co-receptor usage patterns (also time to syncytium induction in U87.CD4 cells and the proportion of cells showing fluorescence in GHOST(3) cells) should be highly repeatable between experiments. Another way of testing receptor expression is by flow cytometry using monoclonal antibodies to CD4 or the chemokine receptors (anti-CCR3, -CCR5, -CXCR4, and -CXCR6 are commercially available) ***(20)***.

Should the proportion of receptor-positive cells decrease with time the cell lines may be reselected by culturing in selective medium for two weeks. Expression of the different markers can be selected according to the following scheme:

a. CD4: 300 μg/mL of G418 for U87.CD4 and 500 μg/mL for GHOST(3).
b. CCR1, CCR2, CCR3, CCR5, CXCR4, BOB, Bonzo: 1 μg/mL puromycin.
c. GFP: 100 μg/mL hygromycin for GHOST(3) (reduce to 50 μg/mL if cells appear too sensitive).

Following selection, passage cells at least once in 10% DMEM (or 7.5% DMEM for GHOST(3) cells), then freeze stocks and use for experiments.

GHOST(3) cells show a background expression of CXCR4. Virus isolates able to use this receptor show a background expression at various levels more or less across the entire panel of GHOST(3) cells. Use of a specific receptor need sometimes be verified by using the CXCR4 antagonist, AMD3100 (*see* Chapter 26).

2. Blue cell assay (for staining of intracellular HIV-1 p24 antigen ***[19]***): cell monolayers fixed with methanol:acetone (1:1) are washed three times with wash buffer (WB) and exposed for 1 h at 37°C to 150 μL of a mouse anti-HIV-1 p24 monoclonal antibody (EF-7, 1:600 dilution). Subsequently, plates are washed three times in WB, followed by further incubation with 150 μL of β-galactosidase-conjugated anti-mouse IgG (diluted 1:400). The washing procedure is then repeated and 100 μL substrate added (5-bromo-4-chloro-3-indolyl-β-D-galactopyranoside, X-gal). After 1–2 h incubation the substrate is removed and PBS added for storage. Infected cells stain blue and can be enumerated under the light microscope.
3. Preparation of GHOST(3) cells for flow cytometry ***(20)***.
   Infection of GHOST(3) cells: *see* **Subheading 3.2.2.**

Day 4:

Check cultures in the fluorescence microscope. If there is an increase in fluorescence from the day before, cultures are ready for testing by flow cytometry.

a. Wash wells with 200 μL PBS.
b. Add 200 μL 5 m*M* EDTA, place trays at 37°C and wait a few minutes until cells detach.
c. Add 200 μL 4% paraformaldehyde to the FACS tube (5 mL polystyrene tubes, sterile).
d. Resuspend cells and transfer into the FACS tube. Mix well. The final concentration of paraformaldehyde is 2%.
e. Keep samples in the dark at 4°C for at least 1 h before FACS analysis. Samples can be kept up to 2–3 d before reading.

Acknowledgments

The U87.CD4 and GHOST(3) cell lines, engineered to express receptors for HIV and SIV, were kindly provided by Dr. Dan R. Littman, Howard Hughes Medical Institute, Skirball Institute for Biomolecular Medicine, New York University, New York. I thank all my colleagues who participated in the development and validation of the assays described herein: Yu Shi, Jan Albert, and Rigmor Thorstensson at the Swedish Institute for Infectious Disease Control/ Karolinska Institute, Stockholm, Sweden; Harvey Holmes and Gary Francis at the National Institute of Biological Standards and Control, South Mimms, Potters Bar, Herts, UK; and Monica Öberg at Lund University, Lund, Sweden. The support of the UNAIDS/WHO Network for HIV Isolation and Characterisation for providing a broad range of genetically characterized isolates is gratefully acknowledged. This work would not have been possible without the grants provided by the Swedish Research Council and the Swedish International Development Cooperation Agency/Department for Research Cooperation (Sida/SAREC).

References

1. Weber, J., Fenyo, E. M., Beddows, S., Kaleebu, P., and Bjorndal, A. (1996) Neutralization serotypes of human immunodeficiency virus type 1 field isolates are not predicted by genetic subtype. The WHO Network for HIV Isolation and Characterization. *J. Virol.* **70,** 7827–7832.
2. Fenyo, E. M., Albert, J., and McKeating, J. (1996) The role of the humoral immune response in HIV infection. *AIDS* **10,** S97–S106.
3. Kwong, P. D., Doyle, M. L., Casper, D. J., Cicala, C., Leavitt, S. A., Majeed, S., et al. (2002) HIV-1 evades antibody-mediated neutralization through conformational masking of receptor-binding sites. *Nature* **420,** 678–682.
4. Wei, X., Decker, J. M., Wang, S., Hui, H., Kappes, J. C., Wu, X., et al. (2003) Antibody neutralization and escape by HIV-1. *Nature* **422,** 307–312.
5. Zhang, Y. J., Fracasso, C., Fiore, J. R., Bjorndal, A., Angarano, G., Gringeri, A., and Fenyo, E. M. (1997) Augmented serum neutralizing activity against primary human immunodeficiency virus type 1 (HIV-1) isolates in two groups of HIV-1-infected long-term nonprogressors. *J. Infect. Dis.* **176,** 1180–1187.
6. Bjorling, E., Scarlatti, G., von Gegerfelt, A., Albert, J., Biberfeld, G., Chiodi, F., et al. (1993) Autologous neutralizing antibodies prevail in HIV-2 but not in HIV-1 infection. *Virology* **193,** 528–530.
7. Zhang, Y. J., Ohman, P., Putkonen, P., Albert, J., Walther, L., Stalhandske, P., et al. (1993) Autologous neutralizing antibodies to SIVsm in cynomolgus monkeys correlate to prognosis. *Virology* **197,** 609–615.
8. Feng, Y., Broder, C. C., Kennedy, P. E., and Berger, E. A. (1996) HIV-1 entry cofactor: functional cDNA cloning of a seven-transmembrane, G protein-coupled receptor. *Science* **272,** 872–877.

9. Berger, E. A., Doms, R. W., Fenyö, E. M., Korber, B. T. M., Littman, D. R., Moore, J. P., et al. (1998) A new classification for HIV-1. *Nature* **391,** 240.
10. Fenyo, E. M. (2001) The role of virus biological phenotype in human iimunodeficiency virus pathogenesis. *AIDS Rev.* **3,** 157–168.
11. Dragic, T., Litwin, V., Allaway, G. P., Martin, S. R., Huang, Y., Nagashima, K. A., et al. (1996) HIV-1 entry into CD4+ cells is mediated by the chemokine receptor CC-CKR-5. *Nature* **381,** 667–673.
12. Deng, H., Liu, R., Ellmeier, W., Choe, S., Unutmaz, D., Burkhart, M., et al. (1996) Identification of a major co-receptor for primary isolates of HIV-1. *Nature* **381,** 661–666.
13. Bjorndal, A., Deng, H., Jansson, M., Fiore, J. R., Colognesi, C., Karlsson, A., et al. (1997) Coreceptor usage of primary human immunodeficiency virus type 1 isolates varies according to biological phenotype. *J. Virol.* **71,** 7478–7487.
14. Deng, H. K., Unutmaz, D., KewalRamani, V. N., and Littman, D. R. (1997) Expression cloning of new receptors used by simian and human immunodeficiency viruses. *Nature* **388,** 296–300.
15. Morner, A., Bjorndal, A., Albert, J., Kewalramani, V. N., Littman, D. R., Inoue, R., et al. (1999) Primary human immunodeficiency virus type 2 (HIV-2) isolates, like HIV- 1 isolates, frequently use CCR5 but show promiscuity in coreceptor usage. *J. Virol.* **73,** 2343–2349.
16. National Institutes of Health (2002) AIDS Research and Reference Reagent Program Catalogue. http://www.aidsreagent.org ed. NIH, pp. 99–1536.
17. National Institute of Biological Standards and Control, Centralised Facility for AIDS Reagents (2002) Catalogue of Reagents. http://www.nibsc.ac.uk/catalog/aids-reagent ed. NIBSC, pp. 8–185.
18. World Health Organisation. (2002) WHO-UNAIDS Guidelines for Standard HIV Isolation and Characterization Procedures. ed. WHO, pp. 1–140.
19. Shi, Y., Albert, J., Francis, G., Holmes, H., and Fenyo, E. M. (2002) A new cell line-based neutralization assay for primary HIV type 1 isolates. *AIDS Res. Hum. Retroviruses* **18,** 957–967.
20. Vodros, D., Tscherning-Casper, C., Navea, L., Schols, D., De Clercq, E., and Fenyo, E. M. (2001) Quantitative evaluation of HIV-1 coreceptor use in the GHOST3 cell assay. *Virology* **291,** 1–11.

23

Detection of Drug-Resistant HIV-1 Strains

Ann A. Kiessling, S. J. Eyre, and B. D. Desmarais

Summary

Human immunodeficiency virus (HIV-1) encodes proteins essential to its replication cycle. Reverse transcriptase, protease, and viral envelope gp120 are three proteins that have been targeted for antiviral drug development. Eleven inhibitors of reverse transcriptase, seven inhibitors of protease, and one inhibitor of viral envelope binding have been approved for use. Antiretroviral therapy has reversed the mortality rate of HIV-infected persons, but over time, therapy-resistant virus variants may outgrow. A large body of information is now available to relate specific amino acid sequences in the resistant variants to specific drug regimens. Designing therapy to compensate for virus resistance results in improved patient outcomes. The advent of microsequencing technologies paved the way for direct sequencing of DNA products generated by polymerase chain reactions, thus dramatically lowering the cost of HIV gene sequencing. Designing therapy according to genetic analysis of HIV variants will not only also improve clinical outcome, but will also deter the transmission of drug-resistant strains.

Key Words: HIV-1; drug resistance; protease; reverse transcriptase; gene sequences; PCR.

1. Introduction

Human immunodeficiency virus (HIV-1) is characterized by genetic diversity. This has historically been attributed to base substitutions introduced during the synthesis of the DNA copy of the viral RNA genome by the viral DNA polymerase, reverse transcriptase, which has no error-proofing activity. The more recent elegant experiments that identified the deaminases, CEM15 (APOBEC3G), that comprise an innate cellular defense mechanism against retrovirus infection have, however, revealed another important cause of HIV genetic diversity, and also explained the relentless accumulation of G-to-A

From: *Methods in Molecular Biology, Vol. 304: Human Retrovirus Protocols: Virology and Molecular Biology*
Edited by: T. Zhu © Humana Press Inc., Totowa, NJ

transitions within HIV quasispecies in an infected individual *(1,2)*. The mutations in the viral genome that result from these processes may weaken it, but they also provide an opportunity for the virus to escape immunological and pharmacological controls.

The complex nature of the HIV life cycle has provided multiple opportunities for drug intervention. HIV reverse transcriptase is sufficiently different in its nucleotide-binding characteristics from cellular DNA replicating and repair enzymes, that nucleotide analogs, such as the adenosine derivative, zidovudine, effectively inhibit reverse transcriptase activity without blocking normal cell replication. Such inhibition prevents synthesis of a proviral DNA copy of the infecting virus RNA genome, thus blocking successful infection. As described in the following section, eight nucleotide-based reverse transcriptase inhibitors have been approved for therapeutic use, in addition to three non-nucleotide reverse transcriptase inhibitors (**Fig. 1**).

The virus-encoded protease converts polypeptide precursors to the proteins required for packaging the virus RNA into infectious particles. The uniqueness of the viral protease has allowed the development of seven protease inhibitors approved for therapeutic use (**Fig. 2**), including the most recent, atazanavir, a prodrug of amprenavir. Two new protease inhibitors that are at the most advanced stage of clinical development are tipranavir and TMC114. Protease inhibitors provide a powerful blockade to infectious virus particle production by otherwise infectious provirus. Viral dynamics studies have shown that blood viral burden is dependent on continual new infections. Hence, the production of defective particles results in a rapid and dramatic drop in plasma viral burden. Importantly, other compartments, such as semen-producing organs, exhibit a marked delay in viral clearance.

In addition to reverse transcriptase and protease inhibitors, polypeptide inhibitors of the process of virus fusion with cell receptors have been developed, including the recently approved Fuseon, which is the first antiretroviral to inhibit new infection cycles.

Mutations within the drug-targeted viral genes can lead to resistance to the inhibitory effects of the drug, either by influencing binding or by lessening the inhibitory effects of the bound drug. Importantly, mutations arise only as a result of new virus infection events. Although lack of strict adherence to a drug regimen is commonly touted as the cause for the development of resistance mutations, this may not be the only cause. Analyses of the kinetics of emergence of resistance mutations indicate the existence of two separate compartments of infection, one smaller with lower concentrations of antiviral drugs, not sufficient to completely inhibit replication, and the other larger with higher concentrations of antiviral drugs, sufficient to suppress replication *(3)*. Over time, resistant virus arising in the smaller compartment migrates to the larger

Nucleoside Analogs

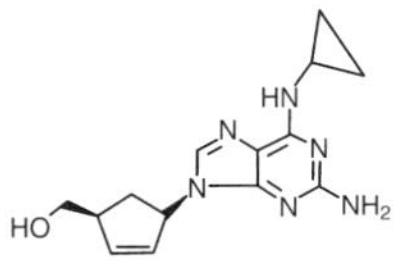 MW 404.42	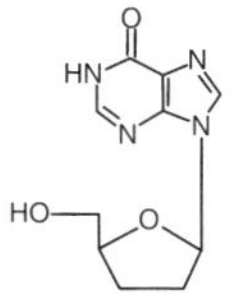 MW 236.23	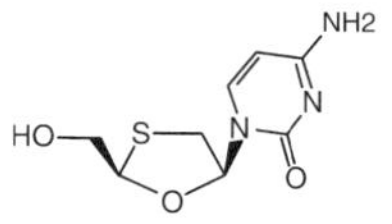 MW 229.26	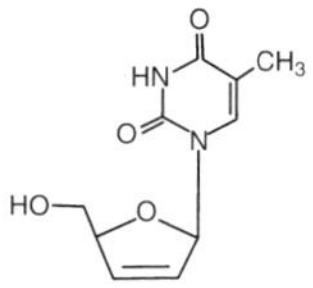 MW
Abacavir *Ziagen®*	Dideoxyinosine *Videx®*	Lamivudine; 3TC *Epivir®*	Stavudine; D4T *Zerit®*

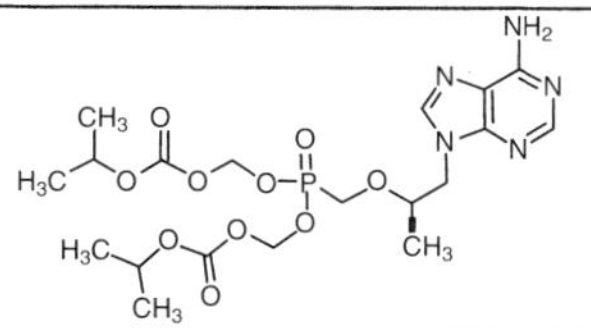 MW 519.45	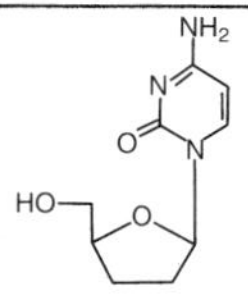 MW 211.22	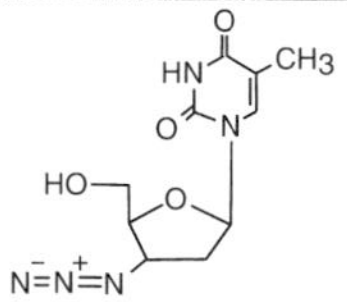 MW 267.24	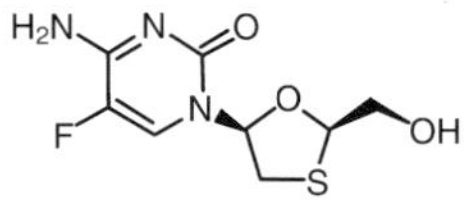 MW 247.24
Tenofovir Disoproxil (Fumarate) *Viread®*	Dideoxycytidine *Hivid®*	Azidothymidine *Retrovir®*	Emtricitabine *Emtriva®*

Non-Nucleoside Reverse Transcriptase Inhibitors (NNRTI's)

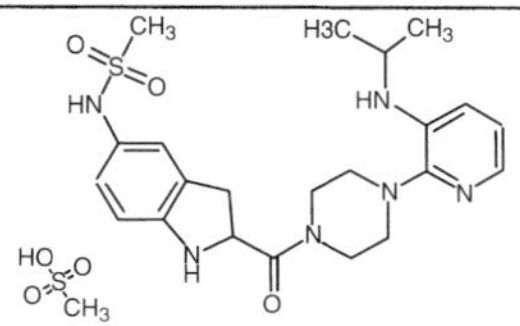 MW 552.68	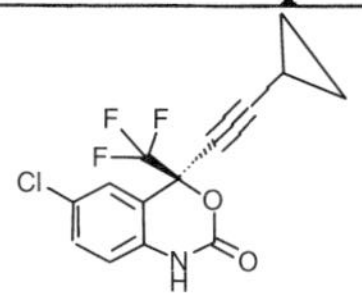 MW 315.68	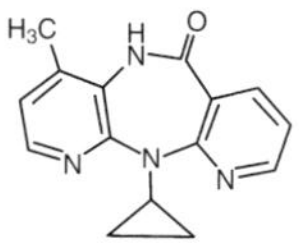 MW 266.31
Delavirdine *Rescriptor®*	Efavirenz *Sustiva®*	Nevirapine *Viramune®*

Fig. 1. Chemical structures of approved reverse transciptase inhibitors.

Protease Inhibitors

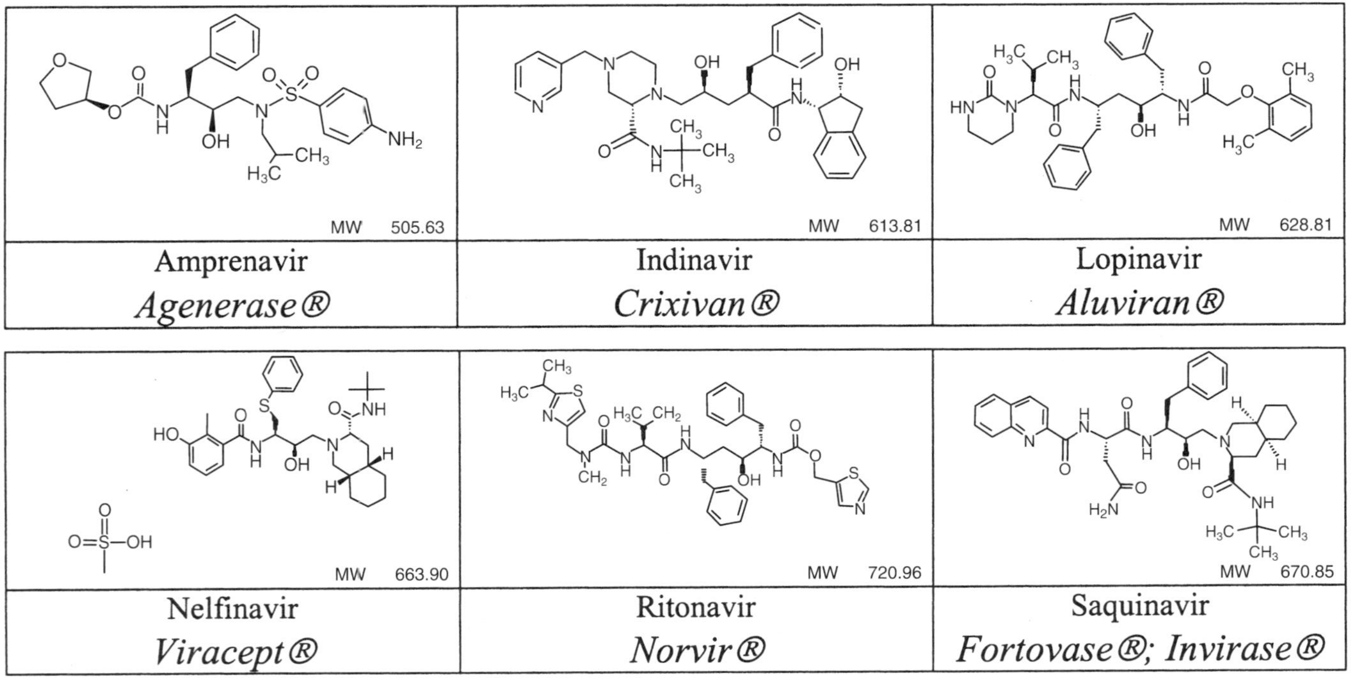

Amprenavir	Indinavir	Lopinavir
MW 505.63	MW 613.81	MW 628.81
Amprenavir *Agenerase®*	Indinavir *Crixivan®*	Lopinavir *Aluviran®*

Nelfinavir	Ritonavir	Saquinavir
MW 663.90	MW 720.96	MW 670.85
Nelfinavir *Viracept®*	Ritonavir *Norvir®*	Saquinavir *Fortovase®; Invirase®*

Fig. 2. Chemical structure of approved protease inhibitors.

compartment, leading to re-emergence of circulating virus resistant to antiviral therapy. The larger compartment in this model is undoubtedly the blood/lymphatic system. There are several candidates for the smaller compartment, including the genital tract, particularly semen-producing organs.

Drug regimens that target more than one viral function have obvious advantages because of the unlikelihood that any one virus particle will contain multiple drug-resistant mutations. Nonetheless, over time, mutations conferring resistance to one or more drugs do accumulate, leading to re-emergence of clinically significant levels of virus replication. The resistance mutations are not caused by the drug therapy per se, and have not been found to increase viral pathogenicity over wild-type, but they represent genetic changes that are consistent with successful viral infection (most mutations are probably lethal to the virus) in the presence of inhibitory drugs.

Identifying viral resistance mutations is highly beneficial in guiding the design of new drug regimens following treatment failure, as well as designing new treatment regimens for individuals at risk for infection by resistant virus. For this reason, several strategies for determining drug resistance have emerged. An early strategy involved phenotypic characterization of infectious virus by measuring the reduction in virus production in cultured cells in the presence of a drug. Although laborious, this strategy laid the groundwork for identifying specific amino acid substitutions associated with resistance to each drug.

Once the resistance-conferring amino acid substitutions were known, less cumbersome methods of detecting drug resistance by direct assay for mutations in the DNA sequence were developed commercially.

Alternatively, sequencing polymerase chain reaction (PCR)-amplified DNA products is available to most clinicians in developed countries and the costs of such analyses have decreased dramatically in the past few years. PCR products can be sequenced directly or following ligation and cloning. There are advantages and disadvantages to each method, as described in the following sections.

2. Materials

2.1. Specimen Collection

1. Laminar flow hood (minimum class 2, type A biosafety hood).
2. Biohazard containers.
3. Centrifuge with capped, swinging bucket rotor.
4. Sterile, bulbed disposable pipets.
5. Bleach (1% household bleach) in a spray bottle and a beaker.
6. 70% Isopropanol or ethanol in a spray bottle.
7. 15-mL sterile, conical tubes.
8. 1.5 mL microfuge tubes.

9. Dulbecco's modified phosphate-buffered saline (Sigma, St. Louis, MO).
10. Guanidium-isothiocyanate solution: 6 *M* guanidium-isothiocyanate (Sigma), 0.5% NP40, 2 m*M* ethylenediaminetetraacetic acid (EDTA).
11. Sterile, bulbed, plastic transfer pipets.

2.2. Isolation of Nucleic Acids

1. QIAamp Viral RNA Mini Kit (Qiagen, Carlsbad, CA).
2. QIAamp DNA Blood Mini Kit (Qiagen).
3. Sterile, embryo culture tested, deionized water (Sigma).
4. Micropipettor.
5. 200 µL aerosol-resistant pipet tips (Molecular BioProducts).
6. 1000 µL aerosol-resistant pipet tips (Molecular BioProducts).
7. 1.5-mL micro test tubes—sterilized with O-ring screwcaps installed.
8. 0.5-mL micro test tubes—sterilized with O-ring screwcaps installed.
9. Fluorometer (Bio-Rad, Hercules, CA).
10. DNA quantitation kit (Bio-Rad).
11. 3 *M* sodium acetate.
12. Absolute ethanol.

2.3. Reverse Transcription

1. 10X assay buffer: embryo-tested water (sterile, Sigma); Ethylene glycol, enzyme-grade (Fisher); 1 *M* KCl; 25% (v/v) NP40 solution in embryo-tested water; 50 mg/mL (w/v) BSA (Fraction V, Sigma) solution; 1 *M* Tris-HCl, pH 8.9, at 25°C; 0.22-µm filter.
   In a sterile laminar flow hood never used for PCR, combine components to form a final solution of 200 m*M* Tris-HCl, 200 m*M* KCl, 1 mg/mL BSA, 0.5% NP40, and 35% ethylene glycol. Sterile-filter and store at –20°C (will not freeze). Solution is stable for up to 6 mo.
2. 75 m*M* $MgCl_2$ (Sigma).
3. Avian myeloblastosis virus (AMV) reverse transcriptase, 20 U/µL (R-9376, Sigma).
4. RNAse Inhibitor, 50 U/µL (RNAse-In, N-2511, Promega, Madison, WI).
5. Random hexamer primers, 2 µg/µL (C-1181, Promega).
6. 1–10 µL glass, disposable capillary pipets (Drummond).
7. Cold block at 0 to 5°C, or ice bath.

2.4. Polymerase Chain Reaction

1. 10X assay buffer (*see* **Subheading 2.3.**).
2. 25 m*M* $MgCl_2$ (Roche).
3. AmpliTaq® DNA Polymerase with GeneAmp® (Roche).
4. Oligonucleotide primers, aliquoted for single use.
5. Thermal cycler with heated lid (GTC-2, iCycler, or MJ).
6. 200-µL thin-wall tubes, DNase- and RNase-free (Bio-Rad).
7. 500-µL PCR tubes.

8. DEAE (diethylaminoethane) chromatography paper (Whatman).
9. PCR primers.

2.5. Gel Electrophoresis

1. Gel electrophoresis bath (Bio-Rad).
2. Adjustable power supply (Bio-Rad).
3. Gel trays (Bio-Rad).
4. Gel combs (Bio-Rad).
5. Agarose (Invitrogen)
6. 10X TBE buffer; 108 g Tris-HCL base, 55 g boric acid, 2 m*M* EDTA.
7. Self-sticking tape (Fisher).
8. Microwave or hot plate.
9. 6X sample loading buffer: 0.25% (w/v) bromophenol blue in sterile, deionized water; 30% (v/v) glycerol.
10. 50-base-pair DNA ladder for molecular weight determination (Sigma).
11. Ethidium bromide solution, 50 μg/mL (w/v) (GibcoBRL).
12. Ultraviolet variable intensity transilluminator.
13. Photodocumentation handheld Polaroid camera with hood (Fisher).
14. Polaroid 667, black-and-white film (Fisher).

2.6. Ligation and Cloning

1. TOPO TA Cloning Kit (containing pCR2.1-TOPO®) (Invitrogen).
2. Bacteriologic plate (1029, Falcon).
3. LB agar (Sigma).
4. Ampicillin (Sigma).
5. Bacterial cell spreader (Fisher).
6. X-gal (Promega).

2.7. PCR Product Clean-Up, Screening

1. Qiagen QIAquick PCR purification kit (Qiagen).
2. 1.5-mL flip-top tubes (Fisher).

2.8. Sequencing and Sequence Editing

1. ABI Prism 377XL automated sequencer.
2. Sequencher 4.2 (Gene Codes Corporation).

3. Methods

3.1. Specimen Collection

3.1.1. Blood

Blood plasma virus is considered the most clinically relevant virus population for the detection of resistance-conferring mutations, although such mutations may first appear in semen in HIV-infected men ***(4)***. Five to 10 mL of blood is collected in EDTA (purple-top tubes, 5 m*M* EDTA) using universal precautions (*see* **Note 1**). All blood-handling procedures are carried out within a laminar flow hood, biosafety level 2 certified, with the biosafety glass in place and the blower on. Lab coats should have sleeves cuffed at the wrist. Two gloves should be worn, one under the cuffed sleeve and one over the cuffed sleeve. All disposable plastic- and glassware is discarded in a biohazard container located within the hood. No scissors, needles, or glass Pasteur pipets are used in the hood.

The rubber stoppers are removed from the vacutainer blood tubes, and the blood transferred to 15-mL conical tubes. One mL is aliquoted to at least 5 vol of hypotonic fixative *(5)* to lyse red blood cells and stabilize leukocytes for quantitation. The conical tubes are then capped and the remaining blood centrifuged at approx 500 to 800*g* for 15 min to separate plasma from cells. Plasma is removed to within 1 cm of the buffy coat and at least 1 mL aliquoted for storage at –80°C, or in liquid nitrogen. The remainder is aliquoted into AVL buffer, according to Qiagen kit instructions, and stored at –80°C until assayed, up to 3 yr.

Blood leukocytes layered above the red blood cells, termed the buffy coat, should also be saved in case genotyping the DNA provirus is of clinical interest. Several studies have demonstrated significant genetic differences between DNA provirus in peripheral blood cells and circulating plasma virus. The evidence indicates that the blood cell virus represents an earlier infection event than the infection focus (presumably lymph nodes) releasing contemporary virus into blood plasma. For this reason, it may be of interest to identify resistance-conferring mutations in blood provirus, as well as blood plasma virus, in patients experiencing multiple therapy failures.

The buffy coat is collected by gentle aspiration of the uppermost 2 to 3 mm of the red blood cell layer and transferred to five equal volumes of the guanidium-isothiocyanate solution to lyse the cells and release the DNA. The buffy coat-guanidium solution may be stored at 4°C for a few weeks, or at –80°C for up to a few years.

3.1.2. Semen

Although it is of less overall clinical importance than the resistance-conferring mutations in blood, determining resistance-conferring mutations in semen may be valuable to guide therapy in men with poor clinical improvement on existing therapy or with a history of multiple drug regimen failures. Statistical modeling approaches to understanding therapy failure point out the need for persistent, low-level virus replication in a relatively small, isolated virus reservoir to give rise to mutant virus *(3)*. Semen-producing organs are such a reservoir and resistance-conferring mutations have been shown to arise first in semen ***(4)***. Not only are such resistance-conferring mutations of potential importance to the clinical status of the patient, but the mutated virus is also a potential candidate for sexual transmission.

A semen specimen is collected either in a condom or by masturbation into a sterile urine container, and allowed to liquefy for no more than 90 min. It is transferred to a 15-mL conical tube, diluted with one equal volume of phosphate-buffered saline, and underlayered with a 19% cushion of Nycodenz in PBS. Following centrifugation at 1000*g* for 20 min, the seminal plasma is removed to within 1 cm of the interface, at least 1 mL transferred to a cryotube for storage at –80°C or in liquid nitrogen, and the remainder aliquoted into Qiagen AVL tubes according to manufacturer's directions, as described for blood plasma.

3.2. Nucleic Acid Isolation

3.2.1. Viral RNA

The Qiagen kit developed for RNA isolation from blood plasma can also be used with seminal plasma and takes advantage of the affinity of RNA for substrates such as silica in the presence of high concentrations of a chaotropic salt, such as guanidium-isothiocyanate (*see* **Note 2**). Proteins and lipids are extracted by high-salt washes containing ethanol. The purified RNA is eluted in sterile buffers, or water, ready for reverse transcription. Unused portions may be stored at –20°C for several weeks, or combined with one-tenth volume of 3 *M* sodium acetate plus two volumes of absolute ethanol and placed at –80°C for longer-term storage. Alcohol-precipitated RNA is recovered by removing an aliquot containing the desired number of RNA copies and centrifuging at 12,000*g* in a cold rotor, and the pellet is washed twice with cold 70% ethanol, dried, and resuspended in sterile water or in the Qiagen kit elution buffer. Alternatively, the RNA in an aliquot of the alcohol precipitate is recovered by reabsorption to the Qiagen viral RNA column following adjustment of salt and alcohol concentrations according to manufacturer's instructions (*see* **Note 3**).

3.2.2. Proviral DNA

The Qiagen kit developed for blood DNA extraction can also be used for semen cell DNA, according to manufacturer's instructions. Eluted DNA is ready for PCR amplification; storage considerations for unused portions are the same as for RNA. DNA concentration in the eluate is determined by fluorometry readings of bound Hoechst fluorescent dye; the Bio-Rad kit contains necessary controls and is reasonably priced. DNA concentration is expressed either in weight/volume (e.g., ng/mL) or in cell equivalents per volume, assuming 5 pg of DNA per diploid cell (*see* **Note 2**).

3.3. Reverse Transcription

3.3.1. Template Copy Number

Several considerations determine the input number of RNA copies for gene sequencing determinations. Although theoretically one RNA copy should be detectable by RT-PCR, input of 10 to 100 RNA copies generally ensures sufficient PCR products for gene sequencing. Under ideal conditions, in which the PCR amplifies each virus species uniformly, all species will be equally represented in the final reaction products. In general, however, PCR amplification is not uniform, and random PCR bias selects species early in the reaction that become overrepresented in the final product mixture. For this reason, it is important to run the reactions at least in duplicate. If only the most abundant viral species are of interest, which is generally the case for clinical decision-making, a higher input of RNA copies, up to 50,000, will not unduly bias the outcome. If the number of input RNA copies cannot be estimated from recent clinical lab measurements of RNA copy number per milliliter of plasma, the reactions should be run at two or three 5- to 10-fold dilutions of the RNA column elutant.

3.3.2. Reverse Transcription Reaction

The reverse transcription (RT) reaction is carried out in the same tube as the first round of the PCR. The total volume is 20 μL. In a PCR tube maintained at 0 to 5°C, 1 to 13 μL of target RNAs are combined with 0.8 μg of random hexamer primers, 50 U of RNase inhibitor, 2 μL of 10X buffer, 2 μL of 75 m*M* $MgCl_2$, and sterile, embryo-tested water to a total volume of 19 μL. The random primers are annealed to the template RNA by heating the tube to 72°C for five min, followed by cooling to room temperature (*see* **Note 4**). Ten to 20 U of AMV RT (*see* **Note 5**) are added, followed by incubation at 37–42°C overnight (*see* **Note 6**).

3.4. Polymerase Chain Reaction

3.4.1. Template Copy Number

One round of PCR under usual conditions can amplify 1000 targets to the 10^{10} to 10^{11} copies needed for ethidium bromide visualization of PCR products following electrophoresis through agarose or acrylamide gels *(5)*. Therefore, a minimum of 500 HIV particles, or 1000 infected cells, is needed to detect protease amplification by standard techniques with one round of PCR (*see* **Note 7**).

Alternatively, two rounds of PCR employing bracket-nested primers can theoretically amplify a single target to the detectable level of 10^{10} to 10^{11} copies. Two rounds of PCR introduce bias for amplification of the most abundant species, but this usually does not pose a problem for identification of clinically significant levels of virus with resistance mutations. A bracket-nested strategy has several advantages: (1) two to three orders of magnitude greater sensitivity, (2) the bracket PCR conditions employ a lower primer concentration, thus reducing nonspecific PCR background, and (3) increased tolerance to mutations within primer binding regions because the greater sensitivity compensates for reduced amplification efficiency secondary to primer mismatch.

Endpoint dilution of cells provides an estimate of the frequency of HIV-infected cells in circulation in the blood, or in an ejaculated semen specimen. Cells stabilized in aldehyde fixative offer many cytologic advantages ***(6,7)***, but the fixation process crosslinks proteins, making it difficult to isolate purified nucleic acids for PCR amplification. To improve the reliability of detection of small numbers (fewer than 10) HIV templates, a triple-bracket nested PCR strategy has been employed ***(5)*** with primers to the *gag* region. Cross-contamination with PCR products is controlled by liberal use of positively charged DEAE paper during all product transfer steps ***(5)***, and the use of disposable, positive displacement glass capillary tubes in place of handheld micropipettors for PCR product transfer (*see* **Note 8**).

3.4.2. PCR Conditions

To estimate viral burden, either viral RNA or proviral DNA, specimens undergo an initial triple bracket-nested analysis for *gag*. Endpoint dilution of cells provides an estimate of proviral burden; endpoint dilution of cDNA products provides an estimate of RNA burden. **Table 1** lists the gag primers, amplifying product sizes of 341, 144, and 105 base pairs, respectively.

3.4.2.1. Bracket PCR

The first round of PCR amplification of cDNAs is added directly to the RT reaction. The PCR mix contains 8 μL 10X buffer, 10 pmol each of forward and reverse bracket primers, 0.5 U *Taq* polymerase, sterile, embryo-tested water to total volume of 80 μL, and no magnesium. Sufficient magnesium for the PCR reaction is contained in the RT reaction. If a single-copy gene, such as β-globin, is co-amplified in the first round of PCR as an internal control, 10 pmol each of forward and reverse β-globin primers are included in the assay mix. If endpoint dilution with the RT reaction products is being carried out, the magnesium concentration in the PCR tubes must be adjusted to 1.5 m*M*.

The first round of PCR amplication of DNA is carried out essentially the same way, but the PCR mix is modified to a total volume of 100 μL including sample DNA volume; the mix contains 10 μL of 10X buffer, 10 pmol each of forward and reverse bracket primers (*see* **Note 9**), 1.25 U *Taq* polymerase, 6 μL 25 m*M* $MgCl_2$, and water to 100 μL, compensating for the volume of DNA sample, which is added last in a location separate from the PCR mix setup location.

Table 1
Primers for PCR Amplification of HIV-1 Genes

Primer	Orientation	Target Gene	Primer Sequence (5' to 3')	Location (HXB2 numbering)
H341-1	Forward	GAG	ttatcagaaggagccacccc	1312-1331
H341-2	Reverse		ccttgtcttatgtccagaatgc	1652-1631
H144-1	Forward		agtgggggacatcaagcagccatgcaaat	1359-1388
H144-2	Reverse		cctgctatgtcacttcccct	1502-1483
H105-1	Forward		gagactatcaatgaggaagc	1396-1415
H105-2	Reverse		tgctatgtcagttccccttggt	1500-1480
PRA-F	Forward	PROTEASE	cctaggaaaaagggctgttggaaatgtgg	2011-2039
RTA-R	Reverse		aacttctgtatgtcattgacagtcca	3328-3303
PRB-F	Forward		actgagagacaggctaattttttaggga	2068-2095
RTB-R	Reverse		catttatcaggatggagttcata	3265-3243
POL-1	Forward		attttcttcagagcagaccagag	2129-2151
NEW PF	Forward		gagcttcaggtctggggtagagac	2172-2195
NEW PR	Reverse		ctgtcaatggccattgtttaacttttgggc	2631-2602
PNY441-1	Forward		gagacaacaactccctctcag	2191-2211
P368-1	Forward		gaagcaggagccgatagacaaggaactg	2211-2238
P368-2	Reverse		actggtacagtctcaatagggctaatg	2578-2552
P365-1B	Forward		agacaaggaactgtatcctttagc	2326-2349
P365-2	Reverse		ggctttaattttactggtacag	2590-2569
MP-F	Forward		ccaaaaatgataggggggaattggagg	2382-2407
MP-R	Reverse	PROTEASE/REVERSE TRANSCRIPTASE	cctccaattcccctatcattttgg	2407-2382

RT NEST-1	Forward	REVERSE TRANSCRIPTASE	ggacataaagctataggtacag	2454-2474
RT NEST-2	Reverse		ctgccagttctagctctgcttc	3462-3441
GRANT RT-1	Forward		ttgggcctgaaaatccatacaatac	2698-2721
GRANT RT-2	Reverse		ctgtatgtcattgacagtccagc	3323-3301
P66-1	Forward		aaaattagtagatttcag	2765-2782
P66-2	Reverse		ccccatattactatgctttc	3699-3681
B268-1	Forward	BETA-GLOBIN	caacttcatccacgttcacc	N/A
B268-2	Reverse		gaagagccaaggacaggtac	N/A
B172-1	Forward		gtcatcacttagacctcacc	N/A
B172-2	Reverse		tggtgtctgtttgaggttgc	N/A

Primer sequence listed is the sequence ordered; reverse primers are the reverse complement of the sequence listed in italics.

To limit amplification of competing, nonspecific product, the first bracket PCR is limited to 20 heating-cooling cycles. The strategy is: two cycles at 98°C (2 min)/55°C (30 s)/72°C (30 s), followed by 18 cycles of 95°C (30 s)/55°C (30 s)/72°C (30 s).

3.4.2.2. First Nested PCR

The PCR assay mix contains 10 pmol of nested PCR primers in 1 µL, 5 µL of 10X assay buffer, 0.5 µL (2.5 U) of *Taq* polymerase, 1 µL of 25 m*M* $MgCl_2$, 37.5 µL sterile, embryo-tested water, and 2 µL of bracket PCR products transferred directly from the bracket PCR assay tube. The heating–cooling cycling strategy is 29 cycles of: 95°C (30 s)/55°C (30 s)/72°C (30 s), followed by 1 cycle of 95°C (30 s)/55°C (30 s)/72°C (7 min).

3.4.2.3. Second Nested PCR

The PCR assay mix contains 50 pmol of nested PCR primers in 5 µL, 5 µL of 10X assay buffer, 0.5 µL (2.5 U) of *Taq* polymerase, 1 µL of 25 m*M* $MgCl_2$, 29.5 µL sterile, embryo-tested water, and 2 µL of bracket PCR products transferred directly from the first nested PCR (*see* **Note 10**). The heating–cooling cycling strategy is the same as the first nested PCR.

3.4.3. Primer Strategies

Table 1 lists sets of bracket-nested primers employed for amplifying *gag* and *pol* regions. The *gag* primers are the triple bracket-nested strategy found to reliably amplify a conserved region of *gag* from all patients tested for over a decade. It should be noted that one primer in the second nested set (H105-2) substantially overlaps one of the primers in the first nested set (H144-2). This is a form of what is sometimes referred to as "PCR-anchoring" because the product formed will contain a sequence certain to be recognized by the subsequent set of primers.

Table 1 also lists sets of bracket-nested primers for amplification of reverse transcriptase and protease gene sequences. Multiple strategies are listed because patients with highly mutated virus may require testing with multiple sets of primers to achieve gene sequence amplification.

One strategy involves amplification of a region of *pol* that spans all of protease and the first two thirds of reverse transcriptase in which the known resistance-conferring mutations occur. One primer set used for this strategy is Pol-1 and P66-2 for the bracket PCR; this amplifies a 1870-base-pair region. To further amplify protease, primers P368-1 with P365-2 in the first nest, followed by p365-1B with P368-2 for the second nest. Each of the second nested primers overlaps slightly with the first nested set, thus providing the PCR anchor (*see* **Note 11**).

The initial portion of reverse transcriptase can also be amplified from the Pol-1/P66-2 bracket amplicon with primers PNY441-1, Nest RT-2 for the first nested reaction, and GRANT RT-1, GRANT RT-2 for the second nested reaction. Other combinations are possible if these fail to yield sufficient PCR products for sequencing.

To provide an internal control for the PCR and template conditions, a portion of a single copy gene, such as β-globin, may be coamplified. Some primer sets can be used together in a single, multiplex reaction. An example of such a reaction is combining β-globin primers B268-1 and B268-2 with H341-1 and H341-2 in the bracket reaction, followed by B172-1 and B172-2 with H144-1 and H144-2 in the first nested set. Two rounds of amplification of a single copy gene will yield sufficient PCR product for visualization on a gel.

3.5. Gel Electrophoresis

In this procedure, 2.3% (w/v) agarose is dissolved in 1X TBE buffer by heating in a microwave oven or on a hot plate. The dissolved agarose is poured into a gel tray fitted with a well-forming comb approximately one-half the depth of the gel thickness. The wells of the cooled gel are loaded with 10 μL PCR products combined with 3 μL of 6X sample loading buffer in individual wells of a 96-well plate. Electrophoretic resolution of PCR products is carried out in a gel bath filled with 1X TBE buffer attached to a power supply set at 50 mA, 120 V. The bromophenol blue dye migrates toward the anode slightly in front of leftover PCR primers (*see* **Note 12**).

DNA bands are visualized on an ultraviolet light gel box following ethidium bromide staining (10 min) in a bath containing 50 ng/mL ethidium bromide, followed by 20 min destaining in sterile, deionized water (*see* **Note 13**). The gel is photodocumented with a Polaroid camera attached to a fixed-distance hood; the resulting photo is scanned into a computer file (*see* **Note 14**).

3.6. Ligation and Cloning

PCR products are ligated directly into the vector supplied in the TOPO TA cloning kit. The kit vector takes advantage of the fact that *Taq* polymerase terminates DNA strands by adding thymidine-adenosine residues that overhang the opposite strand. Transformation of *E. coli* is performed according to kit instructions. The transformed bacteria are plated onto LB agar plates prepared with with ampicillin, 50 μg/mL. Just before plating the bacteria, the plates are coated with 10 μL of X-gal, a chromophore-substituted β-galactose that is hydrolyzed only by β-galactosidase, an enzyme encoded in the TOPO TA vector (*see* **Note 15**).

PCR products are ligated to the cloning vector according to kit instructions. Competent *E. coli* for the transformation reaction is also supplied in the kit.

The transformed *E. coli* cells are spread onto the ampicillin-X-gal plates and incubated overnight at 37°C (*see* **Note 16**).

Blue colonies contain the vector with ampicillin-resistance gene, but not integrated DNA products; white colonies contain the vector with ampicillin-resistance gene, and ligated PCR products that have interrupted the β-galactosidase-encoding sequences. Twenty white colonies are screened for HIV DNA sequences by adding a partial scraping of individual colonies with a sterile pipet tip to PCR assays set up as for the nested products that were ligated.

3.7. PCR Product Cleanup and Sequencing

PCR DNA products are prepared for sequencing by treating positive PCR assay mixtures with the Qiagen PCR cleanup column, according to manufacturer's directions. The ABI Prism 377XL automated sequencer is an example of recent-generation microsequencers that yield reliable sequence data from only 50 to 150 ng template. Approximately 4 ng of PCR amplicons of 400 base pairs is the threshold required for visualization by ethidium bromide; 40 to 400 ng yield a bright ethidium bromide band under most conditions, although staining intensity is not linear. Microsequencing technology employs *Taq* polymerase and generates reliable sequence data from 50 to 150 ng of template plus 2 to 5 pmol of one primer. Sequences derived from sequencing in both directions are more reliable than one-directional sequencing.

3.8. Sequence Editing

Routine gene sequencing requires a well thought-out strategy for storing and retrieving the sequences. Raw, unedited sequences should be stored in two ways, such as on the computer and on Zip or CD disks. Raw sequences are imported into Sequencher, sequence-manipulating software that supports editing of multiple sequences at the same time. Sequences with more than 3% ambiguous bases are not used for the determination of resustance-conferring mutations. Sequences are trimmed to the reading frame of the protein of interest and translated to identify amino acid substitutions.

3.9. Resistance Mutations to Inhibitors of HIV Protease

HIV-1 protease is a 99-amino-acid protein encoded by 297 bases at the 5' end of the *pol* gene (**Table 2a**). The reading frame begins at base 2354 according to the hxb2 numbering system. It can readily be sequenced in its entirety following PCR amplification.

Elegant crystallographic studies have determined the three-dimensional structure of wild-type HIV-1 protease *(8,9)*. It is an aspartic protease, the functional unit of which is a dimer, folded in mirror-image fashion to form an ac-

Table 2a
HIV-1 Protease Amino Sequence (Consensus B)[a]

1	2	3	4	5	6	7	8	9	10	11	12	13	14	15
P	Q	I	T	L	W	Q	R	P	L	V	T	I	K	I
16	17	18	19	20	21	22	23	24	25	26	27	28	29	**30**
G	G	Q	L	K	E	A	L	L	D	T	G	A	D	D
31	32	33	34	35	36	37	38	39	40	41	42	43	44	45
T	V	L	E	E	M	N	L	P	G	R	W	K	P	K
46	**47**	**48**	**49**	**50**	51	52	53	**54**	55	56	57	58	59	60
M	I	G	G	I	G	G	F	I	K	V	R	Q	Y	D
61	62	63	64	65	66	67	68	69	70	71	72	73	74	75
Q	I	L	I	E	I	C	G	H	K	A	I	G	T	V
76	77	78	79	80	81	**82**	83	**84**	85	86	87	**88**	89	**90**
L	V	G	P	T	P	V	N	I	I	G	R	N	L	L
91	92	93	94	95	96	97	98	99						
T	Q	I	G	C	T	L	N	F						

[a]Bold items are high- or intermediate-level resistance; non-bold are low level

tive site, Asp-Thr-Gly, at positions 25–27. The substrate cleft is located just above the active site, and a flexible flap region formed by amino acids 33–62 closes down upon substrate binding. The substrate cleft itself is formed by a loop-back of amino acid residues 78 to 90, stabilized by dimer crossover near residues 97 and 98 (*see* **Note 17**).

HIV-1 protease cleaves both *gag* and *gag-pol* polyproteins to produce capsid, nucleocapsid, protease itself, reverse transcriptase, and integrase, proteins essential to the infectivity of new virions. The seven FDA-approved protease inhibitors block binding of the polyproteins to the active site of protease, resulting in immature virions with uncleaved polyproteins that are incapable of carrying out successful infection.

Mutations that decrease the affinity of protease for drugs, but do not render the enzyme completely inactive, can escape pharmacologic control and produce infectious virions. Such mutated virions are less infectious than wild-type virus, but do promote HIV disease progression. Several approaches have been used to catalog resistance mutations and determine the degree of resistance they confer, including culturing virus with known mutations in the presence of drug, and characterizing protease gene sequences in patients who have failed protease inhibitor therapy ***(10,11)***. Taken together, the data are most consistent with **Table 2b**, which lists resistance mutations to the protease inhibitors.

Mutation at four amino acid residues—54, 82, 84, and 90—confers significant resistance to all inhibitors. Residue 54 is part of the flap, residues 82 and

Table 2b
Resistance-Conferring Mutations to Protease Inhibitors[a]

	HIV-1 Protease Amino Acids (Consensus B)													
	24	30	32	46	47	48	50	53	54	73	82	84	88	90
Protease Inhibitor	L	D	V	M	I	G	I	F	I	G	V	I	N	L
Amprenavir	–	–	**I**	**I,L**	**V**	V,L	**V**	–	**V,L,M,T,A**	–	**A,I,T,F,S,L**	**V**	–	**M**
Atazanavir	–	–	I	**I,L**	–	V,L	**L**	–	**V,L,M,T,A**	–	**A,I,T,F,S,L**	**V**	**D,S**	**M**
Indinavir	I	–	**I**	**I,L**	**V**	V,L	–	**L,Y,I**	**V,L,M,T,A**	S,C,T,A	**A,I,T,F,S,L**	**V**	D,S	**M**
Lopinavir (w/Ritonavir)	–	–	I	**I,L**	**V**	V,L	**V**	–	**V,L,M,T,A**	–	**A,I,T,F,S,L**	**V**	–	**M**
Nelfinavir	–	**N**	–	**I,L**	–	**V,L**	–	–	**V,L,M,T,A**	S,C,T,A	**A,I,T,F,S,L**	**V**	**D,S**	**M**
Ritonavir	–	–	**I**	**I,L**	**V**	V,L	**V**	**L,Y,I**	**V,L,M,T,A**	–	**A,I,T,F,S,L**	**V**	–	**M**
Saquinavir	–	–	–	–	–	**V,L**	–	**L,Y,I**	**V,L,M,T,A**	S,C,T,A	**A,I,T,F,S,L**	**V**	–	**M**

Letters indicate amino acid substitutions relative to wild-type reference virus, Consensus B.

[a]Bolded items represent high- or intermediate-level resistance.

Unshaded boxes represent low-level resistance.

Dots represent no resistance/no mutation reported.

84 are part of the substrate cleft, and residue 90 stabilizes the cleft. These mutations weaken affinity for the drug, while allowing binding and cleavage of the polyproteins, albeit with either slower kinetics, or less precision. Not surprising, in addition to these four, other mutations are more drug-specific, and multiple mutations are additive, leading to increased levels of drug resistance.

As indicated in **Table 2**, the mutation at amino acid 30 occurs only in patients taking indinavir and it does not confer resistance to any other protease inhibitor.

A more recent approach to detecting possible resistance-conferring mutations has been analysis of large databases of protease gene sequences from treated and untreated persons. Those analyses have agreed with the known resistance-conferring mutations and have identified additional amino acid substitutions at residues 11, 22, 23, 45, 58, 66, 74, 75, 76, 79, 83, and 85 that were highly conserved in persons not receiving protease inhibitor therapy. The impact of these 13 amino acid substitutions on overall drug resistance and disease progression has not been determined.

3.10. Resistance Mutations to Inhibitors of HIV-1 Reverse Transcriptase

The active form of HIV reverse transcriptase is a heterodimer of two proteins, the p51 subunit, 440 amino acids encoded at the start of *pol*, and the p66 subunit, the 560 amino acids encoded by the entire *pol* gene. The heterodimer catalyzes RNA-templated DNA synthesis and DNA-templated DNA synthesis, and also carries out degradation of the RNA strand of the RNA–DNA hybrid, an activity termed RNase H.

3.10.1. Nucleotide Reverse Transcriptase Inhibitors

In contrast to protease inhibitors that primarily block binding of the enzyme to its polypeptide substrate, one class of reverse transcriptase inhibitors do not actually inhibit the enzyme at all, but serve to terminate the nascent DNA chain being synthesized from the RNA template. The chain termination feature is accomplished by removing both alcohol groups from the ribose residue of nucleotide triphosphates, thus preventing the addition of the incoming nucleotide. The goal of the design of HIV RT inhibitors has been to design nucleotide triphosphates that are recognized and incorporated by HIV RT, but not by cellular DNA replicating enzymes.

Resistance mutations to the nucleotide RT inhibitors (NRTIs) result from two mechanisms: one is the removal of the inhibitor from the end of the nascent DNA strand to allow the strand to continue growing and the other lessens the affinity of RT for the drug, allowing the enzyme to rely on cellular stores of unmutated nucleotides for DNA strand synthesis. Either class of mutations leads to slower rates of DNA polymerization.

Amino acid substitutions at positions 41, 67, 70, 210, 215, and 219 promote removal of the dideoxynucleotide, thus allowing chain elongation to continue. Interestingly, the T215Y/F mutation has been shown to increase RT affinity for adenosine triphosphate, which is required for excision of the chain-terminating dideoxynucleotide. The other positions are more directly involved in binding the incoming nucleotide, and their mutation selects against the dideoxy-drug, resulting in the incorporation of more nucleotide triphosphates from cellular stores.

Amino acids 65, 69, 74, and 75 make important contacts with the incoming deoxynucleotide triphosphate during polymerization. Amino acid substitutions at these positions generally reduce the affinity for the dideoxy-drug and improve selection for nucleotide triphosphates from cellular stores. An insertion of either one or two amino acids at residue 69 is also found in treated patients, and also leads to discrimination against the drug.

Q151M substitution results from a two-base-pair change in a region associated with binding to the RNA template; interestingly, virus variants with Q151M rarely have the mutations that promote removal of the dideoxynucleotide. This substitution is also common in drug-resistant HIV-2.

A comparison of mutations from a number of patients failing nucleoside reverse transcriptase inhibitor therapy have revealed four patterns that result in resistance to most of the dideoxy-drugs: one is M184V in combination with substitutions at 44, 69, 75, and 118; a second is a dipeptide insertion at residue 69; a third is K65R, alone or in combination; and a fourth is Q151M in combination with V75I, F77L, and F116Y.

3.10.2. Non-Nucleotide Reverse Transcriptase Inhibitors

The three FDA-approved reverse transcriptase inhibitors sit in the hydrophobic polymerase binding groove. This groove is structurally less demanding than the nucleotide binding site and naturally occuring mutations in these locations are common among wild-type virus. HIV-2 virus isolates appear intrinsically resistant to this class of drugs.

K103N is the most common mutation in patients receiving NNRTIs, and it alone can cause significant resistance to all three, resulting in therapy failure. Other amino acid substitutions at K103, K103S, and K103R cause resistance, but not always therapy failure. The location of this amino acid at the entrance to the template binding pocket appears essential to how relaxed, or tightly closed, the pocket is. Tightly closed, the drug cannot gain access once the template is bound.

As listed in **Table 3**, mutations at residues 98, 100, 101, 108, 179, 181, 188, 190, 225, and 238 cause varying degrees of resistance to each of the NNRTIs, but not to all three together.

4. Notes

1. Blood can also be collected in a syringe and transferred directly to the centrifuge tubes into which EDTA has been added to equal a final concentration of 5 m*M*. Blood serum can also be used, but considerable loss of virus occurs during the clotting process. Heparinized blood is not recommended owing to the inhibitory effect of heparin on DNA polymerase activity.
2. Several methods for isolation of RNA and DNA for amplification and sequencing are appropriate for reverse transcription and PCR amplification. Storage up to several months of purified nucleic acids is possible in dilute buffers or sterile water; longer-term storage should include precipitation with one tenth volume of 3 *M* sodium acetate plus two volumes absolute ethanol and storage at –80°C.
3. The Qiagen Viral RNA isolation kit does not select against concomitant DNA isolation; DNA, including genomic DNA, is recovered with approximately the same efficiency as viral RNA.
4. 1st Strand cDNA Synthesis Kit for RT-PCR (AMV) (Roche) also has all the components for the RT step.
5. Many types and forms of reverse transcriptase are commercially available. Higher concentrations of enzyme (20 U or more per reaction) are needed for shorter reaction times, e.g., 1 h or less; lower concentrations of enzyme (5 U) require longer reaction times, e.g., overnight, and may be more reproducible.
6. Longer RT reaction times (3–4 h) at lower enzyme concentrations (10–20 U) may yield a higher percentage of longer cDNA products; the overnight incubation described here works well for collecting and processing specimens and viral RNA purification on the same day.
7. The number of molecules of *Taq* polymerase (approx 10^9 per unit) is rate-limiting for PCR because the number of enzyme molecules available determines the number of DNA copies synthesized with each round of heating and elongation.
8. Several companies manufacture micropipettors for use with disposable, filtered (aerosol resistant) pipet tips; Hamilton pipettors have a self-calibration feature.
9. Oligonucleotide primers have a long shelf-life if stored in water at –80°C, but do not tolerate multiple cycles of freezing and thawing. Aliquoting primers for single use before freezing eliminates problems introduced by freezing and thawing.
10. PCR product cross-contamination is a major problem for HIV gene sequencing. The cationic properties of DEAE chromatography paper allow for immediate adsorption of aerosolized PCR products. The DEAE paper is used as a mat during the transfer of bracket PCR products to nested reaction tubes, and small, single-use swatches of DEAE paper are used to open the caps of PCR product tubes. These precautions can reduce to zero PCR product crossover from bracket to nested reactions.
11. "PCR anchoring" ensures that nested PCR primers have a sequence identity to previously amplified product. One method is primer overlap, as described here; another method is to introduce an unrelated sequence of 10 or 11 bases at the 5' end of the primer. The unrelated sequences will be incorporated into PCR amplicons and can subsequently be recognized by either primers, or probes.

Table 3a
Reverse Transcriptase Amino Acid Sequence (Consensus B)

1	2	3	4	5	6	7	8	9	10	11	12	13	14	15	16	17	18	19	20	21	22	23
P	I	S	P	I	E	T	V	P	V	K	L	K	P	G	M	D	G	P	K	V	K	Q
24	25	26	27	28	29	30	31	32	33	34	35	36	37	38	39	40	41	42	43	44	45	46
W	P	L	T	E	E	K	I	K	A	L	V	E	I	C	T	E	M	E	K	E	G	K
47	48	49	50	51	52	53	54	55	56	57	58	59	60	61	62	63	64	65	66	67	68	69
I	S	K	I	G	P	E	N	P	Y	N	T	P	V	F	A	I	K	K	K	D	S	T
70	71	72	73	74	75	76	77	78	79	80	81	82	83	84	85	86	87	88	89	90	91	92
K	W	R	K	L	V	D	F	R	E	L	N	K	R	T	Q	D	F	W	E	V	Q	L
93	94	95	96	97	98	99	100	101	102	103	104	105	106	107	108	109	110	111	112	113	114	115
G	I	P	H	P	A	G	L	K	K	K	K	S	V	T	V	L	D	V	G	D	A	Y
116	117	118	119	120	121	122	123	124	125	126	127	128	129	130	131	132	133	134	135	136	137	138
F	S	V	P	L	D	K	D	F	R	K	Y	T	A	F	T	I	P	S	I	N	N	E
139	140	141	142	143	144	145	146	147	148	149	150	151	152	153	154	155	156	157	158	159	160	161
T	P	G	I	R	Y	Q	Y	N	V	L	P	Q	G	W	K	G	S	P	A	I	F	Q
162	163	164	165	166	167	168	169	170	171	172	173	174	175	176	177	178	179	180	181	182	183	184
S	M	T	K	I	L	E	P	F	R	K	Q	N	P	D	I	V	I	Y	Q	Y	M	D
185	186	187	188	189	190	191	192	193	194	195	196	197	198	199	200	201	202	203	204	205	206	207
D	D	L	Y	V	G	S	D	L	E	I	G	Q	H	R	T	K	I	E	E	L	R	Q
208	209	210	211	212	213	214	215	216	217	218	219	220	221	222	223	224	225	226	227	228	229	230
H	L	L	R	W	G	F	T	T	P	D	K	K	H	Q	K	E	P	P	F	L	W	M
231	232	233	234	235	236	237	238	239	240	241	242	243	244	245	246	247	248	249	250	251	252	253
G	Y	E	L	H	P	D	K	W	T	V	Q	P	I	V	L	P	E	K	D	S	W	T
254	255	256	257	258	259	260	261	262	263	264	265	266	267	268	269	270	271	272	273	274	275	276
V	N	D	I	Q	K	L	V	G	K	L	N	W	A	S	Q	I	Y	A	G	I	K	V

277	278	279	280	281	282	283	284	285	286	287	288	289	290	291	292	293	294	295	296	297	298	299
K	Q	L	C	K	L	L	R	G	T	K	A	L	T	E	V	I	P	L	T	E	E	A
300	301	302	303	304	305	306	307	308	309	310	311	312	313	314	315	316	317	318	319	320	321	322
E	L	E	L	A	E	N	R	E	I	L	K	E	P	V	H	G	V	Y	Y	D	P	S
323	324	325	326	327	328	329	330	331	332	333	334	335	336	337	338	339	340	341	342	343	344	345
K	D	L	I	A	E	I	Q	K	Q	G	Q	G	Q	W	T	Y	Q	I	Y	Q	E	P
346	347	348	349	350	351	352	353	354	355	356	357	358	359	360	361	362	363	364	365	366	367	368
F	K	N	L	K	T	G	K	Y	A	R	M	R	G	A	H	T	N	D	V	K	Q	L
369	370	371	372	373	374	375	376	377	378	379	380	381	382	383	384	385	386	387	388	389	390	391
T	E	A	V	Q	K	I	A	T	E	S	I	V	I	W	G	K	T	P	K	F	K	L
392	393	394	395	396	397	398	399	400	401	402	403	404	405	406	407	408	409	410	411	412	413	414
P	I	Q	K	E	T	W	E	A	W	W	T	E	Y	W	Q	A	T	W	I	P	E	W
415	416	417	418	419	420	421	422	423	424	425	426	427	428	429	430	431	432	433	434	435	436	437
E	F	V	N	T	P	P	L	V	K	L	W	Y	Q	L	E	K	E	P	I	V	G	A
438	439	440																				
E	T	F																				

Table 3b
Resistance-Conferring Mutations to Nucleotide Analog RT Inhibitors[a]

	HIV-1 Reverse Transcriptase Amino Acids (Consensus B)												
	#	65	67	69	70	74	75	##	151	##	210	215	219
NRTI	M	K	D	T	K	L	V	Y	Q	M	L	T	K
Abacavir	L	**R,N,E**	–	**D,N,S,ins,I,A**	–	V,I	–	F	**M,K**	–	W,F,V	Y,F,S,I,V,C,D,N	–
Azidothymidine	**L**	–	N,G,E	**D,N,S,ins,I,A**	R,I	–	–	–	**M,K**	–	**W,F,V**	**Y,F,S,I,V,C,D,N**	Q,E,N,R
Emtricitabine (FTC)	–	**R,N,E**	–	**D,N,S,ins,I,A**	–	–	–	–	M,K	**V,I**	–	–	–
Dideoxyinosine	L	**R,N,E**	–	**D,N,S,ins,I,A**	–	**V,I**	**I,M,T,A**	–	**M,K**	–	W,F,V	Y,F,S,I,V,C,D,N	–
Lamivudine (3TC)	–	**R,N,E**	–	**D,N,S,ins,I,A**	–	–	–	–	M,K	**V,I**	–	–	–
Stavudine (D4T)	**L**	–	N,G,E	**D,N,S,ins,I,A**	–	–	**I,M,T,A**	–	**M,K**	–	**W,F,V**	**Y,F,S,I,V,C,D,N**	Q,E,N,R
Tenofovir Disoproxil	L	**R,N,E**	–	**D,N,S,ins,I,A**	–	–	–	–	M,K	–	W,F,V	Y,F,S,I,V,C,D,N	–

Letters indicate amino acid substitutions relative to wild-type reference virus, Consensus B.
[a]Bolded items represent high- or intermediate-level resistance.
Unshaded boxes represent low-level resistance. Lines represent no resistance/no mutation reported.

Table 3c
Resistance-Conferring Mutations to Non-Nucleotide Reverse Transcriptase Inhibitors[a]

	HIV-1 Reverse Transcriptase Amino Acids (Consensus B)																
	98	#	100	101	103	106	##	##	##	188	190	225	##	##	236	238	318
Inhibitor	A	G	L	K	K	V	V	V	Y	Y	G	P	F	M	P	K	Y
Delavirdine	G	–	**I**	**E,Q,P,R**	**N,S,Q**	**A,M**	I	D,A	**C**	L	A,S,E	–	–	**L**	**L**	**R,T,N**	**F**
Efavirenz	G	–	**I**	**E,Q,P,R**	**N,S,Q**	A,M	I	D,A	**C**	**L**	**A,S,E**	H	–	**L**	–	R,T,N	F
Nevirapine	G	–	I	**E,Q,P,R**	**N,S,Q**	**A,M**	I	D,A	**C**	**L**	**A,S,E**	H	L	**L**	–	**R,T,N**	F

Letters indicate amino acid substitutions relative to wild-type reference virus, consensus B.
[a]Bolded items represent high- or intermediate-level resistance.
Unshaded boxes represent low-level resistance. Lines represent no resistance/no mutation reported.

12. Many reliable gel electrophoresis systems are available. Tris-borate-EDTA buffers are standard because the borate and EDTA inhibit bacterial growth, promoting stability of electrophoresis baths. Other buffers in the pH range of 8.2 to 8.6 also reliably resolve PCR products.
13. Ethidium bromide intercalated into DNA products fluoresces when exposed to ultraviolet light. Approximately 10^{10} to 10^{11} copies of DNA are the threshold of detection by ethidium-bromide-stained gels exposed to a standard UV light-box. Ethidium bromide is a mutagen and should be handled with extreme caution. Incorporating it into the agarose gel allows the detection of DNA products during the process of electrophoresis, but is accompanied by a markedly increased background that may mask faint bands. Staining the DNA bands by submerging the gel in a bath of ethidium bromide at the end of the electrophoresis, followed by a destaining step in sterile distilled water, is more sensitive.
14. Several gel photodocumentation systems have been developed to provide automatic computerized gel records for archiving and editing purposes. Neither the video-camera nor digital-camera systems has the resolution of Polaroid film. The availability of low-cost optical scanners has streamlined computer archiving and editing of high resolution black-and-white photos developed from low-cost Polaroid cameras fitted to fixed-distance shields.
15. The TOPO TA cloning kit takes advantage of the fact that *Taq* polymerase adds thymidine and adenosine residues to the ends of PCR products that overhang the opposite strand. T-A residues are not stable, however, so the ligation reactions needs to occur within a few hours of the PCR reaction. Ligation products are stable at –20°C for up to 24 h.
16. Bacterial plates with transformants should be removed from 37°C within 18 h to keep the colonies from overgrowing and touching each other. Following a brief equilibration at 37°C, the plates can be incubated at 32°C for up to 24 h, or at room temperature for 2 d.
17. Schematics of the crystallographic structure of HIV-1 protease are available at the University of San Francisco HIV website.

References

1. Gu, Y. and Sundquist, W. I. (2003) Good to CU. *Nature* **424,** 21–22.
2. Vartanian, J. P., Meyerhans, A., Asjo, B., and Wain-Hobson, S. (1991) Selection, recombination, and G-A hypermutation of human immunodeficiency virus type 1 genomes. *J. Virol.* **65,** 1779–1788.
3. Kepler, T. and Perelson, A. (1998) Drug concentration heterogeneity facilitates the evolution of drug resistance. *Proc. Natl. Acad. Sci. USA* **95,** 11,514–11,519.
4. Eyre, R., Zhung, G., and Kiessling, A. (2000) Multiple drug resistance mutations in human immunodeficiency virus in semen but not blood of a man on antiretroviral therapy. *Urology* **55,** 591.
5. Fitzgerald, L. M., Yin, H. Z., and Kiessling, A. A. (1993) PCR amplification of HIV and cellular DNA sequences in formaldehyde-fixed, immunoreactive white blood cells. *Biotechniques* **15,** 128–133.

6. Kiessling, A. A., Crowell, R. C., Brettler, D., Forsberg, A., and Wolf, B. (1993) Human immunodeficiency virus detection and differential leukocyte counts are accurate and safer with formaldehyde-fixed blood [letter]. *Blood* **81,** 864–865.
7. Kiessling, A. A., Yin, H. Z., Purohit, A., Kowal, M., and Wolf, B. (1993) Formaldehyde-fixed semen is suitable and safer for leukocyte detection and DNA amplification. *Fertil. Steril.* **60,** 576–581.
8. Baldwin, E., Bhat, T., Gulnik, S., Liu, B., Topol, I., Kiso, Y., et al. (1995) Structure of HIV-1 protease with KNI-272, a tight-binding transition-state analog containing allophenylnorstatine. *Structure* **3,** 581–590.
9. Chen, Z., Li, Y., Schock, H., Hall, D., Chen, E., and Kuo, L. (1995) Three dimensional structure of a mutant HIV-1 protease displaying cross-resistance to all protease inhibitors in clinical trials. *J. Biol. Chem.* **270,** 21,433–21,436.
10. Condra, J., Petropoulos, C., Ziermann, R., Schleif, W., Shivaprakash, M., and Emini, E. (2000) Drug resistance and predicted virologic responses to HIV-1 protease inhibitor therapy. *J. Infec. Dis.* **182,** 758–765.
11. Molla, A., Korneyeva, M., Gao, Q., Vasavanonda, S., Schipper, P., Ho, H., et al. (1996) Ordered accumulation of mutations in HIV protease confers resistance to ritonavir. *Nat. Med.* **2,** 760–766.

II

Molecular Biology Methods

24

Determination of Cell Tropism of HIV-1

Neeltje A. Kootstra and Hanneke Schuitemaker

Summary

With the discovery that changes in the biological properties of HIV-1 correlate with the progression to disease, it became more and more important to develop assays to distinguish between the viral phenotypes. In this chapter, it is described how the biological phenotype of HIV-1 with regard to cellular tropism can be determined on primary monocyte-derived macrophages, established T-cell lines: MT_2, $SupT_1$, and H9, and promonocytic cell lines: U937, HL-60, and THP-1.

Key Words: HIV-1 tropism; monocyte; macrophage; cell line; T-cell; promonocytic.

1. Introduction

The relevance of changes in the cellular tropism of primary HIV-1 variants during the course of infection has long been demonstrated *(1,2)*.

Macrophage-tropic (M-tropic) HIV-1 variants can be found in all stages of HIV-1 infection but predominate in the asymptomatic phase *(3)*. These virus variants use CCR5 as a coreceptor and are unable to replicate in T-cell lines and promonocytic cell lines.

During the course of infection, HIV-1 variants become more T-cell tropic and lose the ability to replicate in macrophages *(3)*. Some of these viruses are still able to enter macrophages efficiently; however, virus replication is restricted at an early step in reverse transcription *(4)*. In 50% of the HIV-1-infected individuals that progress to AIDS, HIV-1 variants appear that are able to use the alternative coreceptor CXCR4 *(5)*. CXCR4-using HIV-1 variants (X4 HIV-1) are able to replicate efficiently in T-cell lines and promonocytic cell lines, whereas M-tropism is rarely observed among these variants.

This chapter will focus on how the cellular tropism of primary HIV-1 can be determined on primary macrophages and established cell lines.

From: *Methods in Molecular Biology, Vol. 304: Human Retrovirus Protocols: Virology and Molecular Biology*
Edited by: T. Zhu © Humana Press Inc., Totowa, NJ

2. Materials

1. MDM medium: Iscove's modified Dulbecco's medium (IMDM; BioWhittaker) supplemented with 10% of human pooled serum (*see* **Note 1**), 100 U/mL penicillin and 100 U/mL streptomycin (GibcoBRL).
2. Phosphate-buffered saline (PBS; BioWhittaker).
3. EDTA: 10 m*M* EDTA in PBS.
4. Elutriation medium: PBS supplemented with 0.5% bovine serum albumin (BSA) and 1.3 m*M* sodium citrate.
5. Avanti J25 centrifuge (Beckman Coulter).
6. JE-6B elutriation rotor (Beckman).
7. 70% Ethanol.
8. Granulocyte macrophage colony-stimulating factor (GM-CSF).
9. Macrophage colony-stimulating factor (M-CSF).
10. Recombinant interleukin (IL)-3.
11. 175-cm^2 tissue culture flask.
12. 24-well tissue culture plates.
13. DNase (Promega).
14. 3' azido-3' deoxythimymidine (AZT; Sigma).
15. L6-lysis buffer: 120 g guanidine isothiocyanate dissolved in 100 mL 0.1 *M* Tris-HCl, pH 6.4, supplemented with 22 mL 0.2 *M* EDTA, pH 8.0, and 2.6 g Triton X-100.
16. Isopropanol.
17. LightCycler (Roche Diagnostics GmbH).
18. LightCycler FastStart DNA Master SYBR Green I kit (Roche Diagnostics GmbH).
19. Sterile 50-mL tubes.
20. Sterile pipets.
21. T-cell lines: Sup T_1, MT_2, and H9.
22. Promonocytic cell lines: U937, HL-60 and THP-1.
23. Cell-line medium: IMDM supplemented with 10% of fetal calf serum (FCS; Hyclone), 100 U/mL penicillin, and 100 U/mL streptomycin (GibcoBrL).
24. 25-cm^2 tissue culture flasks.

3. Methods

The methods here describe how the cellular tropism of primary HIV-1 for monocyte derived macrophages, established T-cell lines, and promonocytic cell lines can be determined.

3.1. Macrophage Tropism of HIV-1

Here we describe monocyte isolation, culturing monocyte-derived macrophages (MDM), and how M-tropism of HIV-1 can be determined (*see* **Note 2**).

3.1.1. Isolation of Monocytes From Peripheral Blood

Monocytes can be isolated from the peripheral blood mononuclear cell (PBMC) fraction by different methods. Two methods of isolation are described: adherence to plastic and centrifugal elutriation.

3.1.1.1. Monocyte Isolation by Adherence

Prepare PBMC from buffy coats as described in Chapter 1 and resuspend the PBMC after the final wash in MDM medium at a final concentration of 10 × 10^6 cells/mL. Transfer the cells to a tissue culture flask (35–50 mL in a 175-cm^2 flask) and place the cells in an incubator at 37°C for 1–2 h. Remove the nonadherent cells and wash the adherent cell layer two times with PBS. To detach the monocytes from the flask, add 10 m*M* EDTA in PBS to the cells and incubate for 10 min at room temperature. Detach the monocytes from the surface using a disposable cell scraper. Wash the cells and resuspend them in MDM medium, at a density of 0.5–1 × 10^6/mL. Culture the monocytes in 24-well tissue culture plates at 1 mL per well in an incubator with a humidified atmosphere and 5% CO_2.

3.1.1.2. Centrifugal Elutriation

Isolate PBMC from a buffy coat as described in Chapter 1 (*see* **Note 3**) and resuspend the cells in elutriation medium at a density of 20–50 × 10^6 cells/mL. Assemble the elutriation rotor according to the manufacturer's instructions and sterilize the rotor by rinsing with 70% ethanol. Subsequently, rinse the rotor with PBS and elutriation medium for 5 min at maximum flow rate (40 mL/min). Run the elutriator at 1200*g* at 20°C at a constant flow rate of 20 mL/min. Inject the PBMC suspension into the system and let the cells run into the separation chamber at 1200*g*. At this speed all the cells will remain in the chamber and only thrombocytes will flow out of the elutriator. To remove all thrombocytes from the cell suspension in the chamber, lower the centrifugation speed to 1100*g* and let it run for 5 min. For collection of the lymphocyte fraction, lower the centrifugation speed slowly to 800*g* (no brake) and collect the cells for 10 min. For isolation of an ultrapure monocyte fraction, a good separation between the lymphocyte and monocyte cell fractions should be obtained. Lower the centrifugation speed gradually with steps of 50*g* and analyze the lymphocyte/monocyte ratio in the flow-through by a cell counter. When the ratio is around 1:2 in favor of the monocytes, the monocyte fraction can be harvested. The shift in the lymphocyte/monocyte ratio occurs usually at a centrifugation speed between 700 and 500*g*, depending on the donor. Lower the centrifugation speed to 100*g* and collect the monocyte fraction for 5 min. Check the purity of the monocyte fraction by a cell counter. In general this fraction will contain more than 95% monocytes. Clean the elutriator by rinsing with water for 15 min and disassemble the rotor.

Pellet the monocytes by centrifugation (10 min at 425*g*) and resuspend the cells in MDM medium at a density of 0.5–1 × 10^6/mL. Plate the cells in 24-well tissue culture plates at 1 mL per well and place them in an incubator with a humidified atmosphere and 5% CO_2.

3.1.2. Culturing of Monocyte-Derived Macrophages

In the first 2 d after the onset of the culture, the monocytes will spontaneously adhere to the culture plates and differentiation into MDM is initiated. During differentiation, the cells expand in size and multinucleated giant cells are observed after 2 wk of culture.

3.1.2.1. Culture Conditions

During the first 2 wk of the cultures, the medium will acidify rapidly. MDM are relatively resistant to the low pH, but depend on factors secreted into the medium. Therefore, it is recommended to replace the medium only once a week or to replace only part of the medium. MDM cultures can be maintained under these conditions for 4–6 wk.

3.1.2.2. Cytokine Treatment

Although cytokine treatment is not essential to obtain MDM, addition of GM-CSF, M-CSF, or IL-3 has a beneficial effect on cell survival and enhances HIV-1 infection (*see* **Note 4**). Add the cytokines either alone or in combination at the onset of the culture at the following concentrations: 100 U/mL GM-CSF; 100 ng/mL M-CSF; 10 ng/mL IL-3. Cytokines should be present during the first week of culture. Addition after differentiation into MDM has no effect on cell survival and HIV-1 infection.

3.1.3. Determining Macrophage Tropism of HIV-1

M-tropism can be analyzed by two different methods: (1) analysis of the process of reverse transcription or (2) analysis of virus replication. In the first method, M-tropism of HIV-1 variants is analyzed by real-time polymerase chain reaction (PCR) detecting different steps in the reverse transcription process. The advantage of this method is that it allows screening of large panels of HIV-1 variants on multiple MDM donors in a relatively short time. In the second method, M-tropism of HIV-1 is determined by analysis of virus replication in MDM by p24 enzyme-linked immunosorbent assay (ELISA). This method is less accurate and more time-consuming, but it will allow selection and isolation of highly M-tropic HIV-1 variants.

3.1.3.1. Infection of Monocyte-Derived Macrophages

For infection of MDM, high-titer virus stocks should be generated in PHA-stimulated PBMC and titers should be determined by $TCID_{50}$ as described in Chapter 1. Treat the virus stocks with DNase (200 ng/mL) for 45 min and filter it through a 0.22-μm filter, to remove any cell debris and contaminating proviral DNA. Inoculate the MDM at d 5 after the onset of the cultures with an inoculum of 1×10^3 to 1×10^4 $TCID_{50}$.

When PCR is used to analyze M-tropism of HIV-1, it is important to ensure that the inoculum is free of contaminating proviral DNA. Therefore it is recommended to inoculate a duplicate culture in the presence of 3'azido-3'deoxythymidine (AZT; 10 µ*M*). Add the AZT 30 min prior to inoculation.

3.1.3.2. Analysis of Virus Entry and Reverse Transcription of HIV-1 in Monocyte-Derived Macrophages by Real-Time PCR

3.1.3.2.1. DNA Extraction

Remove the culture medium 48 h after inoculation and lyse the cells in 0.9 mL L6 lysis buffer. Transfer the lysate into 2-mL Eppendorf tubes and precipitate the DNA with 1 mL isopropanol. Pellet the DNA by centrifugation (20 min at 14,000*g*) and wash twice with 70% ethanol. Air-dry the DNA for 30 min at room temperature and dissolve the pellet in 100 µL aqua dest.

3.1.3.2.2. Real-Time PCR

To analyze the process of reverse transcription, a real-time PCR utilizing the LightCycler and the LightCycler FastStart DNA Master SYBR Green I kit is used. The primer pairs used to detect early, intermediate, and late products of the reverse transcription process are described in **Table 1**.

Prepare the LightCycler SYBR Green master mix as follows: 2 µL LightCycler FastStart DNA Master SYBR Green I, 0.2 µ*M* of each primer, 3–4 m*M* $MgCl_2$, and add H_2O to a final volume of 18 µL. Pipet 18 µL of the master mix in precooled LightCycler capillaries and add 2 µL DNA (100–300 ng). Close the capillaries and place them in the LightCycler. Run the following programs: (1) preincubation and denaturation: 10 min at 95°C; (2) amplification and quantification: 45 cycles of 10 s at 95°C, 5 s at 55°C, 10 s at 72°C, with the acquisition mode set on single; (3) melting curves: 0 s at 95°C, 15 min at 65°C, 0 s at 95°C with a temperature transition rate of 0.1°C/s and the acquisition mode set on continuous; (4) cool-down: 30 s at 40°C. It is essential to run the melting curve analysis to assess the specificity of the PCR product.

As a standard curve for the HIV-1 proviral DNA content, a serial dilution of the 8E5 cell line, which contains 1 HIV-1 proviral DNA copy per cell in carrier DNA, can be used.

3.1.3.3. HIV-1 Replication in Monocyte-Derived Macrophages

Remove the unabsorbed virus 48 h after inoculation and add 1 mL fresh culture medium. Maintain the cultures for 3 wk and replace the medium once a week as described previously. Twice a week, 100 µL of culture supernatant should be taken and placed at –20°C. Analyze the culture supernatant for the presence of p24 by a p24 antigen capture ELISA.

Table 1
Primers for Analysis of Process of Reverse Transcription by Real-Time PCR

RT product	Primer	Sequence (5'–3') (HxB2)	Target sequence	Reference	$MgCl_2$	Product size (bp)
Early	M667	GGCTAACTAGGGAACCCACTG	42–63	***6***	4 m*M*	140
	AA55	CTGCTAGAGATTTTCCACACTGAC	182–159	***6***		
Intermediate	Pol-E	GATTTAACCTGCCACCTGTAGTAGC	3848–3957	***7***	3 m*M*	129
	Pol-B	ATGTGTACAATCTAGTTGCC	3977–3957	***7***		
Late	M667	GGCTAACTAGGGAACCCACTG	42–63	***6***	3 m*M*	204
	M532	GAGTCCTGCGTCGAGAGAGC	246–176	***8***		

3.2. HIV-1 Infection of Permanent Cell Lines

This section describes HIV-1 infection of the T-cell lines MT_2, $supT_1$, and H9, and promonocytic cell lines U937, HL60, and THP-1.

3.2.1. MT_2 Cell Line

All X4 HIV-1 variants are able to infect the MT_2 cell line efficiently (*see* **Note 5**); therefore this cell line is used frequently to screen for these variants.

3.2.1.1. Infection of MT_2 Cells

The MT_2 cell line can be inoculated either by a cell-free virus stock or by cocultivation with HIV-1-infected PHA-PBL. Dilute MT_2 cells with cell line medium to a density of 6×10^6 cells/mL. Add 0.5 mL of MT_2 cells to 1 mL virus stock or 1×10^6 HIV-1-infected PHA-PBL. Incubate the cells for 2 h at 37°C in a shaking water bath. Add 4 mL cell line medium and transfer the cells into 25-cm^2 tissue culture flasks. Place the cultures in an incubator at 37°C. Replace 4 mL of cell culture with fresh cell line medium twice a week, and maintain the cultures for 3 wk.

3.2.1.2. Virus Replication and Syncytia Formation

HIV-1 replication in the MT_2 cell line is accompanied by formation of large syncytia, which will result in rapid cell death. Analyze the MT_2 cultures for the presence of syncytia, and take culture supernatant samples for p24 capture ELISA twice a week.

3.2.2. Other T-Cell Lines

Other T-cell lines used to determine the tropism of HIV-1 are $SupT_1$ and H9. The procedures for infection of these cell lines are similar to infection of the MT_2 cell line (*see* **Subheading 3.2.1.**). $SupT_1$ cells are susceptible to all X4 HIV-1 variants, and some R5 variants are able to replicate in this cell line as well. Infection of H9 cells is more stringent as compared to $SupT_1$ and MT_2, although the majority of the X4 variants are able to replicate in these cells.

3.2.3. Promonocytic Cell Lines

X4 HIV-1 variants that are able to replicate in established T-cell lines, also replicate in promonocytic cell lines like U937, HL-60, and THP-1. Notably, M-tropic HIV-1 variants do not replicate in these cell lines. The procedure to determine the tropism for promonocytic cell lines is identical to the protocol for HIV-1 infection of the MT_2 cell line (*see* **Subheading 3.2.1.**).

4. Notes

1. *Human pooled serum:* MDM are cultured in medium supplemented with HPS. Batches of HPS are made by pooling serum from at least six donors and batch-to-batch variation has been observed. Some HPS batches only allow very low HIV-1 replication in MDM; therefore, batches should be tested before use. Alternatively, medium supplemented with 10% FCS and 100 ng/mL M-CSF can be used to culture MDM.
2. *Donor variability:* MDM obtained from different donors differ in their susceptibility to HIV-1 infection. Culture conditions and cellular factors that account for these differences are yet unknown. Therefore it is recommended to determine M-tropism of HIV-1 variants on at least three different donors.
3. *Prevention of cell loss during monocyte isolation:* Monocytes adhere easily to plastic and this might lead to cell loss during isolation and purification. This can be prevented by the use of polypropylene tubes instead of polystyrene tubes. Furthermore, it is recommended to use cold PBS supplemented with 0.13 *M* sodium citrate to wash the cells.
4. *Cytokine treatment and HIV-1 infection of MDM:* MDM are susceptible only during a certain differentiation stage. The optimal time point of infection is donor-dependent and is usually between day 5 and day 10 after the onset of the culture. Addition of cytokines to the cultures will influence differentiation and therefore the susceptibility to HIV-1 infections. GM-CSF, M-CSF, and IL-3 treatment will enhance HIV-1 infection, whereas IL-4 and IL-10 treatment reduces the susceptibility to HIV-1.
5. *HIV-1 replication in T-cell lines and cytokine treatment:* T-cell lines described in this chapter are differentially susceptible to HIV-1 infection. The $SupT_1$ cell line supports replication of all X4 variants and some of the R5 variants, whereas the MT_2 cell line can be infected only by X4 variants. The H9 cell line, however, supports replication of only a fraction of the X4 variants. Virus replication in this cell line is sometimes hard to determine. To support virus replication in the H9 cell line, virus replication can be stimulated by the addition of cytokines such as IL-2, IL-6, and/or TNFα.

References

1. Tersmette, M., Gruters, R. A., De Wolf, F., De Goede, R. E. Y., Lange, J. M. A., Schellekens, P. T. A., et al. (1989) Evidence for a role of virulent human immunodeficiency virus (HIV) variants in the pathogenesis of acquired immunodeficiency syndrome: studies on sequential HIV isolates. *J. Virol.* **63,** 2118–2125.
2. Asjo, B., Albert, J., Karlsson, A., Morfeldt-Månson, L., Biberfeld, G., Lidman, K., and Fenyö, E. M. (1986) Replicative capacity of human immunodeficiency virus from patients with varying severity of HIV infection. *Lancet* **2,** 660–662.
3. Schuitemaker, H., Koot, M., Kootstra, N. A., Dercksen, M. W., De Goede, R. E. Y., Van Steenwijk, R. P., et al. (1992) Biological phenotype of human immunodeficiency virus type 1 clones at different stages of infection: progression of disease is associated with a shift from monocytotropic to T-cell-tropic virus populations. *J. Virol.* **66,** 1354–1360.

4. Fouchier, R. A. M., Brouwer, M., Kootstra, N. A., Huisman, J. G., and Schuitemaker, H. (1994) HIV-1 macrophage-tropism is determined at multiple steps of the viral replication cycle. *J. Clin. Invest.* **94,** 1806–1814.
5. Koot, M., Keet, I. P. M., Vos, A. H. V., De Goede, R. E. Y., Roos, M. T. L., Coutinho, R. A., et al. (1993) Prognostic value of human immunodeficiency virus type 1 biological phenotype for rate of $CD4^+$ cell depletion and progression to AIDS. *Ann. Intern. Med.* **118,** 681–688.
6. Zack, J. A., Arrigo, S. J., Weitsman, S. R., Go, A. S., Haislip, A., and Chen, I. S. Y. (1990) HIV-1 entry into quiescent primary lymphocytes: molecular analysis reveals a labile, latent viral structure. *Cell* **61,** 213–222.
7. Bruisten, S. M., Koppelman, M. H. G. M., Van der Poel, C. L., and Huisman, J. G. (1991) Enhanced detection of HIV-1 sequences using PCR and a liquid hybridization technique. *Vox Sang.* **61,** 24–29.
8. Butler, S. L., Hansen, M. S., and Bushman, F. D. (2001) A quantitative assay for HIV DNA integration in vivo. *Nat. Med.* **7,** 631–634.

25

Determination of Co-Receptor Usage of HIV-1

Hanneke Schuitemaker and Neeltje A. Kootstra

Summary

In addition to CD4, HIV-1 uses chemokine receptors for entry in their target cells. The most important chemokine receptors in this respect are β-chemokine receptor 5 (CCR5) and α-chemokine receptor 4 (CXCR4). Coreceptor usage is an important feature of the biological phenotype of HIV-1 variants. In this chapter, methods are described to determine the co-receptor usage of HIV-1 variants.

Key Words: Co-receptor; U87 cell line; CCR5; CXCR4; MT_2 cell line.

1. Introduction

Already in the early days of HIV-1 research, the existence of HIV-1 variants with different biological properties was recognized *(1–3)*. HIV variants that lacked the capacity to induce syncytia, and that were unable to replicate in permanent T-cell lines were called nonsyncytium-inducing (NSI). The presence of these virus variants in an individual was associated with a slow decline of $CD4^+$ T cells *(4)*. HIV-1 variants with the capacity to induce syncytia could infect permanent T-cell lines *(5)*. These virus variants were called syncytium-inducing (SI). The NSI/SI nomenclature almost completely overlapped with the slow/low and rapid/high nomenclature. The underlying mechanism for the differential T-cell line tropism was resolved when the coreceptors for HIV were identified that form the cellular entry complex for HIV-1 in addition to CD4 *(6–9)*. β-Chemokine receptor 5 is the co-receptor for NSI, slow/low HIV-1 variants; these variants are now indicated as R5 HIV-1 *(10)*. α-Chemokine receptor 4 (CXCR4) is the coreceptor for SI, rapid/high HIV-1 variants; these are now named X4 variants. Here, the currently used methods to determine coreceptor usage of HIV-1 variants are provided.

From: *Methods in Molecular Biology, Vol. 304: Human Retrovirus Protocols: Virology and Molecular Biology*
Edited by: T. Zhu © Humana Press Inc., Totowa, NJ

2. Materials

1. Cell line medium: Iscove's modified Dulbecco's medium (IMDM), supplemented with 10% fetal calf serum (FCS), 100 U/mL penicillin, 100 U/mL streptomycin.
2. 96-Well tissue culture microtiter plates.
3. 24-Well tissue culture microtiter plates.
4. 25-cm^2 Tissue culture flasks.
5. Sterile 10-mL pipets.
6. 0.2% Triton X-100 solution in PBS.
7. U87 cells transfected with human CD4.
8. U87-CD4 selection medium (cell line medium supplemented with G418 200 μg/mL).
9. U87 cells transfected with human CCR2.
10. U87 cells transfected with human CCR3.
11. U87 cells transfected with human CCR5.
12. U87 cells transfected with human CXCR4.
13. U87-co-receptor selection medium (cell line medium supplemented with 1 μg/mL puromycin).
14. Antibody 12G5 directed against CXCR4.
15. Antibody 2D7 directed against CCR5.
16. Antibody 7B11 directed against CCR3.
17. Fluorescein isothiocyanate-conjugated goat anti-mouse IgG.
18. PE conjugated anti-CD4 antibody.
19. MT_2 cells.
20. PBMC from a donor with a CCR5 Δ32 homozygous genotype.
21. P24 antigen capture enzyme-linked immunosorbent assay (ELISA).
22. Fluorescence-activated cell sorter (FACS).
23. FACS staining buffer (PBS supplemented with 1% BSA).

3. Methods

The capacity of HIV-1 variants to replicate in cell lines that specifically express a certain coreceptor reflects the capacity of the virus variant to use that coreceptor. For this purpose, U87 cells have been transfected with different coreceptor genes. In addition, permanent T-cell lines express CXCR4 and not CCR5. Replication in these T-cell lines reflects the ability to use CXCR4 (*see* **Note 1**). Conversely, the capacity of an HIV variant to replicate in PBMC from a donor who is homozygous for a 32-base-pair deletion in the CCR5 gene indicates that HIV-1 variant replicates independent of CCR5.

3.1. Determination of Co-Receptor Usage on Stably Transfected U87 Cell Lines

For the determination of coreceptor usage, a panel of transfected U87 cell lines is available from the National Institutes of Health AIDS reagent program.

These U87 cell lines are transfected with human CD4, alone or in combination with human CCR2, CCR3, CCR5, or CXCR4. Selection pressure for maintaining the co-receptor-encoding gene in the U87 cells is provided by the presence of puromycin (1 μg/mL) in the culture medium. Selection for CD4 expression is regularly performed by selection with G418 (200 μg/mL). Cell surface expression of coreceptors and CD4 should be routinely controlled by FACS analysis. For this purpose, a CD4 directed PerCP conjugated antibody is available (BD). In addition, CCR5 directed FITC conjugated antibody 2D7 and CXCR4-directed PE-conjugated antibody 12G5 are available (Pharmingen). For FACS analysis, 2×10^5 U87 cells are incubated with saturating amounts of the relevant antibody for 20 min at 4°C. Cells are then washed with FACS staining buffer and finally resuspended in 200 μL of the same buffer. For each sample, 10^4 cells should be analyzed on the FACS. As a negative control for co-receptor expression, the U87 cell line transfected with only CD4 can be used. As a negative control for the FACS analysis of CD4 expression, the untransfected U87 cell line can be used.

The capacity of an HIV variant to use a co-receptor can be determined by cell-free inoculation of the U87 cell lines with that variant and subsequent monitoring of virus production. U87 cells are seeded in 96-well plates, 10^4 cells per well. After 24 h, medium is removed and cells are inoculated with a minimal inoculum of 10^2 tissue culture infectious dose as established on phytohemagglutinin (PHA)-activated peripheral blood mononuclear cells (PBMC), in a maximum volume of 200 μL. After 24 h, the inoculum is removed and cells are washed with PBS, which is prewarmed to 37°C. Subsequently, 200 mL fresh U87-CD4-coreceptor medium is added. At day 7 after inoculation, 50 μL of culture supernatant is harvested for analysis of HIV production in a p24 antigen capture ELISA. U87 cells are detached by trypsinization and cells from a single well of the original 96-well plates are transferred to a single well of a 24-well plate. At days 10 and 14 after inoculation, 50 μL culture supernatant is harvested for monitoring of virus production.

3.2. Determination of CXCR4 Usage on the MT_2 Cell Line

Already before the discovery of CXCR4 as a coreceptor for SI HIV-1 variants, the MT_2 cell line was used for the identification of SI HIV-1 variants. When an HIV-1 variant has the capacity to replicate in the MT_2 cell line, large syncytia are formed, and ultimately the cells in the culture die as a consequence of the HIV-induced cytopathic effect. We now know that MT_2 cells express CD4 and CXCR4 and only X4 HIV-1 variants replicate in MT_2 cells. Co-cultivation with MT_2 is much less laborious than the U87 infection assays and MT_2 cell infection is often used for the rapid screening of X4 HIV-1 variants. Details on the method of MT_2 infection are described in Chapter 11.

3.3. Determination of CCR5 Dependence on PBMC of a Donor With a CCR5 Δ32 Homozygous Genotype

To measure the capacity of HIV-1 variants to use CXCR4 on primary cells or the incapacity of HIV-1 to use CCR5 on primary cells, PHA-stimulated PBMC from a donor who is homozygous for the previously described 32-base-pair deletion in the CCR5 gene can be used. This deletion results in a frame-shift and a premature stop codon. Consequently, functional CCR5 molecules are absent from the cell membrane and R5 HIV-1 variants cannot infect PBMC from donors with this genotype. Kinetics of X4 HIV-1 variant replication are indifferent from replication on cells of a homozygous CCR5 WT donor. PBMC activation, inoculation, and monitoring of virus replication are identical to that described for HIV infection in CCR5 WT PBMC as described in Chapter 1.

3.4. Determination of Co-Receptor Dependence Using Co-Receptor Antagonists

A panel of CCR5 and CXCR4 directed antagonists is currently available. These agents can be used to inhibit replication of HIV-1 variants. The susceptibility of HIV-1 variants to inhibition by these agents reflects their dependence on the co-receptor that is blocked for their replication. Examples of CXCR4 antagonists are bicyclam AMD3100 ***(11)*** and T22 peptide ***(12)***. Examples of CCR5 antagonists are SCH-C ***(13)*** and TAK779 ***(14)***.

To test the susceptibility of HIV-1 variants for co-receptor antagonist-mediated inhibition, 10^5 PHA-stimulated PBMC or 2.5×10^4 MT_2 cells should be incubated with fivefold serial dilutions of the specific antagonists in a volume of 50 μL for 2 h at 37°C in flat-bottom 96-well plates. To each virus clone, 10 $TCID_{50}$ must be added in a total volume of 100 μL medium (IL-2 medium for PBMC cultures and cell line medium for MT_2 cultures). Every 3 to 4 d, one-third of the MT_2 cultures must be replaced with fresh cell line medium. PBMC cultures can be maintained for 14 d and should be transferred to fresh medium at day 7 after inoculation. Production of p24 in the culture supernatant can be measured by ELISA at days 7 and 14 after inoculation. Each dilution of co-receptor antagonist should be tested at least in triplicate, to allow reliable calculation of inhibitory concentrations. Percent inhibition relative to control infections should be calculated.

4. Note

1. Co-receptor usage as determined on U87 cell lines do not necessarily reflect co-receptor usage on primary cells. HIV-1 variants that are able to use both CCR5 and CXCR4 and/or additionally CCR3 and/or CCR2 on U87 cells in general can only use CXCR4 on primary PBMC, as was demonstrated by 100% inhibition of

replication on primary PBMC when a CXCR4 antagonist was present. These observations also exclude an important role for unidentified co-receptors in HIV-1 infection of primary PBMC.

References

1. Tersmette, M., Gruters, R. A., De Wolf, F., De Goede, R. E. Y., Lange, J. M. A., Schellekens, P. T. A., et al. (1989) Evidence for a role of virulent human immunodeficiency virus (HIV) variants in the pathogenesis of acquired immunodeficiency syndrome: studies on sequential HIV isolates. *J. Virol.* **63,** 2118–2125.
2. Asjo, B., Albert, J., Karlsson, A., Morfeldt-Månson, L., Biberfeld, G., Lidman, K., and Fenyö, E. M. (1986) Replicative capacity of human immunodeficiency virus from patients with varying severity of HIV infection. *Lancet* **2,** 660–662.
3. Schuitemaker, H., Koot, M., Kootstra, N. A., Dercksen, M. W., De Goede, R. E. Y., Van Steenwijk, R. P., et al. (1992) Biological phenotype of human immunodeficiency virus type 1 clones at different stages of infection: progression of disease is associated with a shift from monocytotropic to T-cell-tropic virus populations. *J. Virol.* **66,** 1354–1360.
4. Tersmette, M., Lange, J. M. A., De Goede, R. E. Y., De Wolf, F., Eeftink Schattenkerk, J. K. M., Schellekens, P. T. A., et al. (1989) Association between biological properties of human immunodeficiency virus variants and risk for AIDS and AIDS mortality. *Lancet* **6,** 983–985.
5. Koot, M., Vos, A. H. V., Keet, R. P. M., De Goede, R. E. Y., Dercksen, W., Terpstra, F. G., et al. (1992) HIV-1 biological phenotype in long term infected individuals, evaluated with an MT-2 cocultivation assay. *AIDS* **6,** 49–54.
6. Feng, Y., Broder, C. C., Kennedy, P. E., and Berger, E. A. (1996) HIV-1 entry cofactor: functional cDNA cloning of a seven-transmembrane, G protein-coupled receptor. *Science* **272,** 872–877.
7. Zhang, L., Huang, Y., He, T., Cao, Y., and Ho, D. D. (1996) HIV-1 subtype and second-receptor use. *Nature* **383,** 768.
8. Dragic, T., Litwin, V., Allaway, G. P., Martin, S. R., Huang, Y., Nagashima, K. A., et al. (1996) HIV-1 entry into CD4+ cells is mediated by the chemokine receptor CC-CKR-5. *Nature* **381,** 667–673.
9. Deng, H. K., Liu, R., Ellmeier, W., Choe, S., Unutmaz, D., Burkhart, M., et al. (1996) Identification of the major co-receptor for primary isolates of HIV-1. *Nature* **381,** 661–666.
10. Berger, E. A., Doms, R. W., Fenyo, E. M., Korber, B. T. M., Littman, D. R., Moore, J. P., et al. (1998) A new classification for HIV-1. *Nature* **391,** 240.
11. Donzella, G., Schols, D., Lin, S. W., Este, J. A., Nagashima, K. A., Maddon, P. J., et al. (1998) AMD3100, a small molecule inhibitor of HIV-1 entry via the CXCR4 co-receptor. *Nat. Med.* **4,** 72–77.
12. Murakami, T., Zhang, T. Y., Koyanagi, Y., Tanaka, Y., Kim, J., Suzuki, Y., et al. (1999) Inhibitory mechanism of the CXCR4 antagonist T22 against human immunodeficiency virus type 1 infection. *J. Virol.* **73,** 7489–7496.

13. Strizki, J. M., Xu, S., Wagner, N. E., Wojcik, L., Liu, J., Hou, Y., et al. (2001) SCH-C (SCH 351125), an orally bioavailable, small molecule antagnoist of the chemokine receptor CCR5, is a potent inhibitor of HIV-1 infection in vitro and in vivo. *Proc. Natl. Acad. Sci. USA* **98,** 12,718–12,723.
14. Baba, M., Nishimura, O., Kanzaki, N., Okamoto, M., Sawada, H., Iizawa, Y., et al. (1999) A small-molecule, nonpeptide CCR5 antagonist with highly potent and selective anti-HIV-1 activity. *Proc. Natl. Acad. Sci. USA* **96,** 5698–5703.

26

Quantitative Evaluation of HIV and SIV Co-Receptor Use With GHOST(3) Cell Assay

Dalma Vödrös and Eva Maria Fenyö

Summary

An assay has been established for quantitative evaluation of lentivirus coreceptor use with the help of GHOST(3) cells. GHOST(3) cells were derived from the human osteosarcoma cell line, HOS, and have been engineered to stably express CD4 and one or another of the chemokine receptors CCR3, CCR5, CXCR4, CXCR6/STRL33/Bonzo, or the orphan receptor GPR15/BOB. The indicator cell line carries the HIV-2 long terminal repeat-driven green fluorescence protein (GFP) gene, which becomes activated upon infection with HIV or SIV. Viral entry is followed by Tat activation of transcription and GFP becomes expressed. Infected cells can be detected as early as 2 or 3 d after infection by simple fluorescence microscopic observation. The simplicity of the GHOST(3) cell system makes it particularly suitable for screening of a large number of isolates. In addition, the efficiency of co-receptor use can be accurately quantitated with flow cytometric analysis. Thus, the most efficiently used co-receptor of multitropic isolates can be determined. It is also possible to sensitively determine co-receptor switch of sequential isolates from the same individual.

Key Words: GHOST(3); GFP; co-receptor use; quantitative assay; HIV; SIV.

1. Introduction

1.1. Co-Receptor Use of Primate Lentiviruses

The target cells of HIV and SIV are T lymphocytes. Initial events in the entry into target cells require interactions between the viral envelope glycoprotein and two cellular receptors. The primary cell surface receptor for HIV and SIV is CD4, but CD4 alone is not sufficient for infection. Secondary receptors, called co-receptors, are also involved in virus entry ***(1–15)***. The co-receptors belong to the G protein-coupled seven-transmembrane chemokine receptor

From: *Methods in Molecular Biology, Vol. 304: Human Retrovirus Protocols: Virology and Molecular Biology*
Edited by: T. Zhu © Humana Press Inc., Totowa, NJ

family. CC- and CXC-chemokine receptors are involved in HIV and SIV entry. Some orphan receptors can also serve as co-receptors for virus entry.

First the virus binds to CD4. This leads to conformational changes in the surface envelope glycoprotein (gp120) of the virus, making possible the interaction between the viral envelope and the coreceptor on the cell surface ***(16–19)***. This tight binding reveals the hydrophobic fusion protein (gp41), making it possible to reach the cell surface. Finally, the viral and the cell membrane fuse, releasing the viral core particles in the cytoplasm. With this, the entry is complete, and postentry steps can take place.

Co-receptor use of lentiviruses shows specific characteristics and can be related to viral pathogenesis. CD4-independent co-receptor use of HIV and SIV has also been observed ***(13,20–24)***.

1.2. GHOST(3) Cell Assay

GHOST(3) cells, derived from the human osteosarcoma cell line (HOS), contain the gene of the GFP driven by the HIV-2_{ROD} long terminal repeat (LTR) ***(25)***. The cells have been engineered to stably express CD4, the primary receptor used by HIV and SIV, and one of the chemokine receptors, CCR3, CCR5, CXCR4, CXCR6 (formerly Bonzo/strl33/TYMSTRL), or the orphan receptor gpr15/BOB. The parental cell line expresses CD4 but none of the co-receptors. In case of infection, when the virus enters the cells by using CD4 and the appropriate co-receptor, the viral Tat protein becomes expressed and transactivates transcription of the GFP gene by the LTR (**Fig. 1**). GFP expression of infected cells is easily detected by fluorescence microscope as early as 2 or 3 d after infection. In addition, the proportion of fluorescent cells and their fluorescence intensity can be quantitated by flow cytometry (FACS). It is also possible to monitor productive infection by measuring the viral antigen content of culture supernatants.

1.3. Advantages of the Assay

Advantages of the GHOST(3) cell assay include the following:

- Fast screening of a large material by UV microscopic reading 2–3 d after infection.
- Quantitative evaluation of co-receptor use of multitropic isolates and determination of the most efficiently used co-receptor.
- Sensitive determination of co-receptor switch of sequential isolates ***(26)***.

2. Materials

1. Culture medium: Dulbecco's modified Eagle's medium (DMEM) (Life Technologies) complemented with 7.5% fetal calf serum (FCS) (Life Technologies) and penicillin/streptomycin (25 U/mL penicillin and 25 mg/mL streptomycin; Life Technologies) (*see* **Note 1**).

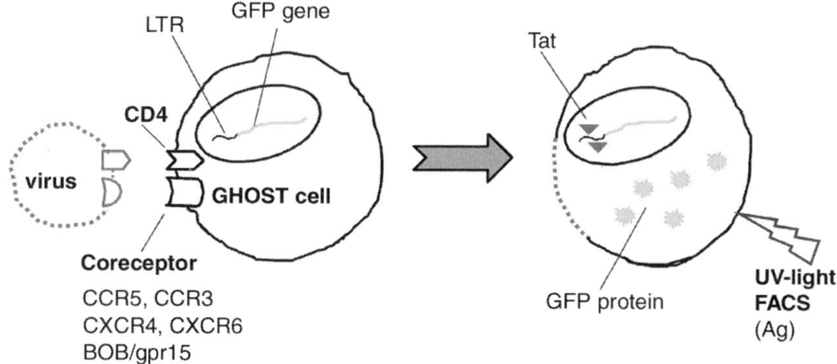

Fig. 1. The GHOST(3) cell system.

2. To enhance infection, polybrene is added to the culture medium (2 μg/mL final concentration) during and after infection.
3. Phosphate-buffered saline (PBS) (Life Technologies).
4. 5 m*M* EDTA (dissolved in PBS).
5. 4% paraformaldehyde (PFA).
6. Equipment: fluorescence microscope, flow cytometer (FACS).

3. Methods

3.1. Maintenance of Cells

The GHOST(3) cell lines are maintained in T25 flasks (*see* **Note 2**) in culture medium and incubated in a humified atmosphere with 5% CO_2 at 37°C. Cells are split 1:15 (1:10–1:25, depending on confluency of cells) twice a week. First remove old culture medium from the flask, wash the cells with 2 mL PBS and then add 0.5 mL 5 m*M* EDTA to the cultures. After 2–3 min treatment at 37°C detach the cells from the bottom of the flask with fresh medium and remove to a centrifuge tube. Refill flask with 7–8 mL fresh culture medium and add an appropriate amount of cells from the centrifuge tube (depending on what rate the cells are split).

3.2. Infection Assay

1. **Day –1:** Prepare 24-well plates with 2–3 × 10^4 cells/well in 1 mL medium 1 d before infection (*see* **Note 3**).
2. **Day 0 (afternoon):** On the day of infection replace medium with 200 μL of fresh medium (from this step on, medium must contain polybrene) and add virus to duplicate wells in a volume of 300 μL/well (*see* **Note 3**). Incubate plates in a humified atmosphere with 5% CO_2 at 37°C for 2 h. After 2 h add culture medium up to 1 mL/well and incubate plates further at 37°C humified atmosphere with 5% CO_2.

3. **Day 1 (morning):** After an overnight incubation remove medium, wash cells with 400 μL of PBS, add 1 mL of fresh medium to each well, and incubate the plates further (humified atmosphere, 37°C, 5% CO_2).
4. **Day 3:** Three days after infection check cultures for
   - Confluence, syncytium, and cytopathicity in normal light microscope.
   - Fluorescence induction in fluorescence microscope.

After observation, prepare cells from one of the parallel wells for flow cytometry and split the remaining wells (*see* **Note 4**). Remove supernatants, wash cells with 400 μL of PBS and add 200 μL of 5 m*M* EDTA to each well. Incubate plates at 37°C for a few minutes. During this incubation period prepare FACS tubes with 200 μL of 4% paraformaldehyde (PFA) in each tube.

When the cells detached from the plate, first split those wells that will be kept for further incubation. For this add 800 μL of culture medium to the wells (the wells now contain 1 mL medium altogether) and detach cells from the wells by pipetting the medium up and down several times. Remove appropriate amount of liquid containing cells. For example: If cells are split 1:5 in a given well, keep 200 μL medium in the well and remove 800 μL. Finally replace the removed medium with fresh medium. All the wells should contain 1 mL medium after split. Incubate plates in humified atmosphere with 5% CO_2 at 37°C.

Next, remove cells for FACS analysis from the appropriate wells. For this use 1 mL blue tip and carefully detach the cells from the wells by pipetting them up and down several times. Move detached cells into tubes containing 200 μL of 4% PFA (this will result 400 μL cell suspension in each tube, with the final concentration of 2% PFA). Check the emptied wells with microscope and make sure that all the cells are removed from the wells. Incubate samples at 4°C in dark for at least 1 h before FACS analysis. (The cells can be analyzed the next day.) **Note 5** details safety precautions for preparing and handling HIV infected cells on FACS.

5. **Day 6:** 3 d after split, check cultures again in both normal and UV light and take supernatants from each well to determine the viral antigen content of each culture.

3.3. FACS Analysis

3.3.1. Running FACS

Select the GHOST cell population on the side scatter-forward scatter diagram (**step I** in **Fig. 2**), and measure the fluorescence intensity of $1–1.5 \times 10^4$ GHOST cells (**step II**).

3.3.2. Calculation

The quadrant location must be adjusted for each cell line in each experiment such that about 0.1% (0.09–0.11%) of the mock-infected cells can be found in the upper right quadrant (region of positive cells) (*see* **Note 6**).

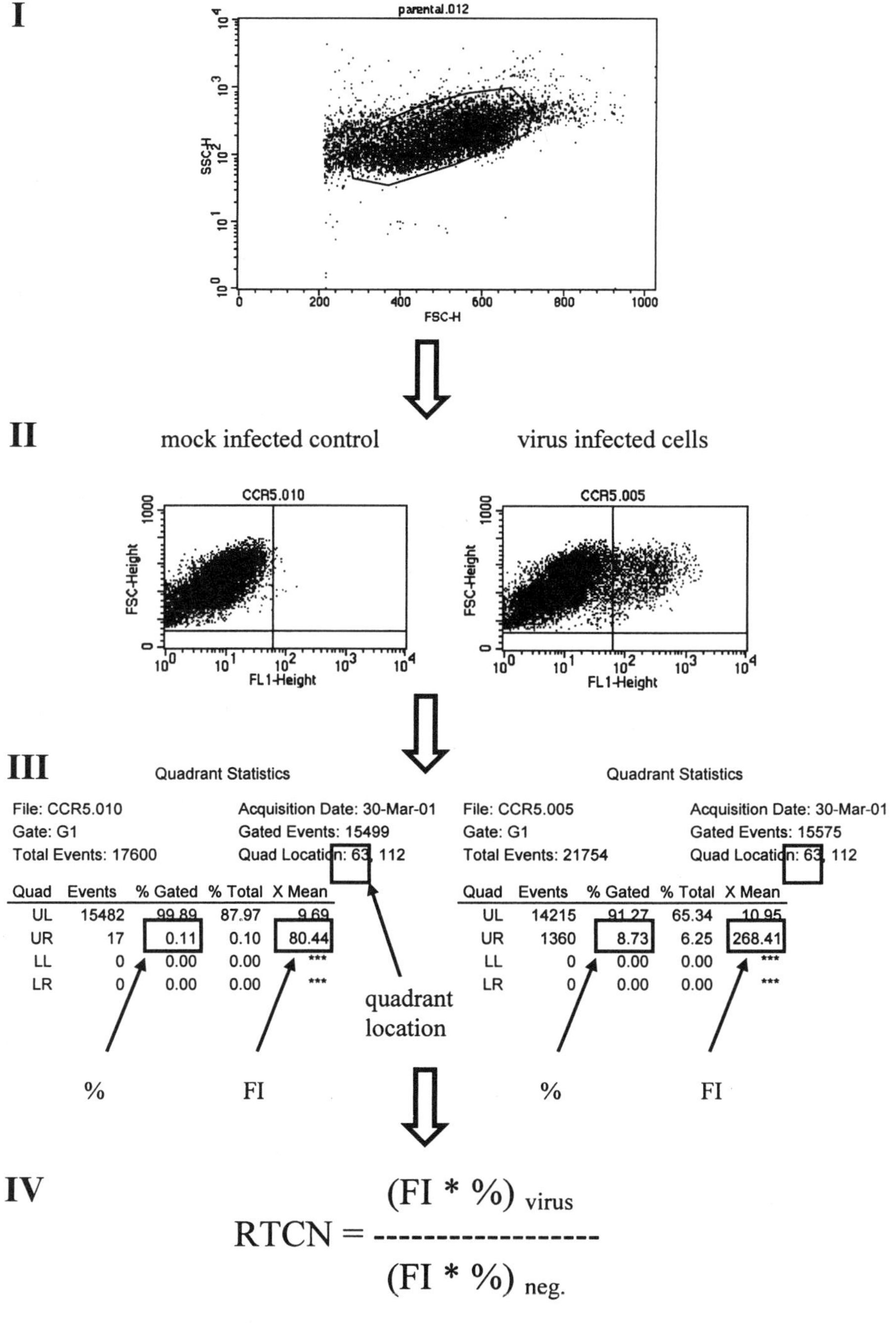

Fig. 2. Steps during FACS analysis.

Determine the proportion of fluorescence-positive cells and their fluorescence intensity (FI) for each infected culture (**step III**). Multiply these two parameters with each other and determine fold difference from mock-infected control cells expressing the same coreceptors (**step IV**):

$$RTCN = (FI{*}\%)_{virus}/(FI{*}\%)_{neg}.$$

This characteristic number, RTCN (**r**atio **t**o **c**ell **n**egative), gives a quantitative measure of the efficiency of infection.

3.3.3. Evaluation

RTCN above 10 indicates that the tested virus used the tested coreceptor.
RTCN below 5 gives a negative result.
RTCN between 5 and 10 is an indeterminate result (*see* **Notes 7** and **8**).

4. Notes

1. Antibiotics can be used for selection of stably transfected cells. CD4 gene can be selected with gentamicin (G418), the co-receptors with puromycin, and the gene of GFP with hygromycin. If large stock of cells is frozen and cells are not grown for more than 4–6 wk, selection of cells with antibiotics is not necessary.
2. Cells can easily be cultured in larger volume (T75 flask) if needed.
3. The amount of infectious virus can influence the RTCN values: the more virus is used for infection, the higher the RTCN *(27)*. On the contrary, if the virus is too weak or diluted too much, the most efficiently used co-receptor(s) can be identified but the use of additional (less efficiently used) co-receptor(s) may be overlooked. The use of at least 50 $TCID_{50}$ is recommended. There is no upper limit for the amount of virus used. If strong virus is tested, however, it is recommended to plate more cells for infection (4–5 × 10^4 cells/well) of those that express the co-receptor used by the virus.
4. At d 3, when cells are split, do not treat cells with EDTA if strong cytopathic effect is observed in any of the wells and the confluence of the cells in that well is less than 40%. In this case, wash cells twice with 400 μL of PBS and then add 1 mL of fresh culture medium to the well. Cells that are going to be FACSed must be treated with EDTA as in any other well.
5. Special care must be taken since HIV-infected cells for FACS analysis are prepared in the safety laboratory (BL-3) and most often FACS machines are outside the safety laboratory. For this reason please follow the following precautions:
   - Change gloves in the safety laboratory before touching the FACS tubes.
   - Incubate HIV-infected cells with PFA for at least 1 h before running cells on FACS.
   - Spray FACS tubes with 70% ethanol before moving them from the safety laboratory for FACS analysis.
   - Try not to touch the keyboard of the FACS machine with the same hand that touches the FACS tubes and loads samples for analysis. Once the machine is set one can change the samples with the left hand and handle the keyboard with the right hand.

- After running samples in FACS machine clean the machine: run FACSclean (contains 1% active Cl) for 5 min at high speed, then run FACSrinse for 5 min; finally, run destilled water for 1 min. Leave the tube with water on and close the machine.

6. The calculated RTCN value depends very much on the cutoff level of fluorescence intensity (the *x*-value of the quadrant location, **Fig. 2**). Therefore it is important to adjust the quadrant carefully for each cell line in each experiment.
7. Indeterminate RTCNs may result when
   - Very low amount of virus ($TCID_{50} < 10$) is tested in cells expressing the coreceptor used by the virus tested, or
   - CXCR4 using viruses are tested on the parental cell line or cells expressing receptors other than CXCR4: background level of the endogenously expressed CXCR4 receptor on GHOST(3) cells may result in the weak induction of GFP production in all cell lines.
8. When CXCR4 using virus is tested for coreceptor use, all cell lines can give positive results since the GHOST(3) cells express endogenous CXCR4 on their cell surface. There are several ways to eliminate the effect of endogenously expressed CXCR4 receptor on the GHOST(3) cell lines:
   a. Isolates can be tested in the presence and absence of specific CXCR4 antagonists or inhibitors such as AMD3100. AMD3100 does not affect positive results obtained with coreceptors other than CXCR4 but inhibits infection via CXCR4 thus eliminates background level of GFP induction by the endogenously expressed CXCR4 receptor on all cell lines *(27)*.
   b. Dilution of CXCR4-using viruses can also decrease CXCR4-background. If virus is used in a $TCID_{50}$ of about 50 in the experiments to infect the GHOST(3) cells the background becomes negligible *(27)*.
   c. Positive RTCN valuess could also be compared with RTCN on the parental cells: if there is at least one log difference between the RTCN on the tested coreceptor expressing cells and the RTCN on parental cells, one can accept that the result is positive.
9. Calculated RTCN values correlate very well with the readout of UV microscopic observations *(27)*.
10. Coreceptor expression: Receptor expression of all GHOST(3) cell lines was examined. CXCR4 expression was similar on each cell line, i.e., with the 12G5 monoclonal antibody. There was no difference in the level of CXCR4 expression between the GHOST(3).CXCR4 cell line engineered to express CXCR4 and the endogenously expressed CXCR4 on all other GHOST(3) cell lines *(27)*. However, GHOST(3).CXCR4 cells, engineered to express CXCR4, were clearly more efficient in supporting replication of CXCR4-using viruses than any other cell line. It is possible that:
    - There might be differences in the expression level of CXCR4 but the 12G5 antibody—which we used to determine the CXCR4 expression on the GHOST(3) cell lines—cannot distinguish between the conformation of the endogenous and the engineered CXCR4 on the surface of the GHOST(3) cells *(28)* or

- Even if the same level of receptors are expressed on the different cell lines, the receptors on the GHOST(3).CXCR4 cells are functioning better than those on the other cell lines.

Interestingly, the CCR3, CCR5, and CXCR6 levels on the GHOST(3).CCR3, GHOST(3).CCR5, and GHOST(3).CXCR6 cell lines, respectively, were comparable to the level of endogenous CXCR4 on the same cell lines *(27)*. However, as mentioned above, endogenous CXCR4 can be inhibited on these cells without any effect on the results obtained with the engineered coreceptors other than CXCR4.

References

1. Alkhatib, G., Combadiere, C., Broder, C. C., Feng, Y., Kennedy, P. E., Murphy, P. M., and Berger, E. A. (1996) CC CKRs—A rantes, MIP-1-alpha, MIP-1-beta receptor as a fusion cofactor for macrophage-tropic HIV-1. *Science* **272,** 1955–1958.
2. Choe, H., Farzan, M., Sun, Y., Sullivan, N., Rollins, B., Ponath, P. D., et al. (1996) The beta-chemokine receptor CCR3 and CCR5 facilitate infection by primary HIV-1 isolates. *Cell* **85,** 1135–1148.
3. Choe, H., Farzan, M., Konkel, M., Martin, K., Sun, Y., Marcon, L., et al. (1998) The orphan seven-transmembrane receptor apj supports the entry of primary T-cell-line-tropic and dualtropic human immunodeficiency virus type 1. *J. Virol.* **72,** 6113–6118.
4. Deng, H. K., Liu, R., Ellmeier, W., et al. (1996) Identification of a major coreceptor for primary isolates of HIV-1. *Nature* **381,** 661–666.
5. Deng, H., Unutmaz, D., KewalRamani, V. N., and Littman, D. R. (1997) Expression cloning of new receptors used by simian and human immunodeficiency viruses. *Nature* **388,** 296–300.
6. Doranz, B. J., Rucker, J., Yi, Y. J., Smyth, R. J., Samson, M., Peiper, S. C., et al. (1996) A dual-tropic primary hiv-1 isolate that uses fusin and the beta-chemokine receptors ckr-5, ckr-3, and ckr-2b as fusion cofactors. *Cell* **85,** 1149–1158.
7. Dragic, T., Litwin, V., Allaway, G. P., Martin, S. R., Huang, Y. X., Nagashima, K. A., et al. (1996) HIV-1 entry into CD4(+) cells is mediated by the chemokine receptor CC-CKR-5. *Nature* **381,** 667–673.
8. Edinger, A. L., Hoffman, T. L., Sharron, M., Lee, B., Yi, Y., Choe, W., et al. (1998) An orphan seven-transmembrane domain receptor expressed widely in the brain functions as a coreceptor for human immunodeficiency virus type 1 and simian immunodeficiency virus. *J. Virol.* **72,** 7934–7940.
9. Farzan, M., Choe, H., Martin, K., Marcon, L., Hofmann, W., Karlsson, G., et al. (1997) Two orphan seven-transmembrane segment receptors which are expressed in CD4-positive cells support simian immunodeficiency virus infection. *J. Exp. Med.* **86,** 405–411.
10. Feng, Y., Broder, C. C., Kennedy, P. E., and Berger, E. A. (1996) Hiv-1 entry cofactor—functional cDNA cloning of a seven-transmembrane, G protein-coupled receptor. *Science* **272,** 872–877.

11. He, J., Chen, Y., Farzan, M., Choe, H., Ohagen, A., Gartner, S., et al. (1997) CCR3 and CCR5 are co-receptors for HIV-1 infection of microglia. *Nature* **385,** 645–649.
12. Horuk, R., Hesselgesser, J., Zhou, Y., Faulds, D., Halks-Miller, M., Harvey, S., et al. (1998) The CC chemokine I-309 inhibits CCR8-dependent infection by diverse HIV-1 strains. *J. Biol. Chem.* **273,** 386–391.
13. Reeves, J. D., McKnight, A., Potempa, S., Simmons, G., Gray, P. W., Power, C. A., et al. (1997) CD4-independent infection by HIV-2 (ROD/B): Use of the 7-transmembrane receptors CXCR-4, CCR3, and V28 for entry. *Virology* **231,** 130–134.
14. Rucker, J., Edinger, A. L., Sharron, M., Samson, M., Lee, B., Berson, J. F., et al. (1997) Utilization of chemokine receptors, orphan receptors, and herpesvirus-encoded receptors by diverse human and simian immunodeficiency viruses. *J. Virol.* **71,** 8999–9007.
15. Samson, M., Edinger, A. L., Stordeur, P., Rucker, J., Verhasselt, V., Sharron, M., et al. (1998) ChemR23, a putative chemoattractant receptor, is expressed in monocyte-derived dendritic cells and macrophages and is a coreceptor for SIV and some primary HIV-1 strains. *Eur. J. Immunol.* **28,** 1689–1700.
16. Kwong, P. D., Wyatt, R., Robinson, J., Sweet, R. W., Sodroski, J., and Hendrickson, W. A. (1998) Structure of an HIV gp120 envelope glycoprotein in complex with the CD4 receptor and a neutralizing human antibody. *Nature* **393,** 648–659.
17. Lapham, C. K., Ouyang, J., Chandrasekhar, B., Nguyen, N. Y., Dimitrov, D. S., and Golding, H. (1996) Evidence for cell-surface association between fusin and the CD4-gp120 complex in human cell lines. *Science* **274,** 602–605.
18. Trkola, A., Dragic, T., Arthos, J., Binley, J. M., Olson, W. C., Allaway, G. P., et al. (1996) CD4-dependent, antibody-sensitive interactions between HIV-1 and its co-receptor CCR-5. *Nature* **384,** 184–187.
19. Wu, L., Gerard, N. P., Wyatt, R., Choe, H., Parolin, C., Ruffing, N., et al. (1996) CD4-induced interaction of primary HIV-1 gp120 glycoproteins with the chemokine receptor CCR-5. *Nature* **384,** 179–183.
20. Edinger, A. L., Blanpain, C., Kunstman, K. J., Wolinsky, S. M., Parmentier, M., and Doms, R.W. (1999) Functional dissection of CCR5 coreceptor function through the use of CD4-independent simian immunodeficiency virus strains. *J. Virol.* **73,** 4062–4073.
21. Edinger, A. L., Mankowski, J. L., Doranz, B. J., Margulies, B. J., Lee, B., Rucker, J., et al. (1997) CD4-independent, CCR5-dependent infection of brain capillary endothelial cells by a neurovirulent simian immunodeficiency virus strain. *Proc. Natl. Acad. Sci. USA* **94,** 14,742–14,747.
22. Endres, M. J., Clapham, P. R., Marsh, M., et al. (1996) CD4-independent infection by HIV-2 is mediated by fusin/CXCR4. *Cell* **87,** 745–756.
23. Liu, H. Y., Soda, Y., Shimizu, N., Haraguchi, Y., Jinno, A., Takeuchi, Y., and Hoshino, H. (2000) CD4-dependent and CD4-independent utilization of coreceptors by human immunodeficiency viruses type 2 and simian immunodeficiency viruses. *Virology* **278,** 276–288.

24. Reeves, J. D., Hibbitts, S., Simmons, G., McKnight, A., Azevedo-Pereira, J. M., Moniz-Pereira, J., and Clapham, P. R. (1999) Primary human immunodeficiency virus type 2 (HIV-2) isolates infect CD4-negative cells via CCR5 and CXCR4: comparison with HIV-1 and simian immunodeficiency virus and relevance to cell tropism in vivo. *J. Virol.* **73,** 7795–7804.
25. KewalRamani, V. N., Volsky, B., Kwon, D. S., Xiang, W.-K., Gao, J., Unutmaz, D., et al. (1998) Unpublished data.
26. Vödrös, D., Thorstensson, R., Biberfeld, G., Schols, D., De Clercq, E., and Fenyö, E. M. (2001) Coreceptor usage of sequential isolates from cynomolgus monkeys experimentally infected with simian immunodeficiency virus (SIVsm). *Virology* **291,** 12–21.
27. Vödrös, D., Tscherning-Casper, C., Navea, L., Schols, D., De Clercq, E., and Fenyö, E. M. (2001) Quantitative evaluation of HIV-1 coreceptor use in the GHOST3 cell assay. *Virology* **291,** 1–11.
28. Baribaud, F., Edwards, T.G., Sharron, M., Brelot, A., Heveker, N., Price, K., et al. (2001) Antigenically distinct conformations of CXCR4. *J. Virol.* **75,** 8957–8967.

27

Phenotypic Characterization of Blood Monocytes From HIV-Infected Individuals

Philip J. Ellery and Suzanne M. Crowe

Summary

Monocytes play an important, yet only partly understood, role in HIV-1 pathogenesis. Two main subsets of peripheral blood monocytes have been described; the major subset of monocytes are phenotypically characterized as being $CD14^{hi}/CD16^{-}$, and a minor subset (5–15% of total monocytes in healthy individuals), which are $CD14^{lo}/CD16^{hi}$, have been reported to be expanded in HIV-infected individuals. These $CD14^{lo}/CD16^{hi}$ monocytes differ from the majority of monocytes in a number of ways, including the molecules expressed on their surface and how they function. Here we describe a flow-cytometric assay to identify and compare the expression of a representative surface molecule (CCR5) on $CD14^{hi}/CD16^{-}$ and $CD14^{lo}/CD16^{hi}$ monocytes in small volumes of whole blood, and methods to isolate monocyte subsets by both fluorescence-activated cell sorting (FACS) and magnetic bead sorting.

Key Words: Flow cytometry; monocyte subsets; CD14; CD16; CCR5; whole blood assay; FACS sorting; magnetic bead sorting.

1. Introduction

Cells of the macrophage lineage at their various stages of differentiation play an important, yet relatively poorly understood, role in the pathogenesis of HIV-1 infection. Monocytes are derived from myelomonocytic stem cells of the bone marrow. They spend up to 3 d circulating in the peripheral blood before they are recruited to tissues where they differentiate predominantly into macrophages *(1)*.

It is also becoming increasingly apparent that monocytes are a heterogeneous population of cells that may represent varying stages of differentiation *(2)* or activation *(3,4)* and may include peripheral blood precursors of dendritic

From: *Methods in Molecular Biology, Vol. 304: Human Retrovirus Protocols: Virology and Molecular Biology*
Edited by: T. Zhu © Humana Press Inc., Totowa, NJ

cells *(5–9)*. Two main subsets of monocytes have been described based on surface receptor expression. The majority of monocytes express high levels of CD14 (a component of the lipopolysachharide recognition complex on monocytes) on their surface, but express little or no CD16 (Fc γ-receptor III) ($CD14^{hi}$/$CD16^{-}$). A minor subpopulation of monocytes (5–15% of total monocytes in healthy individuals) that downregulates surface expression of CD14 and upregulates CD16, and are therefore phenotypically described as being $CD14^{lo}$/$CD16^{hi}$, was first described in 1988 *(10)*. More recently, the $CD14^{lo}$/$CD16^{hi}$ monocyte subset has been shown to be expanded in individuals with sepsis *(11,12)*, arthritis *(13)*, atherosclerosis *(4)*, pulmonary alveolar proteinosis *(14)*, and those undergoing hemodialysis *(15)*. Expansion of this subset to represent up to 50% of total monocytes has also been described in HIV-infected individuals *(16–18)*. They differ from the majority of monocytes in a number of ways: they are slightly smaller *(19)*; they are less phagocytically active *(19)*; cytokine production such as tumor necrosis factor (TNF) may be increased *(3,16,20)*; they are more likely to traffic through endothelial layers in vitro *(5,21,23)*; and they differ in surface receptor phenotype *(4,6,19)*. The $CD14^{lo}$/$CD16^{hi}$ monocytes express significantly more CCR5 than the majority of monocytes *(13,22,24)*, and as a possibly related finding, they are more permissive to HIV-1 infection in vitro *(25)*. Furthermore, HIV-infected $CD14^{lo}$/$CD16^{hi}$ monocytes have been shown to infiltrate the perivascular space of brains of patients with HIV dementia *(23)*, suggesting a potential role in the pathogenesis of HIV-dementia and in the development of tissue reservoirs of HIV.

In this chapter we describe a surface expression assay to identify $CD14^{hi}$/$CD16^{-}$ and $CD14^{lo}$/$CD16^{hi}$ monocytes in small volumes of whole blood obtained from individuals infected with HIV. This method allows further phenotypic characterization of the subsets (as an example, we compare the expression of CCR5 between the two subsets using three-color flow cytometry). We describe the use of a fluorescence quantitative standards kit to determine the mean number of CCR5 molecules on the cell surface of monocyte subsets. We also describe a method to isolate viable $CD14^{hi}$/$CD16^{-}$ and $CD14^{lo}$/$CD16^{hi}$ monocytes from larger volumes of blood by fluorescence-activated cell sorting (FACS) and from smaller volumes of blood by sorting using magnetic bead technology.

2. Materials (*see* Note 1)

1. Phosphate-buffered saline without calcium and magnesium (PBS–): sterile, pH 7.0–7.5, at 4°C.
2. Ficoll-Paque Plus (Amersham Biosciences; Uppsala, Sweden, cat. no. 17-1440-03), at room temperature.
3. FACS wash buffer: PBS– with 1% v/v fetal bovine serum (FBS) or 0.1% w/v bovine serum albumin (BSA), at 4°C.

Table 1
Fluorochrome-Conjugated Monoclonal Antibodies Used

Name	Clone	Antibody isotype	Catalog no.
CD14-PE	M5E2	IgG_{2a}	555398
IgG_{2a}-PE	G155-178	IgG_{2a}	555574
CD16-Cy	3G8	IgG_1	555408
IgG_1-Cy	MOPC-21	IgG_1	555750
CCR5(CD195)-FITC	2D7	IgG_{2a}	555592
IgG_{2a}-FITC	G155-178	IgG_{2a}	555573

PE: phycoerythrin; Cy: Cy-Chrome; FITC: fluoroscein isothiocyanate (all from Becton Dickinson, San Jose, CA).

4. 6-mL polypropylene round-bottom tubes with caps (FACS tubes; Becton Dickinson Labware, Franklin Lakes, NJ, cat. no. 2063).
5. Fluorochrome conjugated monoclonal antibodies (*see* **Table 1**), stored at 4°C, away from light.
6. FACS lysing solution (Becton Dickinson, cat. no. 349202), made fresh to 1X solution according to manufacturer's instructions, at room temperature.
7. Fixing solution: 3% (v/v) ultrapure formaldehyde in PBS–, made fresh, at 4°C away from light.
8. Quantum FITC MESF kit—low level (Bangs Laboratories Inc., Fishers, IN, cat. no. 824).
9. MACS Monocyte Isolation Kit II (Miltenyi Biotec, Bergisch Gladbach, Germany, cat. no. 130-091-153).
10. MACS CD16 microbeads (Miltenyi Biotec, cat. no. 130-045-701).
11. MACS LS columns (Miltenyi Biotec, cat. no. 130-042-401).
12. MidiMACS magnetic separator system (Miltenyi Biotec, cat. no. 130-042-102).

3. Methods

The methods below outline (1) the staining of monocytes within patient whole blood to identify monocyte subsets using fluorochrome-conjugated monoclonal antibodies, (2) flow-cytometric analysis of stained monocyte subsets, (3) the isolation of viable monocyte subsets for culture in vitro by flow-cytometric sorting, and (4) the isolation of monocyte subsets from small volumes of blood using magnetic bead sorting.

3.1. Surface Receptor Staining of Monocytes Within Whole Blood to Identify Monocyte Subsets (see Note 2)

Whole blood (200 µL) was transferred to four FACS tubes and cells were washed with FACS wash and pelletted by centrifugation at 500*g* for 5 min at 10°C; the supernatant was removed to 1 mm above the cell pellet (leaving

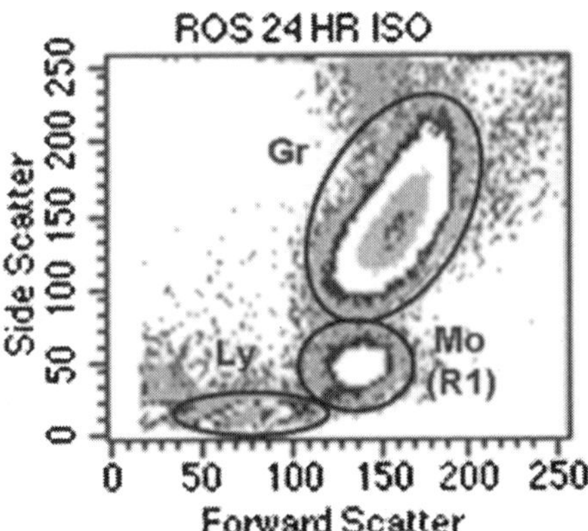

Fig. 1. Major cell populations of blood after erythrocyte lysis. Three distinct cell populations are clearly resolved by flow cytometry based on cell size (forward scatter; FSC) and granularity (side scatter; SSC); lymphocytes (Ly), monocytes (Mo), and granulocytes (Gr). For analysis, the putative monocyte population is gated (R1).

approx 200 μL of buffer). Cells in duplicate tubes were incubated with 20 μL of CD14-PE, CD16-Cy, and CCR5-FITC; 20 μL of the isotype control antibodies IgG_{2a}-PE, IgG_1-Cy, and IgG2a-FITC were added to the third tube; and the fourth tube contained only cells without antibody. Tubes were incubated on ice away from light for 30 min, before unbound antibodies were washed from the cell surface using FACS wash, as above. Erythrocytes were lysed for 10 min at room temperature with 2 mL of FACS lysis buffer. Erythrocyte debris and hemoglobin were removed by washing the cells three times with FACS wash, as above. Cells were fixed with 200 μL of fixing solution at 4°C away from light for 2–24 h before flow-cytometric analysis.

3.2. Flow-Cytometric Acquisition and Analysis

Basic principles of flow cytometry are detailed in ref. ***26***. We used a three-color FACSCalibur and accompanying CELLQuest computer software (Becton Dickinson) (*see* **Note 3**).

Using unstained cells (tube 4; *see* **Subheading 3.1.**), a putative monocyte gate (G1) was set based on forward (size) and side (granularity) scatter characteristics (**Fig. 1**). Background fluorescence and nonspecific staining of monocytes in G1 was then assessed using cells from the same donor stained with the isotype control antibodies (tube 3; *see* **Subheading 3.1.**). A single-parameter histogram, gated on G1, was set on FITC (FL-1) and a "positive cell" marker (M1) was set to include ~5% of cells (**Fig. 3A**). CCR5 expression on monocyte subsets was analysed in duplicate using cells stained with CD14-PE, CD16-

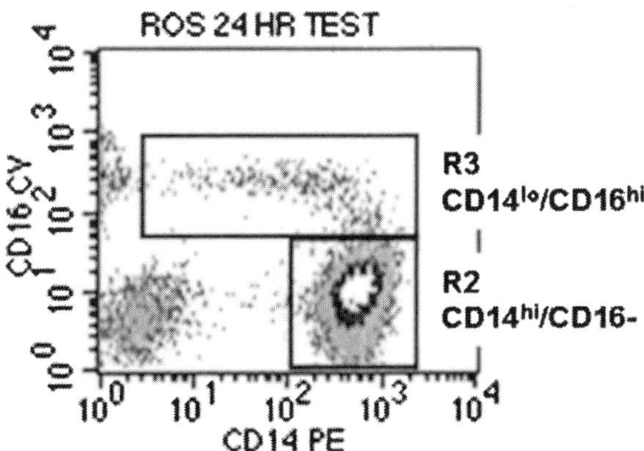

Fig. 2. Gating of monocyte subsets. Gates are set around CD14^{hi}/CD16^{-} (R2), and CD14^{lo}/CD16^{hi} (R3) monocyte populations.

Cy, and CCR5-FITC (tubes 1 and 2; *see* **Subheading 3.1.**). CD14^{hi}/CD16^{-} (G2) and CD14^{lo}/CD16^{hi} (G3) monocytes were gated using a CD14-PE/CD16-Cy dot-plot gated on G1 (**Fig. 2**). Two single-parameter histograms, gated on G2 or G3, were set on FITC (FL-1). A "positive cell" marker (M1) was set at the same level as was set on the isotype control to calculate the net percent positive cells for both of these histograms (**Fig. 3B,C**). Net mean fluorescence intensity (MFI) of CCR5-FITC and net percent CCR5-positive cells was derived from these histograms by subtracting background values of the isotype control-stained cells.

A standard curve between 3000 and 50,000 MESF (molecules of equivalent soluble fluorochrome) values was generated using FITC MESF beads (Quantum) according to the manufacturer's instructions. CCR5 MFI values of CD14^{hi}/CD16^{-} (G2) and CD14^{lo}/CD16^{hi} (G3) monocytes were converted to MESF values using the regression line generated from the standard curve (*see* **Note 4**).

3.3. Isolation of Monocyte Subsets by FACS Sorting

Flow cytometric cell sorting may be used to isolate monocyte subsets from blood. In this example we describe the use of a high-speed FACS sorter (MoFlo; DakoCytomation) to isolate large numbers of CD14^{hi}/CD16^{-} and CD14^{lo}/CD16^{hi} monocytes from HIV-seronegative buffy coats acquired from a blood bank.

Peripheral blood mononuclear cells (PBMC) were isolated from buffy coats by Ficoll density gradient centrifugation *(27)*, which generally yields 5–10 × 10^8 PBMC per buffy coat. Briefly, 50 mL of blood was diluted 1:2 with PBS–

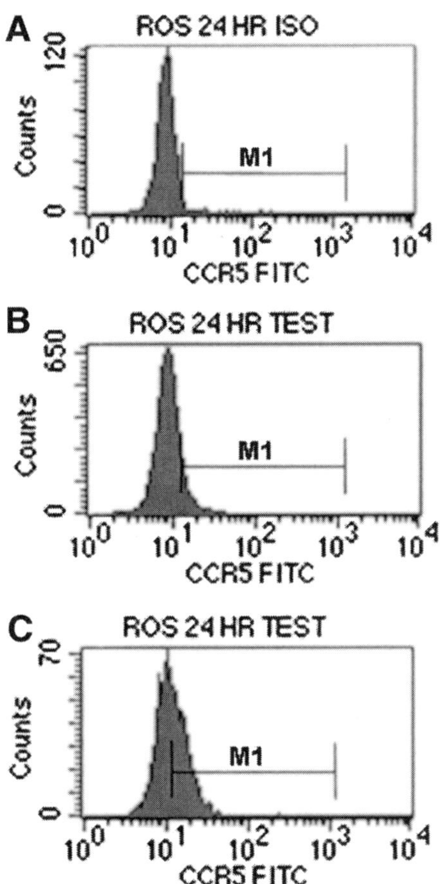

Fig. 3. Comparison of CCR5 expression on monocyte subsets. **(A)** Level of background fluorescence of monocytes (R1) stained with the isotype control antibody. **(B)** $CD14^{hi}/CD16^{-}$ monocytes (R2) show lower CCR5 expression compared to **(C)** $CD14^{lo}/CD16^{hi}$ (R3) monocytes.

at room temperature. Diluted blood (25 mL) was layered over 15 mL of Ficoll and centrifuged at 914*g* for 20 min at room temperature. PBMC were collected from the interface of the Ficoll and transferred to a 50-mL conical tube. PBMC were washed with PBS– (4°C) and centrifuged once at 500*g* for 10 min, then twice at 200*g* for 10 min, at 4°C.

Monocytes were further enriched by countercurrent elutriation to yield 5–10 $\times 10^7$ enriched monocytes. Briefly, PBMC were resuspended in 10 mL FACS wash and elutriated by loading cell into a standard chamber in a Beckman J-Gm/E Centrifuge and JE-5-O RDTOR set at 1550*g* at 4°C with a pump speed calibrated to 28 mL/min. Lymphocytes were depleted by allowing 1 L of PBS– to flow through the chamber and slowly increasing the pump flow rate to 30 mL/min. Monocytes were eluted into six 50-mL tubes coated with 5 mL of fetal bovine serum at 35 mL/min. Cells were combined and washed with FACS wash as above, resuspended in 450 µL FACS wash, and transferred to a FACS tube.

To identify monocyte subsets, elutriated cells were incubated with 25 µL CD14-PE and CD16-Cy for 30 min on ice away from light. Unbound antibody was washed from cells with FACS wash as in **Subheading 3.1.** and cells were resuspended in FACS wash for sorting. A putative monocyte gate (G1) was set based on forward and side scatter characteristics as previously shown (**Fig. 1**). Typically, 90% of cells fell in this gate after the enrichment steps above. $CD14^{hi}/CD16^{-}$ and $CD14^{lo}/CD16^{hi}$ monocytes were gated, as described (**Fig. 2**) and were sorted into FACS tubes on ice which are coated with 500 µL of fetal bovine serum.

This process typically yields $1–4 \times 10^6$ $CD14^{lo}/CD16^{hi}$ monocytes. Sorted monocytes using this method were more than 95% viable as judged by Trypan blue exclusion, and remained adherent to plastic tissue culture plates for at least 14 d when cultured in Iscove's modified Dulbecco medium supplemented with 10% (v/v) heat-inactivated AB+ human serum (*see* **Note 5**), 2 m*M* L-glutamine, and 50 µg/mL gentamicin on plastic tissue culture plates at 37°C in a humidified CO_2 incubator.

3.4. Isolation of Monocyte Subsets by Magnetic Bead Sorting (see Note 6)

Here we describe a method to isolate monocyte subsets from small volumes of blood using MACS magnetic bead technology. The MACS Monocyte Isolation Kit II (Miltenyi Biotec) is a negative isolation kit that consists of a biotinylated antibody cocktail against CD3, CD7, CD16, CD19, CD56, CD123, and CD235a to deplete T cells, B cells, granulocytes, NK cells, dendritic cells, platelets, and erythrocytes. Antibody-bound cells are depleted with antibiotin-coated magnetic beads by placing bead-coated cells through a magnetic column assembly. $CD14^{lo}/CD16^{hi}$ monocytes are then positively selected from the total monocyte fraction using directly conjugated CD16 microbeads (Miltenyi Biotec), leaving the $CD14^{hi}/CD16^{-}$ in the negative fraction.

PBMC (typically $1.5–2.5 \times 10^7$) were isolated from 9 mL of EDTA anticoagulated whole blood by Ficoll density gradient centrifugation (*see* **Subheading 3.3.**). Monocytes were negatively selected using the MACS Monocyte Isolation Kit II (Miltenyi Biotec) according to the manufacturer's directions. A typical yield was $1–4 \times 10^6$ monocytes.

$CD14^{lo}/CD16^{hi}$ monocytes were positively selected from whole monocytes using the CD16 microbeads according to the manufacturer's directions to yield typically 1–4 × 10^5 $CD14^{lo}/CD16^{hi}$ monocytes. Monocyte subsets were shown to be more than 95% pure as judged by flow-cytometric analysis using CD14-PE and CD16-Cy antibodies.

4. Notes

1. Sodium azide (0.1% w/v) may be added to the FACS wash buffer as a preservative and an inhibitor of nonspecific internalization of antibody. Monocytes are highly adherent to plastics; therefore it is important to keep buffers, tubes, and cells cold where possible. Monocytes are less adherent to polypropylene FACS tubes compared with polystyrene tubes and the former is therefore preferable. Fluorochrome-conjugated monoclonal antibodies should be stored at 4°C away from light. Methanol/ethanol-free formaldehyde should be used for fixation to avoid dehydration of cells. **Caution:** formaldehyde is carcinogenic, highly toxic, and an irritant.
2. We suggest collecting blood in tubes containing EDTA as the anticoagulant for surface expression assays of monocyte subsets, as we have found CD16 expression on monocytes is upregulated over 24 h at room temperature if blood is collected in lithium heparin or ACD. As, EDTA has also been shown to inhibit phagocytosis by monocytes *(28)*, the anticoagulant must be chosen to suit the specific requirements of the experiment.
3. Multiparameter flow cytometry using fluorochrome-conjugated monoclonal antibodies, which stain monocyte populations within whole blood, has many advantages over preenrichment steps such as density gradient centrifugation: the assay is quicker, it is cheaper, and it avoids changes that may occur to surface molecule expression by preenrichment steps. Whole-blood analysis of rare cell populations typically generates large data files and investigators should ensure that analysis computers are able to support such files. Flow cytometers may be calibrated and compensated using commercially available fluorescently labeled beads (e.g., CALIBRITE beads; Becton Dickinson), but settings should be checked using single-color-stained cells, as fluorescence characteristics of monocytes, especially from HIV-infected individuals, differs significantly from beads.
4. MFI is a useful tool to compare relative levels of expression of a given molecule between two cell populations, but does not determine the precise number of molecules that are expressed. MESF values may be generated comparing fluorescence levels on cells to beads conjugated with known dilutions of fluorochrome.
5. Alternatively, RPMI-1640 medium and FBS may be used to culture monocytes.
6. The disadvantages of FACS sorting when only a small volume of blood is available are significantly reduced yield due to coincidence/abortion rates associated with all FACS sorting, and only one sample may be sorted at a time. Magnetic bead sorting provides a simple, rapid method in which multiple samples may be processed simultaneously with a yield nearing 100%. The disadvantage of magnetic bead technology is that it is expensive for the isolation of a large numbers of cells, beads cannot be removed from the surface of monocytes once bound, and monocytes may be activated upon contact with beads. The MACS beads have the

advantage over other beads such as Dynabeads (Dynal) as they are small enough (50 nm) to not interfere with downstream applications such as flowcytometry or light microscopy. Despite containing an antibody against CD16, $CD14^{lo}/CD16^{hi}$ monocytes are not depleted by the Monocyte Isolation Kit II. It is presumed that CD16 receptor density on monocytes is too low to be effectively depleted with this kit. $CD14^{hi}/CD16^{-}$ monocytes may still contaminate the $CD14^{lo}/CD16^{hi}$ monocyte fraction. This may be due to a difference in the sensitivity of CD16-binding with MACS beads versus CD16-detection by flow cytometry, or may be due to blocking of CD16 antibody binding due to the binding of the CD16 microbeads.

References

1. Whitelaw, D. M. (1972) Observations on human monocyte kinetics after pulse labeling. *Cell Tissue Kinet.* **5,** 311–317.
2. Ziegler-Heitbrock, H. W., Fingerle, G., Strobel, M., Schraut, W., Stelter, F., Schutt, C., et al. (1993) The novel subset of CD14+/CD16+ blood monocytes exhibits features of tissue macrophages. *Eur. J. Immunol.* **23,** 2053–2058.
3. Belge, K. U., Dayyani, F., Horelt, A., Siedlar, M., Frankenberger, M., Frankenberger, B., et al. (2002) The proinflammatory CD14+CD16+DR++ monocytes are a major source of TNF. *J. Immunol.* **168,** 3536–3542.
4. Draude, G., von Hundelshausen, P., Frankenberger, M., Ziegler-Heitbrock, H. W., and Weber, C. (1999) Distinct scavenger receptor expression and function in the human CD14(+)/CD16(+) monocyte subset. *Am. J. Physiol.* **276,** H1144–H1149.
5. Randolph, G. J., Sanchez-Schmitz, G., Liebman, R. M., and Schakel, K. (2002) The CD16(+) (FcgammaRIII(+)) subset of human monocytes preferentially becomes migratory dendritic cells in a model tissue setting. *J. Exp. Med.* **196,** 517–527.
6. Almeida, J., Bueno, C., Alguero, M. C., Sanchez, M. L., de Santiago, M., Escribano, L., et al. (2001) Comparative analysis of the morphological, cytochemical, immunophenotypical, and functional characteristics of normal human peripheral blood lineage(–)/CD16(+)/HLA-DR(+)/CD14(–/lo) cells, CD14(+) monocytes, and CD16(–) dendritic cells. *Clin. Immunol.* **100,** 325–338.
7. MacDonald, K. P., Munster, D. J., Clark, G. J., Dzionek, A., Schmitz, J., and Hart, D. N. (2002) Characterization of human blood dendritic cell subsets. *Blood* **100,** 4512–4520.
8. Engering, A., Van Vliet, S. J., Geijtenbeek, T. B., and Van Kooyk, Y. (2002) Subset of DC-SIGN(+) dendritic cells in human blood transmits HIV-1 to T lymphocytes. *Blood* **100,** 1780–1786.
9. Sanchez-Torres, C., Garcia-Romo, G. S., Cornejo-Cortes, M. A., Rivas-Carvalho, A., and Sanchez-Schmitz, G. (2001) CD16(+) and CD16(–) human blood monocyte subsets differentiate in vitro to dendritic cells with different abilities to stimulate CD4(+) T cells. *Int. Immunol.* **13,** 1571–1581.
10. Ziegler-Heitbrock, H. W., Passlick, B., and Flieger, D. (1988) The monoclonal antimonocyte antibody My4 stains B lymphocytes and two distinct monocyte subsets in human peripheral blood. *Hybridoma* **7,** 521–527.

11. Fingerle, G., Pforte, A., Passlick, B., Blumenstein, M., Strobel, M., and Ziegler-Heitbrock, H.W. (1993) The no vel subset of CD14+/CD16+ blood monocytes is expanded in sepsis patients. *Blood* **82,** 3170–3176.
12. Blumenstein, M., Boekstegers, P., Fraunberger, P., Andreesen, R., Ziegler-Heitbrock, H. W., and Fingerle-Rowson, G. (1997) Cytokine production precedes the expansion of CD14+CD16+ monocytes in human sepsis: a case report of a patient with self-induced septicemia. *Shock* **8,** 73–75.
13. Kawanaka, N., Yamamura, M., Aita, T., Morita, Y., Okamoto, A., Kawashima, M., et al. (2002) CD14+,CD16+ blood monocytes and joint inflammation in rheumatoid arthritis. *Arthritis Rheum.* **46,** 2578–2586.
14. Yoshioka, Y., Ohwada, A., Harada, N., Satoh, N., Sakuraba, S., Dambara, T., and Fukuchi, Y. (2002) Increased circulating CD16+ CD14dim monocytes in a patient with pulmonary alveolar proteinosis. *Respirology* **7,** 273–279.
15. Nockher, W. A. and Scherberich, J. E. (1998) Expanded CD14+ CD16+ monocyte subpopulation in patients with acute and chronic infections undergoing hemodialysis. *Infect. Immun.* **66,** 2782–2790.
16. Thieblemont, N., Weiss, L., Sadeghi, H. M., Estcourt, C., and Haeffner-Cavaillon, N. (1995) CD14lowCD16high: a cytokine-producing monocyte subset which expands during human immunodeficiency virus infection. *Eur. J. Immunol.* **25,** 3418–3424.
17. Pulliam, L., Gascon, R., Stubblebine, M., McGuire, D., and McGrath, M. S. (1997) Unique monocyte subset in patients with AIDS dementia. *Lancet* **349,** 692–695.
18. Amirayan-Chevillard, N., Tissot-Dupont, H., Capo, C., Brunet, C., Dignat-George, F., Obadia, Y., et al. (2000) Impact of highly active anti-retroviral therapy (HAART) on cytokine production and monocyte subsets in HIV-infected patients. *Clin. Exp. Immunol.* **120,** 107–112.
19. Passlick, B., Flieger, D., and Ziegler-Heitbrock, H. W. (1989) Identification and characterization of a novel monocyte subpopulation in human peripheral blood. *Blood* **74,** 2527–2534.
20. Ziegler-Heitbrock, H. W., Strobel, M., Kieper, D., Fingerle, G., Schlunck, T., Petersmann, I., et al. (1992) Differential expression of cytokines in human blood monocyte subpopulations. *Blood* **79,** 503–511.
21. Geissmann, F., Jung, S., and Littman, D. R. (2003) Blood monocytes consist of two principal subsets with distinct migratory properties. *Immunity* **19,** 71–82.
22. Ancuta, P., Rao, R., Moses, A., Mehle, A., Shaw, S. K., Luscinskas, F. W., and Gabuzda, D. (2003) Fractalkine preferentially mediates arrest and migration of CD16+ monocytes. *J. Exp. Med.* **197,** 1701–1707.
23. Fischer-Smith, T., Croul, S., Sverstiuk, A. E., Capini, C., L'Heureux, D., Regulier, E. G., et al. (2001) CNS invasion by CD14+/CD16+ peripheral blood-derived monocytes in HIV dementia: perivascular accumulation and reservoir of HIV infection. *J. Neurovirol.* **7,** 528–541.
24. Weber, C., Belge, K. U., von Hundelshausen, P., Draude, G., Steppich, B., Mack, M., et al. (2000) Differential chemokine receptor expression and function in human monocyte subpopulations. *J. Leukoc. Biol.* **67,** 699–704.

25. Ellery, P., Sonza, S., and Crowe, S. M. (2002) Monocyte subsets and HIV reservoirs in patients on HAART in 5th International Workshop on HIV, Cells of Macrophage/Dendritic Lineage and Other Reservoirs, Rome, Italy.
26. Shevach, E. M. (2002) Immunofluorescence and cell sorting, in *Current Protocols in Immunology*, vol. 1 (Coligan, J. E., Kruisbeek, A. M., Margulies, D. H., Shevach, E. M., and Strober, W., eds.) John Wiley and Sons, NY, pp. 5.0.1.–5.8.10.
27. Crowe, S., Mills, J., and McGrath, M. S. (1987) Quantitative immunocytofluorographic analysis of CD4 surface antigen expression and HIV infection of human peripheral blood monocyte/macrophages. *AIDS Res. Hum. Retroviruses* **3,** 135–145.
28. Hewish, M. J., Meikle, A. M., Hunter, S. D., and Crowe, S. M. (1996) Quantifying phagocytosis of *Mycobacterium avium* complex by human monocytes in whole blood. *Immunol. Cell Biol.* **74,** 306–312.

28

Methods to Determine HIV-1 Ex Vivo Fitness

Awet Abraha, Ryan M. Troyer, Miguel E. Quiñones-Mateu, and Eric J. Arts

Summary

The fitness of human immunodeficiency virus type-1 (HIV-1) appears to be an important piece of the puzzle in understanding the global HIV-1 epidemic and disease progression in infected individuals. We have developed a dual infection/competition assay followed by a sensitive heteroduplex tracking assay (HTA) to measure the relative fitness of any HIV-1 isolate. Differences in fitness between wild-type and drug-resistant HIV-1 isolates can also be used as indicators of the emergence of drug-resistant isolates. This chapter describes the methods utilized to measure viral fitness (relative replicative capacity) during growth competition experiments. HTA is used to evaluate the production of HIV-1 variants in a competition, which can then be used to estimate ex vivo viral fitness.

Key Words: HIV-1; ex vivo fitness; quasispecies; HTA; PBMC.

1. Introduction

Many RNA viruses, including human immunodeficiency virus type 1 (HIV-1), have the ability of evading host immune response and antiviral therapy due to rapid turnover (10^{10} viral particles/day in an HIV-infected individual), high mutation rate, and high frequency of recombination. This rapid change in HIV-1 evolution will undoubtedly affect the replication kinetics or "fitness" of the virus. Fitness, defined as an organism's replicative capacity/adaptability in a given environment *(1)*, is a multilayered parameter when applied to the impact of this virus on disease or the epidemic. The most rigorous tests of relative viral fitness involve competitions between two or more viral strains in tissue culture and were initially utilized to study fitness/evolution of various RNA viruses (e.g., vesicular stomatitis virus [VSV]) *(1,2)*. It was recently discovered with HIV-1 that drug-resistant mutations reduced replication efficiency

From: *Methods in Molecular Biology, Vol. 304: Human Retrovirus Protocols: Virology and Molecular Biology*
Edited by: T. Zhu © Humana Press Inc., Totowa, NJ

(3–5) and that ex vivo HIV-1 fitness was directly related to disease progression *(6)*. Based on these studies it appears that HIV-1 fitness varies in different environments; however, the fitness in micro- and macroenvironments may be interrelated but not necessarily consequential *(5)*. For example, the ability of virus to replicate in human peripheral blood mononuclear cells (PBMCs) (defined as ex vivo fitness *[5,6]*) may be related to virus production by these same cells in the blood of an infected human host. However, other factors such as immune response and host genetics will influence viral loads, select for different HIV-1 clones, and thus, alter fitness during disease. Nonetheless it appears that HIV-1 fitness is one of the strongest predictors of disease progression *(6,7)*.

In the context of HIV-1 evolution and distribution in the epidemic, it appears that HIV-1 subtype C isolates are less fit than any other group M strain in PBMC cultures *(6)*. However, HIV-1 subtype C has emerged as the predominant subtype in the world. HIV can be classified into three groups, M, N, and O, of which group M is further subdivided into at least nine subtypes (A through J) and 15 circulating recombinant forms (CRF01–CRF15). Based on early reports, it is important to study the role of HIV fitness and evolution in the AIDS epidemic.

This chapter describes and illustrates the competition assay for PBMCs used to measure the ex vivo fitness of any HIV-1 isolates. PBMC can also be substituted for other primary cells (e.g., Langerhans cells, macrophages, T cells) *(8)* or cell lines (e.g., MT_2) *(6)*. Production of two HIV-1 isolates in a competition is analyzed by a heteroduplex tracking assay (HTA) and compared to initial inocula to calculate the relative fitness value for each isolate in a competition. **Figure 1** illustrates the strategies employed for the growth competition, polymerase chain reaction (PCR), and HTA analysis. The techniques have been optimized to provide reliable and accurate estimates of ex vivo fitness for nearly any primary or laboratory HIV strain that is classified as group M strain.

Fig. 1. *(opposite page)* Illustrations of HIV-1 competition experiments and HTA method for dual virus detection and fitness calculations. **(A)** Demonstrates growth competition assay, in which virus is added alone or in pairs to phytohemagglutinin (PHA) and interleukin-2-treated peripheral blood mononuclear cells at the desired multiplicity of infection (MOI). **(B)** Extracted DNA polymerase chain reaction (PCR) amplified from dual or mono-infections with conserved HIV-1 *env* primers. **(C)** DNA heteroduplexes derived from the monoinfections are resolved on a nondenaturing polyacrylamide gel and used to identify isolates found in the HTA derived from the dual infection. Phosphor-imager analysis is used to quantify each heteroduplex and determine the relative production of each virus in the dual infection. Formulas for calculating relative production of each virus in a dual competition and for measuring the fitness difference between the two isolates are also provided.

A GROWTH COMPETITION ASSAY

	MOIs		
● Virus A	0.01	0	0.01
○ Virus B	0	0.01	0.01

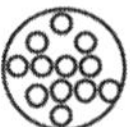
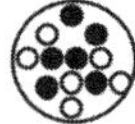

B PCR STRATEGY

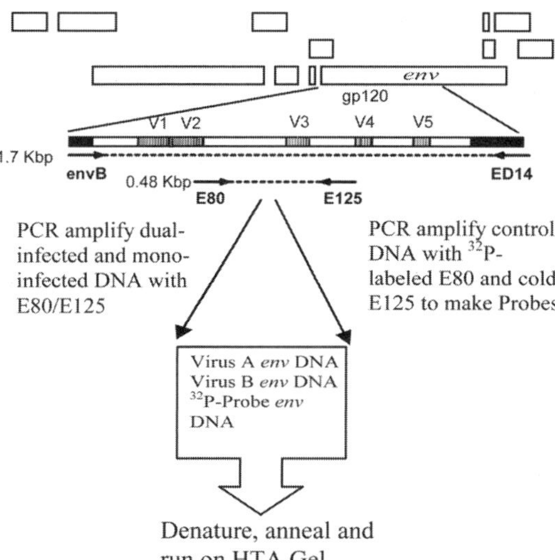

C HETERODUPLEX TRACKING ASSAY (HTA) GEL

Production of Virus A in competition $W_{s\,(A)} = d/a$
Production of Virus A in competition $W_{s\,(B)} = c/a$

Calculate total virus production in a dual infection (W_T):

$$W_{s\,(A)} + W_{s\,(B)}$$

Relative fitness calculation:

$W_{(A)} = [W_{s\,(A)}/W_T]$ / initial inoculum
$W_{(B)} = [W_{s\,(B)}/W_T]$ / initial inoculum

Fitness difference (W_D):

$W_D = W_M/W_L$

W_M and W_L correspond to the relative fitness of the more and less fit virus, respectively.

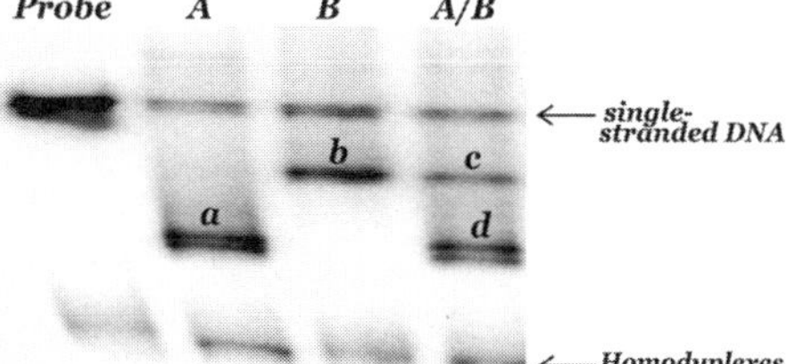

Techniques to study the fitness of more diverse HIV and simian immunodeficiency viruses (SIV) are available on request.

2. Materials

2.1. PMBC Stimulation

1. Biohazard level 2 safety facility.
2. HIV-seronegative whole blood donor (preferably same donor for the entire experiment).
3. Ficoll-Paque.
4. RPMI-1640–2 m*M* L-glutamine supplemented with 10% fetal bovine serum (FBS), 10 m*M* HEPES buffer, 100 U of penicillin/mL, 100 μg of streptomycin/mL, and 2 μg of phytohemagglutinin (PHA; Invitrogen)/mL (complete medium I).
5. 37°C incubator supplemented with 5% CO_2.

2.2. Virus Titration and Growth Competition Assay

1. Virus supernatant to be used for titration.
2. PHA-stimulated PBMCs (need approx 2.4×10^6 cells per virus titration).
3. RMPI-1640–2 m*M* L-Glutamine (medium I).
4. RPMI-1640–2 m*M* L-glutamine supplemented with 10% FBS, 10 m*M* HEPES buffer, 1 ng of recombinant interleukin-2 (IL-2; Invitrogen)/mL, 100 U of penicillin/mL, 100 μg of streptomycin/mL, and 2 μg of PHA (Invitrogen)/mL complete medium II).
5. Phosphate-buffered saline sterile for tissue culture.
6. 48- and 96-well flat-bottom microtiter plates.
7. Multichannel pipet.
8. RT master mix: 50 m*M* Tris-HCl (pH 7.8), 75 m*M* KCl, 2 m*M* dithiothreitol (DTT), 5 m*M* $MgCl_2$, 5 μg/mL of poly(rA), 6.25 μg/mL oligo(dT), 0.5%(v/v) NP40; make 1-mL aliquots and store at –20°C to –70°C indefinitely.
9. 10 mCi/mL α-^{32}P dTTP (Perkin Elmer, cat. no. NEG-505A (>400 Ci/mmol).
10. DEAE Filtermat for use with 1450 MicroBeta (96-well format) (Perkin Elmer, cat. no. 1450-522).
11. 1X saline-sodium citrate (SSC).

2.3. PCR of HIV-1 env Gene (C2–V3 Region 0.48 kb)

1. QIAamp DNA blood kit (Qiagen, cat. no. 51106).
2. Primers:
   - a. External — ENVB AGAAAGAGCAGAAGACAGTGGCAATGA
     ED14 TCTTGCCTGGAGCTGTTTGATGCCCCAGAC
   - b. Nested — E80 CCAATTCCCATACATTATTGTG
     E125 CAATTTCTGGGTCCCCTCCTGAGG
3. 10 m*M* dNTP mix (dATP, dCTP, dGTP, dTTP).
4. *Taq* DNA polymerase (5 U/μL) and buffer.
5. RNase/DNase-free H_2O.
6. Agarose.

7. QIAquick PCR purification kit (Qiagen, cat. no. 28104).

2.4. Primer Labeling and PCR Amplifying env *Gene (C2–V3 Region, 0.48 kb) of HIV-1 Strains to Make Probes*

1. Preferably use up to three specific HIV-1 strain probe DNA for each competition.
2. 10 mCi/mL γ-^{32}P dATP (3000 Ci/mmol) (Perkin Elmer, cat. no. NEG-002A).
3. T4 polynucleotide kinase and 5X forward reaction buffer.
4. Phenol/chloroform/isoamylalcohol.
5. G-25 spin columns (Pharmacia, cat. no. 27-5325-01).

2.5. HTA for Detection of env *Fragments*

1. DNA annealing buffer: 100 m*M* Tris-HCl (pH 7.8), 100 m*M* NaCl, and 2 m*M* EDTA.
2. 10X nondenaturing gel shift buffer: 0.25 *M* Tris-HCl, 1.9 *M* electrophoresis-grade glycine (Bio-Rad, cat. no. 161-0718), 10 m*M* EDTA. Make 1 L solution and store at 4°C.
3. 30% Acrylamide/bis solution 29:1 (3.3%C) (Bio-Rad cat. no. 161-0156).
4. TEMED (Bio-Rad, cat. no. 161-0800).
5. 10% (w/v) Ammonium persulfate (APS). Make 50-mL aliquots and store at 4°C up to 2 mo.
6. PROTEAN II XI electrophoresis system (Bio-Rad).
7. Gel shift HTA loading buffer: 5 mL glycerol, 0.03 g Tris-HCl, 5 µL 1 *M* DTT, 0.25% bromophenol blue, 0.25% xylene cyanol in a final volume of 10 mL. Make 1-mL aliquots and store at –20°C.
8. Whatman chromatography paper (Fisher, cat. no. 05-716-3F).

3. Methods

All incubations are done at 37°C with 5% CO_2 unless stated otherwise.

3.1. PMBC Isolation and Stimulation

Isolate PMBC from heparin-treated venous blood of HIV-seronegative donor by Ficoll-Paque grade centrifugation. Expected yield of PBMC is 1–2 10^6 cells/mL of whole blood (*see* **Note 1**).

1. Resuspend pelleted PBMC in the appropriate volume of complete medium II to yield a final cell concentration of approx 1×10^6 cells/mL and incubate for 48 h.
2. 48 h poststimulation; evaluate cell number and viability of cells by trypan blue exclusion.
3. Centrifuge the cells for 5 min at 1200*g* at room temperature. Discard the supernatant and save the pellet
4. Resuspend the cell pellet with freeze medium (10% DMSO, 70% fetal bovine serum, 20% RMPI-1640–2 m*M* L-glutamine) to achieve a final cell concentration of 2×10^6 cells/mL.

5. Make a 1-mL aliquot of resuspended cells in cryovials.
6. Freeze cells at –80°C for short-term storage and in liquid nitrogen freezer for long-term storage (*see* **Note 2**).

3.2. Virus Titration

Tissue culture dose for 50% infectivity ($TCID_{50}$) of propagated viruses or patient plasma is calculated by utilizing the accumulative Reed and Muench method (**Fig. 2**).

The following steps should be performed in a biohazard level 2 safety facility.

3.2.1. $TCID_{50}$ Infections (see **Note 3**)

1. Retrieve the cryovials containing frozen cells and thaw by immersing the bottom half of the vials in a 37°C water bath and swirl the tube until cells are thawed.
2. In a tissue culture hood, spray the outside of the vials with 70% ethanol. Open vial and resuspend cells with complete medium II by using a sterile pipet to an approximate cell concentration of 1×10^6 cells/mL. Transfer the cells to a sterile conical tube; you may consolidate multiple vials.
3. Centrifuge resuspended cells for 5 min at 1200*g* at room temperature.
4. Discard the supernatant and resuspend pelleted cells in media to an approximate cell concentration of 1×10^6 cells/mL. Transfer to a tissue culture flask and incubate for 2–3 d.
5. On day 2 or 3 after stimulation of cells, while the cells are still in a logarithmic phase of growth, count viable cells by trypan blue exclusion.
6. Using a multichannel pipettor, plate 96-well flat-bottom plates with 100,000 cells per well in a total volume of 150 μL. 24 Wells are required for each virus to be titrated (8 dilutions in triplicate; *see* **Fig. 1**).
7. Once cells are plated, place into incubator until infection.
8. Thaw and make 10-fold serial dilutions of virus supernatants in medium I (*see* **Fig. 1** and **Note 4**).
9. In triplicate: add 100 μL of each serial dilution to the appropriate wells in the 96-well plate containing PBMC (*see* **Fig. 1**). Leave some wells uninfected for negative controls and incubate.
10. 72 h postinfection, spin down plates and carefully harvest 150 μL off of the cell-free supernatant and discard appropriately.
11. Add 150 μL of complete medium II and incubate.
12. Using a multichannel pipettor, harvest 25 μL of cell-free supernatant on days 6, 8, 10, and 12 postinfection and store in –80°C freezer to be used for RT activity assay (*see* **Note 5**).

3.2.2. Reverse Transcriptase Activity Measurement

Reverse transcriptase assay measures the activity of the RT enzyme present in HIV supernatant. The RT mix contains the necessary substrate to drive HIV

RT activity. This step requires the use of both radioactivity and biological agent (HIV) at the same time. You must have the proper training and certification to work with radioactive and biological agents.

1. Thaw harvested virus supernatants at room temperature.
2. Aliquot 10 µL of the 25-µL supernatant into 96-well round-bottom microtiter plates.
3. Aliquot 10 µL of positive control (supernatant previously known to have positive RT activity) and negative control (supernatant from uninfected PBMC) in duplicate.
4. Add 1 µL of fresh 10-mCi/mL [α-^{32}P]-dTTP per mL of RT mix (*see* **Subheading 2.2.**).
5. Using a multichannel pipettor, add 25 µL of the labeled RT mix into each well containing 10 µL cell-free supernatant and incubate at 37°C incubator overnight.

Caution: Tips used to load radioactive RT mix into wells containing supernatant may possess two biohazards (HIV and ^{32}P radionucleotide) and must be discarded properly. Generally it is best to drop the labeled RT mix into the wells and completely avoid touching the virus. The tips can then be discarded as solid radioactive waste. If contaminations by both biohazard agents occur, you must aspirate bleach into the tips and discard the bleach as liquid and the tips as solid radioactive waste ***(1)***.

After more than 2 h incubation with the RT master mix, any HIV will be rendered noninfectious by the NP40 detergent in the mix. The remaining steps can be performed on the bench top employing appropriate radiation safety measures.

6. After the overnight incubation, using the multichannel pipettor, blot 10 µL of the reaction mixture from each well onto the 96-well format DEAE filtermat (*see* **Subheading 2.2.1.**). It is best to avoid touching the paper with the pipet tips to prevent tearing of the filtermat.
7. Allow the filtermat to dry (~10 min). Wash the filtermat five times with 1X SSC solution and twice with 85% ethanol by rocking in a shaker platform for 5 min each. The SSC wash liquid waste must be discarded as radioactive waste, but the ethanol washes can be discarded as regular waste.
8. Allow filters to dry, wrap with Saran wrap and expose overnight onto autoradiography film (Kodak BioMax MR).
9. Develop the film and use along with quantitative data to calculate $TCID_{50}$ (*see* **Note 6**).
10. To generate quantitative data, count the filters with a Matrix 96 β-counter (Packard, Meriden, CT) (*see* **Note 7**).
11. The 50% tissue culture infective dose values can then be calculated for each virus using the Reed-Muench technique ***(1)***.
12. The Reed-Muench accumulative calculation method yields the $TCID_{50}$ value as infectious units per milliliter (IU/mL).

3.3. Growth Competition Assay

Once the infectious titers of the viruses are determined, the next step is to calculate the multiplicity of infection (MOI) for the experimental conditions. MOI is the ratio between the number of infectious units (IU) to the number of cells. Most competitions/dual infections are best performed with an MOI ranging between 0.001 and 0.0001. Recombination between two isolates in a dual infection is lower if the MOI is at the lower range. This is owing to the low-frequency event of recombination within the C2–V3 region and is below the limit of detection by the HTA analysis *(2)*.

$$\text{MOI} = \text{IU/cell}$$

$$\text{TCID}_{50} = \text{IU/mL}$$

1. Stimulate PBMC by following the process described in (*see* **Subheading 3.2.1., steps 1–5** and **Note 8**).
2. Plate 48-well plates with 200,000 cells per well (final volume after infection with virus is 500 μL).
3. Infect PBMC with virus at the desired MOI for the experiment. Each virus can be added in pairs for dual infections and alone for monoinfections.
4. Also leave uninfected cultures to be used as HIV-negative control. All infections should be done in duplicate.
5. Once the infections are performed, add amount needed to bring the final volume of each well to 500 μL (use complete medium II) and incubate.
6. 48 h postinfection, aspirate supernatant and wash cells with 1X PBS.
7. Resuspend cells in complete medium II to final volume of 500 μL per well.
8. 5 d postinfection, harvest 250 μL from each well (save 25 μL and store at –80°C and discard the rest). Replenish with 250 μL of complete medium II.
9. Harvest 25 μL of cell-free supernatant from the cultures on days 7, 9, and, if necessary, 11 and 13 (*see* **Note 9**).
10. Screen for reverse transcriptase activity by using the procedure described in **Subheading 3.2.2.** for each harvest day.
11. Once peak virus production is observed by RT activity assay, harvest cells and supernatant as indicated below.
12. Transfer supernatant containing cells into 1.5-mL Eppendorf tubes and centrifuge in a microcentrifuge at 12,000*g* for 5 min (*see* **Note 10**).
13. Transfer the supernatant into a cryovial and store at –80°C.
14. You may freeze the pelleted cells at –80°C or can proceed to the subsequent steps right away.

3.4. PCR of HIV-1 env Gene (C2–V3 Region, 0.48 kb)

Store isolated DNA and PCR amplified DNA in –20°C freezer between experiments and for long-term storage.

1. Use QIAamp DNA blood kit to isolate proviral DNA from the pelleted cells of the competition assay.

2. Perform external and nested PCR for each dual and mono infected proviral DNA. *See* **Subheading 2.3.** for primer sequences.
   a. External PCR: 50 μL PCR reaction
      Prepare a master mix containing 5 μL 10X PCR buffer (with 1.5 m*M* final $MgCl_2$), 0.75 μL 10 m*M* dNTPs, 1.5 μL EnvB (10 pmol/μL), 1.5 μL ED14 (10 pmol/μL), 0.25 μL *Taq* DNA polymerase (5 U/μL), and 36 μL water. Add 45 μL of master mix to each tube containing 5 μL of proviral DNA. PCR-amplify with the following conditions: 94°C 4 min, [94°C 30 s, 55°C 30 s, 72°C 1 min] × 35 cycles, and 4°C hold.
   b. Nested PCR: 100 μL PCR reaction.
      Prepare a master mix containing 10 μL 10X PCR buffer (with 1.5 m*M* final $MgCl_2$), 1.5 μL 10 m*M* dNTPs, 3 μL E80 (10 pmol/μL), 3 μL E125 (10 pmol/μL), 0.5 μL *Taq* DNA polymerase (5 U/μL), and 72 μL water. Add 90 μL of mastermix to each tube containing 10 μL of external PCR product. Amplify with the same PCR conditions as in the external PCR.
3. Run 5 μL of each nested PCR product (plus loading dye) on a 1% denaturing agarose gel alongside a molecular weight marker.
4. Once the correct size PCR product is identified (0.48 kb), purify the remaining 95 μL of nested PCR product by using QIAquick PCR purification kit (Qiagen) (*see* **Note 11**).

3.5. Primer Labeling and PCR Amplifying env *Gene (C2–V3 Region, 0.48 kb) of HIV-1 Strains to Make Probes*

The next step is radiolabeling the nested sense primer (E80) to make probes for heteroduplex tracking assay.

1. Prepare a reaction mixture of 20 μL E80 (10 pmol/μL), 10 μL of 5X forward reaction buffer, 3 μL T4 PNK, 3 μL γ-^{32}P dATP (3000 Ci/mmol, 10 mCi/mL), and 14 μL of water. Incubate at 37°C for 60 min.
2. Vortex G-25 column resin, loosen cap, and snap off the bottom.
3. Place column onto 1.5-mL tube and spin at 2000*g* for 1 minute. Place the column into a new 1.5-mL tube, being careful not to disturb the resin. Discard the tube containing the eluted solution from the column.
4. After the 60-min primer labeling incubation, add 100 μL of phenol/chloroform/isoamylalcohol, vortex, and spin in a microcentrifuge for 5 min at ~12,000*g*.
5. Harvest aqueous (top) phase and apply onto the prepared G-25 column.
6. Spin column at 2000*g* for 2 min.
7. Discard column and save the eluted primer reaction. Wave the tube over a Geiger hand counter to make sure there is labeled primer in the tube.

The next step is the utilization of the labeled primer to make HIV DNA probes for HTA. It is best to have up to three HIV strain DNA to use as probes. If possible, use plasmids containing the *env* region from various HIV strains to make probes.

Table 1
Recipes for Making Nondenaturing Gel Shift Acrylamide Gels for HTAs

Stock solution	Final acrylamide concentration in gel (%)[a] 6%	8%
30% Acrylamide/*bis* solution 29:1	5 mL	6.7 mL
10X nondenaturing gel shift buffer	2.5 mL	2.5 mL
Deionized H_2O	17.5 mL	15.8 mL
10% (w/v) APS	500 µL	500 µL
TEMED	30 µL	30 µL

[a]This recipe is sufficient for pouring 0.75 mm, 16 × 14 gel plates.

8. Make a 100-µL PCR reaction in a tube for each HIV plasmid or HIV proviral DNA to be used as probe. Each reaction tube should contain 10 µL 10X PCR buffer (with 1.5 m*M* final $MgCl_2$), 1.5 µL 10 m*M* dNTPs, 7 µL γ-^{32}P dATP labeled E80 (4 pmol/µL), 3 µL E125 (10 pmol/µL), 0.5 µL *Taq* DNA polymerase (5 U/µL), and 72 µL water plus 2–10 µL of HIV DNA. (Adjust the volumes accordingly.) Use same PCR-amplifying conditions used previously.
9. Run radiolabeled PCR-amplified probes (0.48 kb) on 1% denaturing gel and purify using a QIAquick gel extraction kit (Qiagen). Use a Geiger hand counter to measure the radioactive probe.

3.6. Heteroduplex Tracking Assay for Detecting env *Fragments*

1. Before starting the HTA analysis, the amplified nested PCR products from the competitions must be equalized. Quantitate each band on the agarose gel using a program such as Kodak 1D. Adjust PCR product concentration with water to be able to add equal volume per HTA reaction.
2. Using the Protean II xi (Bio-Rad) or similar electrophoresis system, pour a 0.75-mm-thick 6% gel (8% for higher resolution; *see* **Table 1** for recipe). Use 15-well comb.
3. Prewarm gel for 2 h at 200 V and make the reactions for HTA during the pre-run.
4. Each HTA reaction should contain 10 µL of nested PCR product, 1 µL of 10X annealing buffer, 1 µL of probe reaction (~300–400 CPM final or 0.1 pmol per reaction). You should also have a probe-alone reaction in which you add water instead of PCR product.
5. Heat samples at 95°C 3 min, 37°C 5 min, 4°C hold.
6. Place the reaction mixtures on wet ice (*see* **Note 12**).
7. Add 3 µL of HTA loading dye into each tube, vortex and spin at 12,000*g* and place on wet ice until loading.
8. After the gel has prewarmed, pull out the comb and thoroughly clean out the wells (*see* **Note 13**).
9. Load the entire HTA reaction mix into the wells and let the gel settle for about 10 min before running it. For each dual competition, it is best to run the respective monoinfections alongside to make visualization and analysis simpler.

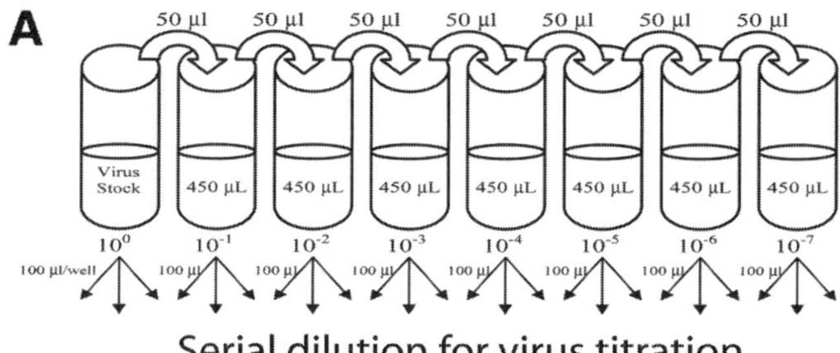

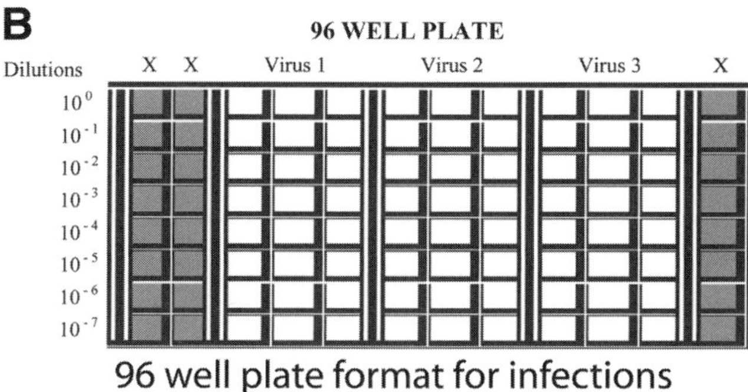

Fig. 2. Determining the tissue culture dose for 50% infectivity of HIV-1 culture supernatants. **(A)** Schematic illustration of the 10-fold serial dilution step for virus titration. **(B)** 100,000 cells/well are plated as illustrated in the figure. Fill the wells marked X with 250 µL of PBS (to reduce rate of evaporation). Infect cells by adding 100 µL of the serially diluted virus (as well as undiluted) as depicted in the figure.

10. Run gel at 200 V for ~4 h (6% gel) or until the second dye marker (xylene cyanol) gets to the bottom of the plate.
11. Take down the gels and blot onto Whatman chromatography paper.
12. Wrap it with Saran wrap and dry on a gel dryer at 85°C for 45 min.
13. Expose the gels overnight onto a phosphor imager screen.
14. Scan the screen on a phosphor imager and proceed to analysis (*see* **Note 14**).
15. **Optional:** Expose gel onto autoradiograph (Kodak MR) overnight with an intensifying screen.

3.7. HTA Analysis for Estimation of Viral Fitness (see *Note 15*)

This method describes how to estimate fitness (replicative capacity) of a virus based on competition experiments using equal MOIs of each virus. Refer to **Fig. 2** for a representative HTA of one competition and the formula for calculating relative viral fitness.

Suppose we're competing virus A with virus B at equal ratio of MOI.

Initial inoculum = 0.5 (this will change if using various ratios of MOI)

1. Determine production of individual HIV isolate in a dual infection, $W_{s(A)}$ for virus A and $W_{s(B)}$ for virus B. This is done by dividing the production of an isolate in a dual infection by its production in a monoinfection (*see* **Note 16**).
2. Calculate total virus production in a dual infection (W_T) by adding $W_{s(A)} + W_{s(B)}$.
3. Calculate relative fitness for each isolate in a competition as follows.
   - Relative fitness of virus A
     $W_{(A)} = [W_{s(A)}/(W_T)]$/initial inoculum.
   - Relative fitness of virus B
     $W_{(B)} = [W_{s(B)}/(W_T)]$/initial inoculum
4. To measure the fitness difference between two isolates in a competition (W_D), determine the ratio of relative fitness values of each HIV variant in the competition as follows.
   - $W_D = W_M/W_L$.

W_M and W_L correspond to the relative fitness of the more and less fit virus, respectively.

Refer to **Fig. 2** for a demonstration of viral fitness calculation.

4. Notes

1. Before beginning this experiment, it is very important to plan ahead. When isolating PBMC, make sure you estimate total PBMC need for the entire experiment (i.e., virus titration, subsequent competitions).
2. For successful cryopreservation and recovery of cells, it is best to freeze the cells by utilizing freezing containers (Fisher cat. no. 15-350-50) that cool cells at –1°C/min.
3. Prior to performing virus titration, make multiple aliquots of the virus stocks and freeze at –80°C or in liquid nitrogen. It is generally best to make small-volume aliquots (200–500 μL) to avoid wasting.
4. For titrating multiple viruses, you may use 48-well plates for making serial dilutions.
5. Harvesting directly into 96-well plates makes it possible to use a multichannel pipettor when setting up RT activity assay.
6. Visualizing the filter on an X-ray film helps identify any potential irregularities that may have occurred due to spotting or washing (i.e., background level) as well as help verify the numerical data that accompanies it.
7. Alternatively, you can cut out the radioactive areas that represent each spot from the filters and count in a liquid scintillation counter. It's also possible to read it on a phosphor imager screen to generate numerical data.
8. It is important to note that the $TCID_{50}$ assay and competition assays are performed with PBMC from the same donor and blood draw.
9. The competitions may be kept in culture for up to 14 d depending on the type of viruses used. For example, syncytium-inducing (SI) or X4 tropic viruses may take shorter duration to replicate than nonsyncytium-inducing (NSI) or R5 tropic viruses resulting in peak RT activity on days 7–9 postinfection, whereas NSI viruses may exhibit peak RT activity on days 11–14 postinfection.

10. Use 1-mL filter tips to aspirate cells and supernatant and you can use the tip to gently scrape off the cells that may have adhered to the wells.
11. The supernatant may be used for further analysis such as viral RNA purification.
12. It is possible to skip the PCR purification step if the nested PCR products look clean. If you observe nonspecific bands, QiaQuick gel purification step should be performed.
13. Rapid cooling facilitates the formation of stable hereroduplexes between highly divergent *env* fragments. This aids in the formation of more heteroduplexed fragments compared to the more stable homoduplexes.
14. Excess gel particles in the wells can alter DNA migration and it is important to level the wells. To clean out excess gels that might have remained in the wells, cut out a spacer (0.75 mm) into a size that can fit in the wells. You can use this "gel cleaning tool" to peel off gel particles stuck in the wells and follow with a syringe and a 21-gage needle to wash off the gel particles.
15. We found it easier to visualize the gels on an X-ray film as supposed to phosphor-imager. You can use the X-ray film to aid in the analysis of the phosphor-imager scan.
16. Evaluate virus production in competition relative to production in the monoinfection controls for any differences in probe binding.

References

1. Domingo, E. and Holland, J. J. (1997) RNA virus mutations and fitness for survival. *Annu. Rev. Microbiol.* **51,** 151–178.
2. Novella, I. S., Duarte, E. A., Elena, S. F., Moya, A., Domingo, E., and Holland, J. J. (1995) Exponential increases of RNA virus fitness during large population transmissions. *Proc. Natl. Acad. Sci. USA* **92,** 5841–5844.
3. Harrigan, P. R., Bloor, S., and Larder, B. A. (1998) Relative replicative fitness of zidovudine-resistant human immunodeficiency virus type 1 isolates in vitro. *J. Virol.* **72,** 3773–3778.
4. Martinez-Picado, J., Savara, A. V., Sutton, L., and D'aquila, R. T. (1999) Replicative fitness of protease inhibitor-resistant mutants of human immunodeficiency virus type 1. *J. Virol.* **73,** 3744–3752.
5. Quinones-Mateu, M. E. and Arts, E. J. (2001) HIV-1 fitness: implications for drug resistance, disease progression, and global epidemic evolution, in *HIV Sequence Compendium* (Kuiken, C., Foley, B., Hahn, B. H., Marx, P., McCutchan, F. E., Mellors, J., et al., eds.), Theoretical Biology and Biophysics Group, Los Alamos National Laboratory, Los Alamos, CA.
6. Quinones-Mateu, M. E., Ball, S. C., Marozsan, A. J., Torre, V. S., Albright, J. L., Vanham, G., et al. (2000) A dual infection/competition assay shows a correlation between ex vivo human immunodeficiency virus type 1 fitness and disease progression. *J. Virol.* **74,** 9222–9233.
7. Blaak, H., Brouwer, M., Ran, L. J., dc Wolf, F., and Schuitemaker, H. (1998) In vitro replication kinetics of human immunodeficiency virus type 1 (HIV-1) variants in relation to virus load in long-term survivors of HIV-1 infection. *J. Infect. Dis.* **177,** 600–610.

8. Ball, S. C., Abraha, A., Collins, K. R., Marozsan, A. J., Baird, H., Quinones-Mateu, M. E., et al. (2003) Comparing the ex vivo fitness of CCR5-tropic human immunodeficiency virus type 1 isolates of subtypes B and C. *J. Virol.* **77,** 1021–1038.

29

A Yeast Recombination-Based Cloning System to Produce Chimeric HIV-1 Viruses and Express HIV-1 Genes

Dawn M. Moore, Eric J. Arts, Yong Gao, and Andre J. Marozsan

Summary

Differential phenotypes or properties of HIV-1 gene products in primary virus isolates are difficult to assess due to interference by the high degree of sequence variation across the entire genome. Thus, chimeric viruses provide a powerful tool to study the function of single gene products or genetic elements in the context of a neutral viral genomic backbone. In this chapter, we describe how to produce HIV-1 chimeric viruses utilizing a yeast-based homologous recombination cloning technique to insert env sequences first into a yeast cloning vector and then into the common pNL4-3 virus backbone. This technique is not limited to the env gene, but can be used to build chimeric viruses with any HIV-1 gene or genetic element. This cloning technique involves the use of a shuttle vector that can replicate in yeast and bacterial cells. Along with acting as a shuttle vector for subsequent subcloning into pNL4-3, this construct pRec/env can also be used to express to the env gene product, gp120/gp41, on the surface of mammalian cells. The chimeric viruses produced by this cloning method are capable of undergoing multiple rounds of replication and are therefore very useful to study drug sensitivity, coreceptor usage, and viral fitness as influenced by a single gene or gene fragment of a primary HIV-1 isolate from any group M subtype.

Key Words: Yeast; recombination; transformation; chimeric virus; HIV-1; shuttle vector.

1. Introduction

Chimeric viruses provide a powerful tool to study the function of single gene products or genetic elements in the context of a neutral viral genomic backbone. HIV-1 isolates can be classified as one of three groups, M, N, and O, of which M is further subdivided into 10 subtypes (A–J) and at least 15

From: *Methods in Molecular Biology, Vol. 304: Human Retrovirus Protocols: Virology and Molecular Biology*
Edited by: T. Zhu © Humana Press Inc., Totowa, NJ

stable recombinant forms (CRF01–CRF 15). HIV-1 subtypes differ the most in the envelope (*env*) gene that encodes for the glycoproteins and mediates HIV-1 entry into the host cell. As described here, the yeast recombination/gap repair technique was adopted by our laboratory to clone diverse genetic elements into a common HIV-1 backbone, i.e., a technique more amenable to cloning diverse genetic elements than restriction endonuclease-based cloning methods.

For HIV-1, utilization of chimeric viruses is often a necessity when trying to access the effects of specific genetic elements on phenotype, as genomic sequence diversity can vary up to 30%. For example, the best way to examine how the sequence diversity between envelope genes from different isolates affects HIV-1 entry is to build chimeric viruses containing these diverse *env* genes. The chimeric virus, as just mentioned, provides a common viral backbone and should eliminate influences of other genomic regions on the phenotype of the introduced gene. Unfortunately, HIV-1 sequence heterogeneity leaves few conserved restriction endonuclease sites that are required for traditional cloning techniques. Restriction endonuclease sites are also not located in convenient places for cloning specific genes. Furthermore, the introduction of restriction endonuclease sites into the genome may also cause problems by interrupting multiple and often overlapping open reading frames of HIV-1 ***(1)***. Other chimeric virus systems have been developed, but most rely on pseudotyping the virus with a *trans* gene product (i.e., on a separate genetic element and not part of the viral genome) and thus, limit analyses of a particular phenotype in a single and often incomplete round of replication. Although single-cycle assays can be useful, studying HIV-1 in the context of multiple rounds of replication more closely mimics infection in vivo and is often required to study coreceptor usage, relative or competitive viral fitness, and sensitivity to some inhibitors (e.g., drugs inhibiting viral assembly). The cloning technique described herein produces chimeric viruses suitable for multiple rounds of replication using a yeast recombination/gap repair technique. Yeast recombination is typically much more efficient than recombination in a mammalian cell system, permits isolation and purification of the recombined vector, and thus allows for further manipulation of the introduced gene or genetic element.

Homologous recombination is a natural mechanism used by yeast to repair its chromosomal DNA by performing double-stranded gap repair when a break in the DNA is detected. The yeast will use a homologous gene located in another chromosome as a template to repair the gap in the damaged DNA by utilizing regions of homology between the damaged DNA and the chromosome ***(2)***. Manipulation of the yeast cell's capability to perform homologous recombination between two double-stranded pieces of DNA has been used for many years to modify yeast chromosomes and more recently to clone an insert into a vector. When a plasmid with a double-stranded gap is transformed into a yeast cell along with an insert that contains regions of homology flanking the gap in the plasmid, the yeast will perform homologous recombination to integrate the insert into the plasmid. This method works best when the genes to be inserted into the plasmid do not match a yeast gene; otherwise, this genetic element will also be inserted into the yeast chromosome *(3)*. Yeast homologous recombination is not only extremely efficient at inserting a gene into a plasmid, but it also accommodates large fragments, maintains correct open reading frames, and requires only a small amount of insert DNA (200–300 ng). We have been able to utilize this cloning procedure to produce chimeric viruses containing different HIV-1 genes, or portions of genes, inserted into a common HIV-1 lab strain called pNL4-3. This method also allows for easy bulk cloning of genes representing the entire HIV-1 population (also referred to as quasispecies) found within an infected patient. Finally, the product of this cloning method can also be used to express an HIV-1 gene in mammalian cells.

In this chapter we describe how to produce HIV-1 chimeric viruses utilizing a yeast-based homologous recombination cloning technique to insert *env* sequences into the common pNL4-3 virus backbone. **Figure 1** outlines the main steps of this protocol. The first step is to prepare a shuttle vector with the necessary selectable marker(s) and containing the genes required to replicate in yeast and bacteria. The selectable marker is generally a gene encoding for enzyme(s) involved in amino acid biosynthetic pathway that compensates for this defect in mutated yeasts and permits growth on media lacking the specific amino acid. Ultimately this vector will contain the gene of interest for subsequent subcloning as well as for direct expression in mammalian cells. We have produced such a shuttle vector, pRec/env, that is designed specifically for the insertion of the HIV-1 *env* gene (*see* **Fig. 2**). This vector has also been modified to insert the *gag*, *pol*, accessory genes, or subdomains of these genes by replacing the HXB2 *env* region of pRec/env with these alternative genes from the HXB2 HIV-1 strain (*see* **Fig. 2**).

The *URA3* gene is used for negative selection and in order to select specifically for the yeast cells containing plasmids with the correct insert following the correct homologous recombination event. The second step to produce a chimeric virus is to insert *URA3* into pRec/env in place of the HXB2 *env* gene by using yeast homologous recombination (*see* **Fig. 1, step 4**). Growth on plates lacking uracil and leucine ensures the correct insertion of *URA3* and the maintenance of the shuttle vector in the yeast cells containing the leucine gene (*see* **Fig. 1, step 5**). Following the insertion of *URA3*, the *env* gene of interest (*env* II) can now be inserted in place of *URA3* by yeast homologous recombination

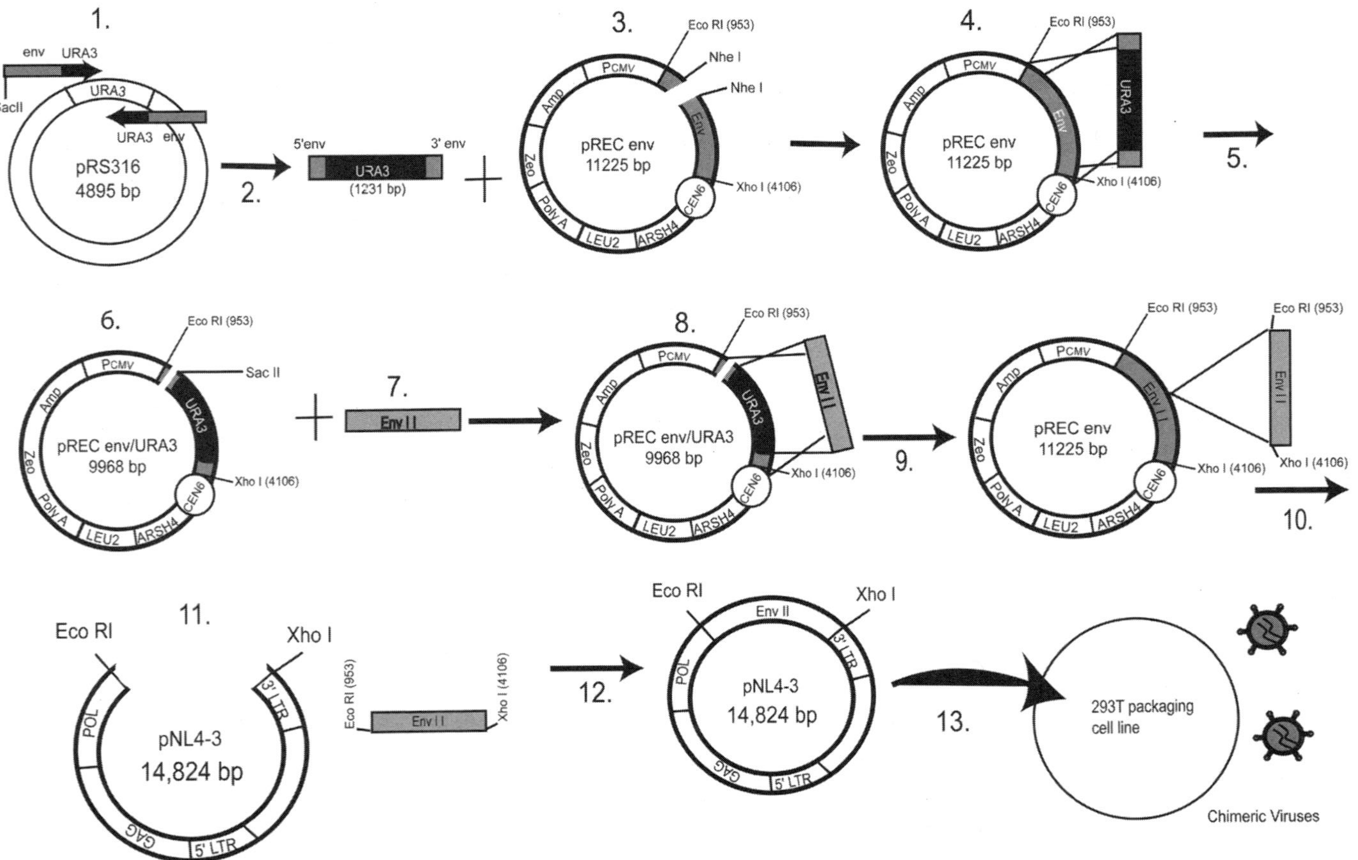
1.
env URA3
SacII
URA3
URA3 env
pRS316
4895 bp
2.
5'env
3' env
URA3
(1231 bp)
3.
Eco RI (953)
Nhe I
Nhe I
PCMV
Amp
Zeo
Poly A
LEU2
ARSH4
CEN6
Env
pREC env
11225 bp
Xho I (4106)
4.
Eco RI (953)
URA3
Env
pREC env
11225 bp
Xho I (4106)
5.
6.
Eco RI (953)
Sac II
URA3
pREC env/URA3
9968 bp
Xho I (4106)
7.
Env I I
8.
Eco RI (953)
Env I I
URA3
pREC env/URA3
9968 bp
Xho I (4106)
9.
Eco RI (953)
Eco RI (953)
Env I I
pREC env
11225 bp
Xho I (4106)
Xho I (4106)
10.
11.
Eco RI
Xho I
POL
3' LTR
pNL4-3
14,824 bp
GAG
5' LTR
Eco RI (953)
Env I I
Xho I (4106)
12.
Eco RI
Xho I
Env II
POL
3' LTR
pNL4-3
14,824 bp
GAG
5' LTR
13.
293T packaging
cell line
Chimeric Viruses

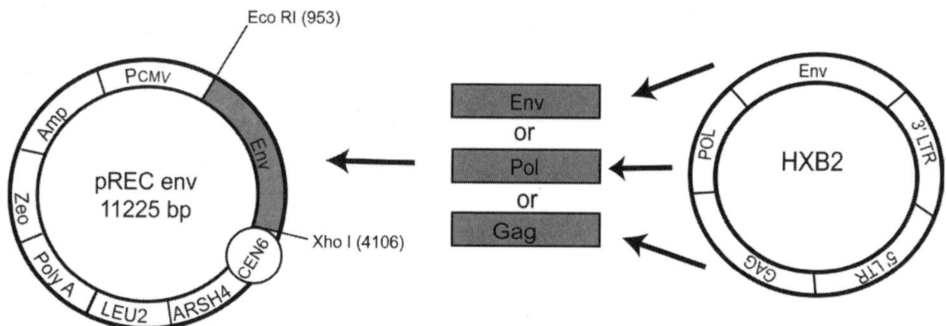

Fig. 2. pRec/env shuttle vector can be used to clone any HIV-1 gene. The Prec/env vector contains *env* from HXB2 that is used to recombine an *env* gene from an HIV-1 isolate of interest. *gag*, *pol*, or any HIV-1 gene or subdomain of a gene from HXB2 may be placed into the pRec vector in order to allow the insertion of that gene of interest from a primary isolate via yeast recombination.

(*see* **Fig. 1, steps 6–8**). This recombination is facilitated on yeast plates containing 5-fluoroorotic acid (5-FOA), which is toxic to cells expressing *URA3* (*see* **Fig. 1, step 9**). Once the *env* II is inserted into pRec/env/URA3, the plasmid is easily extracted from the yeast and shuttled into *E. coli* to produce many

---

Fig. 1. *(opposite page)* Schematic representation of yeast recombination technique to produce chimeric viruses. (1) The first step in the yeast recombination technique is to create primers that will amplify the *URA3* gene from the pRS316 vector and have 40 nucleotides of homology to the regions in the pRec/env vector flanking the HXB2 gene that will be replaced (in this example it is *env*). The sense primer must also contain the SacII restriction endonuclease site at the 5' end of the primer. (2) PCR-amplify the URA3 gene with the designed primers. (3) Digest the pRec/env vector with NheI to create a gap in the vector. (4) Perform a yeast transformation with the *URA3* gene and pRec/env to recombine the *URA3* gene into the pRec/env vector. (5) Select for correct recombinants on C–Leu–Ura plates. (6) Digest the newly created pRec/Env/Ura vector with SacII to linearize the vector for another yeast transformation. (7) PCR-amplify the Env II gene from an isolate of interest, making sure to include the region of the gene with homology to the Env region left in the pRec/env/Ura3 vector following the insertion of *URA3*. (8) Transform *env* II and the digested pRec/env/URA3 into yeast and allow yeast recombination by (9) selection on C–Leu 5–FOA plates. (10) Digest pRec/env II with EcoRI and XhoI to cut out the newly inserted *env* II gene. (11) Digest pNL4-3 with EcoRI and XhoI to prepare the vector for ligation with *env* II while removing *env* from the pNL4-3 vector. (12) Ligate *env* II into pNL4-3. (13) Transfect the ligated pNL4-3 into 293T cells to produce chimeric virus expressing *env* II in the pNL4-3 HIV-1 backbone.

copies of the plasmid. The *env* II gene is now flanked by specific restriction endonuclease sites in the pRec/env plasmid that permits the shuttling of the env gene from pRec/env into pNL4-3 (*see* **Fig. 1**, **steps 10–12**). The pNL4-3 containing the exogenous *env* gene is then transfected into a packaging cell line to produce chimeric viruses containing *env* II in a common HIV-1 backbone (*see* **Fig. 1**, **step 13**). Alternatively, the shuttle vector pRec/env II can be modified and transfected into mammalian cells to express *env* II glycoproteins on the surface of the cell. This is particularly useful to study the effects of *env* II on the fusion and entry steps of the virus.

2. Materials

2.1. Preparation of the Shuttle Vector and Insert

1. pRec/env shuttle vector (*see* **Fig. 2** and **Note 1**).
2. Primers to amplify the *URA3* gene from pRS316 with overhangs homologous to *env* (ENV-START-RECOM and ENV-ENV-RECOM) (*see* **Table 1**).
3. *Taq* DNA polymerase (5 U/μL) and buffer containing a final magnesium concentration of 1.5 m*M*.
4. 10X dNTP mix (dATP, dCTP, dGTP, dTTP).
5. pRS316 plasmid *(4)*.
6. Restriction endonuclease NheI and buffer.
7. Dialysis tubing and clamps (Spectra/Por molecular porous membrane tubing from Spectrum Laboratories, Inc.).
8. 100% cold ethanol and 70% ethanol at room temperature.
9. 3 *M* sodium acetate, pH 5.2.

2.2. Yeast Transformation/Recombination

1. Yeast strain SL1463: *Saccharomyces cerevesiae MAT α leu2 ura3-52 his3-Δ200 trp1*.
2. YPD media: Dissolve 10 g yeast extract, 20 g peptone, and 20 g dextrose in 1 L of water. Filter-sterilize. For plates, add 20 g of Difco agar and boil to sterilize. Plate 25 mL per 100-mm plate.
3. Yeast complete minimal medium (CMM)–Leu–URA: 0.67 g CSM-Leu-Ura (Bio 101 Systems), 1 L water. Filter-sterilize.
4. Yeast complete minimal medium–Leu: 0.67 g CSM-Leu-Ura, 0.024 g uracil (Bio 101 Systems), 1 L water. Filter-sterilize.
5. 10X TE buffer (pH 7.5): 0.1 *M* Tris-HCl (pH 7.5), 0.01 *M* EDTA (pH 8.0). Filter-sterilize.
6. 10X lithium acetate: 1 *M* lithium acetate (pH 7.5). Filter-sterilize.
7. 50% PEG 4000 w/v in water. Filter-sterilize.
8. PEG solution: 8 vol 50% PEG 4000, 1 vol 10X TE buffer (pH 7.5), 1 vol 10X lithium acetate. Make this solution fresh for each use.

Table 1
Primer Sequences, Restriction Enzymes, and Selection Media Used for Each Stage of Yeast Transformation

Transformation	Primer sequence to produce insert[a]	Vector and digestion	Selection medium	Step in Fig. 1
Insert *URA3* into pRec/env	ENV-START-RECOM- tgtgggtcacagtctattatggggtacctgtgtg gaagga*ccgcgg*agattgtactgagagtgcac ENV-END-RECOM- cttttgaccacttgccacccattttatagcaaag ccctttcctgtgcggtatttcacaccg	pRec/env *Sac II*	CMM-Leu-URA	Steps 1–5
Insert *env* II into pRec/env/URA3	ENV-START- tgtgggtcacagtctattatgg ENV-END- cttttgaccacttgccacccat	pRec/env/URA3	CMM-Leu 5-FOA	Steps 6–9
Test for correct insertion of *env* II into pRec/env II	E80- ccaattcccatacattattgtg E125- caatttctgggtcccctcctgagg			After step 12

[a]Underlined sequence indicates Sac II restriction site.

9. Single-stranded sheared salmon sperm carrier DNA denatured at 88°C for 5 min (BD Biosciences Clontech, Palo Alto, CA).
10. 30°C incubator suitable for yeast culture.
11. Shaker at 30°C.
12. Water bath at 42°C.
13. Sterile Millipore water.

2.3. Crude Plasmid DNA Extraction From Yeast

1. Glass beads (425–600 μm, acid-washed; Sigma).
2. Breaking buffer: 2% (v/v) Triton X-100, 1% (v/v) sodium dodecyl sulfate (SDS), 100 m*M* NaCl, 10 m*M* Tris-HCl (pH 8.0), and 1 m*M* EDTA (pH 8.0).
3. Phenol/chloroform/isoamyl alcohol.
4. DH10B electrocompetent bacterial cells (Invitrogen), electroporator, and 1-mm electroporation cuvets.
5. Luria broth (LB): In 1 L of water add 10 g NaCl, 5 g yeast extract, and 10 g tryptone peptone. Autoclave the broth before use. For ampicillin plates add 15 g of agar to the other ingredients as well as ampicillin to a final concentration of 50 μg/mL.
6. Means to extract a plasmid out of bacteria (Qiagen Midiprep kit).

2.4. Homologous Recombination to Replace URA3 With Gene of Interest

1. Restriction endonuclease SacII and buffer.
2. Polymerase chain reaction (PCR) primers to amplify the HIV-1 gene of interest to insert into pRec/env/URA3 (*env* II in this case). The primers should contain 40 nt of homology to the region in the pRec/env/URA3 vector that flanks the region where *URA3* was inserted (*see* **Note 1**). The primers for *env* II are ENV START and ENV END (*see* **Table 1**).
3. 5-FOA/C–Leu plates (18 plates): 10 g dextrose, 0.85 g yeast nitrogen base, 2.5 g $(NH_4)_2SO_4$, 0.335 g CSM-Leu-Ura, 12 mg uracil, 0.5 g 5-fluoroorotic Acid (5-FOA), 15 g Bacto-Agar (Difco), and 470.5 mL water. All materials used to produce plates must be sterilized. 5-FOA is heat-sensitive and will be inactivated when heated above 65°C. Thus, dissolve the Bacto-Agar in 285 mL of water and either boil in a microwave or autoclave. Dissolve the remaining ingredients in 185.5 mL water and filter-sterilize. The agar must be cooled to below 65°C before mixing in the solution containing 5-FOA. Pour 25 mL into a standard 100-mm plate. (**Note:** 5-FOA is harmful and inhalation and exposure to skin must be avoided.)

2.5. Ligation of env Into pNL4-3

1. Restriction endonucleases EcoRI and XhoI and buffers. It is useful to obtain both enzymes from the same company to improve the success of this double digest.
2. Shrimp alkaline phosphatase (Fermentas).
3. T4 DNA ligase and buffer (New England Biolabs, Beverly, MA).
4. pNL4-3 plasmid.
5. PCR primers to amplify the C2-V3 loop of the *env* gene. The primers are E80 and E125 (*see* **Table 1**).
6. Gel purification kit (Qiagen).

2.6. Transfection of pNL4-3 Into 293T Cells to Produce Virus

1. Biohazard level 2 safety facility.
2. 293T packaging cell line (National Institutes of Health [NIH] AIDS Research and Reference Reagent Program).
3. Cell media: Dulbecco's modified Eagle's medium (DMEM) supplemented with 10% fetal calf serum (FCS), 100 U/mL penicillin, and 100 μg/mL streptomycin.
4. Effectene Transfection Kit (Qiagen). Buffer EC and enhancer solution from the kit.
5. Phosphate-buffered saline, sterile for tissue culture.
6. Cell line to passage chimeric viruses: U87 CD4+/CCR5 or U87 CD4+/CXCR4 cells are recommended from the NIH AIDS Research and Reference Reagent Program.

2.7. Expression of env in Mammalian Cells Using Shuttle Vector

1. Restriction endonuclease XbaI and buffer.
2. HeLa cell line from NIH AIDS Research and Reference Reagent Program to ex-

press the HIV-1 *env* gene on the surface of the cell.

3. Cell media to select for transfected cells: DMEM supplemented with 10% FCS, 100 U/mL penicillin, 100 mg/mL streptomycin, and 100 mg/mL zeocin (Invitrogen).

3. Methods

3.1. Preparation of Shuttle Vector and Insert

This method describes specifically the insertion of *env* into a pNL4-3 backbone. However, the same protocols may be used to insert other genes or even parts of HIV-1 genes into pNL4-3. This may be done by changing the HIV-1 gene in the shuttle vector and PCR-amplifying the gene of interest with regions homologous to that gene in the shuttle vector (*see* **Fig. 2**). The following protocol describes the first step to prepare the pRec/env shuttle vector for the insertion of a particular *env* gene-of-interest (*env* II), which involves replacing the HXB2 *env* gene in pRec/env with the *URA3* gene via yeast homologous recombination (*see* **Fig. 1**, **steps 1–5**). This will allow for selection of the insertion of *env* II later.

1. PCR-amplify the *URA3* gene from the pRS316 vector with primers that contain 20 nucleotides homologous to *URA3* for PCR amplification and 40 nucleotides homologous to the flanking region of *env* in the pRec/env vector (ENV-START-RECOM and ENV-END-RECOM) (*see* **Note 2** and **Fig. 1**, **step 1**). For primer sequences, *see* **Table 1**. Make a master mix containing the following reagents/PCR reaction: 5 μL 10X PCR buffer, 1 μL 10 m*M* dNTPs, 1 μL ENV-START-RECOM primer (20 pmol/μL), 1 μL ENV-END-RECOM primer (20 pmol/μL), 0.25 μL *Taq* DNA polymerase, and 41.75 μL water. PCR-amplify with the following program: 95°C 4 min, 95°C 30 s, 63°C 1 min, 72°C 1 min 30 s, 4°C hold. Amplify for 35 cycles.
2. Create a gap in the pRec/env shuttle vector by digesting the plasmid with Nhe I (*see* **Fig. 1**, **step 3**).
3. Separate the digestion products on an agarose gel and obtain the digested vector by cutting out the 9652-base-pair region from the gel and purifying the DNA (*see* **Note 3**).

3.2. Yeast Transformation/Recombination

This protocol describes how to transform yeast using the lithium acetate-based approach. This is a general method that can be used for the insertion of any PCR product into a vector that contains the machinery and selection markers to grow as an episome in yeast cells (*see* **Note 4**). The following protocol describes the insertion of URA3 into pRec/env.

1. Streak a YEPD plate with SL1463 and grow the yeast for 2–4 d at 30°C (*see* **Note 5**).
2. Pick a single colony from the yeast plate and inoculate cells into 5 mL of YPD medium and grow overnight to stationary phase on a shaker at 30°C.

3. Add 500 µL of the overnight culture to 50 mL of YPD medium. Grow the cells on a shaker at 30°C for 4.5–6 h. This is the time frame in which the yeast cells will be in log-phase growth.
4. Pellet the cells by centrifugation for 5 min at 4000*g* at room temperature.
5. Resuspend the cells in 1.0 mL of sterile water and transfer the cells to a 1.5-mL microfuge tube.
6. Centrifuge the cells for 5 min at 5000–6000*g* at room temperature and resuspend the cells in 1.0 mL of freshly prepared TE/LiAc solution (1 vol of 10X TE, 1 vol of 10X LiAc, 8 vol of sterile water). The yeast cells are now competent for transformation and may be incubated for up to 1 h at 30°C before the next step.
7. Mix 50 µL of competent yeast cells with 1 µg of the digested transforming pRec/env vector, 3 µg of the insert PCR product (*URA3* or *env*), and 50 µg of denatured salmon sperm carrier DNA in a 1.5-mL tube. For efficient yeast transformation the DNA cannot dilute the LiAc by more than 10%. Thus, if the volume of DNA exceeds 20 µL, appropriate levels of 10X TE and 10X LiAc should be added to the transformation reaction.
8. Add 300 µL of sterile PEG solution containing TE and LiAc to the transformation reaction and mix thoroughly.
9. Incubate the yeast cells on a shaker at 30°C for 30 min to allow the DNA to enter the competent cells.
10. Heat-shock the cells in a water bath at 42°C for 15 min to close the competent cells. This step is very important to retain the plasmid in the yeast cells.
11. Spin down the cells in a microfuge for 10 s at full speed (~16,000*g*).
12. Resuspend the cell pellet in 500 µL of sterile water and plate 50 µL, 200 µL, and 250 µL onto plates containing the correct selection medium. In the case of the insertion of *URA3* into the pRec/env vector, CMM–Leu–URA must be used to select for the correct insertion of the *URA3* gene into the plasmid and the maintenance of pRec/env in the yeast cells. Plates should be warmed to 25–30°C before plating the yeast cells.
13. Incubate the plates at 30°C for 2–5 d until colonies form on the plate. Colonies should be ~1–2 mm in diameter.

3.3. Crude Plasmid DNA Extraction From Yeast

1. Screen the yeast colonies for cells containing the plasmid with the *URA3* insertion. The same primers used to amplify *URA3* for the insertion can be used to screen for the presence of the *URA3* gene in the yeast cells (*see* **Note 6**).
2. Inoculate 2 mL of selective media with a yeast colony verified to contain the *URA3* insert and grow the culture overnight to stationary phase in a shaking incubator at 30°C. The selective medium used in this step contains the same selection drugs as the plates used for the yeast transformation in **Subheading 3.2., step 12** (CMM–Leu–URA in this case). The selection medium varies depending on what gene is being inserted (*URA3* or *env*) via yeast homologous recombination (*see* **Table 1**).
3. Spin 1.5 mL of the overnight culture for 5 s at high speed in a microfuge.

4. Resuspend the cell pellet in 200 μL of breaking buffer.
5. Add 0.3 g of glass beads to break up the yeast cells and 200 μL of phenol/chloroform/isoamyl alcohol and vortex for 2 min at the highest speed.
6. Microfuge the cells for 5 min at room temperature at the highest speed. Remove approximately 50 μL of the top aqueous layer containing the plasmid DNA.
7. Transform competent bacterial cells with 1–2 μL of the aqueous layer and store the remaining aqueous layer at –20°C for later use. Plate the competent bacteria on LB ampicillin plates and let the bacteria grow overnight until colonies form at 37°C.
8. Pick several bacterial colonies and inoculate LB media to expand the plasmid population. pRec/env/URA3 is a very low copy number plasmid and enough bacteria must be grown up to produce at least 10 μg of DNA for a restriction digest, purification, and another yeast transformation event. Extract the plasmid from the bacteria using a plasmid DNA extraction kit (Qiagen).

3.4. Homologous Recombination to Replace URA3 *With Gene of Interest*

Following the insertion of the *URA3* gene into the shuttle vector pRec/env in place of the HXB2 *env*, the *env* II gene from an individual isolate that will be cloned into pNL4-3 must be PCR-amplified from the isolate of interest. This gene must be amplified with primers that contain about 40 nucleotides of homology to the pRec/env region that flanks the *URA3* gene that is now inserted into that vector. Generally, there will be regions of HXB2 *env* left in the plasmid flanking *URA3* after its insertion that can be used to insert this second *env* gene without creating primers with specific regions of homology. In the system that we have established, we have sequenced the *env* II gene and compared the sequence to HXB2 *env* to ensure that the 40 nt of *env* flanking the *URA3* gene are homologous between the two sequences. Again, a gap must be created in the pRec/env/URA3 vector first and then the new *env* II PCR product can be inserted by homologous recombination.

1. Digest the pRec/env/URA3 plasmid with SacII to create a gap for yeast homologous recombination (*see* **Fig. 1**, **step 6** and **Note 7**).
2. PCR-amplify the *env* gene from an isolate of interest to study (*see* **Fig. 1**, **step 7**). The primers used to amplify this region need to contain 40 nucleotides of homology to the pRec/env/URA3 vector region that flanks the *URA3* gene to promote recombination of the *URA3* gene out of the vector with the insertion of the *env* II gene. An example of the primers for *env* II are ENV-START and ENV-END (*see* **Table 1** for sequences), but homology between these primers and your *env* gene needs to be verified. Prepare a master mix containing 5 μL 10X PCR buffer, 1 μL 10 m*M* dNTPs, 1 μL ENV-START primer (20 pmol/ul), 1 μL ENV-END primer (20 pmol/μL), 0.25 μL *Taq* DNA polymerase (5 U/μL), and 36.75 μL water. Add 45 μL of master mix to each tube containing 5 μL of DNA containing *env* II. PCR amplify with the following program: 95°C 4 min, 95°C 30 s, 55°C 45 s, 72°C 3 min, 70°C for 10 min, and 4°C hold. Amplify 35 cycles (*see* **Note 8**).

3. Complete a yeast transformation as described in **Subheadings 3.2.** and **3.3.** with the newly created and digested pRec/env/URA3 plasmid and new *env* II gene. Select for the correct recombination by plating the transformation on C–Leu 5–FOA plates (*see* **Fig. 1**, **steps 8** and **9** and **Note 9**).
4. Obtain at least 15 μg of pRec/env II plasmid DNA from a plasmid midiprep for very low copy number plasmids.

3.5. Ligation of env Into pNL4-3

The ligation step of *env* II into pNL4-3 requires the introduction of two restriction endonuclease sites flanking the *env* gene that will be compatible with the restriction endonuclease sites used to cut the *env* gene out of pNL4-3. These compatible restriction sites are introduced to *env* II when this gene is inserted into pRec/env that contains these sites. Once compatible restriction endonuclease sites exist between the *env* gene and pNL4-3, the gene can be inserted into pNL4-3 such that the open reading frame of the gene will not be disrupted. The following protocol is optimized for this difficult ligation step of a 3-kb product into a 14-kb vector.

1. Digest pRec/env II with EcoR I and Xho I and obtain the 3154 base pair *env* II insert by gel purification (*see* **Fig. 1**, **step 10**). A Qiagen gel purification kit may be used for this purification.
2. Digest pNL4-3 with EcoR I and Xho I followed directly by dephosphorylation of the plasmid with shrimp alkaline phosphate. Dephosphorylation prevents the plasmid from ligating back together with itself. Dephosphorylation can be performed in the restriction endonuclease buffer following inactivation of the restriction enzymes. Obtain the 11,680-bp vector with the native *env* gene cut out by gel purification. To obtain a sufficient amount of digested product after purification, digest 15 μg of DNA (*see* **Note 2** and **Fig. 1**, **step 11**).
3. Ligate *env* II into the digested pNL4-3 in a ligation reaction using ~100 ng of pNL4-3 and 300 ng of the *env* II insert cut out of the pRec/env II vector. Use T4 DNA ligase from New England Biolabs and ligate in a 30 μL reaction at 16°C overnight followed by inactivation of the enzyme at 65°C for 10 min (*see* **Fig. 1**, **step 12** and **Note 10**).
4. Purify the ligated plasmid from the ligation reaction in chloroform/isoamyl alcohol followed by an ethanol precipitation. Add 50 μL of chloroform/isoamyl alcohol to the ligation reaction and vortex on high speed for 2 min. Centrifuge the reaction for 5 min at about 16,000*g*. Remove the top aqueous layer containing the DNA and add sodium acetate at 10% of the volume of the aqueous layer followed by 100% ethanol at 2.5× the total volume. Precipitate the DNA overnight at –20°C. Centrifuge the DNA for 30 min at approx 16,000*g* and remove the 100% ethanol. Wash the DNA in 1 mL of 70% ethanol, centrifuge for 10 min at about 16,000*g*, remove all the liquid from the DNA pellet, dry the pellet and resuspend the DNA in 20–50 μL of water.

5. Transform 1–2 μL of the purified ligated pNL4-3 plasmid into electrocompetent bacterial cells (Invitrogen) and plate the transformed bacteria onto LB ampicillin plates. Make sure to plate a positive control and negative control for the transformation step. Generally the competent cells will be provided with a positive control and the negative control is competent cells that have undergone electroporation without a plasmid.
6. Screen the bacterial colonies for colonies that contain a plasmid with the *env* II gene by amplifying the V3 loop of the *env* gene. PCR can be done directly from the colonies or may be done on plasmids purified from individual colonies. PCR master mix/colony: 5 μL 10X PCR buffer, 1 μL 10 m*M* dNTPs, 1 μL E80 (20 pmol/μL) (*see* **Table 1** for primer sequences), 1 μL E125 (20 pmol/μL), 0.25 μL *Taq* DNA polymerase (5 U/μL), 21.75 μL of water. Pick a single bacterial colony into 20 μL of water and denature at 95°C for 10 min. Add 30 μL of master mix to each tube containing a denatured colony. PCR with the following conditions: 95°C for 4 min, 95°C for 30 s, 55°C for 30 s, 72°C for 40 s, 4°C hold. Complete 35 cycles.
7. Grow colonies that are positive for pNL4-3 containing an *env* gene and extract the plasmid from the bacteria using a plasmid DNA midiprep kit and the very low copy number protocol (Qiagen) (*see* **Note 11**).

3.6. Transfection of pNL4-3 Into 293T Cells to Produce Virus

1. This protocol utilizes Effectene (Qiagen) for the transfection of pNL4-3 into 293T packaging cells (primary human embryonic kidney cells) to produce chimeric viruses. Other transfection methods may be utilized as well.
2. Culture adherent 293T cells in complete DMEM media containing penicillin/streptomycin and 10% fetal bovine serum. After the cells have been split at least twice and are at confluency again, split the cells to 20–40% confluency in 10 mL of media in a 60-mm plate 24 h before the cells will be used for transfection. The cells need to be 40–80% confluent the day of the transfection. Cells should be incubated under normal growth conditions at 37°C and 5% CO_2.
3. Add 2 μg of pNL4-3 containing *env* II to 300 μL of buffer EC and 16 μL of enhancer from the Effectene kit. Vortex for 1 s and incubate the reaction for 5 min at room temperature.
4. Briefly spin down the reaction and add 60 μL of Effectene transfection reagent. Mix by pipetting up and down five times. Incubate the reaction for 10 min at room temperature.
5. While the reaction is incubating, rinse the 293T cells with PBS. Be careful during this wash step because 293T cells are loosely adherent. After the wash, add 7 mL of complete DMEM medium to the dish.
6. Add 3 mL of medium to the transfection reaction and add that mixture dropwise to the 293T cells swirling after each addition to distribute the transfection mix evenly over the cells. Incubate the cells with the transfection reaction for 12–24 h at 37°C and 5% CO_2.

7. The day after transfection remove the media containing the transfection reagents from the cells and replace with 10 mL of fresh complete DMEM.
8. To harvest the virus, collect the supernatant from the cells 48 h post transfection. Replace the media removed with fresh complete DMEM. Harvest the virus by centrifuging the cells for 15 min at 3000*g*. Aliquot the supernatant into 1.5-mL tubes and store at –140°C, on liquid nitrogen, or at –80°C.
9. At 48 h after harvesting the first round of virus, make a second collection of virus by repeating **step 7**. 24 h after the second harvest, a third harvest can be made, but is not often necessary.
10. The virus harvested from the 293T cells may be up to 60% noninfectious and therefore need to be propagated through one round of replication in U87 cells or human peripheral blood mononuclear cells (PBMCs) to select for infectious virus. Following this passage, the $TCID_{50}$ of the virus should be determined for future application with the chimeric virus.

3.7. Expression of env *in Mammalian Cells Using Shuttle Vector*

1. This protocol utilizes the shuttle vector to express the gene of interest in mammalian cells under the control of the CMV promoter. Basically, the yeast episomal replication machinery *CEN6/ARSH4* along with *LEU2* are removed from the shuttle vector to place the gene of interest directly upstream of the bovine growth hormone polyadenylation sequence. The shuttle vector also contains exons from the *tat* and *rev* genes. *rev* interacts with the rev response element (RRE) within the *env* open reading frame and aids in the transport of unspliced *env* RNA out of the nucleus and enhances the expression of *env* ***(1)***. The pRec/env shuttle vector also contains a zeocin resistance gene so that transfected cells can be placed under zeocin selection to maintain the transfected plasmid.
2. Digest the pRec/env II shuttle vector with Xba I, which removes *CEN6/ARSH4* and *LEU2* from the vector. Separate the digestion products on an agarose gel and gel-purify the 8135-base-pair band by electroelution (*see* **Note 2**).
3. Religate the pRec/envII vector back together with the Xba I compatible ends using NEB T4 DNA ligase. This reaction can be performed at room temperature for 10 min with 1 µL of the ligase in a total volume of 30 µL. Heat inactivate the ligase for 10 min at 65°C.
4. Transform electrocompetent bacterial cells with the ligation product using an electroporator and plate the cells on LB ampicillin plates overnight at 37°C until colonies form.
5. Screen the bacterial colonies for the correct ligation plasmid product using either restriction enzyme digest to verify that there is only one Xba I site left, or by using PCR primers to amplify over the region where the *CEN6/ARSH4* sequence was located.
6. Transform Hela cells (human cervical carcinoma cells) with the plasmid to express the *env* II gene using the Effectene reagent (Qiagen) as described in **Subheading 3.6., steps 2–5**. The pRec/env vector will be transformed in place of pNL4-3.

7. 18–24 h after transfection of the cells, passage the cells 1:5 to 1:10 into selective complete DMEM medium containing 100 mg/mL of zeocin (*see* **Note 12**).
8. Maintain cells under zeocin selection until it is evident that cells that do not contain the pRec/env plasmid have died. At this point the cells should be ready to use and must be maintained under zeocin selection at all times. Verification of the surface expression of the *env* gene by flow cytometry or Western blot analysis is recommended before using these cells in experiments.

4. Notes

1. Each gene in the pRec/env shuttle vector plays an important role in the different cells in which it can replicate (*see* **Fig. 2** for a representation of the vector). The *CEN6/ARSH4* provides a yeast centromere and autonomously replicating sequence to replicate in yeast cells and the leucine gene provides a selectable marker for yeast cells containing this plasmid. The ampicillin resistance gene allows for selection in bacterial cells and the pcDNA3.1 Zeo backbone used to contstruct pRec/env provides the bacterial origin of replication. The CMV promoter and bovine growth hormone polyadenylation sequence allow expression of *env* in mammalian cells and the zeocin gene provides a selectable marker for mammalian cells maintaining the expression plasmid.
2. The longer the region of homology between the PCR product and the pRec/env vector, the more efficient the homologous recombination will be. A study by Fusco et al. found that 30 nucleotides was the minimal amount of homology necessary for yeast recombination and that 40 homologous nucleotides followed by 50 homologous nucleotides yielded even more efficient recombination *(5)*. If a gene other than *env* or a subdomain of a gene has been placed into the pRec vector, the region of homology flanking the *URA3* gene must correspond to the gene that is in the pRec vector.
3. The best method to gel-purify large pieces of DNA following a restriction digestion and subsequent separation of the digestion products on an agarose gel is to electroelute the DNA from the gel. This is done by placing the gel slice containing the plasmid DNA into a small piece of Spectra/Por molecularporous membrane tubing (aka dialysis tubing) (Spectrum Laboratories, Inc.) that is clamped at one end and has been boiled in the buffer used to make the gel (TAE). Place 400 μL of cold gel buffer (TAE) and 3 μL of dilute ethidium bromide into the tubing with the gel slice, remove all bubbles, and clamp the other end of the tubing. Elute the DNA by placing the tubing in an electrophoresis gel box containing buffer such that the gel slices closest to the negative pole to allow the DNA to move out of the gel towards the positive pole. Run the gel box at 70 V for ~30 min and then reverse the poles for 10 s to push the DNA away from the tube wall at the positive end of the tubing. The ethidium bromide allows for visualization of the DNA moving out of the gel under UV light. Once the DNA has electrophoresed out of the gel, it must be ethanol-precipitated to complete the purification. When 400 μL of buffer is used with the gel slice, ethanol precipita-

tion can be done by adding 40 µL of sodium acetate (3 *M*, pH 5.2) (10% of the volume of DNA) and 1 mL of cold 100% ethanol (2.5 × the volume of DNA). Precipitate the DNA overnight at –20°C and spin down the DNA at approx 16,000*g* for 30 min. Wash the DNA with 1 mL of 70% ethanol at room temperature and centrifuge at approx 16,000*g* for 10 min. Dry the DNA and resuspend in 20–50 µL of water.

4. To help troubleshoot potential problems with the yeast transformation/recombination, both a positive and negative control transformation should be performed. The best negative control is to transform the digested shuttle vector along with the salmon sperm DNA but no insert and plate on the same selection plates as the samples. The best positive control is derived from the successful insertion of *URA3* into pRec/env that has been verified by sequence analysis or any plasmid containing both the leucine and uracil genes along with yeast replication machinery.
5. This yeast starter culture can be stored at 4°C for 1 mo and used for subsequent transformations. Longer storage times result in less efficient yeast transformation.
6. Yeast colony PCR can determine if *URA3* was inserted into pRec/env. Make a master mix containing the following reagents/colony to be tested: 5 µL 10X PCR buffer, 1 µL 10 m*M* dNTPs, 1 µL ENV START RECOM primer (20 pmol/µL), 1 µL ENV-END RECOM primer (20 pmol/µL), 0.5 µL *Taq* DNA polymerase (5 U/µL), and 21.75 µL water. Pick a yeast colony and place it into 20 µL of water and denature at 95°C for 10 min. Add 30 µL of master mix to each tube containing a colony. PCR-amplify with the following program: 95°C 4 min, 95°C 30 s, 55°C 40 s, 72°C 40 s, 4°C hold. Amplify for 35 cycles.
7. Sac II is a unique restriction enzyme site that is introduced into pRec/env as part of the primer used to amplify *URA3* (*see* **Table 1**). Sac II is not found in pNL4-3 or in most HIV-1 genomes and is therefore a reasonable site to use for the insertion of any HIV-1 gene into pRec/env to be included as part of the primer to amplify *URA3*.
8. Because the envelope gene is a very large product to PCR-amplify, it is often necessary to perform nested PCR. The ENV-START and ENV-END primers may be used as the nested PCR product and primers that lie outside this region in a given HIV-1 genome may be used to perform the external PCR. These primers may be specific for the isolate to be inserted into pRec/env.
9. The gene product of *URA3* (orotidine-5'-phosphate decarboxylase) converts FOA into the toxic substrate 5-fluorouracil, which will kill any yeast cell expressing *URA3*. This allows for robust selection for the insertion of *env* II and the subsequent removal of the *URA3* gene from pRec/env/URA3.
10. It is best to ligate the *env* into pNL4-3 as soon after purification as possible. Long storage times at –20°C may disrupt the integrity of the sticky ends on the insert and vector and hinder ligation.
11. pNL4-3 plasmids should be sequenced in the *env* gene region to verify that *env* II has indeed been ligated into the plasmid and that the *env* gene from pNL4-3 has been removed.
12. Zeocin is sensitive to light and should be protected from light during usage.

References

1. Marozsan, A. J. and Arts, E. J. (2003) Development of a yeast-based recombination cloning/system for the analysis of gene products from diverse human immunodeficiency virus type 1 isolates. *J. Vir. Methods* **111,** 111–120.
2. Orr-Weaver, T. L. and Szostak, J. W. (1983) Yeast recombination: the association between double-strand gap repair and crossing-over. *Proc. Natl. Acad. Sci. USA* **80,** 4417–4421.
3. Longtine, M. S., McKenzie III, A., Demarini, D. J., Shah, N. G., Wach, A., Brachat, A., et al. (1989) Additional modules for versatile and economical PCR-based gene deletion and modification in *Saccharomyces cerevisiae. Yeast* **14,** 953–961.
4. Sikorski, R. A. and Hieter, P. (1989) A system of shuttle vectors and yeast host strains designed for efficient manipulation of DNA in *Saccharomyces cerevisiae. Genetics* **122,** 19–27.
5. Fusco, C., Guidotti, E., and Zervos, A. S. (1999) In vivo construction of cDNA libraries for use in the yeast two-hybrid system. *Yeast* **15,** 715–720.

30

PCR Amplification, Cloning, and Construction of HIV-1 Infectious Molecular Clones From Virtually Full-Length HIV-1 Genomes

Philip K. Ehrenberg and Nelson L. Michael

Summary

The development of mixtures of highly processive and high-fidelity thermostable DNA polymerases has enabled the routine recovery of DNA sequences in excess of 25 kb generated by polymerase chain reaction. This powerful tool has been instrumental in the ability to recover virtually full-length HIV-1 proviral DNA as a single, contiguous fragment. Such fragments allow for the clean interpretation of the genomic organization of HIV-1 provirus, as they are not confounded by molecular mosaicism that accrues to overlapping subgenomic amplification strategies. We detail here a robust procedure to produce virtually full-length, single contiguous 9.2-kb HIV-1 amplimers whose full-length infectious potential is reconstituted upon cloning into long terminal repeat-replacement vectors. Large numbers of HIV-1 proviral clones can now be quickly generated and screened to identify the fraction of the viral quasispecies with the highest capacity for viral replication. The methods used to construct long terminal repeat-replacement vectors, amplify HIV-1 provirus, reconstitute full-length provirus, and recover viral stocks will be illustrated using a circulating recombinant form 1 (CRF01_AE, formerly known as subtype E) primary isolate.

Key Words: Polymerase chain reaction; provirus; HIV-1; culture; long terminal repeat (LTR); molecular cloning.

1. Introduction

The importance of human immunodeficiency virus type 1 (HIV-1) infectious molecular clones as tools in dissecting the mechanisms of viral replication and pathogenesis cannot be overstated. They have provided a means to systematically study the extraordinary genetic diversity of HIV-1, as well as

From: *Methods in Molecular Biology, Vol. 304: Human Retrovirus Protocols: Virology and Molecular Biology*
Edited by: T. Zhu © Humana Press Inc., Totowa, NJ

the biological and immunological characteristics of the virus as a whole and its component genes. Previous approaches to recovering infectious molecular clones have included both genomic library construction and reconstitution into full-length virus using amplified subgenomic sequences ***(1)***. These approaches are laborious and inefficient, and susceptible to production of mosaic proviral genomes, respectively. These drawbacks, together with the development of stable DNA polymerase cocktails ***(2,3)***, have enabled the design of a more versatile cloning and screening strategy. By introducing a unique Bgl I restriction site into an amplifying primer, virtually full-length, single contiguous 9.2-kb HIV-1 amplimers can be generated whose full-length infectious potential is reconstituted upon cloning into long terminal repeat (LTR)-replacement vectors ***(4,5)***. Large numbers of HIV-1 proviral clones can now be quickly generated and screened. This is important as a high proportion of defective viral genomes are produced during HIV-1 infection ***(6–8)***. The methods used to construct LTR replacement vectors, amplify HIV-1 provirus, reconstitute full-length provirus, and recover viral stocks will be illustrated using a circulating recombinant form 1 (CRF01_AE, formerly known as subtype E) primary isolate.

2. Materials

1. pLTR(E)1/2Δ replacement vector (*see* **Subheading 3.1.**).
2. pBC KS+ (Stratagene, La Jolla, CA).
3. Competent DH5α cells.
4. Oligonucleotide primers.
5. Expand Long Template PCR Kit, Random Prime Labeling Kit, Gel Extraction Kit, Mini-Prep Processing Kit, p24 antigen capture kit.
6. Enzymes T4 DNA ligase, Bbe I (Takara Mirus Bio., Madison, WI), Bgl I.
7. 9600 thermocycler (Perkin-Elmer, Wellesley, MA).
8. Luria-Bertani (LB) medium and plates.
9. SOC medium.
10. Chloramphenicol.
11. 150-mm nitrocellulose disks.
12. 96-prong plate stamper.
13. Toothpicks.
14. Phenol/chloroform/isoamyl alcohol (25:24:1) (toxic; light- and oxygen-sensitive).
15. Chloroform (toxic).
16. Denaturing solution (pH 13.0–14.0): 1.5 *M* NaCl, 0.5 *M* NaOH.
17. Neutralizing solution (pH 7.5): 1.5 *M* NaCl, 1 *M* Tris-HCl, pH 7.5.
18. *E. coli* Pulser (Bior-Rad. Hercules, CA).
19. UV cross-linker (Stratalinker Model 1800, Stratagene).
20. 100X Denhardt's solution.
21. Herring sperm DNA.
22. Formamide (toxic).

23. Electroporation cuvets, 0.1 and 0.4 cm.
24. Fine- and medium-tip plastic Pasteur pipets.
25. Sealable pouches, various sizes.
26. Pouch sealer.
27. Falcon 2059 tubes.
28. 96-well and 6-well plates.
29. 16 × 125 mm TC-treated culture tubes (Corning Life Sciences, Acton, MA).

3. Methods

The following methods describe (1) the construction of the LTR replacement vector, (2) the preparation of virtually full-length HIV-1 sequences, (3) the reconstitution of full-length HIV-1 proviral sequences, and (4) the recovery of viral stocks.

3.1. LTR Replacement Vector

Construction of the CRF01_AE LTR replacement vector is outlined here. For a more detailed description refer to Salminen et al. *(5)*.

The pLTR(E)1/2Δ replacement vector is a modified chloramphenicol-resistant pBC KS+ vector (Stratagene) containing a CRF01_AE (HIV-1$_{CM235}$) LTR-cloning cassette (**Fig. 1**). A complete 670-bp 5' LTR (LTR1) is amplified with E-f13 (GCATGCCCTAGGTGGAWGGGCTARTTYACTCCAAGA) and E-r7 (GCATGCGGATCCGTTCG**GGCGCC**ACTGCTAGAGATT) primers containing Avr II and Bam HI restriction site clamps, respectively (underlined). E-r7 also contains a natural Bbe I site (boldface). A 122-bp 3' U5 region (LTR2Δ) is amplified using E-f15 (GCATGCGGATCCGCCTTGAGgGCTTAAAGTGGTGTG) and E-r8 (GCATGCGAATTCCTGCTAGAGATTTTTACTCAGT) primers containing Bam HI and Eco RI restriction site clamps, respectively (underlined). E-f15 also contains a unique Bgl I restriction site owing to an introduced T → G transversion (lowercase). After deleting a Bgl I site in pBC KS+ by digestion with Nae I plus Pvu I, blunting the ends, and religating, the resulting vector was linearized with Xba I plus Eco RI digestion. Transcriptionally robust LTR1 and LTR2Δ were digested with Avr II plus Bam HI, and Bam HI plus Eco RI, respectively, ligated, and cloned into the linearized vector. The Avr II and Xba I sites are ligation-compatible and result in the loss of both sites. This vector design enables directional cloning and reconstitution of potentially infectious molecular clones of HIV-1. Finally, this pLTR(E)1/2Δ construct is confirmed by sequencing.

3.2. Preparation of Virtually Full-Length HIV-1 Sequences

The steps involved in the preparation of virtually full-length HIV-1 sequences are described in **Subheadings 3.2.1.–3.2.3.** These include (1) amplification of a virtually full-length 9.2-kb portion of the provirus, (2) purifying the

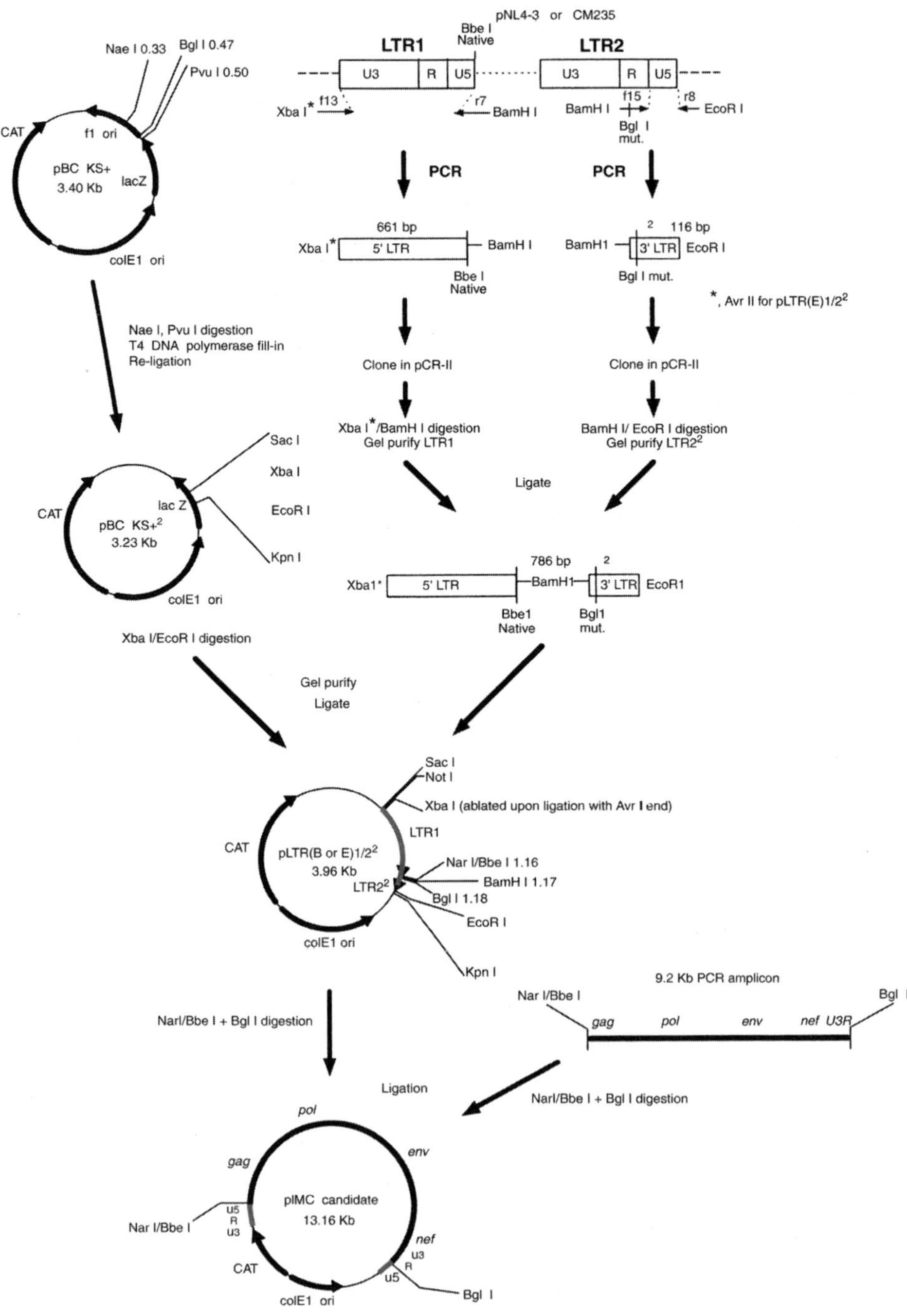
pNL4-3 or CM235
Bbe I
Native
LTR1
LTR2
U3
R
U5
Nae I 0.33
Bgl I 0.47
Pvu I 0.50
Xba I*
f13
r7
BamH I
f15
r8
EcoR I
Bgl I
mut.
CAT
f1 ori
pBC KS+
3.40 Kb
lacZ
colE1 ori
PCR
661 bp
5' LTR
Bbe I
Native
116 bp
BamH1
3' LTR
Bgl I mut.
*, Avr II for pLTR(E)1/2[2]
Nae I, Pvu I digestion
T4 DNA polymerase fill-in
Re-ligation
Clone in pCR-II
Xba I*/BamH I digestion
Gel purify LTR1
BamH I/ EcoR I digestion
Gel purify LTR2[2]
Ligate
Sac I
Xba I
EcoR I
Kpn I
pBC KS+[2]
3.23 Kb
786 bp
Xba1*
BamH1
EcoR1
Bbe1
Native
Bgl1
mut.
Xba I/EcoR I digestion
Gel purify
Ligate
Sac I
Not I
Xba I (ablated upon ligation with Avr I end)
LTR1
pLTR(B or E)1/2[2]
3.96 Kb
LTR2[2]
Nar I/Bbe I 1.16
BamH I 1.17
Bgl I 1.18
EcoR I
Kpn I
9.2 Kb PCR amplicon
Nar I/Bbe I
Bgl I
gag
pol
env
nef
U3R
NarI/Bbe I + Bgl I digestion
Ligation
NarI/Bbe I + Bgl I digestion
pIMC candidate
13.16 Kb
u5
R
u3
nef
Bgl I

amplimer via extraction and precipitation, and (3) restricting the amplimer with Bbe I plus Bgl I and gel purifying.

3.2.1. Amplification

1. Using the Expand Long Template PCR Kit and the MSF12 (AAATCTCTAGCAGTGGCGCCCGAACAG) and MSR5R (ACGTGCcCTCAAGGCAAGCTTTATTGAGGCT) primers titrate the genomic DNA template in 0.2-mL reaction tubes with dNTP, $MgCl_2$, and each primer having final concentrations of 1.4 m*M*, 1.75 m*M*, and 0.5 μ*M*, respectively (*see* **Note 1**). These primers contain Bbe I and Bgl I sites (underlined), respectively. The Bgl I site is introduced via an A → C mutation (lowercase).
2. Amplify in a 9600 thermocycler using the following parameters:
   a. 94°C, 2 min.
   b. 94°C, 10 s; 60°C, 30 s; 68°C, 8 min (10 cycles).
   c. 94°C, 10 s; 55°C, 30 s; 68°C, 8 min (20 cycles).
   d. 68°C, 7 min.
   e. 4°C soak.
3. Mix 3 mL of the polymerase chain reaction (PCR) product with 5 mL of dH_2O and 2 mL of 5X loading buffer.
4. Resolve through a 0.8% TBE gel to confirm the presence of the 9.2-kb amplimer.

3.2.2. Extraction and Precipitation

1. Into a 0.6-mL tube add the 9.2-kb reaction product and an equal volume of phenol/chlorophorm/isoamyl alcohol (25:24:1) (*see* **Note 2**).
2. Vigorously vortex for 20 s.
3. Centrifuge at 14,000*g* at room temperature (RT) for 5 min.
4. Remove the upper aqueous phase to a fresh 0.6-mL tube and repeat **steps 1–3**.
5. Extract once with an equal volume of chloroform (as in **steps 2** and **3**).
6. Transfer the upper aqueous phase to a fresh 0.6-mL tube.
7. Add 0.1 vol of 3 *M* sodium acetate (pH 5.2) and 2 vol of ice-cold 100% ethanol.
8. Gently mix and incubate on dry ice for 20 min (*see* **Note 3**).
9. Centrifuge at 14,000*g* at 4°C for 15 min.
10. Discard the supernatant and wash the pellet with 400 μL of ice-cold 70% ethanol.
11. Centrifuge as in **step 9** for 10 min.
12. Without disturbing the pellet, remove all of the supernatant.
13. Respin for approx 10 s and remove the residual supernatant.
14. Suspend the pellet in dH_2O to a final concentration of about 200–900 ng/μL.

Fig. 1. *(opposite page)* Construction of the long terminal repeat (LTR) replacement vector pLTR(E)1/2Δ is described in **Subheading 3.1.** This cloning strategy enables directional cloning and reconstitution of potentially infectious molecular clones of HIV-1. Finally, this pLTR(E)1/2Δ construct is confirmed by sequencing.

3.2.3. Digestion and Gel Purification

1. In a total volume of 10 μL, digest 1 μg of the amplimer (from **Subheading 3.2.2., step 14**) with 5 to 10 units of Bbe I at 37°C for 5–8 h. Digest the pLTR(E)1/2Δ vector the same way (*see* **Note 4**).
2. Add 0.7 μL of 10X concentrated Bgl I reaction buffer and 5 to 10 units of Bgl I and BbeI to 12 μL total volume with dH_2O (*see* **Note 5**).
3. Digest overnight at 37°C.
4. Add 68 μL of dH_2O to bring the total volume to 80 μL.
5. Extract once each with phenol/chlorophorm/isoamyl alcohol (25:24:1) and chloroform alone (as in **Subheading 3.2.2., steps 1–6**).
6. Add 5X loading buffer to a final concentration of 1X.
7. Resolve this digested amplimer through a 0.8% TAE gel at 80 V for ~2 h (*see* **Note 6**).
8. Use a gel extraction kit to purify this amplimer away from the agarose.

3.3. Reconstitution of Full-Length HIV-1 Sequences

The steps in **Subheadings 3.3.1.** and **3.3.2.** outline the procedures involved in cloning the 9.2-kb HIV-1 amplimer into the LTR replacement vector, and isolating and preparing the resultant colonies for insert screening.

3.3.1. Cloning and *E. coli* Transformation

1. On ice add the following reagents in order to a 0.6-mL tube (*see* **Note 7**):

dH_2O (QS to 12 μL total volume)	
10X ligation buffer	1.2 μL
pLTR(E)1/2Δ (BbeI/BglI)	75 ng
9.2 kb HIV-1 amplimer (Bbe I/Bgl I)	200 ng
T4 ligase	1 μL

2. Mix, spin down to collect reagent at bottom of the tube, and incubate overnight at 14°C.
3. Dilute the ligation product 1:1 with dH_2O.
4. On ice, add 2 μL of this diluted product to 45 μL of ice-thawed competent DH5α cells in a 0.6-mL tube.
5. Gently tap tube to mix, and incubate for on ice for at least 1 min.
6. Transfer 40 μL of this vector/cell mixture into an ice-cooled 0.1-cm electroporation cuvet.
7. Electroporate with an *E. coli* pulser at 1.8 kV.
8. Immediately add 1 mL of SOC medium directly into the cuvet at room temperature.
9. Transfer this mixture to a Falcon 2059 tube using a fine-tip plastic Pasteur pipet.
10. In an orbital shaker/incubator shake at 225 rpm at 32°C for 1.5 h.
11. Spread 350 μL of this culture over a 150-mm LB plate containing 20 μg/mL of chloramphenicol (LB/chloramphenicol plate). The balance of the culture can be stored at 4°C for ~2 wk (*see* **Note 8**).

12. Allow the culture to soak into the plate completely.
13. Invert and incubate plate at 30–32°C, overnight (*see* **Note 9**).

3.3.2. Isolation and Preparation of Clones for Screening

1. Add 200 µL of LB/chloramphenicol medium to wells of 96-well plates.
2. Pick single colonies with sterile toothpicks and inoculate such that each well contains a single colony. Multiple 96-well plates will likely need to be inoculated.
3. Incubate the plates at 32°C for 4 h.
4. Overlay 150 mm LB/chloramphenicol plates with dry 150-mm-diameter nitrocellulose disks.
5. Dip a 96-well plate stamper into the cultures, swirl to mix, and stamp a nitrocellulose-topped plate gently yet firmly to facilitate culture transfer. The 96-well plates can be stored at 4°C for approx 2 wk.
6. Allow cultures to soak in at RT.
7. Invert and incubate the plates at 32°C, overnight.
8. Using forceps remove the colony-bearing disk from the LB plate and place, growth-side up, on the following sequence of buffer-soaked 3MM papers for 5 min each:
   - 10% SDS
   - Denaturing solution
   - Neutralizing solution
   - 2X SSC
9. Lay disks growth-side up on dry 3MM paper for approx 2 min. Do not dry completely; leave somewhat moist.
10. With a pencil draw a pea-sized circle on the disks away from the 96 stamped clones to mark the location of a positive control.
11. Within these circles add 1 mL (~30 ng) of the 9.2-kb HIV-1 amplimer after denaturing at 95°C for 2 min and immediately snap-cooling on ice for about 5 min.
12. UV crosslink the filters, growth-side up, using 1200 mJ. These filters can subsequently be stored at 4°C if they are carefully wrapped in plastic wrap.

3.4. Recovery of Viral Stocks

The final series of steps are outlined in **Subheadings 3.4.1.–3.4.3.** These include sequential screens for (1) reconstituted HIV-1 proviral clones, (2) p24 antigen production, and (3) finally, infectious molecular clones.

3.4.1. Screen for LTR-Reconstituted Proviral Clones

1. Use a Random Prime Labeling Kit to produce a ^{32}P-labeled PCR-amplified 9.2-kb double-strand HIV-1 probe (*see* **Note 10**).
2. Prewet the UV crosslinked disks in dH_2O for about 5 min.
3. Allow excess dH_2O to drip off the disk and insert, back to back, into a 1-pint pouch. Cut the pouch so that it is slightly larger then the disk. Larger pouches can accommodate additional disks.

4. Denature herring sperm (HS) DNA and ^{32}P-labeled probe at 95°C for 2 min and directly snap-cool on ice for at least 5 min.
5. Add 4 mL of the following hybridization solution, prewarmed to 42°C, to a 1-pint pouch:

Formamide	5 mL
20X SSC	2.5 mL
100X Denhardt's solution	0.5 mL
10% SDS	0.5 mL
HS DNA, denatured (10 mg/mL)	0.1 mL
dH_2O	1.4 mL

6. Remove air bubbles and seal the pouch.
7. Prehybridize in a 42°C water bath for at least 1 h.
8. Cut a corner of the pouch and add the denatured probe from **step 4** into the hybridization solution (not directly onto the disk) to a final concentration of 1.5×10^5 cpm/mL of hybridization solution.
9. Remove air bubbles and reseal the pouch.
10. Incubate in a 42°C water bath overnight.
11. Wash the disk twice with gentle rocking in 300 mL of 2X SSC/0.1% SDS at RT for 15 min. Dispose of radioactive waste in an appropriate manner.
12. Wash the disk twice with gentle rocking in 300 mL of 0.1X SSC/0.1% SDS at 56°C for 15 min.
13. Dry the disk growth-side up on 3MM paper.
14. Transfer the disk growth-side up to a new precut piece of 3MM paper, wrap both together in Saran wrap, and expose to a Phosphorimage plate or X-ray film.
15. For hybridization-positive spots (in addition to the positive control) streak-out cultures from the well on the 96-well plate corresponding to these "hits" onto LB/chloramphenicol plates.
16. Allow the culture streaks to soak into the plates for approx 30 min.
17. Invert and incubate the plates at 32°C overnight.
18. Pick two to three colonies from each clone and inoculate corresponding 50-mL conical tubes containing 10 mL LB/chloramphenicol.
19. Grow in an orbital shaker/incubator at 32°C while shaking ~300 rpm overnight.
20. Use a Mini-Prep Processing Kit to purify plasmid clones from these mini-preps.
21. Digest approx 300 ng of each plasmid with Bbe I plus Bgl I (*see* **Subheading 3.2.3., steps 1–3**).
22. Add 3 µL of 5X loading buffer to each reaction and resolve on a 0.8% TBE gel run at 80 V for about 2 h. Potential HIV-1 infectious molecular clones are present if 9.2- and 4.1-kb bands are present. These bands represent the virtually full-length HIV-1 provirus and the pLTR(E)1/2Δ vector, respectively.

3.4.2. Screen for p24 Antigen Production

1. Pellet 15×10^6 COS-1 cells for each reaction at 300*g*, at RT, for 8 min.
2. Dispose of the supernatant and resuspend the pellet in 300 µL of fresh, prewarmed

DMEM supplemented with 20% FCS, 100 U/ml of penicillin, 100 μg/mL of streptomycin, and 2 m*M* of L-glutamine per 15 × 10^6 cells.

3. Into a 1.6-mL Eppendorf tube, add 300 μL of resuspended cells followed by 20–25 μg of an LTR-reconstituted proviral clone.
4. Tap to mix and incubate on ice for 20 min.
5. Pipet gently with a 1-mL tip to mix, and transfer the DNA/cell mixture to an ice-cooled, 0.4-cm electroporation cuvet.
6. Electroporate at 0.22 kV, 960 μF.
7. Incubate on ice for 20 min.
8. Add 3 mL of fresh, prewarmed DMEM (*see* **step 2**) to each well of a 6-well plate, corresponding to the numbering of transfected samples.
9. Using medium-tip plastic Pasteur pipets gently mix the electroporation products and transfer to cognate wells from **step 8**. Rinse the cuvet to maximize product transfer.
10. Maintain these 6-well plates in a 37°C, humidified, 5% CO_2 environment.
11. At 48 and 72 h posttransfection transfer all the supernatant from each well to 15-mL conical tubes. After the 48-h transfer, add back 3 ml of fresh, prewarmed DMEM (*see* **step 2**) to each well and return the plate to the incubator. Terminate cells after the 72-h transfer.
12. Clear the supernatants by centrifugation at 300*g* followed by filtration through a 0.22-μm syringe filter system.
13. To minimize freeze–thaw cycles, subaliquot approx 200 μL of each supernatant to a screw-capped tube prior to storing at –80°C. This fraction will be used to determine p24 antigen concentration.
14. Perform p24 assay per manufacturer's instructions on samples from **step 13**.

3.4.3. Screen for Infectious Molecular Clones

1. In a 16 × 125 mm culture tube, pellet 2 × 10^6 PHA/IL-2-stimulated seronegative donor PBMCs at 300*g* at RT for 8 min.
2. Remove the supernatant and resuspend the pellet in 1 mL of p24 antigen-positive COS-1 cell supernatant.
3. Maintain at 37°C overnight in a humidified, 5% CO_2 environment.
4. Centrifuge at 300*g* for 8 min.
5. Resuspend and maintain cells in 2 mL of fresh prewarmed RPMI-1640 medium supplemented with 15% fetal calf serum, 100 U/mL of penicillin, 100 μg/mL of streptomycin, 2 m*M* of L-glutamine, and 20 U/mL of human recombinant IL-2.
6. Every 2–3 d over a span of about 30 d centrifuge the culture and transfer the supernatant to a fresh 15-mL conical tube. Resuspend the pellet in 2 mL of fresh, prewarmed medium as in **step 5**.
7. Clear and store supernatants as described in **Subheading 3.4.2.**, **steps 12** and **13**.
8. After supernatant collection over a span of about 30 d, perform a p24 antigen assay to determine if virus is replication competent. Typically, p24 antigen concentration increases rapidly over time, peaks, then slowly decreases (*see* **Note 11**).

4. Notes

1. The amplification product yield can vary widely between gDNA sources. These differences can often be attributed to the quality and purity of the gDNA. Extract gDNA from fresh cells, when possible, and always use sterile technique. Titrating the gDNA template over a range of 50 ng to 1 µg will increase the likelihood of at least one high-yield amplification.
2. The products from two or more amplification reactions can be pooled together prior to extraction to increase the final yield. A 1.6-mL tube can be used.
3. This is a good stopping point as the precipitating product can be stored at –20°C for several days, if necessary.
4. Bbe I and Bgl I are tricky restriction enzymes requiring close attention to the possibility of incomplete digestion. This is particularly relevant to the pLTR(E)1/2Δ vector, as its incomplete digestion will result in intramolecular ligation and, ultimately, isolation of clones with no 9.2-kb insert. Thus is important to set up single-enzyme digestion controls. Even better is to subclone an approx 800–1500-bp control insert into the Bbe I/Bgl I restriction site of pLTR(E)1/2Δ to ensure that only the completely digested fraction will be extracted from the gel.
5. The buffer requirements of Bbe I and Bgl I are distinct enough to preclude simultaneous double-digestion. However, successive BbeI and BglI digestion works well when performed as described.
6. To prevent gel overloading, multiple wells of a comb may be taped together to create a larger contiguous well. Do not create seams by overlapping pieces of tape. The hot prepolymerized gel will cause separation. Instead, cut the tape to a length defined by the midpoints of the comb teeth to be taped. It is important to use white tape, as coloring agents can diffuse into the gel. Similarly it is necessary to wipe the tape with 70% EtOH and a tissue to prevent diffusion of the tape's coating into the gel. These impurities absorb UV light and thus can obscure visualization of the bands. They will also coextract with these bands.
7. The most efficient ligation reactions theoretically occur when the insert and vector are present in stoichiometrically equivalent ratios. However, as this is not always easy to control for, varying the insert to vector ratio may result in increased colony yield. Thus, in addition to a ligation reaction containing a 1:1 ratio of 200 ng insert plus 75 ng vector, set up a parallel reaction containing a 3:1 ratio (600 ng plus 75 ng). It is essential to have a 0:1 insert:vector ligation control reaction to establish the proportion of colonies attributable to contaminating undigested or self-ligated vector.
8. To increase the total number of colonies, two other plates can be spread with 325 µL of the culture. Alternatively, by concentrating the remaining 650 µL of culture and plating, the colony yield per plate can be almost doubled.
9. Some insert/vector combinations are inherently unstable. Instability can be manifested by recombination, insert disposal, and translation of proteins toxic to the cell. One way to increase the stability of a vector containing a very large 9.2-kb insert is to decrease the incubation temperature to 30–32°C. This will, of course, require additional growth time (usually about 18 h).

10. There are many nonradioactive labeling options with comparable sensitivity to ^{32}P. This protocol could be easily modified to substitute the use of such labeling techniques for ^{32}P.
11. The growth kinetics of viral stocks prepared from HIV-1 infectious molecular clones can be variable. The time to, and levels of, peak supernatant p24 antigen production will differ between viral isolates.

Acknowledgments

We wish to thank Drs. Mika Salminen, Deborah Birx, Francine McCutchan, and Merlin Robb for efforts to realize these techniques. The views expressed here are the private opinions of the authors and are not to be considered as official or reflecting the views of the US Army or the US Department of Defense. This work was supported in part by Cooperative Agreement No. DAMD17-93-V-3004, between the US Army Medical Research and Materiel Command and the Henry Jackson Foundation for the Advancement of Military Medicine.

References

1. Gibbs, J. S., Regier, D. A., and Desrosiers, R. C. (1994) Construction and in vitro properties of SIVmac mutants with deletions in "nonessential" genes. *AIDS Res. Hum. Retroviruses* **10,** 607–616.
2. Barnes, W. M. (1994) PCR amplification of up to 35-kb DNA with high fidelity and high yield from lambda bacteriophage templates. *Proc. Natl. Acad. Sci. USA* **91,** 2216–2220.
3. Cheng, S., Fockler, C., Barnes, W. M., and Higuchi, R. (1994) Effective amplification of long targets from cloned inserts and human genomic DNA. *Proc. Natl. Acad. Sci. USA* **91,** 5695–5699.
4. Salminen, M.O., Koch, C., Sanders-Buell, E., Ehrenberg, P. K., Michael, N. L., Carr, J. K., et al. (1995) Recovery of virtually full-length HIV-1 provirus of diverse subtypes from primary virus cultures using the polymerase chain reaction. *Virology* **213,** 80–86.
5. Salminen, M. O., Ehrenberg, P. K., Mascola, J. R., Dayhoff, D. E., Merling, R., Blake, B., et al. (2000) Construction and biological characterization of infectious molecular clones of HIV-1 subtypes B and E (CRF01_AE) generated by the polymerase chain reaction. *Virology* **278,** 103–110.
6. Li, Y., Hui, H., Burgess, C. J., Price, R. W., Sharp, P. M., Hahn, B. H., and Shaw, G. M. (1992) Complete nucleotide sequence, genome organization, and biological properties of human immunodeficiency virus type 1 in vivo: Evidence for limited defectiveness and complementation. *J. Virol.* **66,** 6587–6600.
7. Li, Y., Kappas, J. C., Conway, J. A., Price, R. W., Shaw, G. M., and Hahn, B. H. (1991) Molecular characterization of human immunodeficiency virus type 1 cloned directly from uncultured human brain tissue: identification of replication-competent and -defective viral genomes. *J. Virol.* **65,** 3973–3985.

8. Meyerhans, A., Cheynier, R., Albert, J., Seth, M., Kwok, S., Sninsky, J., et al. (1989) Temporal fluctuations in HIV quasispecies in vivo are not reflected by sequential HIV isolations. *Cell* **58,** 901–910.

31

Amplification and Cloning of Near Full-Length HIV-2 Genomes

Feng Gao

Summary

The genomes of human immunodeficiency virus type 2 (HIV-2), like those of HIV-1, are not only extremely variable but are also highly recombinogenic *(1–5)*. Determination of subtypes based on partial genomes cannot predict the subtype classification of other regions of the genome owing to the frequent occurrence of recombinant genomes among subtypes. To fully understand the genetic variation and evolution of HIV-2s, full-length viral genomes need to be obtained for genetic analysis. Full-length HIV-2 genomes can also be used as infectious clones to study viral biological characteristics and as reference sequences for phylogenetic analysis. More important, all genes in the obtained genomes can be cloned into expression vectors to produce proteins that can be used as antigens for diagnostic reagents or as immunogens for vaccine development. The long-range polymerase chain reaction technique has recently become a more preferred method than the λ phage cloning method to obtain near-full-length HIV-2 genomes.

Key Words: Polymerase chain reaction (PCR); full-length genome; HIV-2; amplification; cloning; subtype.

1. Introduction

Previously, full-length HIV genomes were obtained exclusively by cloning the complete or partial viral genomes into λ phage vectors *(6,7)*. The advantage of this method is the generation of infectious full-length clones that can be used to study the viral biological characteristics. However, this method is very time-consuming and labor-intensive. It also requires successful isolation of HIV strains in T-cell lines or normal donor peripheral mononuclear cells (PBMCs) to obtain a high copy number of viral genomes. Since the long-range polymerase chain reaction (PCR) technique was developed *(8)*, it has been

From: *Methods in Molecular Biology, Vol. 304: Human Retrovirus Protocols: Virology and Molecular Biology*
Edited by: T. Zhu © Humana Press Inc., Totowa, NJ

quickly adapted to amplify near full-length HIV genomes ***(9–11)***. The method is much faster than the λ phage cloning method and does not require isolation of viruses in vitro. Since its use in obtaining near-full-length HIV genomes, the number of full-length genomes has significantly increased. The templates for long-range PCR are mainly DNA samples from either patient PBMC or HIV-infected cell cultures. Using RNA as templates to obtain full-length HIV genomes has not been very successful. Because of the nature of two identical long terminal repeats (LTRs) at each end of the integrated HIV genome or two identical R regions in the RNA genome, it is impossible to design primers to amplify full-length HIV genomes. As a result, the amplified products miss a part of the genome (about 100–200 bp). However, the near full-length HIV genomes contain coding regions for all HIV genes and infectious clones can be reconstructed by adding the missing LTR region at the 5' end ***(11,12)***.

2. Materials

1. Genomic DNA extraction kit (Qiagen, Valencia, CA).
2. Agarose.
3. DNA marker: λ DNA/*Hind*III markers (Promega, Madison, WI).
4. dNTP mix: 10 m*M* dATP, dGTP, dCTP, and dTTP each.
5. TBE buffer: 90 m*M* Tris-borate and 2 m*M* EDTA.
6. Oligonucleotide primers.
7. Ethanol (96–100%).
8. Microcentrifuge tubes (1.5-mL).
9. DTT.
10. $MgCl_2$.
11. *E. coli* strains: TOP10 or Mach1™-T1®.
12. Restriction enzyme.
13. Kanamycin.
14. Expand Long Template PCR system (Roche Applied Science, Indianapolis, IN).
15. TOPO XL PCR cloning kit (Invitrogen, Carlsbad, CA).
16. SOC medium: 2% tryptone, 0.5% yeast extract, 10 m*M* NaCl, 2.5 m*M* KCl, 10 m*M* $MgCl_2$, 10 m*M* $MgSO_4$, and 20 m*M* glucose.

3. Methods

3.1. DNA Extraction

DNA quality is critical for success of amplification of near full-length HIV-2 genomes. DNA samples should not be sheared and should be free of proteins. Although the traditional phenol:chloroform method can yield high-quality DNA, the use of toxic chemicals has made it not preferred. Many new methods are available for preparation of high molecular weight genomic DNA without using toxic chemicals. Among these kits, the Qiagen Blood and Cell Culture

DNA Kit routinely yields high-quality genomic DNA (*see* **Note 1**). The detailed protocol can be obtained from the manufacturer. A brief protocol for the Qiagen Blood and Cell Culture DNA Kit is described as follows:

1. Prepare the reagents as suggested in the Qiagen Blood and Cell Culture DNA Kit protocol (*see* **Note 2**).
2. Mix 0.5 mL of fresh whole blood, patient PBMC (5×10^6) or cultured cells (5×10^6) with 0.5 mL of ice-cold buffer C1 and 3 vol of ice-cold distilled water (1.5 mL). Invert the tube several times until the suspension becomes translucent. Incubate on ice for 10 min (*see* **Note 3**).
3. Centrifuge the lysed cells at 4°C for 15 min at 1300*g*. Discard the supernatant.
4. Add 0.25 mL of ice-cold buffer C1 and 0.75 mL of ice-cold distilled water. Resuspend the pelleted nuclei by vortexing. Centrifuge again at 4°C for 15 min at 1300*g*. Discard the supernatant.
5. Add 1 mL of buffer G2, and completely resuspend the nuclei by vortexing for 10–30 s at maximum speed.
6. Add 25 mL of Qiagen protease and incubate at 50°C for 30–60 min.
7. Equilibrate a Qiagen Genomic-tip 20/G with 1 mL of Buffer QGT, and allow the Qiagen Genomic-tip to empty by gravity flow (*see* **Note 4**).
8. Vortex the samples (**step 6**) for 10 s at maximum speed and apply it to the equilibrated Qiagen Genomic-tip. Allow it to enter the resin by gravity flow.
9. Wash the Qiagen Genomic-tip with 3 × 1 mL of buffer QC.
10. Elute the genomic DNA with 2 × 1 mL of buffer QF into a 10-mL centrifuge tube.
11. Precipitate the DNA by adding 1.4 mL room-temperature isopropanol to the eluted DNA.
12. Mix and centrifuge immediately at >5000*g* for at least 15 min at 4°C. Carefully remove the supernatant.
13. Wash the DNA pellet with 1 mL of cold 70% ethanol. Vortex briefly and centrifuge at >5000*g* for 10 min at 4°C. Carefully remove the supernatant without disturbing the pellet. Air-dry for 5–10 min, and resuspend the DNA pellet in 200 µL of Tris-HCl buffer (10 m*M* Tris-HCl, pH 8.5) or water. Dissolve the DNA overnight on a shaker.
14. Determination of DNA concentration: Add 20 mL of DNA samples from **step 13** to 80 µL water (1:5 dilution) and measure the absorbance at 260 nm. The concentration of the DNA samples (µg/mL) equals OD value × 50 × 5 (dilution factor). Normally, the preparation will yield a total of 20–30 µg genomic DNA (*see* **Note 5**).
15. Determination of DNA length: Load 1 µg extracted genomic DNA and DNA marker to a 0.5% agarose gel. Run the gel with Tris-Borate-EDTA (TBE) buffer at 40 V until the bromophenol blue reaches the bottom of the gel (about 6 cm) and stain the gel with ethidium bromide solution (0.5 µg/mL) for 10 min. The eluted genomic DNA should be 50–100 kb long on average (up to 150 kb; *see* **Note 6**).
16. The DNA samples are ready to be used for PCR amplification or stored at –20°C for later use.

3.2. PCR Amplification

To amplify near full-length genomes of diverse genetic variants, the primers are designed at the highly conserved nontranslation region. If DNA is extracted from HIV-2-infected cell cultures, a single round of PCR is sufficient to obtain enough PCR products for cloning, and the successful rate of amplification of near full-length HIV-2 genomes will be much higher. It is highly recommended to establish short-term cultures of primary HIV-2 isolates in vitro for long-range PCR amplification. If the DNA is extracted directly from patient PBMC samples without cell culture, the number of proviral copies in the samples is low and one round of PCR amplification usually does not yield enough products to be cloned. Therefore, a second round of PCR (nested PCR) needs to be carried out to obtain sufficient PCR products. Because nested PCRs are often supersensitive (detection level as low as a few copies of templates in each reaction), a clean working environment is required to avoid sample contamination.

1. Prepare the master PCR reaction mix in an autoclaved tube at ambient temperature or on ice. Always prepare one extra reaction to have enough master mix for all tested samples. The Expand Long Template PCR system from Roche Applied Science is used for the PCR amplification (*see* **Note 7**). The following recipe is for one reaction:

Component	Volume	Final concentration
10X PCR buffer with $MgCl_2$	5 μL	buffer 2 (2.75 m*M* $MgCl_2$)
10 m*M* dNTP mixture	1.75 μL	0.35 m*M* each
Primer HIV2upA	1 μL	20 pmol
Primer HIV2lowA	1 μL	20 pmol
Expand Long Template Enzyme Mix	0.75 μL	3.75 units
Autoclaved distilled water	to 45 μL (*see* **Note 8**)	

2. Vortex the mixture and transfer 45 μL master mix to each 200-μL thin-wall PCR tube.
3. Add 5 μL DNA sample (0.2–1 μg) to each tube.
4. Cap the tubes and centrifuge briefly to collect the contents in the bottom of the tubes.
5. Put tubes into the heat blocks in the thermocycler (with the heating block on the top).
6. Denature the templates for 2 min at 94°C and perform 10 cycles of PCR amplification as follows:

   Denature: 94°C for 15 s
   Anneal: 55°C for 30 s (*see* **Note 9**)
   Extend: 68°C for 10 min

Then perform 20 cycles of PCR amplification as follows:

Denature: 94°C for 15 s

Anneal: 55°C for 30 s
Extend: 68°C for 10 min + extra 15 s for each successive cycle

7. Extend for an additional 10 min at 68°C after the last cycle of the PCR amplification.
8. Maintain the reaction at 4°C after cycling. Samples can be stored at –20°C for later use.
9. Prepare master PCR reaction mixture as in **step 1** with nested primers HIV2upA1 and HIV2lowB.
10. Follow **steps 2–8** for second-round PCR amplification using 5 µL of first-round PCR products (*see* **Note 10**).
11. Load 5 µL PCR products on a 0.5% agarose gel with ethidium bromide (0.5 µg/mL). In a separated well, load DNA marker for estimation of the size of the PCR products. The expected size of PCR products is approx 9 kb.
12. Take photos or digital images for record.
13. Primer sequences:

    First-round PCR primers
    HIV2upA: 5'-GTAAGGGCGGCAGGAACAAACC-3'
    HIV2lowA: 5'-GCGGCGACTAGGAGAGATGGGA-3'
    The length of PCR products is approx 9 kb.
    Second-round PCR primers
    HIV2upA1: 5'-GGCGGCAGGAACAAACCACG-3'
    HIV2lowB: 5'-GAGAACCTCCCAGGGCTCAATCT-3'
    The length of PCR products is approx 9 kb.
14. For troubleshooting and precautions of PCR amplification *see* **Notes 14** and **15**.

3.3. Purification of PCR Products With QIAquick Gel Extraction Kit

1. Load all PCR products on a 0.5% agarose gel containing 0.5 µg/mL ethidium bromide.
2. Run the gel in TBE buffer at 40 V until the blue dye reaches the bottom of the gel (about 6 cm).
3. Excise the DNA fragment from the gel with a clean scalpel under long-wave-length UV light (312 nm).
4. Weight the gel slice in a colorless 1.5-mL centrifuge tube. Add 3 vol of buffer QG to 1 vol of gel (1 mg = 1 µL).
5. Incubate at 50°C for 10 min (or until the gel slice has completely dissolved). Mix by vortexing the tube every 2–3 min during the incubation.
6. Add one gel volume of isopropanol to the sample and mix.
7. Place a QIAquick spin column in a 2-mL collection tube.
8. Apply the sample to the QIAquick column and centrifuge for 1 min.
9. Discard the flow-through and place QIAquick column back in the same collection tube.
10. Add 0.5 mL of buffer QG and centrifuge for 1 min.
11. Add 0.75 mL of buffer PE and centrifuge for 1 min.
12. Discard the flow-through and centrifuge for an additional 1 min at ~17,900*g*.
13. Place the QIAquick column into a clean 1.5-mL tube.

14. Add 50 µL of buffer EB (10 m*M* Tris-HCl, pH 8.5) or water to the center of the membrance and centrifuge for 1 min (*see* **Note 11**).
15. Load 5 µL PCR products on a 0.5% agarose gel with ethidium bromide (0.5 µg/mL). Estimate the concentration of recovered DNA fragment by comparing with known concentration of DNA bands in the DNA marker (*see* **Note 12**).

3.4. Cloning of Near Full-Length HIV-2 PCR Products Using TOPO XL PCR Cloning Kit

1. Combine 4 µL of purified PCR products (**Subheading 3.3.**, **step 14**) and 1 µL of pCR-XL-TOPO vector in a sterile 1.5-mL tube.
2. Mix gently and incubate for 5 min at room temperature. Do not leave it at room temperature for more than 5 min.
3. Add 1 µL of the 6X TOPO Cloning Stop Solution and mix for several seconds at room temperature.
4. Briefly centrifuge the tube and place on ice.
5. Thaw on ice one vial of One Shot competent cells for each transformation.
6. Add 2 µL of the TOPO Cloning reaction (**step 3**) to one vial of One Shot competent cells and mix gently by flipping the end of the tubes with the fingers.
7. Incubate on ice for 30 min.
8. Heat-shock the cells for 30 s at 42°C without shaking.
9. Immediately transfer the tubes to ice and incubate for 2 min.
10. Add 250 mL of room temperature SOC medium.
11. Cap the tube tightly and shake the tube at 37°C for 1 h.
12. Spread 50–150 µL of transformed bacteria on prewarmed plates (37°C) containing 50 µg/mL kanamycin.
13. Incubate plates overnight at 37°C.

3.5. Analysis of Transformants

pCR-XL-TOPO vector contains an *E. coli* lethal *ccd*B gene that is fused to the C-terminus of the LacZα fragment. Insertion of a long PCR fragment disrupts the expression of the LacZα-*ccd*B fusion gene and permits growth of only positive recombinants upon transformation. Bacteria that contain nonrecombinant vector are killed upon plating. Therefore, blue-white screening is not required.

1. Pick up 10 colonies with sterile toothpicks and inoculate them into 15-mL tubes containing 3 mL LB medium with 50 mg/mL kanamycin.
2. Incubate at 37°C overnight by shaking at 220 rpm.
3. Prepare miniprep DNA using QIAprep Spin Miniprep Kit or any other alternative methods.
4. Digest DNA with EcoR I in a total of 20 µL volume.
5. Run 10 µL of digestion solution in a 0.5% agarose gel. Two bands are expected: a 3.5-kb band for the vector and a 9-kb band for the insert (*see* **Note 13**).
6. After the right size inserts are confirmed by restriction enzyme mapping, the

recombinant plasmids can be sequenced with M13 forward and/or reverse primers. The insert sequences then can be used to search against the GenBank sequence database to determine if the insert is HIV-2-related genomes.

4. Notes

1. Although the Qiagen Blood and Cell Culture DNA Kit has been widely used for extraction of high molecular weight DNA, the following kits from other manufacturers can also be used: GenElute Mammalian Genomic DNA Miniprep kit from Sigma, St. Louis, MO; Easy-DNA™ Kit from Invitrogen, Carlsbad, CA; and Wizard® Genomic DNA Purification Kit from Promega, Madison, WI.
2. The buffers and protease K are supplied in the kit. There is no need to prepare them. If needed, however, they can be prepared based on the instructions from the protocol.
3. All infectious materials (whole blood, plasma, or infected cells) that contain live HIV should be prepared under biosafety level 2 (BSL-2) laminar flow hoods. Extreme caution should be taken to avoid any direct contacts with the infectious materials.
4. It is critical to use the Qiagen Genomic-tips to prepare high molecular weight genomic DNA (50–100 kb). Other kits that yield only 20–30 kb genomic DNA should not be used for amplification of 10-kb HIV genomes.
5. When more DNA samples are needed, DNA extraction kits for larger amounts of samples are available from the manufacturer.
6. If the length of genomic DNA is less than 30 kb, the DNA sample should be discarded and new DNA should be prepared.
7. A number of long-range PCR kits have been developed to amplify target DNA up to 20 kb. If the Expand Long Template PCR system is not available, other systems (AccuTaq LA DNA Polymerase from Sigma and PCR SuperMix High Fidelity from Invitrogen) can be used as alternative methods for the amplification of near full-length HIV-2 genomes.
8. The total volume of the PCR reaction is 50 μL. Five μL of each sample were subtracted from the master mix.
9. Annealing temperature is critical for successful PCR amplification. The temperature should be adjusted for each PCR reaction to achieve the optimal amplification.
10. In general, 5 μL of the first-round PCR reaction is a good starting point for the second-round PCR amplification. However, depending on the DNA sample quality and the copy numbers of target molecules, more or less of the first-round PCR reaction may be needed for the optimal second-round PCR reaction.
11. Promega also manufactures a similar plasmid DNA kit (Wizard PCR Preps DNA Purification System), which can yield the same quality DNA fragments for the subsequent cloning step.
12. The specificity of the PCR amplification will have a huge impact on the cloning efficiency of long PCR products. If too many unspecific bands are amplified, PCR conditions should be adjusted to obtain more specific amplification prod-

ucts (*see* **Notes 14** and **15**).

13. There are two EcoR I sites flanking the cloning site in the pCR-XL-TOPO vector. After digestion with EcoR I, a 9-kb insert will be released from the vector if there are no internal EcoR I sites in the amplified HIV-2 near full-length genomes. Sometimes, the amplified genomes may contain internal EcoR I site(s), and the inserts will be digested into smaller fragments. However, the sum of the smaller fragments should be about 9 kb if a correct HIV-2 genome is cloned.
14. Troubleshooting:

Problem	Solution
No band	Lower annealing temperature
	Increase amounts of DNA templates
	Increase $MgCl_2$ concentration
	Increase cycle number
Too many bands	Increase annealing temperature
	Decrease $MgCl_2$ concentration
	Decrease amount of template
	Decrease primer concentration
Wrong size bands	Increase annealing temperature
Primer-dimer	Decrease primer concentration
	Increase annealing temperature
Band in negative control	Contaminated; prepare new reagents, buffers, and water

15. Precautions
    a. Never use PCR products or work with molecular clones in PCR reaction preparation areas or rooms.
    b. Always use the following controls:
        i. Low copy positive control.
        ii. Negative control (normal genomic DNA).
    c. Always use aerosol-resistant tips.
    d. Always wear new gloves.

References

1. Gao, F., Yue, L., Robertson, D. L., Hill, S. C., Hui, H., Biggar, R. J., et al. (1994) Genetic diversity of human immunodeficiency virus type 2: evidence for distinct sequence subtypes with differences in virus biology. *J. Virol.* **68,** 7433–7447.
2. Chen, Z., Luckay, A., Sodora, D. L., Telfer, P., Reed, P., Gettie, A., et al. (1997) Human immunodeficiency virus type 2 (HIV-2) seroprevalence and characterization of a distinct HIV-2 genetic subtype from the natural range of simian immunodeficiency virus-infected sooty mangabeys. *J. Virol.* **71,** 3953–3960
3. Gao, F., Yue, L., White, A. T., Pappas, P. G., Barchue, J., Hanson, A. P., et al. (1992) Human infection by genetically diverse SIVsm-related HIV-2 in west Africa. *Nature* **358,** 495–499.

4. Yamaguchi, J., Devare, S. G., and Brennan, C. A. (2000) Identification of a new HIV-2 subtype based on phylogenetic analysis of full-length genomic sequence. *AIDS Res. Hum. Retroviruses* **16,** 925–930.
5. Robertson, D. L. and Gao, F. (1998) Recombinantion of HIV genomes, in *Human Immunodeficiency Viruses: Biology, Immunology and Molecular Biology* (Saksena, N. K., ed.), Medical System S.p.A., Genoa, Italy, pp. 183–208.
6. Clavel, F., Guyader, M., Guetard, D., Salle, M., Montagnier, L., and Alizon, M. (1986) Molecular cloning and polymorphism of the human immune deficiency virus type 2. *Nature* **324,** 691–695.
7. Hahn, B. H., Shaw, G. M., Arya, S. K., Popovic, M., Gallo, R. C., and Wong-Staal, F. (1984) Molecular cloning and characterization of the HTLV-III virus associated with AIDS. *Nature* **312,** 166–169.
8. Cheng, S., Fockler, C., Barnes, W. M., and Higuchi, R. (1994) Effective amplification of long targets from cloned inserts and human genomic DNA. *Proc. Natl. Acad. Sci. USA* **91,** 5695–5699.
9. Salminen, M. O., Koch, C., Sanders-Buell, E., Ehrenberg, P. K., Michael, N. L., Carr, J. K., et al. (1995) Recovery of virtually full-length HIV-1 provirus of diverse subtypes from primary virus cultures using the polymerase chain reaction. *Virology* **213,** 80–86.
10. Gao, F., Robertson, D. L., Morrison, S. G., Hui, H., Craig, S., Decker, J., et al. (1996) The heterosexual human immunodeficiency virus type 1 epidemic in Thailand is caused by an intersubtype (A/E) recombinant of African origin. *J. Virol.* **70,** 7013–7029.
11. Gao, F., Robertson, D. L., Carruthers, C. D., Morrison, S. G., Jian, B., Chen, Y., et al. (1998) A comprehensive panel of near-full-length clones and reference sequences for non-subtype B isolates of human immunodeficiency virus type 1. *J. Virol.* **72,** 5680–5698.
12. Salminen, M. O., Ehrenberg, P. K., Mascola, J. R., Dayhoff, D. E., Merling, R., Blake, B., et al. (2000) Construction and biological characterization of infectious molecular clones of HIV-1 subtypes B and E (CRF01_AE) generated by the polymerase chain reaction. *Virology* **278,** 103–110.

32

Growth and Manipulation of a Human T-Cell Leukemia Virus Type 2 Full-Length Molecular Clone

Matthew Anderson and Patrick L. Green

Summary

In retrovirus research, the generation of an infectious molecular clone is a landmark event, opening up new avenues of research using the cloned virus. A full-length proviral plasmid clone of the human T-cell leukemia virus (HTLV) makes possible reproducible viral genetic studies. However, the growth of full-length infectious HTLV proviral plasmid clones in bacteria, their manipulation using molecular techniques, and further characterization of replication capacity and other biological properties are not trivial. This chapter describes successful methods used for the preparation and manipulation of the full-length HTLV-2 proviral plasmid clone pH6neo. The plasmid-borne full-length clone of HTLV-2 permits the study of the interactions and contributions of viral proteins in viral replication and cellular transformation in vitro and in animal models in vivo. These types of studies have provided and will continue to provide critical insight into understanding the virus–host interactions and ultimately the contribution of viral genes and elements to the pathogenesis of HTLV.

Key Words: HTLV-2; pH6neo; immortalization; transformation; human T-lymphocytes; HTLV-associated myelopathy; adult T-cell leukemia.

1. Introduction

Human T-cell leukemia virus type 1 (HTLV-1) and type 2 (HTLV-2) are closely related retroviruses that have been causally associated with a variety of human diseases. HTLV-1 is associated with adult T-cell leukemia (ATL), an aggressive $CD4^+$ T-cell malignancy, and a chronic neurodegenerative disorder termed HTLV-associated myelopathy/tropical spastic paraparesis (HAM/TSP) *(1,2)*. HTLV-2 is less clearly associated with disease, with only a few cases of leukemia or neurological disease reported *(3–6)*. The genetic basis for the dif-

From: *Methods in Molecular Biology, Vol. 304: Human Retrovirus Protocols: Virology and Molecular Biology*
Edited by: T. Zhu © Humana Press Inc., Totowa, NJ

ference in pathobiology of HTLV-1 and HTLV-2 is not yet clear, but likely resides in the activities of the regulatory and/or accessory proteins, thus highlighting the importance of comparative structure/function studies of the viral elements and gene products.

Initial HTLV experimental studies were restricted to examination of infected patients, overexpression of individual viral genes using reporter assays in cell lines, or characterization of infected cell lines or animals using viral isolates obtained directly from patients. Although these types of studies have been very informative, the understanding of HTLV biology and pathogenesis has further benefited from the isolation and manipulation of infectious proviral clones and development and refinement of methodologies for characterization of these clones in primary human T-lymphocytes and relevant animal models.

This chapter focuses on the methods used in the preparation of the infectious HTLV-2 proviral plasmid clone, pH6neo; the introduction of pH6neo or pH6neo variants into cells to generate stably transfected virus producer cell lines; the use of these viral producer cell lines in cell-to-cell transmission of virus to primary human T-lymphocytes through co-culture; and the subsequent care and management of those primary human T-lymphocyte cultures. Several HTLV-2 studies will be highlighted to illustrate the methods utilized, the experimental results observed, and the contribution to our understanding of HTLV.

2. Materials

1. Bio-Rad Genepulser, with capacitance extender (Bio-Rad).
2. Sorvall ultracentrifuge with GSA and HB4 rotors.
3. Beckman G-80 ultracentrifuge with vTi90 and SW55 rotors.
4. Bacterial culture equipment and media.
5. Maxiprep solution I: 50 m*M* glucose, 25 m*M* Tris-HCl, pH 8.0, and 10 m*M* EDTA.
6. Maxiprep solution II: 0.2 *M* NaOH, 1% SDS.
7. Maxiprep solution III: 3 *M* sodium acetate, 2 *M* acetic acid.
8. Rotary shaking incubator, 33°C.
9. Ampicillin, 50 μg/mL stock.
10. Phenol (Tris-buffered)/chloroform solution, 1:1 ratio.
11. Sterile cheesecloth.
12. Dialysis buffer: 10 m*M* Tris-HCl, pH 8.0, 20 m*M* NaCl, 1 m*M* EDTA, 1 m*M* dithiothreitol.
13. 3 *M* sodium acetate, pH 5.6.
14. Butanol, 100%.
15. Ethanol, 100%.
16. Isopropanol, 100%.
17. Ethidium bromide, 10 mg/mL stock.
18. Cesium chloride.

19. TE buffer: 10 m*M* Tris-HCl, pH 8.0, 0.1 m*M* EDTA.
20. Agarose and agarose gel electrophoresis equipment.
21. HTLVI/II p19 ELISA kit.
22. Sodium dodecyl sulfate polyacrylamide gel electrophoresis (SDS-PAGE) equipment.
23. Protein A-Sepharose.
24. Whatman filter paper.
25. Gammacell-40 cesium-source irradiator.
26. 729-6 cell line.
27. BJAB cell line.
28. Primary human peripheral blood mononuclear cells (PMBCs).
29. Cell culture media and cell culture equipment.
30. Human interleukin-2, 10 U/mL.
31. G418, 100 mg/mL active drug.
32. Ficoll.
33. FITC-conjugated α-mouse IgG.
34. α-HTLV p19 Gag antibody (murine).
35. Fluorescence microscope.

3. Methods

The methods below describe (1) the structure of the HTLV-2 infectious proviral clone, pH6neo, (2) expression of pH6neo-encoded virus in stably transfected cell lines, and (3) the infection and transformation of primary T lymphocytes by co-culture.

3.1. HTLV-2 Proviral Clone, pH6neo

The pH6neo plasmid contains a complete proviral sequence of HTLV-2 subtype A isolate. Its source was a cDNA isolated from CEM cells infected by coculture with the HTLV-2 infected Mo cell line. This cDNA, λH-6, contained a complete 8952 base sequence of HTLV-2 and approx 1.3 kb of cellular flanking sequences. A HindIII digest of the λH-6 cDNA was end-filled and cloned into the EcoRI site of the expression vector SV2neo, resulting in a plasmid with an approximate size of 15.9 kb (**Fig. 1A**) *(7–10)*. pH6neo encodes ampicillin and kanamycin resistance in bacteria, and G418 resistance in mammalian cells. It is also a low copy number plasmid; consequently, DNA plasmid yields from bacterial preparations are lower than would be possible if propagated in a high copy number plasmid backbone such as pUC or Bluescript. However, propagation in a low copy number plasmid backbone does provide a strategic advantage. As with all retrovirus proviral clones, HTLV-2 pH6neo contains two direct long terminal repeats (LTRs) at each end. These large direct repeats contribute to increased recombination and sequence deletion facilitated by high plasmid DNA copy number in bacteria. Low copy number plasmid growth in Rec^- *E. coli* significantly reduces the generation of deleted clones.

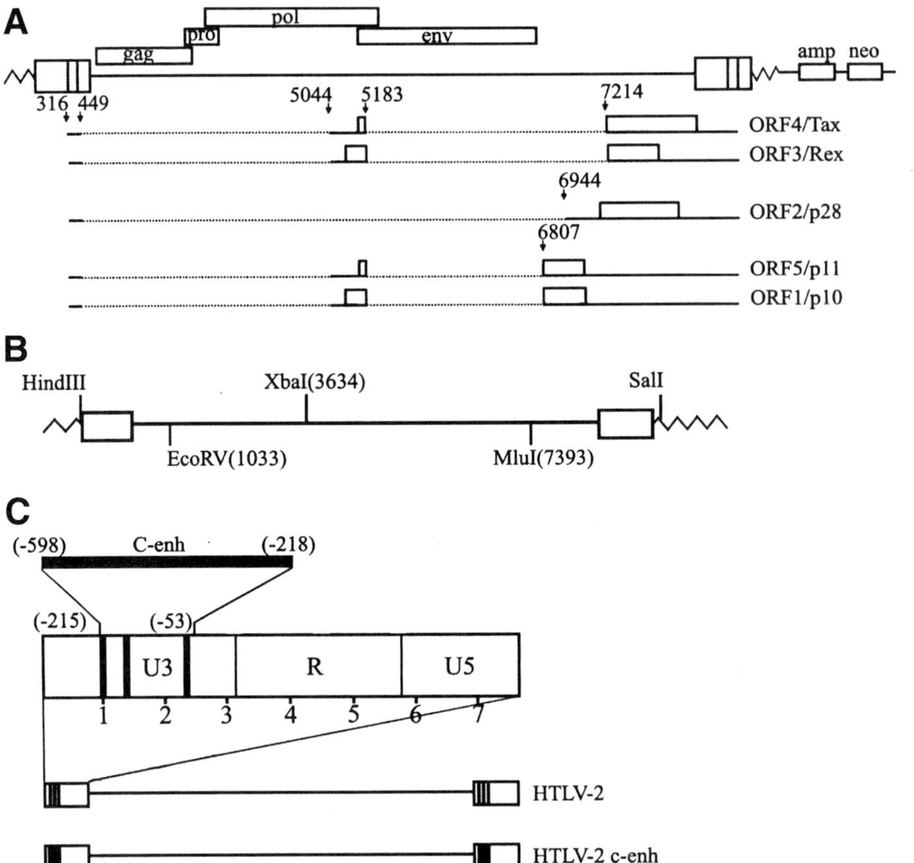

Fig. 1. (**A**) pH6neo plasmid and RNA transcripts. The pH6neo plasmid contains a gene encoding ampicillin resistance and a neomycin-resistance gene that provides kanamycin resistance in bacteria and G418 resistance in mammalian cells. The provirus itself is flanked by short stretches of cellular genomic DNA (sawtooth lines). *Gag*, *pro*, and *pol* genes are expressed from the unspliced, full-length RNA. *Env* is expressed from an mRNA spliced singly, between a splice donor (SD) at nt449 and a splice acceptor (SA) at nt5044. There are three potential SA sites within the pX region, one at nt7214 and two alternates at nt6807 and 6944. *Tax* and *Rex* are expressed from a doubly spliced mRNA created from splicing first between the SD at nt449 and a SA at nt5044, and a second splice between an SD at nt5183 and the third pX region SA at nt7214. p28 can be produced from a singly spliced mRNA, bypassing the SA at 5044 and instead utilizing the SA at nt6944. The putative accessory proteins p11 and p10 are most likely produced from doubly spliced mRNAs using the SA at nt6807. Other potential transcripts could express truncated forms of these accessory proteins in the same reading frame using internal methionine codons for initiation

The pH6neo plasmid has been modified throughout the years to facilitate numerous experimental studies. Most notable modifications include deletion of the majority of the 5' and 3' cellular flanking sequences to the provirus and introduction of unique restriction enzyme sites immediately 5' and 3' to the viral proviral genome (**Fig. 1B**). These modifications reduced the overall size of the plasmid clone and facilitated DNA fragment exchanges containing specific gene alterations. In an effort to perform Tax structure-function analyses in the context of a replicating virus, pH6neo was further modified by exchanging the Tax-response element in the U3 region of both the 5' and 3' LTRs with the cytomegalovirus immediate early gene enhancer (c-enh) (**Fig. 1C**). This chimeric provirus resulted in an HTLV-2 that replicates independently of Tax *(11)* (*see* **Note 1**).

3.2. Generation of pH6neo DNA

The pH6neo plasmid and its derivatives can be generated in useful quantities through standard maxi-preparation techniques, with only slight modifications. Plasmids are best grown after transformation into competent *E. coli* ($recA^-$, $endA^-$) overnight at a temperature of 33°C. Plasmid DNA can be isolated using filter columns such as those produced by Qiagen, but cesium chloride-banded plasmid preparations result in DNA stocks that consistently have higher transfection efficiency in mammalian cells.

3.2.1. Preparation of pH6neo Plasmid DNA

LB media (400 mL) supplemented with 30 mg/mL of ampicillin is inoculated with a single freshly transformed bacterial colony and grown overnight approx 16 h at 33°C (*see* **Note 2**). Overnight cultures are pelleted at 2600*g* for 5 min in 250-mL centrifuge bottles in a Sorvall GSA rotor and resuspended in a 25-mL solution of 50 m*M* glucose, 25 m*M* Tris-HCl, pH 8.0, and 10 m*M*

Fig. 1. *(continued)* ***(14)***. **(B)** Unique restriction sites within the pH6neo plasmid. The majority of cellular flank has been deleted in pH6neo and unique restriction sites (HindIII and SalI) have been introduced. EcoRV, XbaI, and MluI are naturally occuring unique restriction sites within the provirus genome. **(C)** Schematic representation of the HTLV-2 LTR showing the locations of the U3, R, and U5 regions. The three hatched boxes in the U3 region represent the three imperfectly conserved 21-bp repeats contained within the TRE. The enhancer for the cytomegalovirus immediate early gene (C-enh, -598, and -218 relative to the start of transcription) is depicted above (black box). Numbers below the long terminal repeat (LTR) represent bases (hundreds). wtLTRs containing the three 21-bp repeats (hatched boxes) are also indicated. The proviral clone $HTLV_{C\text{-}enh}$ contains both a 5' and a 3' chimeric $LTR_{C\text{-}enh}$ *(10)*.

EDTA (Maxiprep solution I) and kept on ice. Add 50 mL of 0.2 *M* NaOH, 1% SDS lysis buffer (Maxiprep solution II) and mix well. Following lysis, 37.5 mL of 3 *M* sodium acetate: 2 *M* acetic acid (Maxiprep solution III) is added and mixed again. The entire mixture is again centrifuged at a speed of 10,000*g* for 5 min at 4°C. The supernatant is filtered through a few layers of cheesecloth into 67.5 mL of 100% isopropanol and incubated at room temperature for 30 min. The precipitated DNA is then pelleted for 15 min at 5900*g*, 15°C, and resuspended in 3 mL of TE buffer (10 m*M* Tris-HCl, 0.1 m*M* EDTA, pH 8.0) in a 15-mL Falcon tube. Next, 0.5 mL of ethidium bromide (10 mg/mL stock) is added, the entire volume is weighed, and an equal weight of cesium chloride (CsCl) is dissolved in the mixture. The resulting solution (now containing 1 g/mL CsCl) is then loaded into a 5-mL Optiseal tube (Beckman) and centrifuged at 450,000*g* in a Vti90 vertical rotor in a Beckman G80 ultracentrifuge at room temperature for 4 h, without braking. Plasmid bands are visualized in the Optiseal tubes by handheld long-wave UV light. Banded plasmid DNA is removed by first uncapping the tube, and then piercing just beneath the band with an 18-gage needle attached to a 10-mL syringe and removing approx 1–2 mL of solution containing the plasmid band (*see* **Note 3**). The plasmid-containing solution is then diluted to 3 mL total volume with TE and extracted three to four times with 3-mL volumes of H_2O-saturated butanol to remove excess ethidium bromide. The plasmid solution is then dialyzed overnight at 4°C in a large volume (~3 L) of 20 m*M* Tris-HCl (pH 8.0), 0.1 m*M* EDTA, 10 m*M* NaCl, and 1 m*M* DTT (*see* **Note 4**). Following dialysis, the DNA is phenol/chloroform (1:1 ratio) extracted in a 30-mL Corex tube and centrifuged for 5 min at 16,000*g*, the clear fraction (top) is removed, and the DNA is then precipitated by adding 1/10 vol of 3 *M* sodium acetate, pH 5.6, and 2.5 vol of 100% ethanol in a 30-mL Corex centrifuge tube. After 30 min at –80°C, the DNA is pelleted from solution in a HB-4 Sorvall rotor at 16,400*g* for 30 min at 4°C, and resuspended in 250–500 µL of distilled deionized water. DNA concentration is determined by 260/280 spectrophotometry and the integrity of the plasmid is confirmed by restriction digest analysis on 0.7% agarose gels.

3.2.2. Introduction of Plasmid DNA Into Mammalian Cells

The pH6neo plasmid DNA can be introduced transiently into mammalian cell lines (729-6, 293, 293T, Jurkat, CEM) using a number of transfection methods including electroporation, Lipofectamine, and calcium phosphate precipitation allowing the assessment of various viral gene activities (*see* **Note 5**). However, this approach rarely results in efficient expression of virus particles with the capacity to infect a significant proportion of target cells. We have found that the B-cell line 729-6 supports transient as well as long-term stable expression of HTLV. 729-6 is a 6-thioguanine resistant human lymphoblastoid

B cell line isolated by treating parent WIL-2 cells with 6-thioguanine. 729-6 expresses the Epstein-Barr virus nuclear antigen (EBNA) ***(12)***. Stably transfected 729-6 cells are best-isolated following electroporation of pH6neo plasmid DNA and growth selection. Ten million 729-6 cells and 5 μg of plasmid DNA are suspended in 250 μL of growth medium and electroporated in a Bio-Rad Genepulser (with Capacitance Extender) at settings of 250 V, 950 μF, and then placed in 5 mL of growth medium in a tissue culture incubator for 2 d. After 2 d, the cells are centrifuged and resuspended in 25 mL of growth medium supplemented with 1 mg/mL of G418 to select for neomycin-resistant cells. These cells are then divided into 1-mL volumes in a 24-well plate. After selection for 3–5 wk, outgrowth of G418-resistant cells are seen in some wells. Wells are then expanded for further characterization.

3.2.3. Screening for 729 Virus-Producer Cell Lines

Wells containing G418-resistant cells are originally screened for production of HTLV-2 p24 Gag by metabolic labeling with ^{35}-S methionine/cysteine followed by radioimmunoprecipitation with HTLV-2 patient antiserum that detects primarily p24 Gag (**Fig. 2A**). Currently, a commercial HTLV-1/2 p19 Gag ELISA kit (Zeptometrix) is also available and can be used to screen the supernatants of stable cell lines for virions containing p19 Gag, allowing a much larger quantity of wells to be screened at one time, with only slightly less sensitivity than radioimmunoprecipitation (RIPA) provides (**Fig. 2B**). Testing for viral gene activity following selection for G418-resistant cells is important, as the integration process necessitates a breakage of the plasmid at some point in its length, much of which is taken up by proviral sequences. An intact integration of the neomycin-resistance gene does not exclude the possibility of plasmid breakage at some point within the provirus, which would prevent it from producing full-length viral RNA. Because the Gag gene products p19 and p24 are produced from full-length viral RNA, and efficient expression is dependent on both Tax and Rex (3' end of the genome), these proteins are ideal for use in screening stable HTLV-2 producer cell lines. Following confirmation of Gag production, cells are single-cell cloned by limiting dilution to obtain a cloned virus-producer cell line.

Further confirmation of virus production and the ability of virus to spread throughout a culture can be quantitated by syncytia-forming assays if necessary. Syncytia-forming assays are performed as follows: 729 producer cells are irradiated with 10,000 rads, and 10-fold dilutions, starting with 5×10^5 cells, are cocultivated with 10^5 BJAB cells in 24-well culture plates. Cells are fed twice a week with RPMI 1640 supplemented to contain 10% FCS, 100 U/mL of penicillin and streptomycin, and 2 m*M* glutamine. Syncytia are scored in BJAB co-cultures microscopically 3 to 5 d after plating (**Fig. 2C**). If there is

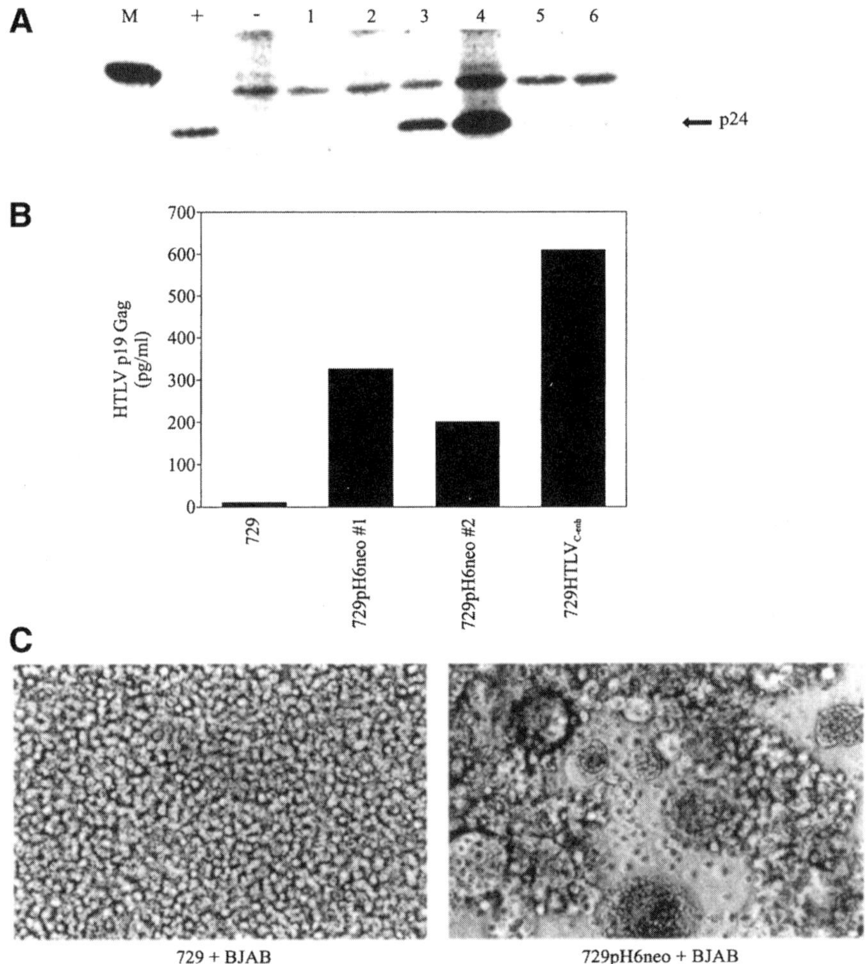

Fig. 2. **(A)** Immunoprecipitation of [35S]methionine-cysteine-labeled 729 producer cells. 729 producer cells (5×10^6) were metabolically labeled, and cell lysates were prepared. Stable transfectant cell lysates were immunoprecipitated with human antisera directed against the HTLV-2 p24 Gag capsid protein. Lane M contains a 30-kD marker band. Uninfected 729 cells (–) and 729pH6neo cells (+) are positive and negative controls, respectively. Six separately isolated neomycin-resistant samples are in lanes 1–6; lanes 3 and 4 are positive. **(B)** HTLV p19 Gag ELISA results from supernatants of uninfected 729 cells, two clones of the 729pH6neo infected cell line, and one clone stably transfected with HTLV-$2_{\text{C-enh}}$. p19 Gag levels vary between different cell clones, even those expressing the same virus. **(C)** HTLV-II syncytium induction in BJAB cells. 729 or stable transfectants were irradiated with 10,000 rads and cocultured with BJAB cells. Syncytia were scored in BJAB cell cocultures microscopically 3 to 5 d postplating. Cells were photographed at 72 h postplating.

efficient virus infection and spread, syncytia can be induced with as few as 100 irradiated viral producer cells.

Infection of the BJAB or 729 cells can be confirmed by detection of the viral p19 Gag by immunofluorescence. Cells are washed in PBS, spotted onto slides, and allowed to air-dry. Cells are fixed for 10 min in acetone-methanol (1:1) and allowed to air-dry. Following fixing, cells are incubated with anti-p19 Gag antibody (Zeptometrix) for 30 min at room temperature under a humidified chamber. Cells are washed with PBS and incubated with fluorescein-conjugated rabbit anti-mouse immunoglobulin G for 30 min at room temperature in a humidified chamber. These cells are then washed three times with PBS and examined immediately under a fluorescence microscope.

3.3. Infection of Primary Human T-Lymphocytes

Once a stable producer line has been generated, it can be irradiated and co-cultured with freshly isolated PBMCs to produce infected T cells, the natural targets of HTLV-1 and -2 infection, immortalization, and transformation. Cell-free HTLV-1 and -2, unlike HIV, is poorly infectious. As a result, it was necessary to develop a reproducible co-culture system to allow direct cell-to-cell transmission of virus, the natural route of infection in vivo.

3.3.1. Irradiation of Virus-Producer Cells

Producer cells are irradiated in their normal growth medium, using a cesium-source irradiator (Gammacell 40) with 10,000 rads of γ-radiation (*see* **Note 6**). These cells cannot divide after irradiation, but continue to produce infectious virus until cell death, which usually occurs in approx 1–2 wk (*see* **Note 7**).

3.3.2. Co-culture of Irradiated Virus-Producer Cells With Human PBMCs

Irradiated producer cells are co-cultured in a 1:1 or up to a 1:4 ratio with freshly isolated human PBMCs. The co-cultured cells are plated into 24-well plates in 1-mL volumes with PBMC total numbers in the range of $1–2 \times 10^6$ cells/mL (well). These experiments can be used to study short- or long-term effects of alterations made within the pH6neo clone (*see* **Note 8**). **Figure 3** shows representative data of a typical transformation assay. In this particular experiment the transforming capacity of a chimeric Tax-independent HTLV-2 ($HTLV_{C\text{-}enh}$), and various Tax mutant viruses were compared to wild-type HTLV-2. The chimeric HTLV-2 had the capacity to transform primary human T-lymphocytes with an efficiency similar to wild-type HTLV-2. A *tax*-knock-out virus, termed $HTLV_{C\text{-}enh}\Delta Tax$, was shown to be replication-competent, but failed to transform primary human T-lymphocytes, therefore providing direct

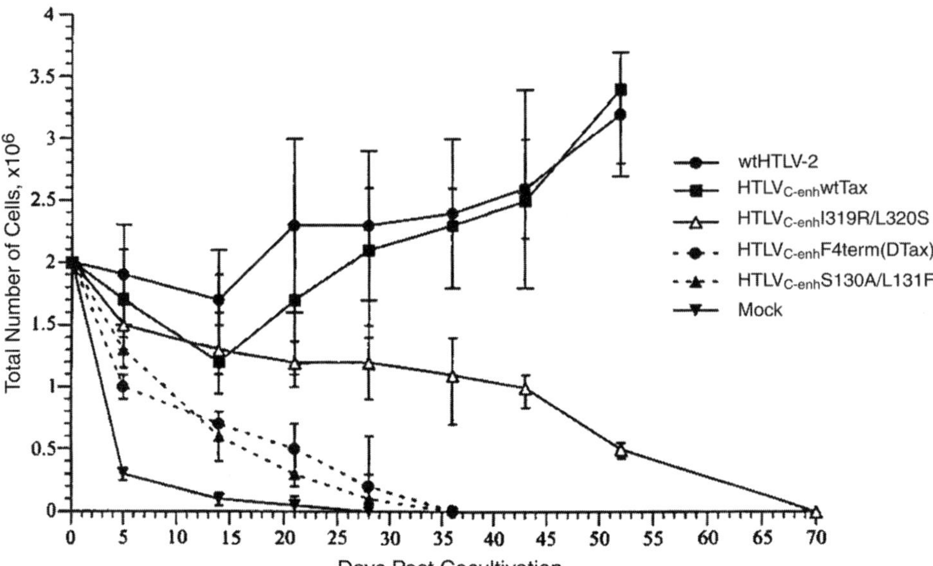

Fig. 3. Growth curve of $HTLV_{C\text{-}enh}$ T-lymphocyte transformation assay. Cell viability was determined by Trypan blue exclusion staining at 0, 5, 14, 21, 28, 36, 43, 53, and 70 days postcocultivation. Four distinct growth patterns were observed. The mean and standard deviation was determined from three independent samples of each coculture: pattern 1 contains PBL/729 negative control (Mock) coculture; pattern 2 contains PBL/729-$HTLV_{C\text{-}enh}$F4Term (ΔTax), and PBL/729-$HTLV_{C\text{-}enh}$S130A/L131F coculture showing the critical importance of NFκB/Rel activation; pattern 3 contains the PBL/729-$HTLV_{C\text{-}enh}$I320R/L320S coculture showing the requirement for activation of CREB/ATF in addition to NFκB/Rel; and pattern 4 contains PBL/729-wtHTLV-2 and PBL/729-$HTLV_{C\text{-}enh}$wtTax, or fully transformed cells ***(13)***.

evidence that Tax is essential for HTLV-mediated transformation of human T-lymphocytes ***(11)***. The creation of a proviral clone that could replicate efficiently in the absence of Tax provided a unique opportunity to study the effects of Tax-2 mutants on transformation without concern for those mutations that affect viral transcription and replication. A series of Tax-2 cDNA mutants were created and evaluated for their ability to stimulate NFκB/Rel and CREB/ATF activity in T-lymphocytes. These mutants were then cloned into $HTLV_{C\text{-}enh}$ proviral clone to determine the respective roles of the CREB/ATF or NFκB/Rel signaling pathway in viral replication and cellular transformation. Distinct *tax*-2 mutant virus phenotypes were identified, but Tax activation of NFκB/Rel and CREB/ATF both contribute to induction of IL-2 independent T-lymphocyte transformation (**Fig. 3**). The activation of NFκB/Rel provides a critical proliferative signal early in the cellular transformation process, whereas CREB/ATF activation is required to promote the IL-2 independent transformed state ***(8,11,13)***. This is a major step in the elucidation of the mechanism of T-lymphocyte transformation.

4. Notes

1. There are a number of useful restriction sites located within the HTLV-2 genome that are useful for creating mutants or chimeras for experimental study. In addition to the use of preexisting restriction sites within the proviral genome, it is sometimes necessary to use site-directed mutagenesis either to add restriction sites or to create deliberate mutations within viral genes. When adding restriction sites to assist in cloning, extra care should be taken to avoid introducing mutations in overlapping viral genes. In situations in which it is necessary to create mutations in coding regions, it is important to minimize disruption of genes with overlapping reading frames, as is the case in the pX region of the provirus. In early studies, mutagenesis was carried out using M13-based methods. More recently, mutagenesis has been performed with Stratagene's PCR-based site directed mutagenesis kits. PCR-based mutagenesis is best carried out on small templates, so it is advantageous to use a smaller subclone of the provirus to conduct the mutagenesis, and then clone that portion back into the full-length plasmid.
2. Large plasmid deletion can be detected at this early stage by first performing a small-scale plasmid preparation on a small aliquot of the 400-mL culture followed by visualization of undigested and diagnostically digested (BamHI or EcoRI) DNA on a 0.7% agarose gel prior to large-scale plasmid preparation.
3. Extreme care should be taken when piercing the tubes with a syringe needle, particularly in how the tube is handled. The tube should always be gripped at a point well away from where the syringe needle will pierce it. It is fairly easy to pierce completely through the opposite side of the tube if too much pressure is applied. There are two drawbacks to this. The first and most important is that it places the health and safety of the handler at risk. The tube at this point of the experiment contains a concentrated solution of ethidium bromide (a mutagen) and cesium chloride, as well as proviral DNA plasmid. The consequences of an accidental needle stick with the mixture are best left unexplored. The risk to the handler is negligible so long as the handler remains aware of the danger and handles the tube correctly. Second, when a second hole is made in the barrel of the tube, the tube will begin to leak as soon as the needle is withdrawn from the hole, usually causing loss of the sample before it can be collected.
4. RNA and protein contamination of plasmid preparations is occasionally a problem, and is undesirable when that plasmid is to be used for transfection. RNA contamination is usually visible after electrophoresis on an ethidium-stained agarose gel as a smear in the lower molecular weight region. Protein contamination is best indicated by a 260/280 ratio of the preparation that falls outside the expected range for nucleic acid. In such a case, the following steps should be added

between collection of the DNA from dialysis and the phenol/chloroform extraction steps: a 1-h treatment with RNAse A (10 μg/mL) at 37°C followed by proteinase K treatment (20 μg/mL for another hour at the same temperature). This will minimize RNA and bacterial protein contamination and result in more accurate concentration values when the plasmid DNA is analyzed by UV spectrophotometry.

5. Primary cells directly transfected by electroporation have poor survival rates and often fail to produce detectable infection. Currently we are attempting to achieve efficient transfection of PBMCs with proviral DNA using nucleofection (Amaxa). This technique is a modification of standard electroporation that delivers the DNA into the cell nucleus. It provides much higher transfection efficiency in primary cells and faster onset of expression of the transfected genes.
6. Producer cells are irradiated before co-culture, damaging their DNA to a point where they can no longer divide. The producer cell line will survive for a short time in the co-culture and continue to produce virus that can infect PBMC. The most common endpoint of an HTLV co-culture experiment is immortalization or transformation of primary T-cells. If a source of γ radiation is not available, it is possible to use mitomycin C treatment of the producer cells in place of irradiation. Mitomycin C should be applied to the producer cells at a concentration of 100 μg/mL for 30 min at 37°C. Extensive washing of the producer cells with cell culture medium after treatment is important if this method is to be used, as the primary cells may be sensitive to residual mitomycin C.
7. At approx 2–3 wk post-co-culture the majority of irradiated 729 cells are dead and the co-culture can be littered with cell debris. In some cases, depending on the overall number of viable cells present at this time, it can be beneficial to remove dead cells and debris by Ficoll treatment and replating only the viable cells. To limit cell loss, Ficoll treatment is performed in a sterile microfuge tube.
8. The decision on whether or not to provide exogenous IL-2 in co-culture experiments depends on the goal of the experiment. It is more difficult to generate IL-2-independent transformed cells, but the results are more relevant to the events that occur in natural infections. Although it is easier to produce immortalized, IL-2-dependent cells, there are likely to be differences in the processes that lead to the transformation of these cells in comparison to the normal development of ATL.

References

1. Yoshida, M., Miyoshi, I., and Hinuma, Y. (1982) Isolation and characterization of retrovirus from cell lines of human adult T-cell leukemia and its implication in the disease. *Proc. Natl. Acad. Sci. USA* **79,** 2031–2035.
2. Yoshida, M., Seiki, M., Yamaguchi, K., and Takatsuki, K. (1984) Monoclonal integration of human T-cell leukemia provirus in all primary tumors of adult T-cell leukemia suggests causative role of human T-cell leukemia virus in the disease. *Proc. Natl. Acad. Sci. USA* **81,** 2534–2537.
3. Hjelle, B., Appenzeller, O., Mills, R., Alexander, S., Torrez-Martinez, N., Jahnke, R., and Ross, G. (1992) Chronic neurodegenerative disease associated with HTLV-II infection. *Lancet* **339,** 645–646.

4. Jacobson, S., Lehky, T., Nishimura, M., Robinson, S., McFarlin, D. E., and DhibJalbut, S. (1993). Isolation of HTLV-II from a patient with chronic, progressive neurological disease clinically indistinguishable from HTLV-I-associated myelopathy/tropical spastic paraparesis. *Ann. Neurol.* **33,** 392–396.
5. Kalyanaraman, V. S., Sarngadharan, M. G., Robert-Guroff, M., Miyoshi, I., Blayney, D., Golde, D., and Gallo, R. C. (1982) A new subtype of human T-cell leukemia virus (HTLV-II) associated with a T-cell variant of hairy cell leukemia. *Science* **218,** 571–573.
6. Rosenblatt, J. D., Golde, D. W., Wachsman, W., Jacobs, A., Schmidt, G., Quan, S., et al. (1986) A second HTLV-II isolate associated with atypical hairy-cell leukemia. *N. Engl. J. Med.* **315,** 372–377.
7. Chen, I. S. Y., McLaughlin, J., Gasson, J. C., Clark, S. C., and Golde, D. W. (1983) Molecular characterization of genome of a novel human T-cell leukaemia virus. *Nature* **305,** 502–505.
8. Green, P. L., Ross, T. M., Chen, I. S., and Pettiford, S. (1995) Human T-cell leukemia virus type II nucleotide sequences between env and the last exon of tax/rex are not required for viral replication or cellular transformation. *J. Virol.* **69,** 387–394.
9. Shimotohno, K., Takahashi, Y., Shimizu, N., Goiobori, T., Chen, I. S. Y., Golde, D. W., et al. (1985) Complete nucleotide sequence of an infectious clone of human T-cell leukemia virus type I and type II long terminal repeats for trans-activation of transcription. *Proc. Natl. Acad. Sci. USA* **82,** 3101–3105.
10. Shimotohno, K., Wachsman, W., Takahashi, Y., Golde, D. W., Miwa, M., Sugimura, T., and Chen, I. S. Y. (1984) Nucleotide sequence of the 3' region of an infectious human T-cell leukemia virus type II genome. *Proc. Natl. Acad. Sci. USA* **81,** 6657–6661.
11. Ross, T. M., Pettiford, S. M., and Green, P. L. (1996) The *tax* gene of human T-cell leukemia virus type 2 is essential for transformation of human T lymphocytes. *J. Virol.* **70,** 5194–5202.
12. Glassy, M. C., Handley, H. H., Hagiwara, H., and Royston, I. (1983) UC 729-6, a human lymphoblastoid B-cell line useful for generating antibody-secreting human-human hybridomas. *Proc. Natl. Acad. Sci. USA* **80,** 6327–6331.
13. Ross, T. M., Narayan, M., Fang, Z. Y., Minella, A. C., and Green, P. L. (2000) Human T-cell leukemia virus type 2 tax mutants that selectively abrogate NFkappaB or CREB/ATF activation fail to transform primary human T cells. *J. Virol.* **74,** 2655–2662.
14. Ciminale, V., D'Agostino, D. M., Zotti, L., Franchini, G., Felber, B. K., and Chieco-Bianchi, L. (1995) Expression and characterization of proteins produced by mRNAs spliced into the X region of the human T-cell leukemia/lymphotropic virus type II. *Virology* **209,** 445–456.

33

Construction and Analysis of Genomic, Full-Length Infectious Foamy Virus DNA Clones

Roman Wirtz and Martin Löchelt

Summary

The molecular engineering of recombinant plasmid DNA clones containing the full-length and replication-competent feline foamy (retro)virus (FFV) proviral genome is described. The methods used to combine subgenomic FFV DNA fragments can be applied to other retrovirus genomes, resulting in full-length, bacterially cloned retroviruses. In addition, techniques used to determine the replication competence of the cloned viral genomes are described. Alternative technologies (not described here) are also discussed, together with the application of cloned genomes in modern molecular virology; for example, in functional genetic studies or as starting material for retrovirus vector development.

Key Words: Molecular virology; infectious DNA clone; provirus; full-length genome; retrovirus; feline foamy virus (FFV).

1. Introduction

The introduction of advanced molecular biology methods to study the biology of viruses has established the powerful field of molecular virology. This rapidly developing field provides fundamental new insights into virus replication and pathogenesis in vitro and in vivo *(1)*. Many of these studies rely on the use of recombinant bacterial plasmids and phage genomes carrying full-length viral genomes. These chimeric DNAs encoding all viral functions required for viral gene expression and morphogenesis often also yield fully infectious progeny virus. Such cloned viral genomes can be easily subjected to random or site-specific mutagenesis. Upon introduction into permissive cells, the phenotype produced by genetic changes can be analyzed. In this way, the functions of individual genes may be subjected to direct experimental investigation.

From: *Methods in Molecular Biology, Vol. 304: Human Retrovirus Protocols: Virology and Molecular Biology*
Edited by: T. Zhu © Humana Press Inc., Totowa, NJ

In addition, surrogate in vitro replication systems using cloned viral genomes often provide a measurable experimental system for molecular biology studies. For instance, permissive cell-culture systems are presently not available for hepatitis B and C viruses and thus, only cloned viral genomes can be studied in vitro *(1)*. Full-length replication-competent viral genomes are also the basis for viral vectors used to specifically transduce therapeutic genes or to express vaccine antigens *(2)*.

Here, the methods of constructing bacterial plasmid DNA clones of full-length retrovirus genomes from subgenomic DNA fragments are presented. The cloning procedure will be described for the feline foamy virus (FFV) genome *(3)*. This strategy is universal but has to be adapted to the intrinsic and individual features of each (foamy) retroviral genome.

2. Materials

1. Standard equipment for eukaryotic cell culture.
2. FFV wt stock, Crandell feline kidney cells (CRFK), foamy virus activated β-galactosidase (Fe-FAB) titration cells.
3. Complete CRFK medium: Dulbecco's modified Eagle's medium (DMEM), 7% fetal calf serum (FCS), 1% penicillin/streptomycin.
4. Phosphate-buffered saline (PBS): 8 m*M* Na_2HPO_4, 1.5 m*M* KH_2PO_4/HCl, pH 7.4, 140 m*M* NaCl, 2.6 m*M* KCl.
5. Trypsin solution for cell culture purposes.
6. Hirt lysis buffer: 0.6% sodium dodecyl sulfate (SDS), 10 mM Tris-HCl, pH 7.4, and 10 m*M* EDTA.
7. TE buffer: 10 m*M* Tris-HCl, pH 7.4, 1 m*M* EDTA.
8. RNase A, 10 mg/mL in TE buffer (Roche).
9. Proteinase K 10 mg/mL in TE buffer (Roche).
10. Equilibrated and buffer-saturated phenol.
11. Chloroform/isoamyl alcohol (24:1, v/v).
12. Polymerase chain reaction (PCR) primers.
13. dNTP mix: 0.5 m*M* of each dNTP in water (Roche).
14. Expanded high-fidelity PCR system (Roche).
15. TOPO TA cloning kit including vectors pCRII and pCR2.1 (Invitrogen).
16. *Taq* polymerase.
17. Qiagen Plasmid Midi Kit (Qiagen).
18. QIAquick Gel Extraction Kit (Qiagen).
19. QIAquick PCR Purification Kit (Qiagen).
20. Klenow DNA polymerase (Roche).
21. 10X buffer *M*: 10 m*M* Tris-HCl, 10 m*M* $MgCl_2$, 50 m*M* NaCl, 1 m*M* dithioerythritol (Roche).
22. T4 DNA ligase and 10X reaction buffer (1 and 5 U/μL, Roche).
23. Restriction enzymes.
24. *E. coli* strain JM 109 and plasmids pAT 153 and pBluescript KS (Stratagene).

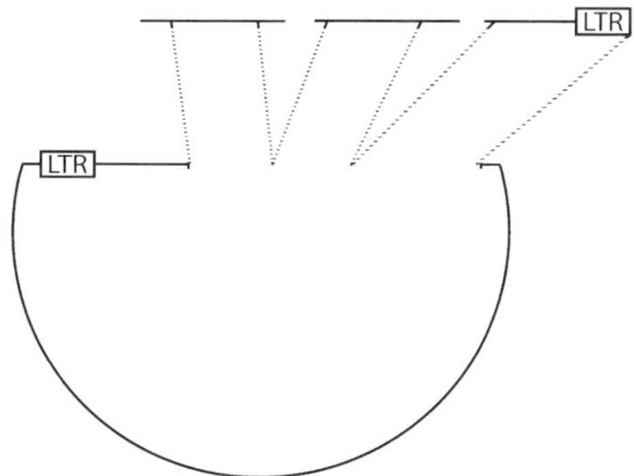

Fig. 1. Schematic outline of the cloning strategy used. In particular, the sequential addition of defined proviral DNA fragments (bars) into the plasmid vector (semicircle) is shown. This strategy results in full-length FFV DNA clones.

25. 2-mm-gap electroporation cuvets.
26. β-gal staining solution: 4 m*M* potassium ferrocyanide, 4 m*M* potassium ferricyanide, 1 m*M* $MgCl_2$, 0.4 mg/mL of 5-bromo-4-chloro-3-indolyl-β-D-galactopyranoside (X-Gal) in PBS.

3. Methods

The full-length cloning strategy described here for a foamy virus (FV) genome relies on the sequential addition of defined DNA fragments by directed, asymmetrical cloning. Owing to the size of the FV genomes, single-cutting restriction enzyme recognition sites required for cloning are rare and not evenly distributed over the provirus. In addition, the subgenomic DNA fragments generated under high-fidelity PCR conditions are limited in size. These features make it necessary to generate partially overlapping FV DNA that share unique or rare restriction sites with the preexisting DNA clones. This technique is schematically shown in **Fig. 1** (*see* **Note 1**). Standard molecular biology techniques are used *(4,5)*. The safety regulations concerning molecular biology experiments must be strictly followed and permission to perform full-length cloning of retroviral genomes may need to be granted by state authorities.

3.1. Construction of Full-Length FFV Genome

The nucleotide sequence of the FFV isolate FUV is known and clone 7 (**Fig. 2**), covering the 5' terminal 5.8 kb of the FFV genome, is available *(6)*. In clone 7, FFV DNA sequences from nt position 17 to 5811 are inserted into the *Eco*RI site of the standard cloning vector pBluescript KSII *(6)*. In order to add the lacking sequences, overlapping subgenomic DNA fragments containing FFV-sequences from nt position 5118 to the 3' end of the proviral FFV genome were

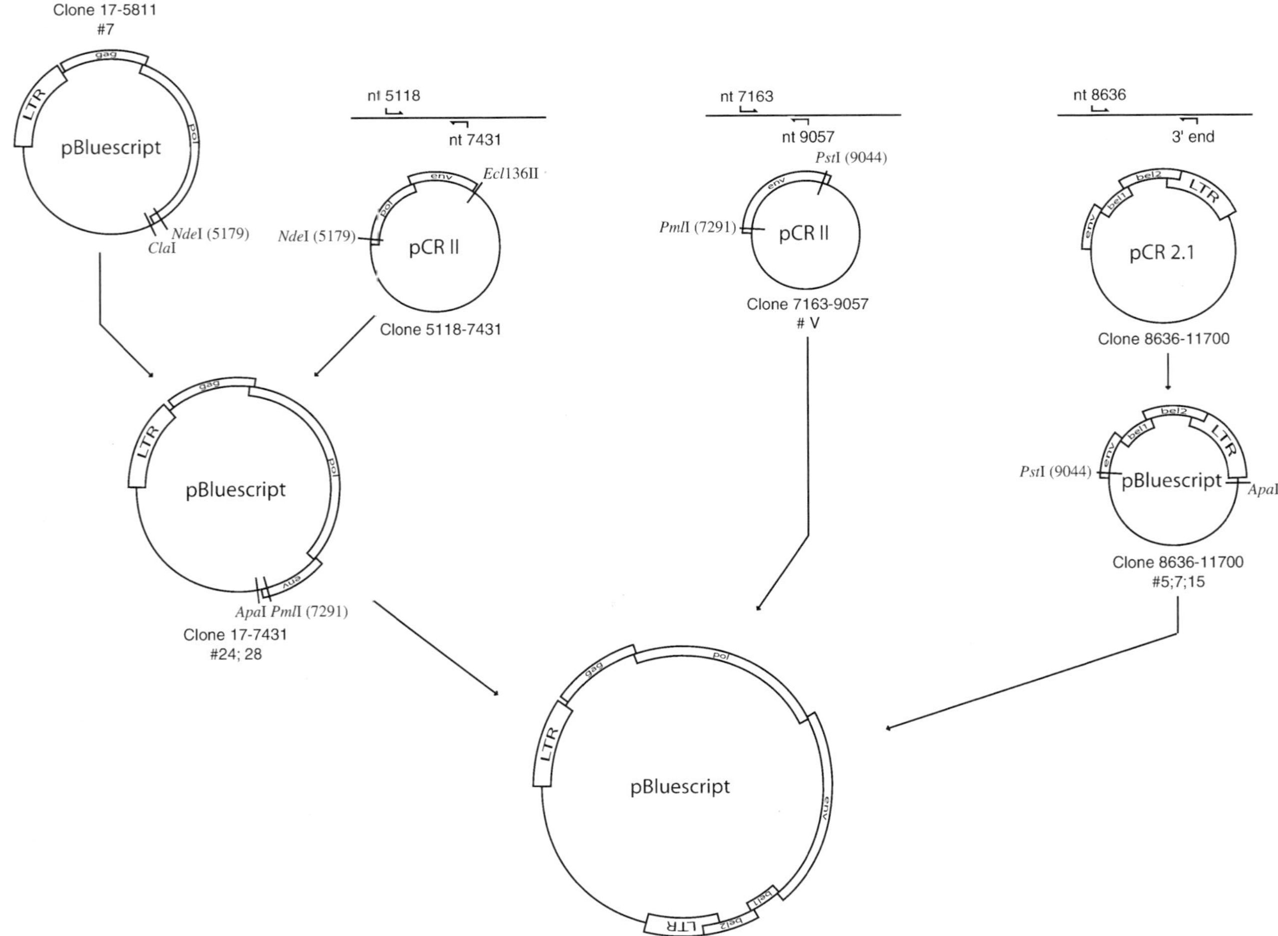

Clone 17-5811
#7
gag
LTR
pol
pBluescript
NdeI (5179)
ClaI
nt 5118
nt 7431
env
Ecl136II
pol
NdeI (5179)
pCR II
Clone 5118-7431
nt 7163
nt 9057
PstI (9044)
env
PmlI (7291)
pCR II
Clone 7163-9057
V
nt 8636
3' end
bel2
bel1
LTR
env
pCR 2.1
Clone 8636-11700
bel2
bel1
LTR
env
PstI (9044)
pBluescript
ApaI
Clone 8636-11700
#5;7;15
gag
LTR
pol
pBluescript
env
ApaI PmlI (7291)
Clone 17-7431
#24; 28
gag
pol
LTR
pBluescript
env
bel1
LTR
bel2

cloned. This was achieved by PCR-mediated amplification of DNA samples from FFV-infected CRFK cells, and cloning of each of the amplicons into TA cloning vectors (Invitrogen) as described below ***(3,6)*** (*see* **Fig. 2**).

3.1.1. Infection of CRFK Cells and Isolation of Proviral FFV DNA

1. Proviral DNA required as a template for PCR-mediated amplification of DNA was derived from FFV-infected cells. CRFK cells were cultured in DMEM supplemented with 7% (vol) FCS and 1% (vol) penicillin/streptomycin at 37°C and 5% CO_2 ***(6)***.
2. One day before FFV infection, cells were washed once with PBS, resuspended after trypsin treatment, and plated at 40% confluence on 60-cm^2 dishes in 10 mL of complete medium. At about 80% confluence, 0.4 mL of an FFV stock was added at a multiplicity of infection of 1.
3. Two days later, nonintegrated FFV DNA was extracted as follows ***(7)***: The FFV-infected cells were lysed in 1 mL of Hirt lysis buffer (0.6% SDS, 10 m*M* Tris-HCl, pH 7.4, and 10 m*M* EDTA) for 20 min at room temperature.
4. The viscous lysate was gently scraped from the dish and poured into 2-mL Eppendorf tubes while carefully avoiding shearing of the DNA.
5. Next, 5 *M* NaCl was added to a final concentration of 1 *M* NaCl, the sample was carefully but thoroughly mixed by inverting the tube several times, and stored at 0 to 4°C overnight.
6. Chromosomal DNA and the majority of cellular proteins were removed by centrifugation at 12,000*g* for 30 min at 4°C. High-molecular-weight, cellular DNA was pelleted, leaving the smaller proviral FFV DNA in solution.
7. The proviral DNA was subsequently precipitated by the addition of three volumes of ethanol, incubated at –20°C for 16 h, and collected by centrifugation at 12,000*g* for 30 min at 4°C.
8. DNA pellets were washed twice with 70% ethanol/water (v/v), dried, and resuspended in 200 µL TE buffer. The samples were sequentially digested with RNase A (5 µL, 10 mg/mL RNase A in TE buffer) for 30 min at 56°C and with proteinase K (5 µL, 10 mg/mL proteinase K in TE buffer) for 30 min, 56°C (Roche).
9. The DNA was twice extracted with equilibrated phenol/chloroform/isoamyl alcohol (25:24:1) and finally once with chloroform/isoamyl alcohol (24:1). DNA

Fig. 2. *(opposite page)* Diagram of the cloning steps required to engineer a full-length FFV proviral DNA genome into a bacterial plasmid. In the upper part of the figure, the preexisting clone 7 and the products of the different PCR reactions (and the primers used) are shown. Below the amplicons, the corresponding molecular clones and the names of the parental vectors are given. FFV genes are indicated as boxes and the plasmid vector backbones as solid lines. Restriction enzyme sites, relevant for the subsequent cloning step (indicated by arrows), are indicated. The length of genes and plasmid backbones is intimated but not to scale. The full-length provirus in the pBluescript vector was finally cloned into pAT153 (not shown).

was precipitated as above except that the solution was adjusted to 0.3 *M* sodium acetate, pH 5.2.

10. The Hirt DNA was collected by centrifugation (as above), dried, resuspended in TE buffer, and stored at –20°C. The DNA concentration was spectrophotometrically determined at 260 nm.

3.1.2. PCR Amplification and Subcloning of FFV Proviral DNA Fragments

1. Each 100 ng of the enriched proviral Hirt DNA served as a template for three individual PCRs to generate FFV DNA fragments from genomic position 5118–7431 (PCR 5118–7431), 7163 to 9057 (PCR 7163–9057) and 8636 to the 3' end of the genome (PCR 8636–11700). The PCRs were performed with the Expand High Fidelity PCR System (Roche). This PCR system contains an enzyme mixture consisting of the strongly processive *Taq* DNA polymerase plus *Tgo* polymerase. The latter is a thermostable DNA polymerase with proofreading activity, thus reducing the level of amplification-dependent errors (*see* **Note 2**). Primers used for the PCR reactions are given in **Table 1**. The PCR conditions, including the extension times and annealing temperatures, were adjusted to fragment length and primer properties. The extension times were calculated to be 2 min per kilobasepair (kb) of target DNA and annealing temperatures were 58°C.
2. After PCR amplification using proofreading polymerases, A-overhangs are added to the 3' ends of the PCR products in order to use the highly efficient TA-cloning strategy (Invitrogen). Thus, 1 unit of *Taq* polymerase was added to each PCR tube and the A-addition reaction was performed at 72°C for 10 min.
3. Directly thereafter, the TA cloning reaction including transformation of electrocompetent *E. coli* was performed according to the manufacturer's protocol. PCR amplicons 5118–7431 and 7163–9057 were cloned into vector pCRII-TOPO and fragment 8636–11700 into vector pCR2.1.
4. Bacterial clones were picked and plasmid DNA isolated according to a rapid lysis protocol *(8)*. The plasmid DNAs were restriction digested and the fragments oriented according to the expected fragment pattern. For the insert 8636–11700, three independent clones were obtained.
5. Plasmid DNA was purified using the Qiagen Plasmid Midi Kit according to the manufacturer's instructions (Qiagen).

3.1.3. Cloning of a Co-Linear, Full-Length FFV Genome From PCR Fragments

1. FFV plasmid DNA clone 7 *(6)* was digested with *Cla*I (in the pBluescript plasmid backbone downstream of the FFV insert), extracted with phenol/chloroform, alcohol-precipitated, and resuspended in 23 µL water.
2. To fill in the recess 3'-ends, 3 µL of 10X buffer M (Roche), 3 µL of a 0.5 m*M* dNTP mix, and 1 µL Klenow DNA polymerase (1 U/µL, Roche) were added and the reaction incubated for 30 min at 37°C. The enzyme was inactivated at 65°C for 10 min and the DNA digested with *Nde*I that cuts at FFV position 5179.
3. The DNA fragment of about 8 kb representing the plasmid backbone plus the 5' half of the FFV genome was isolated by preparative agarose gel electrophoresis.

Table 1
Polymerase Chain Reaction (PCR) Primers and Reaction Conditions Used to Perform PCRs

	Primers	Annealing temperature	Extension time
PCR 5118 to 7431	sense: 5'-CCTCATGCTTACGGGAATAATCTGGCTG-3'		
	antisense: 5'-GAATAGCATACCAGAGCCTACAGGGCTC-3'	58°C	5 min
PCR 7163 to 9057	sense: 5'-CCAATTGGACAAGAGTAGAATCCTATGG-3'		
	antisense: 5'-TTCTCCAAGGAGCTGCAGCCACTCTGG-3'	58°C	4 min
PCR 8636 to 11700	sense: 5'-GTGGAAATGGAACTGGTTCAGACTGCC-3'		
	antisense: 5'-GCCATCGATGTCGGTGCCTATACCTGGGATAAG-3'	58°C	6 min
PCR 5775 to 10522	sense: 5'-TTTGCTCAGTGGGCAAAGGAAAGGAATATACAATTGG-3'		
	antisense: 5'-GTTGACACTGATTTATATGGCACAATAATTCCTCTC-3'	58°C	10 min

The DNA was extracted using a commercially available QIAquick Gel Extraction Kit according to the manufacturer's suggestions (Qiagen).

4. The pCRII clones containing FFV sequences 5118–7432 were digested with *Nde*I (at FFV position 5179) and the blunt-ending enzyme *Ecl*136II (in the pCRII sequence downstream of the FFV insert). The 2.2-kb FFV DNA fragment was gel-purified as above and inserted into the correspondingly cleaved clone 7. For ligation, similar amounts of vector and insert DNA (as judged after gel-electrophoresis of an aliquot of the DNA) were used in 20 μL containing 2 μL of 10X ligation salts and 1 mL T4 DNA ligase (5 U/μL, Roche). Ligations were performed at 15°C overnight and terminated at 65°C for 10 min.
5. Electrocompetent *E. coli* (strain JM109) was prepared and transformated with 1.5 μL of the ligation reaction as described ***(4)***. Ampicillin-resistant colonies were picked and their DNA extracted and analyzed by restriction enzyme digestion. Several independently correct clones containing FFV DNA from nt position 17 to 7431 were identified.
6. Three pCR2.1 inserts 8636–11700 ***(6)*** were subcloned into pBluescript KS using the common restriction enzymes *Eco*RV (in the polylinkers) and *Cla*I introduced by the antisense primer of the PCR. This subcloning was done to attach the pBluescipt KS-derived *Apa*I site to the 3'-end of the FFV genome fragment, as this restriction enzyme site was required for the subsequent cloning step.
7. In summary, the following FFV DNA clones were identified (**Fig. 2**):
   - FFV nt 17 to 7431 in vector pBluescript, subclones 24 and 28.
   - FFV nt 7163 to 9057 in vector pCRII-TOPO, subclone V.
   - FFV nt 8636-11700 in vector pBluescript, subclones 5, 7, and 15.
8. Defined sequences of the three plasmids were combined in three-component ligations in order to obtain full-length, replication-competent FFV DNA clones. In a chessboard fashion, subclones 24 and 28 were combined with inserts 5, 7, and 15 and DNA from the unique clone V. This resulted in six different insert combinations.

   Therefore, DNAs from clones 17–7431 (24 and 28) were digested with *Apa*I and *Pml*I (FFV position 7291), clone 7163-9057 (V) with *Pml*I and *Pst*I (FFV position 9044) and clones 8636–11700 (5, 7, and 15) with *Pst*I and *Apa*I. As described above, the following DNAs were separated by agarose gel electrophoresis and purified (*see* **Note 3**):
   a. Clones 17–7431 *Apa*I-*Pml*I-digested: 10.2 kb
   b. Clone 7163–9057 *Pml*I-*Pst*I-digested: 1.75 kb
   c. Clones 8636–11700 *Pst*I-*Apa*I-digested: 2.3 kb
9. The DNA concentrations were determined by gel electrophoresis and equal amounts of the DNA fragments were ligated for 16 h in a three-component ligation using T4-DNA-ligase (5 U/ μL, see above). The resulting clones contained the FFV-sequence from position 17 to 11,700 in a pBluescript backbone (**Fig. 2**).
10. Finally, different FFV proviral DNAs present between the flanking unique restriction sites *Eag*I and *Cla*I were recloned into the correspondingly digested

plasmid pAT153. This was done since human FV DNA clones were genetically stable in the simple pAT153 backbone but not in some multipurpose vectors containing complex genetic features *(9)*. By these means, different FFV proviral clones derived from different insert combinations were obtained.

3.2. Analysis of Full-Length FFV Clones

The clones containing FFV sequence 17 to 11700 were subsequently analyzed for function by measuring the infectivity of plasmid-derived FFV in parallel with the wild-type, uncloned virus. As outlined below, FFV-permissive CRFK cells were transfected with each of the plasmid DNAs and the amount of infectious virus released into the cell-culture supernatant was determined by titration using Fe-FAB cells.

3.2.1. Transfection of CRFK Cells by Electroporation (6)

1. For each transfection, about 5×10^6 subconfluently grown CRFK cells were harvested by trypsin digestion, sedimented by low-speed centrifugation, washed with PBS, and resuspended in 90 µL PBS.
2. 10 µg plasmid DNA in 10 µL PBS were added, and the mixtures preincubated for 10 min at room tempertature in 2-mm cuvets.
3. Electroporation was carried out at 160 V, 975 µF for 20–30 ms, then 1 mL of complete medium was added and the cells were left at room temperature for another 10 min.
4. Transfected cells were transferred to cell culture dishes, avoiding coagulated and dead cells. Cultures were incubated for 2 d at 37°C until the typical cytopathic effect of syncytia formation was clearly visible by light microscopy in fully permissive cells (*see* **Note 4**).
5. The cell culture medium was collected, cleared by low-speed centrifugation, and used for virus titration employing Fe-FAB-indicator cells *(3)*. The cells were directly harvested in protein sample buffer for immunoblot analysis.

3.2.2. FAB-Assay

Fe-FAB cells are genetically engineered FFV-permissive CRFK cells that carry the FFV long terminal repeat (LTR) promoter directing nuclear expression of β-galactosidase (β-gal) *(3)*. Upon FFV infection, the Bel1 transactivator is expressed, inducing expression of the FFV genome and the engineered LTR-β-gal construct. β-gal expression is detected at the single cell level as a blue nuclear staining allowing specific and very sensitive identification of FFV-positive cells by light microscopy *(3)*.

1. Fe-FAB indicator cells were plated at 40% density on 24-well plates in 1 mL complete medium per well.
2. After 5–16 h of incubation, virus stocks were titrated in 10-fold dilutions. 110 µL of cell-free supernatant was added into the first well, vigorously mixed by

pipetting, and 110 µL were given to the next well. By this method, dilutions up to 10^{-8} were made. Infected cells were incubated for 2 d at 37°C.

3. The cells were washed once with PBS containing 1 m*M* $MgCl_2$, fixed for 5 min at RT in 1 mL of PBS containing 1% formaldehyde, 0.2% glutaraldehyde, and subsequently washed three times with PBS.
4. β-gal activity was detected in 400 µL of staining solution (PBS with 4 m*M* potassium ferrocyanide, 4 m*M* potassium ferricyanide, 1 m*M* $MgCl_2$, and 0.4 mg/mL of 5-bromo-4-chloro-3-indolyl-β-D-galactopyranoside [X-Gal]) for 1 h at 37°C. Reactions were terminated by removing the staining solution and washing the cells with distilled water. Blue cells were identified by light microscopy. Titers of FFV infectivity were quantified as number of (blue) focus-forming units (FFU) per milliliter of virus solution ***(3)***.

The titrations revealed that none of the plasmid clones yielded more than 10^4 FFU/mL progeny virus. As determined by immunoblotting, the overall FFV gene expression was not altered and all detectable FFV proteins appeared to be intact (not shown). Because the moderate infectivity obtained did not allow efficient experimentation with these FFV clones, parts of the present genome were replaced by sequences directly obtained by long-distance PCR using DNA from FFV-infected cells. *Env* sequences were exchanged with the goal of enhancing viral infectivity to the level of wild-type virus.

3.3. Enhancement of Infectivity by Exchange of env *Sequences*

In order to exchange *env* sequences, we took advantage of the unique *env*-flanking *BstZ17I* (FFV position 5980) and *BsaI* (FFV position 10137) sites. PCR fragments covering these sites were generated and following restriction enzyme digestion, inserted into the correspondingly cleaved FFV clone (infectivity of 10^4 FFU/mL).

1. CRFK cells were FFV-infected and FFV DNA extracted as described in **Subheading 3.1.1.** to generate viral DNA as template for PCR.
2. FFV DNA sequences from positions 5775 to 10522 were amplified by long-template PCR, as described using the primers shown in **Table 1**.
3. After PCR amplification, the DNA was purified using the QIAquick PCR Purification Kit according to the manufacturer's instructions (Qiagen).
4. The purified PCR product and the FFV full-length clone (in the vector pAT153) were then digested with *Bst*Z17I (FFV position 5980) and *Bsa*I (FFV position 10137).
5. After gel purification of the 4.2-kb PCR fragment and the plasmid vector fragment, the novel *env* sequences were used to replace the corresponding sequences of the original FFV full-length clone.
6. Importantly, the plasmid DNA clone was genetically stable in bacteria.

Among the different full-length DNA clones, clone pFeFV-7 yielded infectivity of more than 10^6 FFU/mL 3 d after transfection. The infectivity obtained could be serially propagated and the pattern of gene expression was undistinguishable from that of wt FFV-infected cells. Thus, clone pFeFV-7 was fully replication-competent. In addition, virus derived from this plasmid displayed wild-type growth characteristics in experimentally infected cats ***(10)***.

3.4. Application of FFV Plasmid DNA Clone to Molecular Virology of FFV

The FFV DNA clone pFeFV-7 served as the parental clone for the following recombinant genomes and was used to study several questions:

1. Exchange of FFV-isolate-specific *env* sequences required for virus neutralization ***(3)***.
2. Determination of the function of the FFV accessory Bet protein ***(11)***.
3. Construction of replication-competent empty and gfp-transducing FFV vectors ***(10)***.
4. Construction of replication-competent FFV-based vaccine vectors and successful application in cats ***(12)***.
5. Construction of self-inactivating FFV-based replication-deficient vectors ***(13)***.
6. Determination of the Gag-Env interaction for particle formation ***(14,15)***.

4. Notes

1. For the bovine foamy virus (BFV), full-length genomes were directly cloned into lambda vectors after linker addition ***(16)***.
2. At present, other advanced high-fidelity, long-distance PCR systems are commercially available and fully suited for this purpose.
3. In order to optimize the ligation conditions, the DNA fragments can be additionally purified by organic solvent extraction and ethanol precipitation in the presence of 2 μg glycogen (Roche) as a carrier.
4. For DNA transfection, alternative methods may be used depending on cell type and transfection efficiency required. $CaCl_2$ precipitation is often an efficient, inexpensive, and straightforward alternative but fails in some cases. Transfection with lipofectamine 2000 (Invitrogen) or other cationic lipids is comparative expensive but yields good results with a broad range of cells.

Acknowledgments

Part of the presented data is reprinted from **ref. *3*** with permission from Elsevier. We thank Patrizia Bastone, Nadine Kirchner, and Fabian Romen for critically reading the manuscript, and Lutz Gissmann for support.

References

1. Fields, B., Knipe, D., and Howley, P. (eds.) (1996) *Virology*. Philadelphia, PA: Lippincott-Raven.
2. Miller, A. (1997) Development and application of retroviral vectors, in *Retroviruses* (Coffin, J., Hughes, S., and Varmus, H., eds.), Plainview, NY: Cold Spring Harbor Laboratory Press, pp. 437–473.

3. Zemba, M., Alke, A., Bodem, J., Winkler, I. G., Flower, R. L., Pfrepper, K., et al. (2000) Construction of infectious feline foamy virus genomes: cat antisera do not cross-neutralize feline foamy virus chimera with serotype-specific Env sequences. *Virology* **266,** 150–156.
4. Ausubel, F., Brent, R., Kingston, R., Moore, D., Seidman, J., Smith, J., and Struhl, K. (eds.) (1993) *Current protocols in molecular biology.* New York: Greene Publishing Associates and Wiley-Interscience.
5. Sambrook, J., Fritsch, E., and Maniatis, T. (eds.) (1989) *Molecular Cloning—A Laboratory Manual,* 2nd ed. Cold Spring Harbor Laboratory Press, Plainview, NY.
6. Winkler, I., Bodem, J., Haas, L., Zemba, M., Delius, H., Flower, R., et al. (1997) Characterization of the genome of feline foamy virus and its proteins shows distinct features different from primate spumaviruses. *J. Virol.* **71,** 6727–6741.
7. Hirt, B. (1967) Selective extraction of polyoma DNA from infected mouse cell cultures. *J. Mol. Biol.* **26,** 365–369.
8. Zhou, C., Yang, Y., and Jong, A. Y. (1990) Mini-prep in ten minutes. *BioTech.* **8,** 172–173.
9. Löchelt, M., Zentgraf, H., and Flügel, R.-M. (1991) Construction of an infectious DNA clone of the full-length human spuma-retrovirus genome and mutagenesis of the bel1 gene. *Virology* **184,** 43–54.
10. Schwantes, A., Ortlepp, I., and Löchelt, M. (2002) Construction and functional characterization of feline foamy virus-based retroviral vectors. *Virology* **301,** 53–63.
11. Alke, A., Schwantes, A., Kido, K., Flötenmeyer, M., Flügel, R. M., and Löchelt, M. (2001) The *bet* gene of feline foamy virus is required for virus replication. *Virology* **287,** 310–320.
12. Schwantes, A., Truyen, U., Weikel, J., Weiss, C., and Löchelt, M. (2003) Application of chimeric feline foamy virus-based retroviral vectors for the induction of antiviral immunity in cats. *J. Virol.* **77,** 7830–7842.
13. Bastone, P. and Löchelt, M. (2003) Kinetics and characteristics of replication-competent revertants derived from self-inactivating foamy virus vectors. *Gene Therapy* **11,** 465–473.
14. Wilk, T., Geiselhart, V., Frech, M., Fuller, S. D., Flügel, R. M., and Löchelt, M. (2001) Specific interaction of a novel foamy virus Env leader protein with the N-terminal Gag domain. *J. Virol.* **75,** 7995–8007.
15. Geiselhart, V., Schwantes, A., Bastone, P., Frech, M., and Löchelt, M. (2003) Characterization of the feline foamy virus Env leader protein and the N-terminal Gag domain. *Virology* **310,** 235–244.
16. Renshaw, R., Gonda, M., and Casey, J. (1991) Structure and transcriptional status of bovine syncytial virus in cytopathic infections. *Gene* **105,** 179–184.

34

Molecular Characterization of Proteolytic Processing of the Gag Proteins of Human Spumaretrovirus

Klaus-Ingmar Pfrepper and Rolf M. Flügel

Summary

Molecular characterization of proteolytic processing of the human spumaretrovirus (HSRV) Gag proteins and the precise determination of cleavage sites was performed. For in vitro processing of recombinant HSRV Gag proteins, a recombinant enzymatically active HSRV protease was employed. Recombinant Gag proteins and protease were cloned and expressed as hexa-histidine-tagged proteins in pET-32b and pET-22b vectors, respectively, in the *E. coli* BL21 expression strain. The recombinant proteins were purified by affinity chromatography on an immobilized metal ion matrix. To determine the precise processing sites, recombinant Gag proteins or synthetic peptides derived from Gag sequences were cleaved in vitro by the recombinant protease. Proteolytic processing reactions were carried out under optimal reaction conditions of HSRV protease in sodium phosphate buffer, pH 6.0, supplied with 2 *M* NaCl at 37°C. The cleavage sites were determined by amino-terminal amino acid sequencing as well as by matrix-assisted laser desorption/ionization mass spectrometry analysis of the reaction products. Fluorescence spectrophotometry was used to determine cleavage kinetics of peptides mimicking different cleavage sites within the HSRV Gag proteins.

Key Words: Spumaretrovirus; foamy virus; Gag; aspartic protease.

1. Introduction

Proteolytic processing of the human spumaretrovirus (HSRV) Gag proteins is one of the uncommon features that distinguish HSRV as well as foamy viruses in general from other retroviruses *(1,2)*.

In contrast to Gag processing in other retroviruses, spumaretrovirus Gag processing seems to be delayed or incomplete *(3–6)*. Low amounts of processed Gag proteins are found in human cell lines infected with HSRV. With

From: *Methods in Molecular Biology, Vol. 304: Human Retrovirus Protocols: Virology and Molecular Biology*
Edited by: T. Zhu © Humana Press Inc., Totowa, NJ

the exception of p68[Gag], the amounts of processed Gag are insufficient and can hardly be detected by radioimmunoprecipitation (RIPA), Western blotting, or protein sequencing techniques and mass spectrometry (MS) ***(6)***.

For this reason, a recombinant enzymatically active HSRV protease was used with recombinant Gag proteins or synthetic peptides derived from HSRV Gag proteins as substrates in order to determine the precise cleavage sites of the HSRV protease within the Gag precursor protein p71[Gag] of HSRV ***(6–8)***. Edman sequencing MS and fluorescence spectrophotometry were utilized to analyze cleavage products. The resulting cleavage sites found in the HSRV Gag protein match well with the Gag protein patterns found in human cell lines infected with HSRV. This shows the suitability of the recombinant approach ***(6)***.

2. Materials

Materials applied to perform the techniques are listed only once, in the technique mentioned first.

2.1. Eukaryotic Cell Culture, Radioimmunoprecipitation and Western Blotting

1. HEL299 cells.
2. Dulbecco's modified Eagles's medium (DMEM) with Glutamax (GibcoBRL, Eggenstein, Germany).
3. DMEM with Glutamax, methionine and cysteine free (Gibco BRL).
4. Pro-Mix [(^{35}S)-Cys and (^{35}S)-Met], 14.3 µCi/µL (Amersham, Braunschweig, Germany).
5. RIPA buffer: 10 m*M* Tris-HCl, pH 7.5, 150 m*M* NaCl, 1% Triton, 0.5% desoxycholic acid, 0.1% SDS, 2 m*M* EDTA.
6. Protein A agarose (Boehringer Mannhein, Mannheim, Germany).
7. Protein sample buffer: 4X 125 m*M* Tris-HCl, pH 6.8, 40% glycerol, 8% SDS, 2% β-mercaptoethanol, 10 µg/mL bromophenol blue.
8. Sodium dodecyl sulfate polyacrylamide gel electrophoresis (SDS-PAGE) equipment and nitrocellulose membranes (Schleicher & Schuell, Dassel, Germany).
9. Western blot blocking buffer: PBS containing 0.1% Tween-20, 1% skimmed milk.
10. Rabbit anti-HSRV-Gag antisera ***(3,6)***.
11. Peroxidase-conjugated protein A (Boehringer Mannheim).
12. ECL Western blot kit (Amersham).

2.2. Expression Cloning and Recombinant Protein Expression

1. JM109 and BL21 *E. coli* strains.
2. Luria-Bertani (LB) agar supplied with 100 µg/mL ampicillin.
3. LB medium and isopropyl-β-D-thiogalactoside (IPTG).
4. pET expression system (pET-22b and pET-32b vectors; Novagen, Madison, WI).
5. Oligonucleotide primers.
6. *Pfu*-DNA-Polymerase (Stratagene, Heidelberg, Germany).

7. Restriction enzymes (NEB, Frankfurt, Germany; MBI Fermentas, St. Leon Roth, Germany).
8. Agarose gel electrophoresis equipment and DNA sequencing equipment.
9. Coomassie Brilliant Blue R250 (0.25% in 5% acetic acid and 6% methanol).

2.3. Cell Disruption and Immobilized Metal Ion Affinity Chromatography (IMAC)

1. Sonification device.
2. IMAC buffer: 20 m*M* Tris-HCl, pH 8.0, 10% glycerol, 0.5 *M* NaCl; IMAC buffer solutions containing imidazole should be stored at 4°C.
3. Imidazole (light-sensitive).
4. "His-Bind-Resin" (Novagen).
5. "Centricon," "Centriprep" ultra filtration system (Millipore [Amicon], Eschborn, Germany).

2.4. Proteolytic Processing and Analysis of Processed Products

1. Processing buffer: 50 m*M* sodium phosphate buffer, pH 6.0, 2 *M* NaCl.
2. Polyvinylidene difluoride (PVDF) membranes (ProBlot, Perkin Elmer, Rodgau-Jügesheim, Germany).
3. Procise 494A Edman-Sequencer (Applied Biosystems, Weiterstadt, Germany).
4. 2,5-dihydroxybenzoic acid; trifluoroacetic acid.
5. Matrix-assisted laser desorption/ionization mass spectrometry (MALDI-MS; Vison 2000, Finnigan MAT, Bremen, Germany).
6. Sodium borate buffer: 0.2 *M* sodium borate, pH 8.5.
7. Fluorescamine 0.05% w/v in acetone (Serva, Heidelberg, Germany).
8. Sulfosuccinimidyl-acetate (Sulfo-NHS-Acetate; Pierce, Rockford, IL).
9. F-2000 fluorescence spectrophotometer (Hitachi, Japan).

3. Methods

This section discusses the detection of processed Gag proteins in infected human cultured cells, the cloning and expression of recombinant HSRV protease and HSRV Gag proteins, the purification of the recombinant proteins, in vitro enzymatic cleavage reactions, and the analysis of the reaction products by Edman degradation, by MALDI-MS, and by fluorescence spectrophotometry.

Cell culture experiments and molecular cloning were performed according to standard procedures of cellular and molecular biology and are not described in every detail.

3.1. Detection of Processed HSRV Gag Proteins in Infected HEL299 Cells and Purified HSRV Particles

3.1.1. Cell Cultures and Radiolabeling

In order to detect the low amounts of processed Gag proteins that can be found in infected human cell cultures, human embryonic lung fibroblast cells

(HEL299) were selected by their capacity to produce high HSRV titres as well as large amounts of HSRV proteins. Cells were cultivated in DMEM with Glutamax. Please be aware of radiation hazard and protect yourselves by plexiglass.

1. Infect HEL299 cells with 2 mL of an infectious HSRV cell culture supernatant of confluent HEL299 cell showing strong syncytia formation.
2. Incubate cells for 2 d at 37°C in 5% CO_2.
3. Wash two times with 10 mL PBS.
4. Starve cells for 3 h in 7 mL cysteine- and methionine-free medium.
5. Add 50 µCi radiolabeled (^{35}S) cysteine and (^{35}S) methionine per mL of medium.
6. Incubate cells for 24 h.
7. Remove the supernatant, clear it by centrifugation (300g, 5 min), sediment virus particles in the remaining supernatant by ultracentrifugation (76,000*g*, 2 h, 4°C) through a 20% sucrose cushion in 5 mL PBS (careful overlay the sucrose solution with the cell culture supernatant) and lyse precipitated virus particles for 5 min in RIPA buffer.
8. Wash the cells two times with PBS and lyse them for 5 min in RIPA buffer.
9. For control reasons, treat mock-infected cells and their supernatants in parallel.
10. For Western blot analyses, nonradiolabeled infected and mock-infected cells and cell culture supernatants were used; harvest 3 d postinfection as described and lyse in protein sample buffer instead of RIPA buffer.

3.1.2. Radioimmunoprecipitation and Western Blotting

In order to detect sufficient amounts of processed HSRV Gag proteins in human cells, RIPA turned out to be the most efficient method; however, it is easier to detect processed HSRV Gag proteins in virus particles by Western blotting *(6)*. Please be aware of radiation hazard and protect yourselves by plexiglass.

3.1.2.1. Radioimmunoprecipitation

1. Wash 40 µL protein A agarose two times with RIPA buffer.
2. Add 50–100 µL of radiolabeled material (*see* **Subheading 3.1.1.**).
3. Fill up to 300 µL with RIPA buffer.
4. Add 10 µL of the appropriate preimmune rabbit anti-Gag antiserum *(3,6)*.
5. Overhead-shake for 1 h at room temperature.
6. Centrifuge (11,000*g*, 5 min, room temperature).
7. Save the supernatant and mix it again with 40 µL prewashed protein A agarose.
8. Add 4 µL of rabbit anti-Gag serum *(3,6)*.
9. Overhead-shake for 16 h at 4°C.
10. Centrifuge (11,000*g*, 5 min) and remove the supernatant (it can be used for further radioimmunoprecipitations with other antisera).
11. Wash protein A agarose five times with RIPA buffer.

12. Add 25 µL protein sample buffer.
13. Analyze sample by SDS-PAGE, dry the gel, and analyze it by autoradiography.

3.1.2.2. Western Blotting

1. Lyse harvested cells and virus particles in protein sample buffer (*see* **Subheading 3.1.1.**).
2. Separate proteins by SDS-PAGE and electrotransfer to nitrocellulose.
3. Block nitrocellulose with Western blot blocking buffer.
4. Add appropriate rabbit anti-Gag antiserum *(3,6)* in dilutions (made in Western blot blocking buffer) that must be determined empirically for every single antiserum, incubate for 1 h, shaking slightly (longer incubation might be appropriate).
5. Add peroxidase-conjugated protein A (1:10,000 in Western blot blocking buffer as secondary conjugate), incubate for 1 h, shaking slightly.
6. Visualize the antigen-antibody-reaction by staining with ECL Western blot kit.

3.2. Expression Cloning of Enzymatically Active Recombinant HSRV Protease

As polymerase chain reaction (PCR) template pHSRV13 *(9)* was used. The vector pET-22b was chosen for recombinant protein expression because of its high expression potential. This vector allows one-step protein purification of the recombinant expressed protein by immobilized metal ion affinity chromatography owing to the carboxy-terminal fused hexahistidine tag and adds 8 foreign amino acids. For specificity control of downstream cleavage reactions we additionally cloned, expressed, and purified an enzymatically inactive HSRV protease mutant in which the catalytic active aspartic acid residue was mutated to alanine; the corresponding PCR template was pHSRV13/DA *(8,10)*.

1. Perform PCR with primers 5'-CTCCA*CATATG*AATCCTCTTC-3' (*Nde*I) and 5'-TCC*CTCGAG*ATTTTCCCAATGTTGCCATAG-3' (*Xho*I).
2. Ligate PCR product in *Nde*I and *Xho*I sites of pET-22b.
3. Transform 40 mL of electrocompetent JM109 by electroporation (200 Ω, 2.5 kV, 25 µF) with 2 µL of ligation mix.
4. Select postitive clones and transform BL21(DE3) (as described above; *see* **Note 1**).
5. Inoculate 400 mL LB containing ampicillin with 10 mL of an overnight culture.
6. Shake at 37°C to OD_{600} of 0.7 (~2 h).
7. Add 400 µL IPTG solution (1 *M* in H_2O).
8. Shake at 37°C for 3 h.
9. Harvest bacteria by centrifugation (4000*g*, 10 min, 14°C).

3.3. Expression Cloning of Different Recombinant HSRV Gag Protein Sequences

PCR template pHSRV13 *(9)* was used. The vector pET-32b was chosen because of its thioredixin fusion tag. This vector is designed to add thioredoxin and a hexahistidine tag at the amino-terminus of the expressed protein. The thioredoxin fusion enables the isolation of sufficient amounts of soluble HSRV Gag proteins. The hexahistidine tag allows one-step protein purification of the expressed recombinant protein by immobilized metal ion affinity chromatography.

1. Perform PCR with sense primers 5'-TCC*GAATTC*GATGGCTTCAGGAAGTAATG-3' (*Eco*RI) to give total HSRV Gag sequence GagT (aa 1–648) *(6)*, 5'-TCC*GAATTC*GCCTGGACCCTCTCAACCTC-3' (*Eco*RI) to give partial HSRV Gag sequence Gag1 (aa 244–648) *(7)* or 5'-TCC*GAATTC*GCCAATGCATCAGCTTGGAA-3' (*Eco*RI) to give partial HSRV Gag sequence Gag2 (aa 387–648) *(7)* and antisense primer 5'-CTT*GTCGAC*GTCCCTTTGATCTCCGCCG-3' (*Sal*I).
2. Ligate PCR products in *Eco*RI and *Sal*I sites of pET-32b.
3. Continue cloning and express recombinant proteins as described in **Subheading 3.2.**

3.4. Purification of Recombinant HSRV Gag Proteins by Immobilized Metal Ion Affinity Chromatography

The recombinant HSRV proteins were purified in a one-step procedure by immobilized metal ion affinity chromatography enabled by the hexahistidine tag fused to the recombinant proteins. All recombinant HSRV proteins were natively purified from the soluble supernatant resulting from cell disruption, although about 90% of HSRV protease and 50% of HSRV Gag proteins remained insoluble in inclusion bodies (*see* **Note 2**).

1. Suspend sedimented bacteria in 10 mL IMAC buffer containing 5 m*M* imidazole.
2. Disrupt bacteria by sonification while cooling with ice (pulse 30 s with 30 s interuption to allow recooling until the suspension starts to clear up).
3. Centrifuge (15,000*g*, 10 min, 4°C) in a precooled rotor.
4. Apply supernatant on a 1-mL His bind resin column sequentially preequilibrated with 5 mL 100 m*M* EDTA, 5 mL H_2O, 5 mL 50 m*M* $NiSO_4$, and 5 mL IMAC containing 5 m*M* imidazole by gravity flow.
5. Wash column with 10 mL IMAC containing 5 m*M* imidazole.
6. Wash with 5 mL IMAC buffer each while increasing imidazole concentration stepwise to 25 m*M* and 65 m*M* purifying HSRV Gag proteins or 25 m*M* and 80 m*M* purifying HSRV protease.
7. Elute recombinant proteins with 5 mL IMAC buffer containing 200 m*M* imidazole.
8. Dialyse eluted protein against 250 mL 10 m*M* Tris-HCl, pH 8.0; 10% glycerol each three times for 15 min to remove imidazole and NaCl (*see* **Note 2**).
9. Concentrate protein by ultrafiltration using the Centricon or Centriprep system to about 1.5 mL and store small aliquots at –20°C.

3.5. In Vitro Processing of Recombinant HSRV Gag Proteins and Peptides by Recombinant HSRV Protease

Enzymatic activity of HSRV protease is quite poor, so high concentrations of enzyme are needed for efficient cleavage reactions. To ascertain specific cleavage by HSRV protease, every reaction was repeated and compared with the enzymatically inactive HSRV protease mutant in which the catalytically active aspartic acid residue was mutated to alanine ***(8,10)***. Peptides were synthesized, purified, and characterized as described ***(11,12)***.

1. Use 0.2 nmol purified recombinant HSRV protease (*see* **Note 3**).
2. Add 1–10 μg of purified recombinant HSRV Gag protein or 1 nmol of synthetic peptide.
3. Mix into 10 μL reaction buffer (end concentration 50 m*M* sodium phosphate buffer, pH 6.0, containing 2 *M* NaCl).
4. Incubate up to 16 h at 37°C.

3.6. Sequencing of Processed HSRV Gag Proteins by Edman Degradation

Edman sequencing of the blot cartridge device and the Procise 494A protein sequencer (Applied Biosystems, Weiterstadt, Germany) was applied ***(11)***.

1. Separate cleavage products (*see* **Subheading 3.5.**) by SDS-PAGE.
2. Electrotransfer cleavage products to PVDF membranes (in order to electrotransfer proteins completely out of the gel, prolong general blotting time by about 30%).
3. Do not block membrane.
4. Stain PVDF membrane for about 2–5 min with Coomassie Brilliant Blue solution until the protein band to be sequenced can be seen (*see* **Note 4**).
5. Mark protein band.
6. Destain PVDF membrane for about 2–5 min with 5% acetic acid (*see* **Note 4**).
7. Cut marked area out of the membrane and sequence by Edman degradation.

3.7. MALDI-MS Analysis of Processed Gag Peptides

MALDI-MS was performed with a reflex II time-of-flight instrument equiped with a SCOUT multiprobe inlet and a 337 nm nitrogen laser (Vison 2000, Finnigan MAT, Bremen, Germany); ion acceleration voltage was 25 kV and reflector voltage was 26.5 kV. Mass spectra were yielded by accumulation of 10 to 50 individual laser shots. Spectra were calibrated using angiotensin I and insulin β-chain ***(6,7,12)***.

1. Dilute sample (*see* **Subheading 3.5.**) 10-fold with 0.1% aqueous trifluoroacetic acid.
2. Mix 0.5 μL dilution with 0.5 μL matrix (10 mg/mL 2,5-dihydroxybenzoic acid in 0.1% trifluoroacetic acid).
3. Bring the mixture directly onto the target and let dry at room temperature.
4. Perform MALDI-MS (see **Note 5**).

3.8. Quantitative Analysis of Enzymatic Turnover by Fluorescence Spectrophotometry

For fluorescence spectrophotometry the fluorescence spectrophotometer F-2000 (Hitachi) was used according to the method described by Kotler et al. (***13***; *see* **Note 6**). Net increase of relative fluorescence intensity was monitored against zero standards using a calibration curve.

1. Perform cleavage reaction as described (*see* **Subheading 3.5.**).
2. Preincubate HSRV protease in reaction buffer for 10 min before adding the peptide.
3. Start cleavage reaction by adding the peptide.
4. Incubate for 30 min at 37°C.
5. Stop enzymatic cleavage by adding 500 µL 0.2 *M* sodium borate, pH 8.5.
6. Add 80 µL 0.05% fluorescamine (in acetone) and mix thoroughly.
7. Start fluorescence measurement as soon as possible (*see* **Note 6**).
8. Monitor fluorescence at excitation and emission wavelengths of 390 and 485 nm, respectively (*see* **Note 6**).

4. Notes

1. Cloning and expression: It is recommended to use bacterial strains like JM109 that do not allow recombinant protein expression by the T7 RNA polymerase system for initial transformation, selection of positive clones, and storage of plasmid DNA before transformation of BL21(DE3) that allows recombinant protein expression. The reason is that BL21(DE3) strains transformed with pET vectors expressing large amounts of recombinant proteins tend to be unstable.
2. Purification of recombinant proteins: Owing to the fact that recombinant proteins tend to aggregate, purification was continued immediately after cell disruption and completed the same day, including protein dialysis and concentration if necessary. Dialysis is compulsory to remove imidazole that inhibits enzymatic activity needed for downstream cleavage reactions. More extended dialysis did not improve enzymatic activity.
3. In vitro cleavage reactions: In order to exclude side reactions by other proteases that may contaminate the purified recombinant HSRV proteins and to acertain specificity of the cleavage reactions, it is compulsory to repeat every single reaction by another one in which just active HSRV protease is replaced by its enzymatically inactive mutant ***(8,10)***.
4. Edman sequencing: It is important to stain and destain membrane-bound proteins just for a few minutes to avoid side reactions that may interfere with downstream Edman sequencing reactions.
5. MALDI-MS: In addition to the original clevage products, the MALDI-MS spectra contain corresponding Na^+-adducts that are formed due to the high NaCl concentration in the reaction buffer.
6. Fluorescence spectrophotometry:
   a. The chemical principle of the fluorescence detection is based on the reaction of fluorescamine with the released P1' secondary amino group of the cleavage product ***(13)***. Thus, it is indispensable to block the free secondary amino group at the amino-terminus of every single peptide to be analyzed by an additional proline residue. The secondary amino group of proline is nonreactive with fluorescamine.
   b. Lysine residues in peptides must be blocked by acetylation because the ε-amino group of the lysine side chain reacts with fluorescamine. Acetylate lysine residues in peptides by sulfosuccinimidyl-acetate according to the manufacturer's instructions.

c. Fluorescence measuring should be performed immediately after stopping the enzymatic reactions.
d. At least four individual sets per single reaction must be performed and averaged to give consistent results.
e. A reaction mix of preincubated enzyme should be used as zero standard.
f. The absolute concentrations of the reaction products can be determined by comparison to a calibration curve with known concentrations of an uncleaved non-proline-blocked peptide.

Acknowledgments

We thank Hans Heid for carrying out Edman sequencing, Hans-Richard Rachwitz for synthesis of peptides, Martina Schnölzer for MALDI-MS analysis, Wolfgang Weinig and Hajo Delius for DNA sequencing, and Andrea Wagner for providing cell cultures. In particular we are indebted to Helmut Bannert for exellent technical assistance.

We thank David Baldwin and Maxine Linial for providing Gag antisera, Jennifer Reed and Alla Gustchina for helpful discussions, and Alexandra Alke for critical reading of the manuscript.

References

1. Löchelt, M. and Flügel, R. M. (1995) The molecular biology of primate spumaviruses, in *The Retroviridae* (Levy, J., ed.), vol. 4, Plenum Press, NY, pp. 239–292.
2. Linial, M. L. (1999) Foamy viruses are unconventional retroviruses. *J. Virol.* **73,** 1747–1755.
3. Bartholomä, A., Muranyi, W., and Flügel, R. M. (1992) Bacterial expression of the capsid antigen domain and identification of native Gag proteins from spumavirus-infected cells. *Virus Res.* **23,** 27–38.
4. Enssle, J., Fischer, N., Moebes, A., Mauer, B., Smola, U., and Rethwilm, A. (1997) Carboxyterminal cleavage of the human foamy virus Gag precursor molecule is an essential step in the viral life cycle. *J. Virol.* **71,** 7312–7317.
5. Giron, M.-L., Colas, S., Wybier, J., Rozain, F., and Emanoil-Ravier, R. (1997) Expression and maturation of human foamy virus Gag precursor polypeptides. *J. Virol.* **71,** 1635–1639.
6. Pfrepper K.-I., Rackwitz H.-R., Schnölzer M., Heid H., Löchelt, M., and Flügel, R. M. (1999) Molecular characterization of proteolytic processing of the Gag proteins of human spumavirus. *J. Virol.* **73,** 7907–7911.
7. Pfrepper, K.-I., Löchelt, M., Schnölzer, M., and Flügel, R. M. (1997) Expression and molecular characterization of an enzymatically active recombinant human spumaretrovirus protease. *Biochem. Biophys. Res. Commun.* **237,** 548–553.
8. Pfrepper, K.-I., Rackwitz, H.-R., Schnölzer, M., Heid, H., Löchelt, M., and Flügel, R. M. (1998) Molecular characterization of proteolytic processing of the Pol proteins of human foamy virus reveals novel features of the viral protease. *J. Virol.* **72,** 7648–7652.

9. Löchelt, M., Zentgraf, H., and Flügel, R. M. (1991) Construction of an infectious DNA clone of the full-length human spumaretrovirus genome and mutagenesis of the bel1 gene. *Virology* **184,** 43–54.
10. Konvalinka, J., Löchelt, M., Zentgraf, H., Flügel, R. M. and Kräusslich, H.-G. (1995) Active spumavirus proteinase is essential for virus infectivity but not for formation of the Pol polyprotein. *J. Virol.* **69,** 7264–7268.
11. Pfrepper, K.-I., Reed, J., Rackwitz, H.-R., Schnölzer, M., and Flügel, R. M. (2001) Characterization of peptide substrates and viral enzyme that affect the cleavage site specifcity of the human spumaretrovirus proteinase. *Virus Genes* **22,** 61–72.
12. Schnölzer, M., Alewood, P., Jones, A., Alewood, D., and Kent, S. B. H. (1992) In situ neutralization in Boc-chemistry solid phase peptide synthesis. Rapid high yield assambly of difficult sequences. *Int. J. Peptide Protein Res.* **40,** 180–193.
13. Kotler M., Katz, R., Danho, W., Leis, J., and Skalka, A. M. (1988) Synthetic peptides as substrates and inhibitors of a retroviral protease. *Proc. Natl. Acad. Sci. USA* **85,** 4185–4189.

35

Assessing the Relative Efficacy of Antiretroviral Activity of Different Drugs on Macrophages

Stefano Aquaro and Carlo-Federico Perno

Summary

HIV-infected monocyte/macrophage-derived cells are believed to play a major role in the spread of HIV through the body. Not only are fresh monocytes and more differentiated macrophages relatively insensitive to the cytopathic effect of HIV, but once infected they can efficiently infect T-cells. The protocols in this chapter can be used to culture HIV in monocytes/macrophages and to study factors such as drugs and chemokines that influence its replication. Support protocols describe the 6-d-adherence method for preparing mature monocytes/macrophages, a quick and simple means of obtaining mature cells for routine HIV-infection studies, as well as methods for quantitation of HIV in monocytes/macrophages.

Key Words: Macrophages; antivirals; methods of infection; HIV titer; drug activity; chemokines; monocytes.

1. Introduction

Cells of macrophage lineage represent a key target of human immunodeficiency virus (HIV) in addition to $CD4^+$ T lymphocytes. The peculiar dynamics of HIV replication in macrophages, their long-term survival after HIV infection, and their ability to spread virus particles to bystander cells make evident their substantial contribution to the pathogenesis of HIV infection *(1–4)*.

In addition, macrophages are able to activate chemotaxis and start productive HIV replication in latently infected $CD4^+$ T-lymphocytes while inducing HIV-1 infection in resting non-cell-cycling $CD4^+$ T-lymphocytes through a cell-to-cell crosstalk *(5,6)*. Moreover, HIV chronically infected macrophages stimulate apoptotic events in lymphocytes, neurons, and astrocytes *(7–12)*. Finally, monocytes/macrophages sustain the infection in patients undergoing

From: *Methods in Molecular Biology, Vol. 304: Human Retrovirus Protocols: Virology and Molecular Biology*
Edited by: T. Zhu © Humana Press Inc., Totowa, NJ

HAART therapy, since they are productively infected (and not latently infected) by HIV *(13,14)*.

For all these reasons, therapeutic strategies aimed to achieve the greatest and longest control of HIV replication should inhibit HIV not only in $CD4^+$ T lymphocytes, but also in macrophages *(15)*. Testing new and promising antiviral compounds in such cells may provide crucial hints about their efficacy in patients infected with HIV.

The protocols described here can be used to culture HIV in monocytes/macrophages and to study drugs and factors such as chemokines *(16)* that influence its replication in cultured cells. Support protocols describe the 6-d-adherence method for preparing mature monocytes/macrophages for routine HIV-infection studies (*see* **Notes 1** and **2**).

2. Materials

2.1. Isolation and HIV Infection of Cultured Monocytes/Macrophages

1. Ficoll-Hypaque-gradient-separated preparation of peripheral blood mononuclear cells (PBMCs).
2. RPMI-1640 medium (GibcoBRL).
3. Complete RPMI-1640 medium (hereinafter called complete medium), which contains 20% heat-inactivated, low-endotoxin, mycoplasma-free fetal bovine serum (FBS; HyClone Laboratories), 4 m*M* L-glutamine (GibcoBRL), 50 U/mL penicillin, and 50 μg/mL streptomycin (GibcoBRL), both sera endotoxin-free and heat-inactivated (30 min at 56°C).
4. Phosphate-buffered saline (PBS) containing 0.2% EDTA, 4°C.
5. 50-mL conical polypropylene centrifuge tube.
6. 48-well flat-bottom microtiter plate (1 cm^2; Costar) *or* 25- or 75-cm^2 flasks (Corning T-25 or T-75).
7. Reagents and equipment for counting cells (*see* **Note 3**).

2.2. Assessment of Antiviral Drug Activity in Acutely Infected Monocytes/Macrophages

1. Purified macrophages prepared by 6-d-adherence method from PBMCs in either a 48-well flat-bottom microtiter plate or a 25- or 75-cm^2 flasks (*see* **Note 4**).
2. Expanded, tirated clinical isolates or laboratory-adapted stock of monocytotropic strains of HIV.
3. RPMI-1640 medium (*see* **Note 5**).
4. Complete medium.
5. Kit for detecting HIV-p24 gag antigen.

2.3. Assessment of Antiviral Drug Activity in Chronically Infected Monocytes/Macrophages

1. Purified macrophages prepared by 6-d-adherence method from PBMCs in either a 48-well flat-bottom microtiter plate or a 25- or 75-cm^2 flask (*see* **Note 6**).

2. Expanded, titrated clinical isolates or laboratory-adapted stock of monocytotropic strains of HIV.
3. RPMI-1640 medium.
4. Complete medium.
5. Kit for detecting HIV-p24 gag antigen.

2.4. Quantification of HIV Infectivity

1. HIV sample to be titrated.
2. 50-mL conical polypropylene centrifuge tube.
3. Complete medium 48-well flat-bottom microtiter plate containing macrophage cultures derived from PBMCs by 5-d adherence (first support protocol).
4. RPMI-1640 medium (GibcoBRL), 37°C.
5. Kit for HIV-p24 gag antigen detection.

Incubations are performed in a humidified 37°C, 5% CO_2 incubator unless noted otherwise. Except where specified, media may be at either room temperature or 37°C.

3. Methods

3.1. Isolation of Monocytes/Macrophages From PBMC by 6-Day Adherence

This protocol describes the establishment and maintenance of a culture of HIV-infected monocytes/macrophages in either 48-well microtiter plates or flasks. This approach is particularly useful for assessing anti-HIV drug activity. It involves adding various concentrations of each agent being tested to mature monocyte/macrophage cells at different time points before or after viral challenge.

1. Count PBMCs and resuspend in a polypropylene tube at 1×10^6 cells/mL (if using a microtiter plate) or 2.5×10^6 cells/mL (if using a flask) in complete RPMI-20 supplemented with 10% human AB serum (*see* **Notes 7** and **8**).
2. Using a 1000 μL pipettor, place 1 mL of 1×10^6 cells/mL PBMC suspension in the inner 24 wells of a 48-well microtiter plate (filling the outer 24 wells with sterile distilled water); or using a plastic pipet, place 10 or 30 mL of 2.5×10^6 cells/mL PBMC suspension in a 25- or 75-cm^2 flask, respectively (*see* **Note 9**).
3. Incubate plate or flask for 6 d.
4. Remove nonadherent cells by repeated gentle washing with 37°C RPMI 1640 medium (≥4 washes). After each wash, aspirate supernatant using a 2-mL plastic pipet with a sterile yellow tip attached. The pipet's cotton plug should be removed before it is connected to the vacuum system as described in **Subheading 3.2., step 2** of the first basic protocol. After the last aspiration, replenish well or flask with the same original volume of fresh 37°C complete medium (*without* human serum) (*see* **Notes 10–12**).

Estimate the number of macrophages attached to each well as follows.

5. Remove adherent cells from a single representative well by adding cold PBS containing 0.002% EDTA, incubating 10 min, and scraping off cells with a rubber scraper.
6. Count viable cells in a hemocytometer chamber using the trypan blue exclusion method.
7. Determine purity of the macrophages by a flow-cytometric analysis using CD14, CD4, CD8, and CD3 antigen staining (*see* **Note 13**).

3.2. Assessment of Antiviral Drug Activity in Acutely Infected Monocytes/Macrophages

This protocol is important especially for the evaluation of preintegrational stage inhibitors of the HIV replication cycle (e.g., chemokines, entry inhibitors, reverse transcriptase inhibitors, integrase inhibitors, etc.).

1. Add HIV virus stock to a concentration of 100 to 300 $TCID_{50}$/mL to a macrophage culture using a 48-well microtiter plate or 25- or 75-cm^2 flask. Incubate 2 d (*see* **Notes 14–16**).
2. Carefully remove excess virus from the macrophage culture by washing twice with 37°C serum-free RPMI-1640 as follows: Taking care not to disturb the cells, aspirate supernatant using a 2-mL pipet (with cotton plug removed) or equivalent attached to a vacuum system (for 48-well microtiter plate, attach a sterile tip to the pipet when aspirating) (*see* **Note 17**). Add medium (1 mL per microtiter plate well; 10 or 30 mL per 25- or 75-cm^2 flask, respectively). Repeat once.
3. Aspirate RPMI-1640 a third time and add 37°C complete medium (1 mL per well; 10 or 30 mL per 25- or 75-cm^2 flask, respectively).
4. Continue to incubate macrophages, feeding them every 5 d by aspirating old medium and adding fresh complete medium in the amounts described in **step 3**.
5. Starting from d 7, assess virus production by measuring supernatant viral gag antigen using a commercial p24-HIV antigen assay kit (*see* **Note 18**).

3.3. Assessment of Antiviral Drug Activity in Chronically Infected Monocytes/Macrophages

This protocol is especially important for the evaluation of late-stage inhibitors of the HIV replication cycle (e.g., protease inhibitors, inhibitors of oxidative stress, etc.).

1. One day after separation (i.e., 7 d after plating), infect macrophages with 300 $TCID_{50}$/mL of HIV-1_{BaL}.
2. After 24 h, carefully wash all wells with RPMI to remove excess virus and add fresh complete medium.
3. Feed cells every 5 d with fresh complete medium.
4. Assess HIV-1 p24 released in the supernatants every 3 d starting from d 7 after infection (*see* **Note 19**).

5. When HIV-1 p24 production reaches a plateau (e.g., at d 12–14 after infection), carefully wash macrophage cultures with RPMI at least twice to remove any virus present in the supernatants.
6. Replenish with fresh complete medium containing various concentrations of protease inhibitors or control drugs when requested.
7. Keep the macrophage cultures under the same conditions as described before.
8. Starting from the day of drug treatment wash all wells daily and replenish with fresh medium containing the appropriate drug concentration, according to the experimental protocol (*see* **Note 20**).

3.4. Quantification of HIV Infectivity

In this protocol, the number of infectious units of HIV produced by an infected macrophage culture and treated with antiviral compounds is determined by analyzing serial dilutions of viral culture. Titration of a monocytotropic strain of HIV should generally be performed using normal, resting macrophages.

This protocol is especially crucial for the evaluation of late-stage inhibitors of the HIV replication cycle (e.g., protease inhibitors, inhibitors of oxidative stress, etc.).

1. Remove one vial of each lot of virus from the freezer, place vials in a plastic container in a laminar flow hood, and allow them to thaw. Open vials inside the plastic container in the hood after they have thawed, as the contents may be under pressure.
2. Prepare serial dilutions of the virus using complete medium. A useful set of dilutions is 1:10, 1:30, 1:100, 1:300, and so on. Place 7.2 mL medium in a 50-mL conical polypropylene centrifuge tube and add 0.8 mL of virus (this will yield 8 mL of a 1:10 dilution of virus). A subsequent 1:3 dilution can be made by adding 2 mL of this preparation to 4 mL medium (final dilution 1:30), and a 1:10 dilution can be made by adding 1 mL of the preparation to 9 mL medium (final dilution 1:100). Make subsequent 1:3 and 1:10 dilutions in the same way, each time using the virus that was previously diluted 1:10. Alternatively, the first dilution can be 1:100, prepared by adding 10 μL of the original frozen sample to 990 μL medium in wells. Then use cryovials to make subsequent dilutions of the stock virus (1:3, 1:10, 1:30, etc.) and add 10 μL of these dilutions in the cryovials to 900 μL medium to yield final dilutions of 1:300, 1:1000, 1:3000, and so on.
3. Add 1 mL of diluted virus to each cell of a 48-well microtiter plate containing freshly aspirated macrophage culture with six replicate wells for each dilution. Incubate plate 2 d. Avoid cells drying out between medium aspiration.
4. Carefully remove excess virus by aspirating supernatant and adding 1 mL of 37°C RPMI-1640 medium. Repeat wash three times, then add 1 mL of 37°C fresh complete medium to each well.
5. Incubate 14 d, replacing the medium every 5 d.
6. At 14 d after viral challenge, assay supernatants for HIV p24 gag antigen production with p24 kit.

For the purpose of titration, it is sufficient to determine whether HIV p24 gag antigen is present; quantitation of the antigen is not necessary.

4. Notes

1. Procedures described in this chapter involve the use of human blood and HIV virus. Strict attention to appropriate biosafety procedures is therefore mandatory.
2. All solutions and equipment must be sterile, and proper sterile technique should be used accordingly.
3. Incubations are performed in a humidified 37°C, 5% CO_2 incubator unless noted otherwise. Except where specified, media may be at either room temperature or 37°C.
4. Incubations are performed in a humidified 37°C, 5% CO_2 incubator unless noted otherwise.
5. Except where specified, media may be at either room temperature or 37°C.
6. Incubations are performed in a humidified 37°C, 5% CO_2 incubator unless noted otherwise. Except where specified, media may be at either room temperature or 37°C.
7. PBMC may be obtained from a blood bank. Leukapheresis represents the best method for deriving a large number of PBMCs for monocyte/macrophage separation. Buffy coat preparations are another potential source, but usually produces two- to eightfold lower yields. The anticoagulant used should be an endotoxin-free citrate preparation.
8. Use of human serum increases adherence of monocytes to plastic during long-term culture. Both this serum and FCS must be strictly endotoxin-free.
9. T-25 and T-75 Corning flasks and 48-well Costar plates permit good adherence of monocytes/macrophages. However, adherence may vary between lots of flasks or plates.
10. These 6-d-adherent cells are generally used immediately to culture HIV.
11. Avoid leaving cells uncovered by medium for more than a few seconds during this process. If using microtiter plates, the pipet tip should be held at the bottom corner of each well during aspiration to minimize accidental removal of adherent cells. It is critical that the RPMI-1640 medium be at 37°C.
12. After 6 d of culture, the cells will have matured into macrophages, as indicated by their fibroblastoid or polymorphic shape and substantial (>10-fold) increase in volume.
13. If the washing procedure is performed correctly, the number of adherent cells in each well should be similar (i.e., all wells should produce similar levels of virus). Generally, about 10% of PBMCs plated (e.g., 10^5 monocytes/macrophages per microtiter-plate well, or 2.5 or 7.5×10^6 monocytes/macrophages per flask) are recovered, which corresponds to $\geq$50% of the monocytes/macrophages initially present in the PBMCs.
14. Frozen virus stock in cryovials should be rapidly thawed to room temperature just before inoculation, with the vial being placed in a plastic container inside a

laminar flow hood and thawed before being opened, as the contents may be under pressure. Cultures should be set up in microtiter plates when a relatively small number of macrophages per sample is required (e.g., for evaluation of the anti-HIV activity of antiviral drugs or assessment of cytokine production by HIV-infected macrophages); flasks should be used when a larger number of macrophages is required (e.g., for separation of cellular extracts to be used in molecular or biochemical studies).

15. The HIV titer suggested here generally gives optimal infection and virus production, but greater or lesser titers may be appropriate for certain purposes (e.g., a lower titer may be useful in testing agents for the ability to enhance HIV replication). Ideally, the macrophage concentration should be approximately the same as that used in titrating the virus.
16. To study drugs that block infection at the preintegrational steps (entry inhibitors, reverse transciptase inhibitors, integrase inhibitors), the anti-HIV agents should be added to cells about 20 min before the virus is added. This may be varied to accommodate specific properties of each drug. For example, drugs that are slowly activated in cells are best added more than 20 min before infection. Cultures without added HIV and without macrophages should be established as controls. The latter is especially important as a control for residual virus.
17. The vacuum aspiration system should be set up so that aspirated fluid passes into a 4-L flask containing approx 500 mL of 100% bleach. Air from this flask should pass through a second flask (to trap any liquid material) and a 0.45-μm gas filter before entering the vacuum line. Flasks and tubes should be periodically decontaminated with bleach.
18. Limited virus production can usually be detected between days 3 and 7; production increases to a plateau between days 12 and 30. In some experiments, virus production may still be detected up to 120 d after viral challenge. Absolute production levels vary, partly depending on the strain used. With the HIV-1_{Ba-L} strain grown in macrophages, a production peak of 50 to 500 ng/mL HIV p24 antigen is common. Alternatively, other measures of viral production can be monitored (e.g., reverse transcriptase activity), but the p24 assay is generally preferred because it is sensitive and easy to perform.
19. Based on our previous experience *(2,17)*, chronic infection is generally established about 10–12 d after virus challenge (some variation being detectable among different donors and/or different virus strains).
20. At each established time point, about 1 mL of each supernatant is harvested and replaced with new medium with appropriate drugs as before.

References

1. Crowe, S. M., Mills, J., Kirihara, J., Boothman, J., Marshall, J. A., and McGrath, M. S. (1990) Full-length recombinant CD4 and recombinant gp120 inhibit fusion between HIV infected macrophages and uninfected CD4-expressing T-lymphoblastoid cells. *AIDS Res. Hum. Retroviruses* **6,** 1031–1037.

2. Aquaro, S., Bagnarelli, P., De Luca, A., Guenci, T., Balestra, E., Clementi, M., et al. (2002) Long-term dynamic of high-level of human immunodeficiency virus type-1 replication in human macrophages and its modulation by antiviral compounds. *J. Med. Virol.* **68,** 479–488.
3. Garaci, E., Caroleo, M. C., Aloe, L., Aquaro, S., Piacentini, M., Costa, N., et al. (1999) Nerve growth factor is an autocrine factor essential for the survival of macrophages infected with HIV. *Proc. Natl. Acad. Sci. USA* **96,** 14,013–14,018.
4. Garaci, E., Aquaro, S., Lapenta, C., Amendola, A., Spada, M., Covaceuszach, S., et al. (2003) Anti-nerve growth factor Ab abrogates macrophage-mediated HIV-1 infection and depletion of CD4+ T lymphocytes in hu-SCID mice. *Proc. Natl. Acad. Sci. USA* **100,** 8927–8932.
5. Swingler, S., Mann, A., Jacque, J., Brichacek, B., Sasseville, V. G., Williams, K., et al. (1999) HIV-1 Nef mediates lymphocyte chemotaxis and activation by infected macrophages. *Nat. Med.* **5,** 997–1003.
6. Swingler, S., Brichacek, B., Jacque, J. M., Ulich, C., Zhou, J., and Stevenson, M. (2003) HIV-1 Nef intersects the macrophage CD40L signalling pathway to promote resting-cell infection. *Nature* **424,** 213–219.
7. Mastino, A., Grelli, S., Piacentini, M., Oliverio, S., Favalli, C., Perno, C. F., and Garaci, E. (1993) Correlation between induction of lymphocyte apoptosis and prostaglandin E2 production by macrophages infected with HIV. *Cell Immunol.* **152,** 120–130.
8. Badley, A. D., Dockrell, D., Simpson, M., Schut, R., Lynch, D. H., Leibson, P., and Paya, C. V. (1997) Macrophage-dependent apoptosis of CD4+ T lymphocytes from HIV-infected individuals is mediated by FasL and tumor necrosis factor. *J. Exp. Med.* **185,** 55–64.
9. Herbein, G., Mahlknecht, U., Batliwalla, F., Gregersen, P., Pappas, T., Butler, J., et al. (1998) Apoptosis of CD8+ T cells is mediated by macrophages through interaction of HIV gp120 with chemokine receptor CXCR4. *Nature* **395,** 189–194.
10. Shi, B., De Girolami, U., He, J., Wang, S., Lorenzo, A., Busciglio, J., and Gabuzda, D. (1996) Apoptosis induced by HIV-1 infection of the central nervous system. *J. Clin. Invest.* **98,** 1979–1990.
11. Aquaro, S., Panti, S., Caroleo, M. C., Balestra, E., Cenci, A., Forbici, F., et al. (2000) Primary macrophages infected by human immunodeficiency virus trigger CD95-mediated apoptosis of uninfected astrocytes. *J. Leukoc. Biol.* **68,** 429–435.
12. Mollace, V., Salvemini, D., Riley, D. P., Muscoli, C., Iannone, M., Granato, T., et al. (2002) The contribution of oxidative stress in apoptosis of human-cultured astroglial cells induced by supernatants of HIV-1-infected macrophages. *J. Leukoc. Biol.* **71,** 65–72.
13. Sonza, S., Mutimer, H. P., Oelrichs, R., Jardine, D., Harvey, K., Dunne, A., et al. (2001) Monocytes harbour replication-competent, non-latent HIV-1 in patients on highly active antiretroviral therapy. *AIDS* **15,** 17–22.
14. Zhu, T., Muthui, D., Holte, S., Nickle, D., Feng, F., Brodie, S., et al. (2002) Evidence for human immunodeficiency virus type 1 replication in vivo in CD14(+) monocytes and its potential role as a source of virus in patients on highly active antiretroviral therapy. *J. Virol.* **76,** 707–716.

15. Aquaro, S., Caliò, R., Balzarini, J., Bellocchi, M. C., Garaci, E., and Perno, C. F. (2002) Macrophages and HIV infection: therapeutical approaches toward this strategic virus reservoir. *Antiviral Res.* **55,** 209–225.
16. Aquaro, S., Menten, P., Struyf, S., Proost, P., Van Damme, J., De Clercq, E., and Schols, D. (2001) The LD78beta isoform of MIP-1alpha is the most potent CC-chemokine in inhibiting CCR5-dependent human immunodeficiency virus type 1 replication in human macrophages. *J. Virol.* **75,** 4402–4406.
17. Perno, C. F., Newcomb, F. M., Davis, D. A., Aquaro, S., Humphrey, R. W., Caliò, R., and Yarchoan, R. (1998) Relative potency of protease inhibitors in monocytes/macrophages acutely and chronically infected with human immunodeficiency virus. *J. Infect. Dis.* **178,** 413–422.

36

Gene Expression Profiling of HIV-1 Infection Using cDNA Microarrays

Angélique B. van 't Wout

Summary

To illustrate the methods employed in gene expression profiling using cDNA microarrays, infection of CD4+ T cell lines with HIV-1_{LAI} is used to identify expression changes relevant to in vitro HIV-1 infection. Cell lines are infected at a high multiplicity of infection to ensure a population of near-synchronously infected cells to be compared to uninfected cells. Infection status is verified using flow cytometry to determine the intracellular expression of the viral gag p24 protein before samples are harvested for total RNA extraction. Total RNA is extracted and amplified using commercially available kits, and RNA quality is verified using Bioanalyzer technology. To obtain fluorescently labeled cDNA probes, the amplified RNA is reverse-transcribed to yield cDNA, using random nonamers in the presence of dye-labeled dCTP. After first-strand cDNA synthesis, RNA is degraded and the probes are purified. For each infection condition (LAI and mock), two slides are hybridized with identical probes generated from the same RNAs, but with fluorescent labels reversed on one of the slides to control for dye-specific effects. Troubleshooting strategies and issues to consider prior to starting the experiment are discussed in detail in the notes section.

Key Words: Microarrays; cDNA; HIV-1; RNA.

1. Introduction

Since the advent of microarray technology in 1995 *(1)*, microarrays have been applied to the gene expression profiling of a wide range of biological systems (reviewed in **ref.** *2*). Microarrays provide a powerful genomic approach to the identification of new targets for drug development and diagnostic screening. These studies involve a complex multistep process (**Fig. 1**), in which each step influences the reliability and significance of the results. This chapter addresses the experimental portion of HIV-1-related microarray analysis using

From: *Methods in Molecular Biology, Vol. 304: Human Retrovirus Protocols: Virology and Molecular Biology*
Edited by: T. Zhu © Humana Press Inc., Totowa, NJ

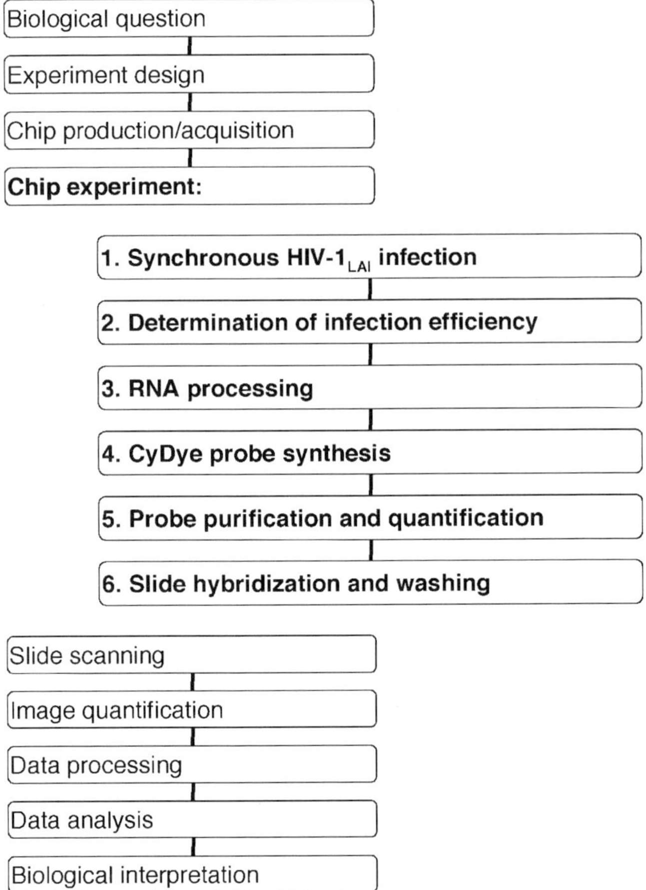

Fig. 1. Workflow of microarray experiment. Gene expression profiling involves a complex multistep process, from initial biological question, through experimental design and application, to data analysis and biological interpretation. Each step influences the reliability and significance of the results. This chapter addresses the experimental portion (delineated in bold) of HIV-1-related microarray analysis using cDNA microarrays.

cDNA microarrays. Several excellent recent reviews are available that describe other aspects, including spotting, experiment design, complex vs simple biological systems, other array formats and array data analysis *(3–5)*.

HIV/SIV-related gene expression profiling has been used to study changes in response to both in vitro and in vivo infection and upon expression of individual virus proteins (**Table 1**). However, interpretation of the results of these studies is complicated by experimental protocols that result in heterogeneous

Table 1
Published HIV/SIV-Related Microarray Studies

First author	Experiment	Platform
Geiss *(6)*	HIV LAI on CEM	cDNA/glass 1500 genes
Corbeil *(7)*	HIV LAI on CEM	Affymetrix 6800 genes
Vahey *(8)*	HIV RF on PBMC	Affymetrix 5600 genes
Van 't Wout *(9)*	HIV LAI on CEM, Jurkat, SupT1	cDNA/glass 4600 genes
Mitchell *(10)*	HIV-vector in Nalm6	Affymetrix 12000 genes
Simmons *(11)*	SF2 Nef in Jurkat	cDNA/membrane 3500 genes
Shaheduzzaman *(12)*	NL43 Nef in HeLa	cDNA/glass 6000 genes
Galey *(13)*	HIV or gp120 on astrocytes	Oligo/glass 2300 genes
Cicala *(14)*	JRFL gp120 on PBMC and MDM	Affymetrix 12,000 genes
De la Fuente *(15)*	LAI Tat in H9	cDNA/glass 2400 genes
Gibellini *(16)*	Tat in Jurkat	cDNA/membrane 1200 genes
Izmailova *(17)*	Tat in immature DC	Affymetrix 6800 genes
Patel *(18)*	Vpr in neurons	cDNA/membrane 200 genes
Chun *(19)*	Resting $CD4^+$ T cells from viremics vs aviremics	Affymetrix 12,000 genes
Guadalupe *(20)*	Jejunal biopsies HIV-1 infected patients	Affymetrix 10,000 genes
George *(21)*	Jejunal biopsies SIV infected macaques	Affymetrix 6800 genes
Vahey *(22)*	PBMC from SIVmac251 infected macaques	Affymetrix 6800 genes
Roberts *(23)*	Brain tissue SIVmac182 infected macaques	Affymetrix 12,000 genes
Sui *(24)*	Brain from SHIV89.6P infected macaques	cDNA/membrane 277 genes
Miller *(25)*	HIV JD or NL4-3 on SCIDhu Thy/Live	cDNA/glass 10,000 genes

samples, nonsynchronous infection and/or selection of cells resistant to infection. This reduces sensitivity and specificity for the detection of changes in gene expression in infected cells. To increase the likelihood of detecting relevant changes induced by HIV/SIV, comparison between homogeneous samples (e.g., 100% infected versus uninfected cell lines) is essential (*see* **Note 1**). Gene expression profiling of HIV-1_{LAI}-infected $CD4^+$ T-cell lines on cDNA microarrays will be used to illustrate the methods employed in the identification of changes relevant to HIV-1 infection that could serve as new drug targets (*see* **Note 2**).

2. Materials

All reagents should be molecular biology-grade and free of contaminating nucleases. Use distilled water for the preparation of solutions unless instructed otherwise and sterilize by filtration with a 0.22-µm filter. Solutions are stable at room temperature unless otherwise indicated.

2.1. Reagents

1. $CD4^+$ T-cell lines CCRF-CEM, Jurkat Clone E6-1 or SupT1 (American Type Culture Collection).
2. HIV-1_{LAI} (AIDS Research and Reference Reagent Program, Division of AIDS, National Institute of Allergy and Infectious Diseases, National Institutes of Health). Frozen aliquots stored at –70°C are stable for more than 1 yr.
3. Antibody anti-HIV-1 gag p24 KC57-FITC (Beckman-Coulter, Fullerton, CA). Store at 4°C, protect from light.
4. Antibody FITC-conjugated mouse IgG2a, kappa isotype control (Becton Dickinson, San Jose, CA). Store at 4°C, protect from light.
5. Fix and Perm kit (CALTAG, Burlingame, CA).
6. RNeasy total RNA isolation kit (Qiagen, Valencia, CA).
7. T7 RNA amplification kit (Arcturus, Mountain View, CA).
8. Random nonamers (1 µg/µL). Frozen aliquots stored at –20°C are stable for more than 1 yr.
9. Superscript II (Invitrogen, Carlsbad, CA). Store at –20°C, return to freezer immediately after use.
10. Nucleotides (Amersham Biosciences, Piscataway, NJ). When combining nucleotides, use 10 m*M* dGTP/dATP/dTTP and 1 m*M* dCTP (the unlabeled dCTP is used in a 1:1 ratio with Cy-labeled dCTP). Frozen aliquots stored at –20°C are stable for approx 1 mo.
11. Cy3-dCTP and Cy5-dCTP (Amersham Biosciences). Protect from light at all times, dispense into single-use aliquots, and store at –20°C.
12. RNase Inhibitor (Roche, Indianapolis, IN). Store at –20°C, return to freezer immediately after use.

2.2. Solutions

1. Tissue culture medium: RPMI-1640 supplemented with penicillin (100 IU/mL), streptomycin (100 µg/mL), L-glutamine (2 m*M*) and fetal bovine serum (10%). Store at 4°C, protect from light.
2. Phosphate-buffered saline (PBS).
3. PBS with 1% bovine serum albumin (BSA). Store at 4°C.
4. Fluorescence-activated cell sorter (FACS) fixation buffer: 1% formaldehyde and 0.025% glutaraldehyde in PBS. Store at 4°C, protect from light.
5. Trypan blue.
6. 3 *M* NaOAc, pH 5.2.
7. 2.5 *M* NaOH.
8. 2 *M* 3-(*N*-morpholino) propane sulfonic acid (MOPS).
9. Binding buffer: 5.3 *M* Gua-HCl in 150 m*M* Kac, pH 4.8; adjust with glacial acetic acid.
10. 10 m*M* Tris-HCl, pH 8.0.
11. Ethanol (80% and 100%).
12. Hybridization solution: 50% formamide, 5X sodium chloride-sodium citrate (SSC), 5X Denhardt's, 0.1% sodium dodecyl sulfate (SDS), 100 µg/mL poly A_{72}, 100 µg/mL human CotI DNA. Frozen aliquots stored at –20°C are stable for more than 1 yr.
13. Wash buffers (1X SSC/0.2% SDS, 0.1X SSC/0.2% SDS, 0.1X SSC).

2.3. Disposables

1. 5-mL FACS tubes.
2. Glass fiber filter plates (Millipore, Billerica, MA).
3. 96-well catch plates.
4. ProbeQuant G50 columns (Amersham Biosciences).
5. Human cDNA microarrays (Agilent, Palo Alto, CA).
6. Cover slips.

2.4. Instruments

1. Flow cytometer.
2. 2100 Bioanalyzer (Agilent).
3. Vacu system (Millipore).
4. SpeedVac concentrator.

3. Methods

The methods described below outline (1) synchronous infection of CD4+ T-cell lines with HIV-1_{LAI}, (2) determination of infection efficiency by flow cytometry, (3) extraction, amplification, and quality control of total RNA samples, (4) generation of fluorescently labeled probes, (5) probe purification and quantification, and (6) slide hybridization and washing.

3.1. Synchronous HIV-1$_{LAI}$ Infection

To ensure that the population of cells under study is undergoing a near-synchronous infection, samples are harvested from CD4$^+$ T-cell lines within the first 24 h of infection with a high multiplicity of infection (MOI). Cells are grown in RPMI-1640 supplemented with penicillin, streptomycin, L-glutamine, and FBS.

1. Generate high-titer HIV-1$_{LAI}$ stock by infecting CCRF-CEM cells at an MOI of 0.01 on day 0, replacing the culture media on day 3 and harvesting the culture supernatant on day 4. Titer will be approx 10^7 TCID$_{50}$/mL. Collect mock supernatant from uninfected CCRF-CEM cells treated as above, excluding the addition of virus. Store at –70°C until use.
2. Expand CD4$^+$ T-cell lines (CCRF-CEM, Jurkat or SupT1) until the appropriate number of cells has been obtained (the cell suspension will be ~10^6 cells/mL 3–4 d after passage). One million cells per condition (infected and mock-infected) are needed to yield sufficient total RNA (>1 μg), and another 1×10^6 cells are needed for flow cytometry. So, 4×10^6 cells (2×10^6 for LAI and 2×10^6 for mock) are needed for one time point, 8×10^6 cells for two time points, etc.
3. Spin cells down at 250*g* for 5 min at room temperature (RT).
4. For each condition, aliquot cells at 10×10^6 cells/mL into a 50-mL tube.
5. Add HIV-1$_{LAI}$ stock at a multiplicity of infection of 2 (200 μL stock = 20×10^6 TCID$_{50}$ for 10×10^6 cells) to one aliquot and the same volume of mock-infected supernatant to the other aliquot.
6. Incubate at 37°C for 1 h in shaking water bath.
7. Wash cells three times with PBS.
8. Resuspend cells in culture media to yield 1×10^6 cells/mL.
9. Incubate at 37°C, 5% CO_2 until harvesting for total RNA extraction.

3.2. Determination of Infection Efficiency

To ensure that the samples are homogeneously infected, the intracellular expression of viral gag p24 protein is determined by flow cytometry using the CALTAG Fix and Perm kit.

1. For each sample to be analyzed, use 1×10^6 cells: 5×10^5 for staining with the specific antibody and 5×10^5 for the isotype control stain. Protect samples from light.
2. Spin cells down in 5-mL FACS tubes at 250*g* for 5 min at RT.
3. Resuspend in 100 μL of reagent A (fixation medium).
4. Incubate at RT for 15 min.
5. Wash cells with 4 mL of PBS/1% BSA.
6. Remove supernatant and resuspend cell pellet in 100 μL of reagent B (permeabilization medium) and 1 μL anti-HIV-1 gag p24 KC57-FITC antibody or 1 μL FITC-labeled isotype control.
7. Vortex at low speed for 1–2 s.
8. Incubate at RT for 15 min.

9. Wash cells with 4 mL of PBS/1% BSA.
10. Remove supernatant and resuspend cells in 0.5 mL of FACS fixation and store them at 2–8°C in the dark. Analyze fixed cells within 24 h using standard flow cytometry procedures (**Fig. 2**).

3.3. RNA Processing

This section describes total RNA isolation, amplification, and quality control, using specific kits that have yielded the best results with these sample types in our experience (*see* **Note 3**). Experimental sampling and extraction of RNA are vitally important components, as successful microarray studies are dependent upon the consistent extraction of high-quality RNA. The 2100 Bioanalyzer system (Agilent) is a very sensitive and rapid tool now commonly employed to verify the quality and consistency of RNA samples.

1. Harvest cells from both HIV-1_{LAI}- and mock-infected cultures at the appropriate time points. Transfer each cell suspension to a separate 50-mL tube.
2. Spin cells down at 250*g* for 5 min at RT.
3. Wash cells twice with PBS.
4. Remove cell sample for trypan blue cell counting and viability assessment.
5. Wash once more with PBS.
6. Discard the supernatant. Remove any remaining drops of fluid.
7. Use Qiagen RNeasy kit to extract total RNA from the cell pellets according to the manufacturer's instructions. It is imperative to thoroughly homogenize the cell lysates to avoid DNA and/or protein contamination and RNA degradation and to optimize total RNA yield.
8. Determine RNA quality on a 2100 Bioanalyzer total RNA chip or its equivalent. Good-quality RNA free of contaminating DNA (*see* **Note 4**) has two distinct ribosomal RNA peaks with a 28S/18S ratio of around 2 and very low background (**Fig. 3A**).
9. Total RNA can be stored at –70°C after adding 0.1 vol 3 *M* NaOAc, pH 5.2, and 2.5 volumes 100% ethanol until it is used for amplification.
10. Amplify mRNA from 1 to 5 µg of total RNA using the Arcturus T7 amplification kit according to the manufacturer's instructions. Five micrograms of total RNA yields approx 100 µg amplified RNA (aRNA).
11. Determine aRNA quality on a 2100 Bioanalyzer mRNA chip or its equivalent. The bulk of the aRNA product is 250–1800 bases in length after one round of amplification (**Fig. 3B**).
12. Amplified RNA can be stored at –70°C after addition of 0.1 vol 3 *M* NaOAc, pH 5.2, and 2.5 vol 100% ethanol until it is used for probe synthesis.

3.4. Cy-Dye Probe Synthesis

To obtain fluorescently labeled cDNA probes, the amplified RNA is reverse-transcribed to yield cDNA using random nonamers in the presence of dye-labeled dCTP (*see* **Note 5**). For each RNA sample, both Cy3- and Cy5-labeled

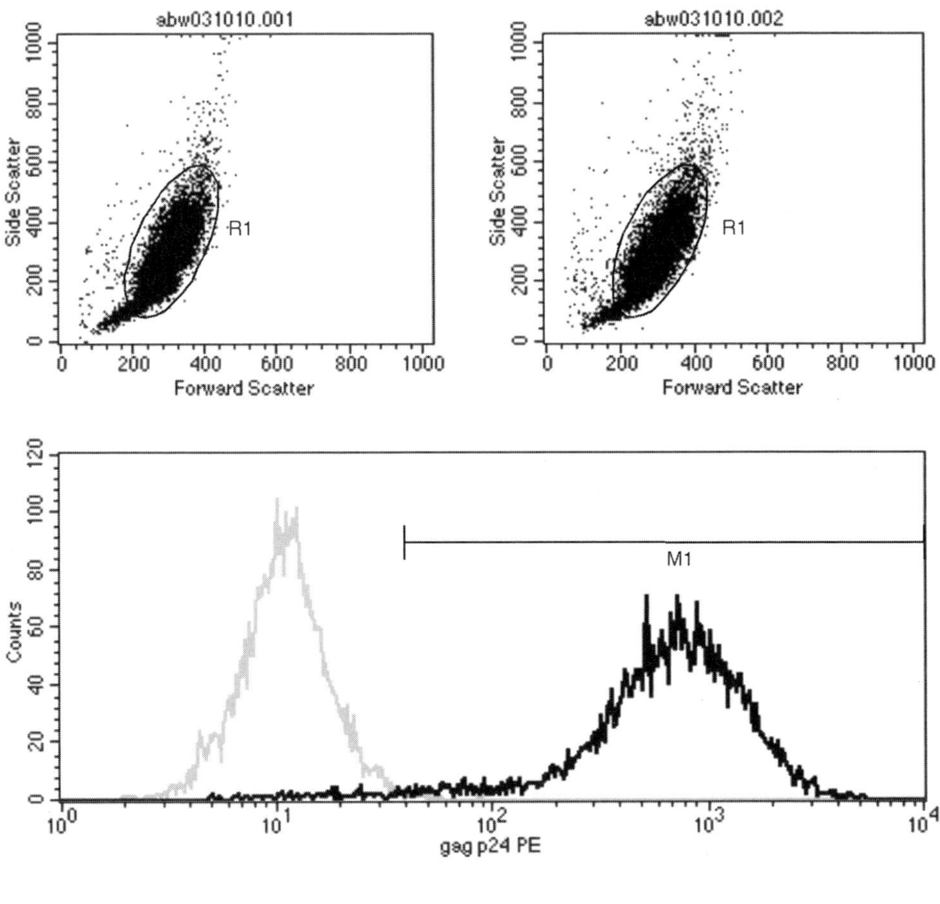

File: abw031010.001
Sample ID: CEM 24hr
Patient ID: MOCK (5-Apr-03)
Acquisition Date: 10-Oct-03
Gate: G1
Gated Events: 9488
Total Events: 10219
X Parameter: FL2-H gag p24 PE (Log)

Marker	Events	% Gated	Mean	Median
All	9488	100.00	11.92	10.94
M1	24	0.25	46.13	42.94

File: abw031010.002
Sample ID: CEM 24hr
Patient ID: LAI (5-Apr-03)
Acquisition Date: 10-Oct-03
Gate: G1
Gated Events: 9462
Total Events: 10224
X Parameter: FL2-H gag p24 PE (Log)

Marker	Events	% Gated	Mean	Median
All	9462	100.00	806.64	691.58
M1	9298	98.27	820.49	697.83

Fig. 2. Intracellular HIV-1 gag p24 expression in mock- and HIV-1_{LAI}-infected cells measured by flow cytometry. Cells were exposed to infectious virus at a multiplicity of infection of 2 (black) or mock-infected supernatant (gray). At 24 h postexposure, cells were fixed and permeabilized for intracellular p24 staining. Only cells exposed to the infectious stock showed detectable levels of p24.

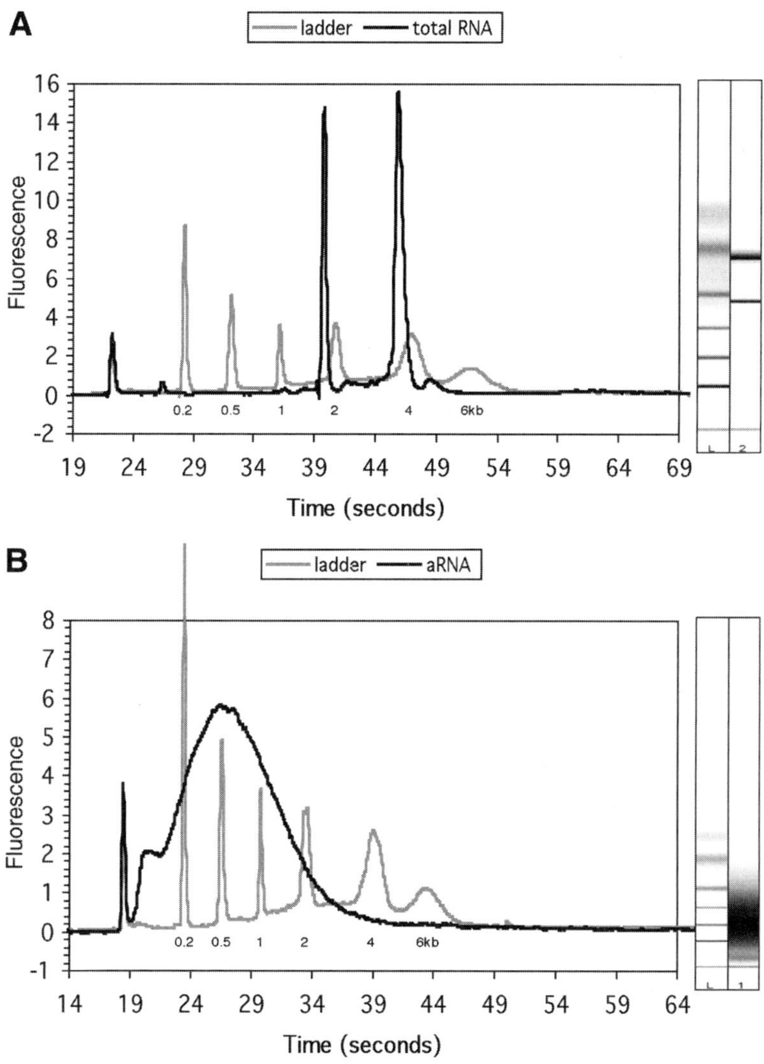

Fig. 3. Agilent 2100 Bioanalyzer RNA analysis. **(A)** Electropherogram of total RNA sample (black) and RNA ladder (gray). Ladder sizes are indicated below each peak. Note distinct 18S and 28S ribosomal RNA peaks with a 28S/18S ratio around 2 and very low background in the total RNA sample. Indications for RNA degradation are: decreasing ratio of ribosomal bands, additional peaks below the ribosomal bands, decrease in overall RNA signal and a shift toward shorter fragments. **(B)** Electropherogram of amplified RNA (black) and RNA ladder (gray). Ladder sizes are indicated below each peak. The bulk of the aRNA product is 250–1800 bases in length after one round of amplification.

probes are generated to control for dye-specific effects (*see* **Subheading 3.6.**). Perform all labeling reactions in amber microcentrifuge tubes and keep light exposure of Cy-labeled nucleotides and cDNA to a minimum. After addition of each reagent, mix contents of tube gently and centrifuge briefly.

1. Combine 5 μg aRNA and 2 μg random nonamers. Add water to bring the total volume to 10.5 μL.
2. Heat at 70°C for 10 min. Chill on ice for 30 s. Spin briefly.
3. Add the following: 4 μL 5X First Strand Buffer, 2 μL DTT (0.1 M), 1 μL nucleotide mix (10 m*M* dGTP/dATP/dTTP, 1 m*M* dCTP), 1 μL Cy3- or Cy5-dCTP (1 m*M*) and 0.5 μL RNase inhibitor. When performing multiple reactions, combine everything excluding Cy-dyes, aliquot 7.5 μL premix to each tube, and then add Cy-dyes.
4. Mix gently and incubate at RT for 10 min.
5. Add 1 μL SuperScript II (to yield 20 μL total reaction volume).
6. Mix gently and incubate at 42°C for 3 h.
7. Unless proceeding immediately to purification, probes can be stored at –20°C.

3.5. Probe Purification and Quantification

After first-strand cDNA synthesis, RNA is degraded and the probe is purified. To remove the degraded RNA template, short oligomers and unincorporated CyDye-nucleotides, two purification steps, glass fiber filter plate and G50 columns, are used to clean up the probes. This greatly reduces background fluorescence and nonspecific binding.

1. Denature cDNA/mRNA hybrid as follows: Add 2 μL 2.5 *M* NaOH. Mix gently (but thoroughly) and incubate at 37°C for 10 min. Add 10 μL 2 *M* MOPS to neutralize the reaction.
2. Add 200 μL binding buffer to probe and mix gently.
3. Dispense into glass fiber filter plate.
4. Place filter plate atop vacu-system and apply vacuum.
5. Wash 6 times with 200 μL 80% ethanol.
6. Place filter plate atop catch plate along with centrifuge alignment frame.
7. Spin to remove residual ethanol at 2000*g* for 5 min at RT.
8. Add 50 μL 10 m*M* Tris-HCl, pH 8.0. Incubate at RT for 1 min.
9. Place filter plate atop a clean catch plate along with centrifuge alignment frame.
10. Spin at 2000*g* for 5 min at RT.
11. Repeat spin after adding another 50 μL 10 m*M* Tris-HCl.
12. Measure probe absorbance at 260 nm, 550 nm, and 650 nm (**Fig. 4**). Use a clean cuvet for each probe, as even minute amounts of contaminating probes will result in detectable microarray signals.
13. Calculate probe yield, amount of incorporated dye and specific activity (see Note 6). Probe yield: OD 260 nm × 33 × probe volume (mL) = μg of cDNA. Incorporated Cy3: OD 550 nm × probe volume (μL)/0.15 = pmol Cy3. Incorporated Cy5:

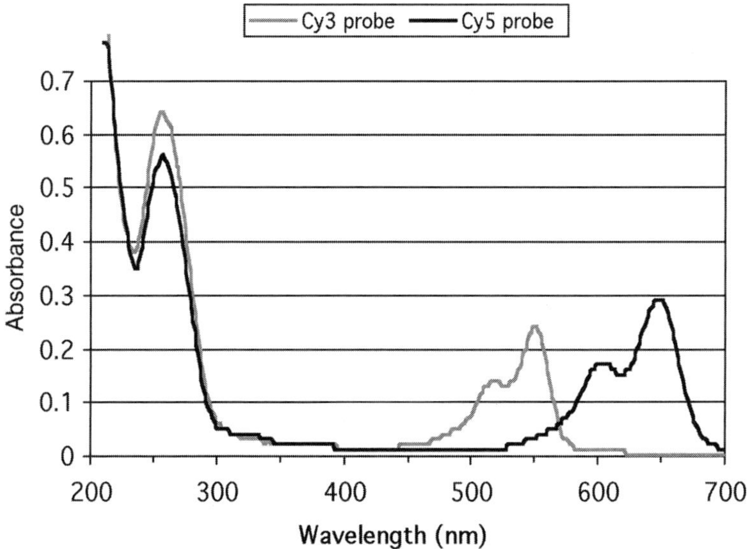

Fig. 4. Probe quantification. Wavelength scan of Cy3- (gray) and Cy5- (black) labeled cDNA probes. The OD values obtained at 260 nm, 550 nm, and 650 nm combined with probe volume are used to calculate probe yield and amount of dye incorporated (1.9 µg DNA and 152 pmol for Cy3 probe, 1.7 µg and 110 pmol for Cy5 probe). Typical probe yields are 1.5–2 µg cDNA and 100–150 pmol incorporated dye.

OD 650 nm × probe volume (µL)/0.25 = pmol Cy5. Specific activity: (µg cDNA/ 324.5)/pmol Cy-dye = number of nucleotides per Cy-labeled nucleotide.

14. Purify probe through a G50 column using standard procedures.
15. Dry purified probe in SpeedVac concentrator (1 h at 50°C).
16. Unless proceeding immediately to slide hybridization, probes can be stored at –20°C.

3.6. Slide Hybridization and Washing

For each infection condition, two slides are hybridized with identical probes generated from the same RNAs, but with fluorescent labels reversed on one of the slides to control for dye-specific effects (*see* **Note 7**). Protect probes and slides from light.

1. Rinse slides and cover slips with two quick dips in distilled water and dry with compressed air.
2. Resuspend dried-down probe in 25 µL of hybridization solution.
3. Boil probe for 3 min. Chill on ice for 30 s. Spin briefly.
4. Combine with appropriate reaction (to yield 50 µL total reaction volume).
5. Apply probe mixture to an area of the slide that does not contain targets and cover with cover slip.

6. Incubate at 42°C overnight (14–16 h) in humid hybridization chamber.
7. Preheat wash buffers to 55°C.
8. Remove cover slip by immersing slide in 1X SSC/0.2% SDS. All subsequent washing steps are performed on a rocking platform.
9. Wash once in 1X SSC/0.2% SDS at RT for 10 min.
10. Wash twice in 0.1X SSC/0.2% SDS at RT for 10 min.
11. Wash twice in 0.1X SSC at RT for 1 min.
12. Rinse with two quick dips in distilled water and dry with compressed air.
13. Store desiccated in the dark until scanning (Cy3 emits at 532 nm and Cy5 at 633 nm).

4. Notes

1. Some systems are not capable of achieving homogeneous target cell populations for study; this needs to be taken into account when analyzing the results. For example, in a heterogeneous collection of cells, the degree of pathway regulation may be commensurate to the proportion of target cells present. In the case of primary patient peripheral blood mononuclear cells, the ratio of cell subsets should be accounted for in order to avoid selecting genes that reflect merely the changes in subset ratios, rather than actual effects of HIV-1 on the cells.
2. The selection of microarray platforms is rapidly expanding and the breadth of genes interrogated per array is starting to approach the full genome. Irrespective of the platform and test system, the general principles and methods described in this chapter will be applicable up to the amplification step. The Affymetrix GeneChip platform uses a different labeling strategy involving dye incorporation during the RNA amplification step and subsequent fragmentation of the probe (*see* **Note 5**). Detailed protocols are available from the manufacturer.
3. Good-quality total RNA upon extraction is essential to all microarray applications and many excellent methods and kits are available toward that end. Any RNA extraction and amplification kits can be used provided they yield ample, high-quality total or amplified RNA. In our experience, the Qiagen RNeasy kit has given very consistent results when working with $CD4^+$ T-cell lines or primary white blood cells. When extracting total RNA from more complex samples, such as tissue blocks, other extraction methods are recommended. In general, non-column-, but phenol-based extractions such as Trizol work well for tissue samples. An additional column-based purification is then performed afterward.
4. Total RNA samples should be checked for genomic DNA contamination (e.g., by polymerase chain reaction). When contaminating DNA is present, DNase treatment after total RNA isolation (e.g., using Ambion DNA-free kit) has proven superior to on-column DNase treatment. Extreme care should be taken to avoid RNA degradation during the DNase step.
5. This chapter describes the generation of fluorescent probes using direct Cy-dCTP labeling of T7 amplified RNA. Many alternatives to this strategy exist. Total cellular RNA (using anchored oligo(dT) primers), poly(A)+ mRNA, or polysomal fractions of poly(A)+ mRNA can be used. T7 amplified RNA works especially well for small samples because as little as 1 μg of total RNA is sufficient. Many

fluorochromes, such as fluorescein, lissamine, rhodamine, and phycoerythrin, have been used. The cyanine dyes Cy3 and Cy5 are currently the most popular due to their brightness, stability, and spectral separation. Fluorescent nucleotides (either dCTP or dUTP) can be incorporated directly or indirectly, for example, using incorporation of aminoallyl-conjugated nucleotides followed by reaction with reactive fluorochromes. Labeled probes for Affymetrix are prepared as cRNA. Biotinylated nucleotides are incorporated during cRNA transcription, followed by staining with streptavidin-fluorochrome conjugates.

6. When probe yields are lower than expected, degraded and/or contaminated RNA are usually responsible. It is crucial to use standard procedures to protect RNA from ribonucleases and to test RNA quality before continuing on to probe synthesis. In addition, protect Cy-labeled dNTPs from light at all times and dispense into single-use aliquots to minimize freeze-thaw cycles.
7. Glass microarrays (whether spotted with cDNA or oligomers) use dual-color hybridization in contrast to Affymetrix GeneChip hybridization, in which each sample is hybridized on an individual chip. Although many factors influence the choice of platform, approaches that test treatment versus control samples are especially amenable to dual color studies, whereas patient studies, in which baseline samples or reference RNA pools are used, can be performed using either platform. In each case, experimental design is as important as subsequent implementation and should be given ample consideration.

Acknowledgments

These protocols were developed in close collaboration with the members of the Microbiology Array Group at the University of Washington. The author thanks Katie A. Davis for editorial assistance. This work was supported by grants from the National Institutes of Health (AI37984, AI52028, HL072631, and HG002360).

References

1. Schena, M., Shalon, D., Davis, R. W., and Brown, P. O. (1995) Quantitative monitoring of gene expression patterns with a complementary DNA microarray. *Science* **270,** 467–470.
2. Xiang, C. C. and Chen, Y. (2000) cDNA microarray technology and its applications. *Biotechnol. Adv.* **18,** 35–46.
3. Duggan, D. J., Bittner, M., Chen, Y., Meltzer, P., and Trent, J. M. (1999). Expression profiling using cDNA microarrays. *Nat. Genet.* **1,** 10–14.
4. Deyholos, M. K. and Galbraith, D. W. (2001) High-density microarrays for gene expression analysis. *Cytometry* **43,** 229–238.
5. Quackenbush, J. (2001) Computational analysis of microarray data. *Nat. Rev. Genet.* **2,** 418–427.
6. Geiss, G. K., Bumgarner, R. E., An, M. C., Agy, M. B., van 't Wout, A. B., Hammersmark, E., et al. (2000) Large-scale monitoring of host cell gene expression during HIV-1 infection using cDNA microarrays. *Virology* **266,** 8–16.

7. Corbeil, J., Sheeter, D., Genini, D., Rought, S., Leoni, L., Du, P., et al. (2001) Temporal gene regulation during HIV-1 infection of human CD4+ T cells. *Genome Res.* **11,** 1198–1204.
8. Vahey, M. T., Nau, M. E., Jagodzinski, L. L., Yalley-Ogunro, J., Taubman, M., Michael, N. L., and Lewis, M. G. (2002) Impact of viral infection on the gene expression profiles of proliferating normal human peripheral blood mononuclear cells infected with HIV type 1 RF. *AIDS Res. Hum. Retroviruses* **18,** 179–192.
9. van 't Wout, A. B., Lehrman, G. K., Mikheeva, S. A., O'Keeffe, G. C., Katze, M. G., Bumgarner, R. E., et al. (2003) Cellular gene expression upon human immunodeficiency virus type 1 infection of CD4(+)-T-cell lines. *J. Virol.* **77,** 1392–1402.
10. Mitchell, R., Chiang, C. Y., Berry, C., and Bushman, F. (2003) Global analysis of cellular transcription following infection with an HIV-based vector. *Mol. Ther.* **8,** 674–687.
11. Simmons, A., Aluvihare, V., and McMichael, A. (2001) Nef triggers a transcriptional program in T cells imitating single-signal T cell activation and inducing HIV virulence mediators. *Immunity* **14,** 763–777.
12. Shaheduzzaman, S., Krishnan, V., Petrovic, A., Bittner, M., Meltzer, P., Trent, J., et al. (2002) Effects of HIV-1 Nef on cellular gene expression profiles. *J. Biomed. Sci.* **9,** 82–96.
13. Galey, D., Becker, K., Haughey, N., Kalehua, A., Taub, D., Woodward, J., et al. (2003) Differential transcriptional regulation by human immunodeficiency virus type 1 and gp120 in human astrocytes. *J. Neurovirol.* **9,** 358–371.
14. Cicala, C., Arthos, J., Selig, S. M., Dennis, G., Jr., Hosack, D. A., Van Ryk, D., et al. (2002) HIV envelope induces a cascade of cell signals in non-proliferating target cells that favor virus replication. *Proc. Natl. Acad. Sci. USA* **99,** 9380–9385.
15. de la Fuente, C., Santiago, F., Deng, L., Eadie, C., Zilberman, I., Kehn, K., et al. (2002) Gene expression profile of HIV-1 Tat expressing cells: a close interplay between proliferative and differentiation signals. *BMC Biochem.* **3,** 14.
16. Gibellini, D., Re, M. C., La Placa, M., and Zauli, G. (2002) Differentially expressed genes in HIV-1 tat-expressing CD4(+) T-cell line. *Virus Res.* **90,** 337–345.
17. Izmailova, E., Bertley, F. M., Huang, Q., Makori, N., Miller, C. J., Young, R. A., and Aldovini, A. (2003) HIV-1 Tat reprograms immature dendritic cells to express chemoattractants for activated T cells and macrophages. *Nat. Med.* **9,** 191–197.
18. Patel, C. A., Mukhtar, M., Harley, S., Kulkosky, J., and Pomerantz, R. J. (2002) Lentiviral expression of HIV-1 Vpr induces apoptosis in human neurons. *J. Neurovirol.* **8,** 86–99.
19. Chun, T. W., Justement, J. S., Lempicki, R. A., Yang, J., Dennis, G., Jr., Hallahan, C. W., et al. (2003) Gene expression and viral prodution in latently infected, resting CD4+ T cells in viremic versus aviremic HIV-infected individuals. *Proc. Natl. Acad. Sci. USA* **100,** 1908–1913.
20. Guadalupe, M., Reay, E., Sankaran, S., Prindiville, T., Flamm, J., McNeil, A., and Dandekar, S. (2003) Severe CD4+ T-cell depletion in gut lymphoid tissue during primary human immunodeficiency virus type 1 infection and substantial delay in restoration following highly active antiretroviral therapy. *J. Virol.* **77,** 11,708–11,717.

21. George, M. D., Sankaran, S., Reay, E., Gelli, A. C., and Dandekar, S. (2003) High-throughput gene expression profiling indicates dysregulation of intestinal cell cycle mediators and growth factors during primary simian immunodeficiency virus infection. *Virology* **312,** 84–94.
22. Vahey, M. T., Nau, M. E., Taubman, M., Yalley-Ogunro, J., Silvera, P., and Lewis, M. G. (2003) Patterns of gene expression in peripheral blood mononuclear cells of rhesus macaques infected with SIVmac251 and exhibiting differential rates of disease progression. *AIDS Res. Hum. Retroviruses* **19,** 369–387.
23. Roberts, E. S., Zandonatti, M. A., Watry, D. D., Madden, L. J., Henriksen, S. J., Taffe, M. A., and Fox, H. S. (2003) Induction of pathogenic sets of genes in macrophages and neurons in NeuroAIDS. *Am. J. Pathol.* **162,** 2041–2057.
24. Sui, Y., Potula, R., Pinson, D., Adany, I., Li, Z., Day, J., et al. (2003) Microarray analysis of cytokine and chemokine genes in the brains of macaques with SHIV-encephalitis. *J. Med. Primatol.* **32,** 229–239.
25. Miller, E. D., Smith, J. A., Lichtinger, M., Wang, L., and Su, L. (2003) Activation of the signal transducer and activator of transcription 1 signaling pathway in thymocytes from HIV-1-infected human thymus. *AIDS* **17,** 1269–1277.

37

Determination of DC-SIGN and DC-SIGNR Repeat Region Variations

Huanliang Liu and Tuofu Zhu

Summary

DC-SIGN and DC-SIGNR efficiently bind HIV-1 and other viral as well as nonviral pathogens and assist either *cis* or *trans* infection. Both are type II transmembrane proteins that consist of an N-terminal cytoplasmic domain, a repeat region consisting of seven 23-amino-acid tandem repeats, and a C-terminal C-type carbohydrate recognition domain that binds mannose-enriched carbohydrate modifications of host and pathogen proteins. The normal functions of DC-SIGN and DC-SIGNR include binding to ICAM-2 and ICAM-3. Binding of DC-SIGN to ICAM-2 on endothelial cells facilitates chemokine-induced dendritic cell extravasation; binding to ICAM-3 on T lymphocytes provides the initial step for establishing cell-mediated immunity. Based on the number of tandem repeats, DC-SIGNR is highly polymorphic in the repeat region, while variations in DC-SIGN repeat region are rare. A change in the number of DC-SIGN and DC-SIGNR repeats may influence their normal functions as well as their binding capacity to viral and nonviral pathogens. This chapter describes the methods for detection of DC-SIGN and DC-SIGNR repeat region variations by polymerase chain reaction.

Key Words: DC-SIGN; DC-SIGNR; allele; genotype; HIV-1; transmission.

1. Introduction

Dendritic cells (DC) are among the first cells encountered by HIV-1 during sexual transmission and migrate from mucosal sites to the secondary lymphoid organs upon capturing an antigen *(1,2)*. DC-SIGN (DC-specific ICAM-3 [intercellular adhesion molecular 3]—grabbing nonintegrin), a C-type lectin, is expressed on the surface of immature and, to a lesser extent, mature DCs *(3)* as well as on some types of tissue macrophages, including Hofbauer cells in human placenta *(4,5)*. Mucosa such as the rectum, vagina, uterus, and cervix also contain DC-SIGN-expressing DC within the mucosal lamina propria *(4,6)*. The normal functions of DC-SIGN include binding to ICAM-2 and ICAM-3. The

From: *Methods in Molecular Biology, Vol. 304: Human Retrovirus Protocols: Virology and Molecular Biology*
Edited by: T. Zhu © Humana Press Inc., Totowa, NJ

DC-SIGN–ICAM-2 interaction facilitates chemokine induced DC transendothelial migration *(3,7,8)*, whereas the DC-SIGN–ICAM-3 interaction supports initial antigen-nonspecific contact between DC and T-lymphocytes *(3)*. DC-SIGN has also been shown to mediate antigen uptake *(9,10)*. DC-SIGN and its homolog, DC-SIGNR (DC-SIGN-related) or L-SIGN (liver/lymph node-specific ICAM3 grabbing nonintregrin) can capture HIV-1 at low inoculation and enhance the viral infection of T-cells in *trans* *(4,11–14)*. Other viral pathogens, such as Ebola virus *(15–17)*, hepatitis C virus (HCV) *(18,19)*, human cytomegalovirus *(20)*, and alphavirus *(21)*, as well as nonviral pathogens such as *Leishmania amastigotes* **(22)** and *Mycobacterium tuberculosis* *(23,24)*, can also bind to DC-SIGN/DC-SIGNR-expressing cells with the interaction facilitating *trans* infection of distant permissive cells and/or enhancing *cis* infection of the lectin-bearing cells. Both DC-SIGN and DC-SIGNR are organized into three domains: an N-terminal cytoplasmic region, a neck region containing seven repeats of the 23-amino-acid sequence (69 nucleotides), and a C-terminal domain homologous to C-type lectins—all located on chromosome 19 *(14,25,26)*.

It has been reported that DC-SIGNR is highly polymorphic in the repeat region based on the number of repeats ranging from three to nine (alleles 3 to 9) *(14,27)*, whereas no similar polymorphisms in DC-SIGN were found in a study of 150 Caucasian participants *(14)*. In one of our studies, we demonstrated that the repeat region of DC-SIGN can also exhibit variation based on numbers of repeats ranging from six to eight (alleles 6 to 8), although much less frequently *(28)*. Furthermore, cloning and sequencing of each repeat region allele of both DC-SIGN and DC-SIGNR revealed that the difference among alleles was the result of a deletion or insertion of 69 nucleotides, which encodes the multiple 23-amino-acid repeats. We also found that individuals with heterozygous DC-SIGN repeat alleles may have a reduced probability of acquiring HIV-1 infection *(28)*. We further demonstrated that DC-SIGNR repeat region polymorphisms may affect the susceptibility to HIV-1 infection *(29)*. It is also possible that these variations may alter the normal functions of DC-SIGN and DC-SIGNR.

The protocol described here can be divided into five general steps: (1) extraction of genomic DNA; (2) preparation of the polymerase chain reaction (PCR) reaction mixture; (3) PCR amplification; (4) detection and analysis of the reaction products; and (5) recording of the genotypes.

2. Materials

2.1. Primers

The primers 1F1/1R *(28)* and L28/L32 *(14)* are used to specifically amplify the DC-SIGN and DC-SIGNR repeat regions from genomic DNA, respectively. Primers are listed in **Table 1**.

Table 1
Primers for Amplifying DC-SIGN and DC-SIGNR Repeat Regions

Primer	Sequence (5'-3')	Target sequence	GenBank access no.
1F1	CCACTTTAGGGCAGGAC	1096–1112	AF209479
1R	AGCAAACTCACACCACACAA	1948–1929	AF209479
L28	TGTCCAAGGTCCCCAGCTCCC	539–559	AF209481
L32	GAACTCACCAAATGCAGTCTTCAAATC	1116–1090	AF209481

2.2. Solutions and Reagents

1. Genomic DNA extraction kit.
2. Ethanol (96–100%).
3. PCR buffer: 100 m*M* Tris-HCl, pH 8.3, 500 m*M* KCl, 15 m*M* $MgCl_2$.
4. *Taq* DNA polymerase.
5. dNTP (2.5 m*M* each).
6. Agarose.
7. Ethidium bromide.
8. Molecular-weight markers.
9. PCR thermal cycler.
10. Centrifuge.
11. Gel electrophoresis equipment.
12. Tubes and tips for DNA extraction and PCR.

3. Methods

3.1. Genomic DNA Extraction

Genomic DNA extracted from traditional methods and commercial kits is suitable for genotyping. Because most traditional methods of DNA isolation are time-consuming, labs are increasingly using DNA purification kits for their speed and convenience. In addition, traditional methods of DNA preparation involve hazardous materials requiring special disposal protocols; kits generally use nontoxic reagents and columns that do not require special handling procedures. Today, nearly all biotech companies sell a genomic DNA isolation kit and many of today's kits are designed to provide high-quality PCR-ready genomic DNA from varied sample sources. The QIAamp DNA Blood Mini Kit (Qiagen, Valencia, CA) is one of them. It simplifies isolation of genomic DNA from whole blood, plasma, serum, buffy coat, body fluids, bone marrow, lymphocytes, platelets, cultured cells, tissue, buccal swabs, and dried blood spots with fast spin-column or vacuum procedures. No phenol–chloroform extraction is required. DNA binds specifically to the QIAamp silica-gel membrane while contaminants pass through. PCR inhibitors such as divalent cations and proteins are completely removed in two efficient wash steps, leaving pure nucleic acid to be eluted in either water or a buffer provided with the kit. DNA extraction spin protocol from blood or cells using this kit is taken as an example (revised from copyrighted material by Qiagen: QIAamp DNA Mini Kit and QIAamp DNA Blood Mini Kit Handbook). DNA extraction methods for the vacuum protocol or from other sample sources are detailed in the handbook (www.qiagen.com).

1. Equilibrate samples and all the reagents to room temperature (*see* **Note 1**).
2. Ensure that buffer AW1, Buffer AW2, and Qiagen protease have been prepared according to the instructions.
3. Heat a water bath or heating block to 56°C for use in **step 7**.
4. Pipet 20 μL of Qiagen protease into the bottom of a 1.5-mL microcentrifuge tube (*see* **Note 2**).
5. Add 200 μL of sample to the microcentrifuge tube. Use up to 200 μL of whole blood or cells in 200 μL of PBS (*see* **Note 3**).
6. Add 200 μL of buffer AL to the sample. Mix by pulse-vortexing for 15 s (*see* **Note 4**).
7. Incubate at 56°C for 10 min (*see* **Note 5**).
8. Briefly centrifuge the 1.5-mL microcentrifuge tube to remove drops from the inside of the lid.
9. Add 200 μL of ethanol (96–100%) to the sample, and mix again by pulse-vortexing for 15 s. After mixing, briefly centrifuge the 1.5-mL microcentrifuge tube to remove drops from the inside of the lid.
10. Carefully apply the mixture from **step 9** to the QIAamp Spin Column (in a 2-mL collection tube) without wetting the rim, close the cap, and centrifuge at 6000*g* for 1 min. Place the QIAamp Spin Column in a clean 2-mL collection tube (provided), and discard the tube containing the filtrate (*see* **Note 6**).
11. Carefully open the QIAamp Spin Column and add 500 μL of buffer AW1 without wetting the rim. Close the cap and centrifuge at 6000*g* for 1 min. Place the QIAamp Spin Column in a clean 2-mL collection tube (provided) and discard the collection tube containing the filtrate.
12. Carefully open the QIAamp Spin Column and add 500 μL of buffer AW2 without wetting the rim. Close the cap and centrifuge at 20,000*g* for 3 min. Continue directly with **step 13**, or to eliminate any chance of possible buffer AW2 carryover, perform **step 12a**, and then continue with **step 13** (*see* **Note 7**).

12a. (*Optional*): Place the QIAamp Spin Column in a new 2-mL collection tube (not provided) and discard the collection tube with the filtrate. Centrifuge at 20,000*g* for 1 min.

13. Place the QIAamp Spin Column in a clean 1.5-mL microcentrifuge tube (not provided), and discard the collection tube containing the filtrate. Carefully open the QIAamp Spin Column and add 200 μL of buffer AE. Incubate at room temperature for 1 min, and then centrifuge at 6000*g* for 1 min (*see* **Note 8**).
14. Determine the concentration and purity of DNA. Add 2 μL of DNA product from **step 13** to 98 μL of water, mix thoroughly, centrifuge gently, and measure the

Table 2
PCR Master Mix Reagent Volume and Concentration

Reagent with initial concentration	Quantity for 50 mL of reaction mixture	Final concentration
Sterile water	38.5–x µL[a]	–
10X PCR buffer	5 µL	1X
2.5 mM dNTP mix	4 µL	0.2 mM of each
50 µM forward primer	1 µL	1 µM
50 µM reverse primer	1 µL	1 µM
Taq DNA Polymerase	0.5 µL	1 unit
Genomic DNA	x µL[a]	50–300 ng

[a]The volume of genomic DNA and sterile water should be adjusted according to the DNA concentration. You can also modify the reaction volume proportionally.

absorbance at 260 and 280 nm. The DNA concentration is calculated using the following formula: DNA concentration (µg/mL) = OD value at 260 nm × 50 (1 OD 260 = 50 µg/mL DNA) × 50 (dilution factor). The DNA purity is determined by calculating the ratio of absorbance at 260 nm to absorbance at 280 nm (*see* **Note 9**).

15. DNA is ready for PCR or storing at –20°C for future use.

3.2. Reaction Mixture

Master mix containing water, buffer, dNTPs, $MgCl_2$, primers and *Taq* DNA polymerase is prepared in a single tube for each pair of primers specific for each gene (**Table 1**) and then aliquoted into individual PCR tubes (**Table 2**). Genomic DNA solutions are added next (*see* **Note 10**); **Table 2** shows the amount of individual reagent solutions and template necessary per PCR reaction. You can increase the quantities accordingly when a large number of parallel samples are needed to be genotyped (*see* **Note 11**). This method of reaction mixture setup minimizes the possibility of pipetting errors and saves time by reducing the number of reagent transfers. A set of water and positive controls should always accompany each set of PCR to confirm both the absence of contamination and success of PCR (*see* **Note 12**). After setup, place samples in a thermocycler and perform PCR.

3.3. PCR Reaction

The cycle conditions for DC-SIGN repeat region genotyping are: denaturation at 94°C for 5 min (*see* **Note 13**), followed by 40 cycles of 94°C for 15 s, 60°C for 30 s and 72°C for 30 s, then followed by 72°C for 7 min for final extension. The PCR conditions for DC-SIGNR repeat region genotyping are:

denaturation at 94°C for 5 min, followed by 35 cycles for 94°C for 5 s and 70°C for 1 min, then followed by incubation at 70°C for 10 min (**Fig. 1**). The PCR conditions for DC-SIGNR repeat region are modified from Bashirova et al. ***(14)***.

3.4. Analysis of PCR Products

After the PCR reaction, DC-SIGN and DC-SIGNR repeat region alleles are analyzed by 3.0% agarose gel electrophoresis and ethidium bromide staining. Samples taken from the reaction product, along with appropriate molecular-weight markers, are loaded onto an agarose gel that contains 3.0% ethidium bromide. DNA bands on the gel can then be visualized and pictured under ultraviolet transillumination (**Fig. 1**).

3.5. Genotype Recording

Recording genotypes relies on the pictures taken from the agarose gel. By comparing reaction product bands with bands from the known molecular-weight markers, the alleles and genotypes can be distinguished for each sample (*see* **Note 14**). The difference among alleles is the multiple of 69 nucleotides, which represents the length of repeats. Numbers after allele indicate the number of repeats of DC-SIGN or DC-SIGNR. The size of the PCR products of DC-SIGN repeat region allele 8 is 922 bp, 7 is 853 bp, and 6 is 684 bp. The fragment length of DC-SIGNR repeat region allele 9 is 715 bp, 8 is 647 bp, 7 is 578 bp, 6 is 509 bp, 5 is 440 bp, 4 is 371 bp, and 3 is 302 bp (**Fig. 1**) (*see* **Note 15**).

4. Notes

1. All centrifugation steps should be carried out at room temperature. If a precipitate has formed in buffer AL, dissolve by incubating at 56°C.
2. It is optional to add Qiagen protease to samples that have already been dispensed into microcentrifuge tubes. In this case, it is important to ensure proper mixing after adding the enzyme. However, do not add Qiagen protease directly to buffer AL.
3. If the sample volume is less than 200 μL, add the appropriate volume of PBS.
4. In order to ensure efficient lysis, it is essential that the sample and buffer AL are mixed thoroughly to yield a homogeneous solution. For some cell pellet samples, a longer vortex time is needed.
5. Before heating, a gentle centrifugation to remove all trace amounts of mixture between the microcentrifuge tube cap and wall is highly recommended for reducing the chance of contamination. Otherwise, solutions remaining between them may be expelled after heating due to the pressure produced from heat expansion. Heating at 90°C for another 10 min can inactivate possible infectious pathogens.
6. Close each spin column in order to avoid aerosol formation during centrifugation. Centrifugation is performed at 6000*g* in order to reduce noise. Centrifugation at full speed will not affect the yield or purity of the DNA. If the lysate has

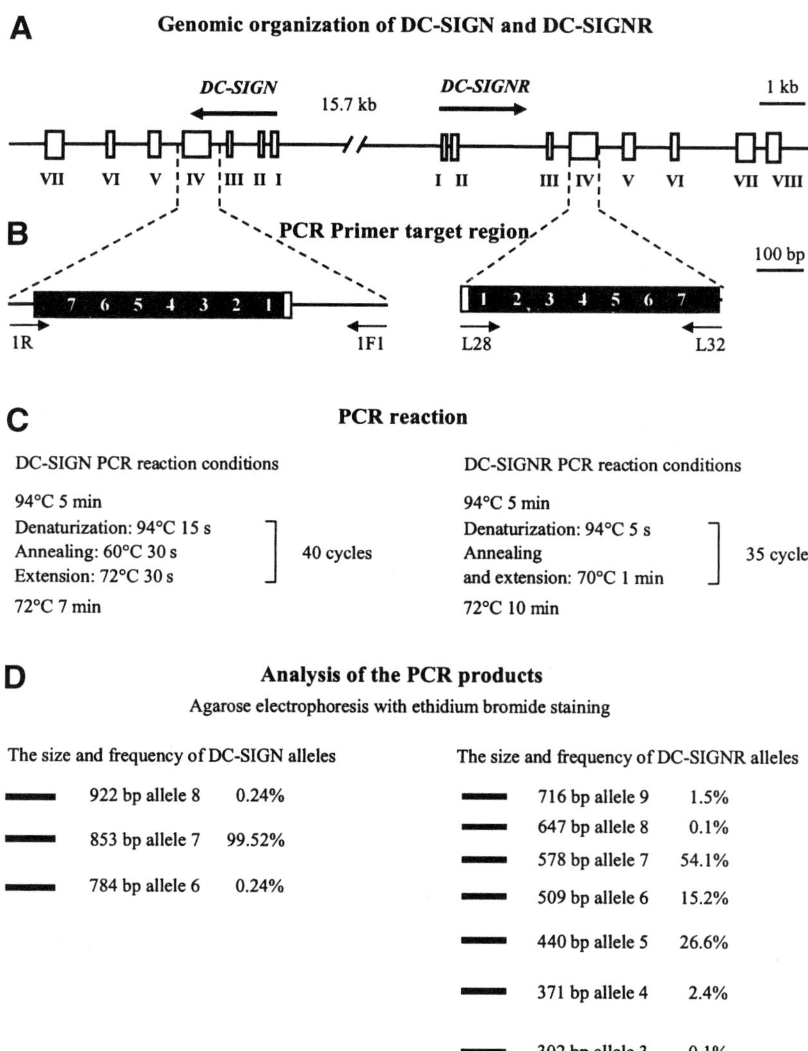

Fig. 1. Schematic representation of the method. **(A)** Genomic organization of DC-SIGN and DC-SIGNR. DC-SIGN and DC-SIGNR are located on chromosome 19p13.2-3 and positioned in a head-to-head orientation ***(14,25,26)***. Boxes are exons and lines between boxes are introns. Roman numerals boxes denote the exons. **(B)** PCR primer target region. Primers used for PCR amplification are labeled and their directions are indicated with arrows. The Arabic numbers in blacked boxes denote the seven full repeats in exon IV of both DC-SIGN and DC-SIGNR. **(C)** PCR reaction conditions. **(D)** Schematic representation of the size of DC-SIGN and DC-SIGNR PCR products. We genotyped 835 individuals for DC-SIGN and 822 for DC-SIGNR repeat region, respectively. The allele frequencies from these individuals are shown after each allele.

not completely passed through the column after centrifugation, centrifuge again at higher speed until the QIAamp Spin Column is empty.

7. Residual buffer AW2 in the eluate may cause problems in downstream PCR amplification. Some centrifuge rotors may vibrate upon deceleration, resulting in the flowthrough, which contains buffer AW2, coming into contact with the QIAamp Spin Column. Removing the QIAamp Spin Column and collection tube from the rotor may also cause flowthrough to come into contact with the QIAamp Spin Column. In these cases, the optional **step 12a** should be performed.
8. Incubating the QIAamp Spin Column loaded with buffer AE for 5 min at room temperature before centrifugation generally increases DNA yield. A second elution step with a further 200 μL of buffer AE into a second 1.5-mL tube will increase the total yields by up to 15%. Volumes of more than 200 μL should not be eluted into a 1.5-mL microcentrifuge tube because the spin column will come into contact with the eluate, leading to possible aerosol formation during centrifugation. Elution with volumes of less than 200 μL increases the final DNA concentration in the eluate significantly, but slightly reduces the overall DNA yield. For samples containing less than 1 μg of DNA, elution in 50 μL of buffer AE is recommended. Eluting with 2 × 100 μL instead of 1 × 200 μL does not increase elution efficiency. For long-term storage of DNA, eluting in buffer AE and storing at –20°C is recommended, since DNA stored in water is subject to acid hydrolysis.
9. A 200-μL sample of whole human blood (~5×10^6 leukocytes/mL) typically yields 6 μg of DNA in 200 μL of buffer AE (30 ng/μL) with an *A*260/*A*280 ratio of 1.7–1.9.
10. Make sure all the reagents and DNA samples are thawed completely before use. Always briefly centrifuge all solutions and samples before opening the cap. Genomic DNA can be prepared from any cell and tissue source. Usually the template DNA amount is in the range of 50–300 ng of genomic DNA in a total reaction mixture of 50 μL. Increased genomic DNA quantities may harmfully affect PCR efficiency and may even inhibit PCR completely. Nearly all routine methods are suitable for genomic DNA purification. Although even minute amounts of agents used in DNA purification procedures (phenol, EDTA, proteinase K, etc.) strongly inhibit *Taq* DNA polymerase, ethanol precipitation of DNA and repetitive treatments of DNA pellets with 70% ethanol is usually effective in removing traces of contaminants. Periodically, the PCR may not work for certain samples. In this case, diluting the genomic DNA may resolve the problem.
11. Stock reagents for PCR can be obtained from a variety of companies. Often the initial concentration of the reagent will differ depending on which company produced it. The amount of a given stock reagent to be used in the master mixture can be easily calculated through the following formula: volume needed = [(final concentration) × (volume per PCR) × (number of PCR reactions)] ÷ (stock concentration).
12. Necessary precautions for avoiding PCR contamination should be taken seriously. The PCR reaction mixture should always be set up in a different location

than where genomic DNA or PCR products are worked with. The following points should also be considered to avoid the problems of PCR contamination from sources: cross-contamination among samples, products from previous PCR, laboratory benches, equipment, thermal cycler, and pipetting devices, which could be contaminated by previous DNA preparation and plasmid DNA. The samples should be handled and DNA isolated in a hood with filtered airflow, and in an area not used for the preparation of template DNA and post-PCR analysis.

13. Complete denaturation of the genomic DNA before the start of the amplification cycle is a key point for PCR success. Incomplete denaturation of DNA results in the inefficient utilization of template in the first amplification cycle, a poor yield of PCR product, or even failed PCR.
14. There are a number of things that can disrupt your PCR, causing your gels to turn up blank after electrophoresis and staining. When you encounter this problem, the following causes occur more often than others: no or incomplete denaturation before the start of the PCR amplification cycle (*see* **Note 13**); excess, insufficient, or impure genomic DNA; too much or degraded enzyme or dNTP; $MgCl_2$ not mixing thoroughly before use or wrong $MgCl_2$ concentration; wrong primer concentration; or simply wrong cycle conditions. If only a small number of samples fail in a large batch of PCR, the most likely source of error is the genomic DNA (*see* **Note 10**).
15. The DC-SIGN repeat region allele frequencies we found in 835 individuals are 0.24% for allele 8, 99.52% for allele 7, and 0.24% for allele 6 ***(28)***. The DC-SIGNR repeat region allele frequencies we identified among 822 individuals are 1.5% for allele 9, 0.1% for allele 8, 54.1 % for allele 7, 15.2% for allele 6, 26.6% for allele 5, 2.4% for allele 4, and 0.1% for allele 3. You may find other repeat region alleles in your investigations. The frequency of these alleles, however, is likely to be extremely low.

Acknowledgments

We thank T. Andrus and B. Greene for editing the manuscript. This work was supported by Public Health Service grants AI 27757 (the University of Washington/Fred Hutchinson Cancer Research Center CFAR Clinical Research Core Award to H.L.), AI 56994, AI 45402, and AI 49109 (T.Z).

References

1. Stahl-Hennig, C., Steinman, R. M., Tenner-Racz, K., Pope, M., Stolte, N., Matz-Rensing, K., et al. (1999) Rapid infection of oral mucosal-associated lymphoid tissue with simian immunodeficiency virus. *Science* **285,** 1261–1265.
2. Barratt-Boyes, S., Watkins, S., and Finn, O. (1997) In vivo migration of dendritic cells differentiated in vitro: a chimpanzee model. *J. Immunol.* **158,** 4543–4547.
3. Geijtenbeek, T. B., Torensma, R., van Vliet, S. J., van Duijnhoven, G. C., Adema, G. J., van Kooyk, Y., and Figdor, C. G. (2000) Identification of DC-SIGN, a novel dendritic cell-specific ICAM-3 receptor that supports primary immune responses. *Cell* **100,** 575–585.

4. Geijtenbeek, T. B., Kwon, D. S., Torensma, R., van Vliet, S. J., van Duijnhoven, G. C., Middel, J., et al. (2000) DC-SIGN, a dendritic cell-specific HIV-1-binding protein that enhances trans-infection of T cells. *Cell* **100,** 587–597.
5. Soilleux, E. J., Morris, L. S., Lee, B., Pohlmann, S., Trowsdale, J., Doms, R. W., and Coleman, N. (2001) Placental expression of DC-SIGN may mediate intrauterine vertical transmission of HIV. *J. Pathol.* **195,** 586–592.
6. Jameson, B., Baribaud, F., Pohlmann, S., Ghavimi, D., Mortari, F., Doms, R. W., and Iwasaki, A. (2002) Expression of DC-SIGN by dendritic cells of intestinal and genital mucosae in humans and rhesus macaques. *J. Virol.* **76,** 1866–1875.
7. Geijtenbeek, T. B., Krooshoop, D. J., Bleijs, D. A., van Vliet, S. J., van Duijnhoven, G. C., Grabovsky, V., et al. (2000) DC-SIGN-ICAM-2 interaction mediates dendritic cell trafficking. *Nat. Immunol.* **1,** 353–357.
8. van Kooyk, Y. and Geijtenbeek, T. B. (2002) A novel adhesion pathway that regulates dendritic cell trafficking and T cell interactions. *Immunol. Rev.* **186,** 47–56.
9. Engering, A., Geijtenbeek, T. B., van Vliet, S. J., Wijers, M., van Liempt, E., Demaurex, N., et al. (2002) The dendritic cell-specific adhesion receptor DC-SIGN internalizes antigen for presentation to T cells. *J. Immunol.* **168,** 2118–2126.
10. Schjetne, K. W., Thompson, K. M., Aarvak, T., Fleckenstein, B., Sollid, L. M., and Bogen, B. (2002) A mouse C kappa-specific T cell clone indicates that DC-SIGN is an efficient target for antibody-mediated delivery of T cell epitopes for MHC class II presentation. *Int. Immunol.* **14,** 1423–130.
11. Curtis, B., Scharnowski, S., and Watson, A. (1992) Sequence and Expression of a Membrane-Associated C-Type Lectin that Exhibits CD4-Independent Binding of Human Immunodeficiency Virus Envelope Glycoprotein gp120. *Proc. Natl. Acad. Sci. USA* **89,** 8356–8360.
12. Pohlmann, S., Baribaud, F., Lee, B., Leslie, G. J., Sanchez, M. D., Hiebenthal-Millow, K., et al. (2001) DC-SIGN Interactions with human immunodeficiency virus type 1 and 2 and simian immunodeficiency virus. *J. Virol.* **75,** 4664–4672.
13. Pohlmann, S., Soilleux, E. J., Baribaud, F., Leslie, G. J., Morris, L. S., Trowsdale, J., et al. (2001) DC-SIGNR, a DC-SIGN homologue expressed in endothelial cells, binds to human and simian immunodeficiency viruses and activates infection in trans. *Proc. Natl. Acad. Sci. USA* **98,** 2670–2675.
14. Bashirova, A. A., Geijtenbeek, T. B. H., van Duijnhoven, G. C. F., van Vliet, S. J., Eilering, J. B. G., Martin, M. P., et al. (2001) A dendritic cell-specific intercellular adhesion molecule 3-grabbing nonintegrin (DC-SIGN)-related protein is highly expressed on human liver sinusoidal endothelial cells and promotes HIV-1 infection. *J. Exp. Med.* **193,** 671–678.
15. Alvarez, C. P., Lasala, F., Carrillo, J., Muniz, O., Corbi, A. L., and Delgado, R. (2002) C-type lectins DC-SIGN and L-SIGN mediate cellular entry by Ebola virus in cis and in trans. *J. Virol.* **76,** 6841–6844.
16. Baribaud, F., Pohlmann, S., Leslie, G., Mortari, F., and Doms, R. W. (2002) Quantitative expression and virus transmission analysis of DC-SIGN on monocyte-derived dendritic cells. *J. Virol.* **76,** 9135–9142.
17. Simmons, G., Reeves, J. D., Grogan, C. C., Vandenberghe, L. H., Baribaud, F.,

Whitbeck, J. C., et al. (2003) DC-SIGN and DC-SIGNR bind ebola glycoproteins and enhance infection of macrophages and endothelial cells. *Virology* **305,** 115–123.

18. Pohlmann, S., Zhang, J., Baribaud, F., Chen, Z., Leslie, G. J., Lin, G., et al. (2003) Hepatitis C Virus Glycoproteins Interact with DC-SIGN and DC-SIGNR. *J. Virol.* **77,** 4070–4080.
19. Lozach, P.-Y., Lortat-Jacob, H., De Lacroix De Lavalette, A., Staropoli, I., Foung, S., Amara, A., et al. (2003) DC-SIGN and L-SIGN are high affinity binding receptors for hepatitis C virus glycoprotein E2. *J. Biol. Chem.* **278,** 20,358–20,366.
20. Halary, F., Amara, A., Lortat-Jacob, H., Messerle, M., Delaunay, T., Houles, C., et al. (2002) Human cytomegalovirus binding to DC-SIGN is required for dendritic cell infection and target cell trans-infection. *Immunity* **17,** 653–664.
21. Klimstra, W. B., Nangle, E. M., Smith, M. S., Yurochko, A. D., and Ryman, K. D. (2003) DC-SIGN and L-SIGN can act as attachment receptors for alphaviruses and distinguish between mosquito cell- and mammalian cell-derived viruses. *J. Virol.* **77,** 12,022–12,032.
22. Colmenares, M., Puig-Kroger, A., Pello, O. M., Corbi, A. L., and Rivas, L. (2002) Dendritic cell (DC)-specific intercellular adhesion molecule 3 (ICAM-3)-grabbing nonintegrin (DC-SIGN, CD209), a C-type surface lectin in human DCs, is a receptor for Leishmania amastigotes. *J. Biol. Chem.* **277,** 36,766–36,799.
23. Geijtenbeek, T. B., Van Vliet, S. J., Koppel, E. A., Sanchez-Hernandez, M., Vandenbroucke-Grauls, C. M., Appelmelk, B., and Van Kooyk, Y. (2003) Mycobacteria target DC-SIGN to suppress dendritic cell function. *J. Exp. Med.* **197,** 7–17.
24. Tailleux, L., Schwartz, O., Herrmann, J. L., Pivert, E., Jackson, M., Amara, A., et al. (2003) DC-SIGN is the major Mycobacterium tuberculosis receptor on human dendritic cells. *J. Exp. Med.* **197,** 121–127.
25. Soilleux, E. J., Barten, R., and Trowsdale, J. (2000) Cutting edge: DC-SIGN; a related gene, DC-SIGNR; and CD23 form a cluster on 19p13. *J. Immunol.* **165,** 2937–2942.
26. Mummidi, S., Catano, G., Lam, L., Hoefle, A., Telles, V., Begum, K., et al. (2001) Extensive repertoire of membrane-bound and soluble dendritic cell-specific ICAM-3-grabbing nonintegrin 1 (DC-SIGN1) and DC-SIGN2 isoforms. Inter-individual variation in expression of DC-SIGN transcripts. *J. Biol. Chem.* **276,** 33,196–33,212.
27. Kobayashi, N., Nakamura, H. T., Goto, M., Nakamura, T., Nakamura, K., Sugiura, W., et al. (2002) Polymorphisms and haplotypes of the CD209L gene and their association with the clinical courses of HIV-positive Japanese patients. *Jpn. J. Infect. Dis.* **55,** 131–133.
28. Liu, H., Hwangbo, Y., Holte, S., Lee, J., Wang, W., Kaupp, N., et al. (2004) Analysis of genetic polymorphisms in CCR5, CCR2, stromal cell-derived factor 1, RANTES, and dendritic cell-specific intercellular adhesion molecule-3-grabbing nonintegrin in seronegative individuals repeated exposed to HIV-1. *J. Infect. Dis.* **190,** 1055–1058.
29. Liu, H., McElrath, M. J., Holte, S., Celum, C., Lee, J., Corey, L., and Zhu, T. (2003) Genetic polymorphisms in DC-SIGNR repeat region affect HIV-1 transmission. *10th Conference on Retroviruses and Opportunistic Infections* abstract # 373.

Index

D

H